Jean Chrysostome K. Kiyala · Geoff Thomas Harris
Editors

Civil Society and Peacebuilding in Sub-Saharan Africa in the Anthropocene

An Overview

DUT
DURBAN UNIVERSITY OF TECHNOLOGY
INYUVESI YASETHEKWINI YEZOBUCHWEPHESHE

INTERNATIONAL CENTRE OF NONVIOLENCE

Springer

Editors
Jean Chrysostome K. Kiyala
International Centre of Nonviolence
Durban University of Technology
Durban, South Africa

Geoff Thomas Harris
International Centre of Nonviolence
Durban University of Technology
Durban, South Africa

Acknowledgement: This contributed volume was supported by The International Center of Nonviolence, Durban University of Technology, South Africa. The cover photo was taken by Hans Günter Brauch who granted the permission to use it here. The photo on the internal title page was taken by Chrys Kiyala, Kwilu River, in Bandundu Town, Democratic Republic of Congo, 27 July 2021.

More on this book is at: http://afes-press-books.de/html/APESS_34.htm

ISSN 2367-4024 ISSN 2367-4032 (electronic)
The Anthropocene: Politik—Economics— Society—Science
ISBN 978-3-030-95178-8 ISBN 978-3-030-95179-5 (eBook)
https://doi.org/10.1007/978-3-030-95179-5

Copyediting: PD Dr. Hans Günter Brauch, AFES-PRESS e.V., Mosbach, Germany
Copy and Language Editors: Dr. Vanessa Greatorex, UK; Ms Elizabeth Norton-Amor and Mr Alan Bell, South Africa

This Springer imprint is published by the registered company Springer Nature Switzerland AG
The registered company address is: Gewerbestrasse 11, 6330 Cham, Switzerland

The Anthropocene: Politik—Economics—Society—Science

Volume 34

Series Editor

Hans Günter Brauch, Peace Research and European Security Studies (AFES-PRESS), Mosbach, Baden-Württemberg, Germany

More information about this series at https://link.springer.com/bookseries/15232
http://www.afes-press-books.de/html/APESS.htm
http://afes-press-books.de/html/APESS_34.htm

To peace-lovers, nature-lovers,
environmentalists,

ecologists, conservationists,
environment-preservationits,

Greenpeace movements and activists,

The Green Teen Team (GTT) generation and
friends

of environment and global peace;

to Environment and Peace Nobel Prize
Laureates,

researchers, engineers, sciences and
technology institutions, think tanks,

various civil society organisations, social
movements;

to all of you who have developed new
paradigms in

environmental and peace studies,

and demonstrated human resilience to

guiding humanity towards the path of
sustainable development,

social justice, environmental stability and global peace;

your commitment and engagement to keep the world green and

mitigate the Earth systems' disintegration are acknowledged here

and will continue to inspire future generations.

Foreword by Urmila Bob

In the context of notions of peace being an elusive concept and attaining a state of peace being a challenge in many parts of the world, the book provides timely and valuable insights in this critical area of civil society and peacebuilding. The focus on Sub-Saharan Africa is also notable given widespread vulnerabilities linked to persistent poverty and inequalities, discrimination, poor service delivery and infrastructural backlogs, governance and political issues, resource scarcity and poor environmental management practices; that are among the many factors that pose several challenges for meaningful civil society participation and peacebuilding efforts. The COVID-19 pandemic, climate change and future disruptions will severely strain existing (albeit limited) capacity to further peace interests as well as increase conflicts, especially environmental conflicts. These are threat multipliers that reinforce existing challenges and worsen conditions on the ground in contexts where there is limited resources, infrastructure, capacity and capabilities to respond. The importance of conflict-sensitive adaptation and responsiveness is significant.

The number of case studies in the book, from different parts of Sub-Saharan Africa, underscore the range of contexts and experiences in this region. The range of thematic foci also highlights a key contribution of the book, which is the interdisciplinary and transdisciplinary orientation that permits varied perspectives, experiences and voices to be examined. The focus on human rights and environmental justice is particularly relevant given that there is widespread consensus that Sub-Saharan Africa will experience the brunt of extreme climate events and natural disasters. The conceptualisation and centrality of peace ecology and indigenous knowledge (that informs African social philosophies) provide a critical and previously neglected lens to better understand civil society and peacebuilding in Sub-Saharan Africa. Noteworthy also is the attention to gender aspects and the focus on children. Women and children are vulnerable segments in society and understanding impacts and their agency to affect/influence change is important in peacebuilding efforts that are empowering, transformational and embedded in promoting resiliency and sustainability. The examination of the role of faith-based organisations, schools and community collectives as well as non-governmental organisations (such as the African Centre for the Constructive Resolution of Disputes), for example, illustrate how different parts of civil society can contribute to peacebuilding initiatives and environmental education.

The perspectives of African-based scholars is particularly notable. Additionally, the diverse methodological approaches used (from integrating and examining secondary sources of information to primary data collection methods) also reveal the need for adopting interdisciplinary approaches. Furthermore, the assessment of peacebuilding initiatives (including campaigns and training programmes) reveals the critical stance embraced. The case studies traverse local community contexts to country-level analyses to regional and continental assessments. Furthermore, challenges associated with undertaking global peace research, especially in relation to scholars from the global South, reveal that more attention needs to be paid to what types of research is being undertaken in Sub-Saharan Africa, by whom and for what purpose. The discussions are rooted in examining specific dynamics while extracting lessons that reflect on implications more broadly and incorporate a holistic approach. The 20 Chapters give voices to many scholars from the Global South, especially from Sub-Saharan Africa. The authors not only focus on how to promote peacebuilding but also the causes of conflict, emphasising the need to empower and enable the agency of individuals, communities, nations and regions to be proactive and responsive. Enduring peace must include all members of society (especially vulnerable groups) as well as place environmental sustainability and justice at the centre of peacebuilding efforts.

I commend the editors and authors for compiling valuable research that will be used by scholars, practitioners, decision- and policymakers, and students on the African continent and globally. The book is an important scholarly contribution, growing the field of research and generating knowledge in an area where limited scholarly resources are available.

Durban, South Africa
November 2021

Prof. Urmila Bob

Prof. Urmila Bob is a Full Professor of Geography in the School of Agriculture, Earth and Environmental Sciences and University Dean of Research at the University of KwaZulu-Natal, South Africa. She has a Masters and Ph.D. in Geography from West Virginia University, USA and a Masters in Tertiary Education Management from the University of Melbourne, Australia. She conducts research on a range of developmental and environmental issues, including socio-economic impact assessments of developmental projects in relation to conservation and tourism projects as well as sustainable livelihoods in both rural and urban contexts. She has published in these fields in both nationally and internationally recognised academic books and journals as well as been involved in consultancy-based projects. She has supervised to completion more than 70 Masters and Ph.D. students in these areas of research as well. She has training expertise in quantitative and qualitative methodologies and techniques, monitoring and evaluation, development of indicators, social and environmental impact assessments, gender analytical methodologies, research planning and the development of academics and postgraduate students.

Address: Research Office, N-Block, Govan Mbeki Building, Westville Campus, University of KwaZulu-Natal, Durban, South Africa.

Email: bobu@ukzn.ac.za.

Websites: https://www.linkedin.com/in/urmilla-bob-0725b09/;
https://saees.ukzn.ac.za/urmilla-bob/;
https://www.accord.org.za/people/urmilla-bob/;
https://www.researchgate.net/profile/Urmilla-Bob.

Foreword by Jan Cillié (Jannie) Malan

Every reader taking up this book in her/his hands should do so with some justified trembling. The key word in its title is, and is meant to be, a shocking term. After 4 600 000 000 years of Planet Earth's history, geologists for the very first time found it necessary to name an epoch after some of its inhabitants. The compelling reason for this naming was not to honour this species of living beings, but to blame and shame them for intruding into the systems sustaining the well-being of all living beings on Earth. And, this species are *us*, you and me!

The detrimental intrusion already began with the Industrial Revolution of around 1800, but the devastating invasion came with the atomic bomb in 1945. The alarming result is that the climate-changing gap in the ozone layer has come about in less than two centuries, after the billions of years during which it had remained undamaged. If so much harm has been done in such a short period, how rapidly may the disaster escalate in an even shorter time?

It should therefore be very obvious that the Anthropocene-causing Anthropos must be accosted, and that the consequences of the Anthropocene must be counteracted with determination. But how?

The problem is that in the overwhelming majority of cases Anthropos happens to be a self-centred being. The Latin designation of Homo Sapiens Sapiens seems to be complimentary, but it may be true to reality to take the one 'sapiens' as 'wise' and the other as 'cunning'. This self-centredness may very well be seen as an underlying reason for the writing of this book. It is the background to all the case studies that were researched. Whether the case was one of broken peace between human beings themselves or between humans and their environment, the root cause appears to have been some form of human self-interest. What seems to be very common among us as human beings, is the singular version, egotism—where an individual ego is preoccupied with own status, prestige or power. But what can apparently be even more prevailing, is the plural version, own-groupishness—where individual egos are obsessed with the pre-eminence of their own ethnic group. On the analogy of 'egotism' this may aptly be called 'nostricism', from Latin 'nostri' for 'our people'. In this book we indeed find that ethnic factors are specifically referred to in the case studies of Burundi, the Democratic Republic of the Congo, South Africa, South Sudan, Uganda and Zimbabwe.

What we also find in this book, however, are inspiring examples of faith-based and other civil society organisations in which fellow human beings have experienced an inner change of attitude and behaviour. We read how three Eastern Religions, the three Abrahamic Religions and African Traditional Religions encourage their followers to appreciate and respect nature, and to take care of our common planetary home. A mind-shift *is* possible from thinking that the Earth belongs to us to rather thinking that we belong to the Earth. Apart from religious organisations, there are also secular civil society organisations who model and propagate rethinking about the purpose of human life. From a United Nations document the following optimistic sentence is quoted: 'If people have the power to create an entirely new geological epoch, then people also have the power to choose to change' (p. 50). The urgent need is therefore discussed for educating the exploding global population in peace ecology (p. 70). Large numbers of publications have already appeared, and further debate should follow, but the important goal should be to integrate research and action—locally, regionally and globally. The findings should be used to educate not only *about* transformation, but in an actually transformative way. The belief that satisfying individual needs is the basic purpose of life should be discarded, and conscious stewardship—which is not ownership—should be adopted (p. 115).

However, for human-natured people to change their attitude and behaviour, both towards fellow-humans and to the human environment, *willingness* is needed. And willingness cannot be induced from outside; it has to emerge from within. There *are* a few ways which *might* play a role to bring this about, for instance to share stories of people who have indeed become willing to experience an inner change. That is precisely what we have in most, if not all, of the case studies. The founders and members of the civil societies were inspired by inward urges and they remained committed to their calling. I would therefore suggest reading the case studies in an exploring mode, trying to distinguish between people adhering to an environment-damaging and/or conflict-causing status quo, and those willing to change themselves and the structures concerned.

Pointing the way towards the changing of attitudes and behaviours may serve as a reply to the above question on how to address the stubborn self-centredness of us, human beings.

I hope that you will find the reading of this significant publication rewarding, and transforming if necessary. The contents are comprehensive and complicated, but the large number of captioned sections and sub-sections helps to make it reader-friendly.

October 2021

Prof. em. Jan Cillié (Jannie) Malan
University of the Western Cape
Capetown, South Africa

Prof. em. Jan Cillié (Jannie) Malan born in 1931, deviated from apartheid ideology and fundamentalist tendencies in Christianity in the 1950s. He worked as Minister of Religion in Xhosa-speaking congregations in Eastern and Western Cape, and also taught Science and Mathematics in high schools. In 1973 he began teaching Biblical Studies at the University of the Western Cape (UWC), Cape Town, South Africa, and took part in the anti-Apartheid 'struggle'. He also presented a post-graduate module in Conflict Studies from 1991 to 2007. He became Professor in 1981 and emeritus Professor in 1999. After retiring at UWC, he was senior researcher at the African Centre for the Constructive Resolution of Disputes (ACCORD), Durban, South Africa (1993–2021). He focused on education towards justice and peace, culture- and diversity-friendliness, mutual understanding and conflict resolution wisdom from Africa. He edited ACCORD's *African Journal on Conflict Resolution* (1999–2021). He and Magriet have been happily married for almost 64 years.

Address: Prof. em. Jan Cillié (Jannie) Malan, Starckwood Retirement Village, Frans Conradie Drive, Bellville, 7530, South Africa.

Email: magrietmalan@live.co.za.

Preface by Will Steffen

The Anthropocene is one of the most important—and confronting—concepts that has arisen in the academic world over the past few decades, and possibly much longer. Based on a vast and rapidly growing body of observations, modelling and other studies, there is overwhelming evidence that human activities have now driven the Earth System out of the stability of the 11,700-year Holocene epoch and onto an accelerating trajectory into a rapidly changing and very risky future. It is literally a change of geological proportions.

Climate change is arguably the most dominant feature of the Anthropocene. Global temperature is now over 1 degree C higher than the pre-industrial level and it is increasing at an ever greater rate. The destabilisation of the climate is already profound—melting ice sheets, acidifying oceans, rising sea levels, massive wildfires, intense storms and heavy rainfall. The current rates at which atmospheric carbon dioxide concentration and global temperature are rising are almost unprecedented in the entire 4.5 billion year history of the Earth.

But climate change is not the only disconcerting feature of the Anthropocene. The biosphere—the life on this planet that makes it unique in the solar system and probably far beyond—is now under a growing threat. A recent study, published in

2019, that assessed the condition of the biosphere revealed some startling observations. Around 1 million animal and plant species are threatened with extinction, many within decades. Extinction rates now rival those of the great extinction events in Earth history. In general, nature is declining globally at rates unprecedented in human history. The web of life on Earth is getter smaller and increasingly frayed.

The fundamental cause of this unfolding global tragedy is absolutely clear—human activities. We are waging war on the Earth, destroying our own planetary life support system.

We are now approaching 8 billion in number, but even more importantly, our economic and technological systems are expanding at unprecedented rates. Our energy consumption is growing, as is our production of chemicals and other novel entities. Our transportation and telecommunication systems are expanding and now span the globe, and our consumption of natural resources continues to rise. The rates at which our activities have expanded increased sharply around the middle of the 20th century, a period that is sometimes called the Great Acceleration.

Although the Great Acceleration has generated unparalleled wealth for humanity as a whole, it is being distributed in a highly unequal manner. Many more people have indeed been brought out of poverty, but the growing differential between the wealthy and the others, both between countries and within countries, is driving escalating health and social problems. Around 800 million humans still suffer from malnutrition and lack of food, but over 2 billion humans are obese or suffer from micronutrient deficiencies, outcomes of having too much food or the wrong types of food. Well-being within most wealthy countries has stagnated or declined since the 1970s or 1980s. Growing income inequalities within wealthy countries are linked to increases in mental illness, drug and alcohol addiction, obesity, homicides, teenage births, infant mortality and other health and social problems.

What is the future trajectory of the Anthropocene? Because the Anthropocene is driven by humans, its future trajectory is—at least at present—in our hands. But that may not be the case much longer. The Earth System has so-called tipping points—features that once pushed past a critical threshold can change abruptly and/or irreversibly, taking the Earth System out of human control and towards very different conditions. Tipping points include the melting of large polar ice sheets, the onset of uncontrollable wildfires in large ecosystems like the Amazon rainforest, and changes in large circulation systems like the North Atlantic ocean current. Ominously, there are already signs that we are approaching several of these tipping points.

How to meet this unprecedented challenge for humanity?

There is much research already on what we might call the ‘proximate’ responses to the threats of the Anthropocene. Technologies for dealing with climate change are advancing rapidly—renewable energy systems, electrified transport solutions, carbon-neutral buildings and other infrastructure, and so on. Research is underway on ways to draw carbon out of the atmosphere and store it safely for hundreds or thousands of years. Some scientists are working on ways to ‘geoengineer’ the climate system, such as dimming incoming solar radiation or brightening clouds to reflect more sunlight.

Solutions to the biosphere crisis are also being proposed, including the translocation of some endangered species, the preservation of critical habitats and building corridors of protected areas to connect nature reserves. More innovative approaches include 'eco-engineering' techniques that deliberately modify or reconstruct ecosystems to help them cope with a rapidly changing climate.

The Anthropocene presents enormous and accelerating challenges for humanity. While a plethora of responses are being considered, nearly all of them neglect the fact that the Anthropocene demands fundamental changes in our thinking and our values. Instead of waging war on our planetary home, we need to make peace with the Earth.

The overall message of the Anthropocene to humanity is embedded in the word itself. The 'cene' refers to a new geological interval in the long history of Earth. That is, the current changes to the Earth System are so profound that they are changing the long-term trajectory of the planetary system as a whole. The 'anthropos', of course, refers to us. It places the spotlight clearly on ourselves—our societies, how they operate internally and with other societies and—critically—their relationship to the rest of life on Earth, and to the Earth itself.

Our focus should not be on how we manage or change the structure or functioning of the Earth System, but rather on ourselves, and on our relationships to each other and to the rest of life on Earth. There is a growing number of approaches to the human dimension of the Anthropocene. These include approaches such as circular economies, the promotion and valuing of social capital, the creation of well-being societies and life-centric rather than human-centric philosophies to guide us through the Anthropocene.

Central to all of these approaches, though, is the need for peaceful and respectful relationships amongst ourselves and with the rest of life on Earth. The concepts of peace ecology and environmental peace consciousness need to be at the centre of our journey into the Anthropocene. The building and maintenance of social capital—not the accumulation of individual material wealth—need to be at the centre of societies that can build peace with each other and with the natural world.

Sub-Saharan Africa, the birthplace of humanity, offers an excellent opportunity to provide the leadership needed to steer humanity away from this present headlong dash towards a catastrophic Anthropocene. It is an epicentre for many of the urgent challenges that face humanity more generally—social division and violent conflicts, exploitation and degradation of the biosphere, changing regional climates driving severe changes in water availability and a lack of effective governance. But if there is any place on Earth that is truly humanity's Common Home, it is Africa. Meeting the deep and urgent challenges that the Anthropocene poses for civil societies in Africa would provide hope and guidance for the rest of us around Earth.

Drawing on their common African origins, indigenous cultures around the planet offer wisdom that can not only guide the healing of relationships amongst our human societies, but also the building of peace between humans and the rest of life on Earth. I leave the last word of this Preface to an elder from the Noongar people of southwestern Australia, one of the hundreds of clans of indigenous people of my home continent:

> We're only here for a short amount of time to do what we've been put here to do, which is to look after Country. We're only a tool in the cycle of things…… we go out into the world and help keep the balance of nature. It's a big cycle of living with the land, and then eventually going back to it…..

September 2021

Prof. em. Will Steffen
The Australian National University
Canberra, Australia

Prof. em. Will Steffen is an Earth System scientist. He is a Councillor on the publicly-funded Climate Council of Australia that delivers independent expert information about climate change, an Emeritus Professor at the Australian National University (ANU); Canberra, a Senior Fellow at the Stockholm Resilience Centre, Sweden; and a Fellow at the Beijer Institute of Ecological Economics, Stockholm. He is the chair of the jury for the Volvo Environment Prize; a member of the International Advisory Board for the Centre for Collective Action Research, Gothenburg University, Sweden; and a member of the Anthropocene Working Group of the Sub-committee on Quaternary Stratigraphy. From 1998 to mid-2004, Steffen was Executive Director of the International Geosphere-Biosphere Programme, based in Stockholm. His research interests span a broad range within climate and Earth System science, with an emphasis on incorporation of human processes in Earth System modelling and analysis; and on sustainability and climate change.

Address: Prof. em. Will Steffen, The Australian National University (ANU); Canberra, Australia.

Email: will.steffen@anu.edu.au.

Websites: https://researchers.anu.edu.au/researchers/steffen-wl;
https://iceds.anu.edu.au/people/academics/professor-will-steffen;
https://en.wikipedia.org/wiki/Will_Steffen;
https://www.stockholmresilience.org/meet-our-team/staff/2009-08-24-steffen.html.

Preface by Luc Reychler

Building sustainable peace is not an option; it is the last chance to ensure our survival and the quality of nature. The deterioration of the natural environment and the negative consequences, such as droughts, floods, wildfires, food insecurity and air pollution, are well documented. The latest UN report gave the climate change 'code red' for humanity. Sustainable peacebuilding aims to prevent and reduce all forms of violence that reduce the length and quality of the lifetime of citizens. In addition to environmental violence, other forms of violence include physical and armed violence, structural or systemic violence, psychological violence, cultural violence, gender violence, bad governance, corruption, epistemic violence, organised crime and temporal violence. Despite progress being made, much armed and unarmed violence continues to exist between people.

Without advances in sustainable peace between humans, sustainable development and environmental peace, it will be close to impossible to prevent and reduce violence. Sustainable peacebuilding among humans and towards nature are closely interlocked processes. A lack of progress can retard or block sustainable development and environmental peace. It can create vicious circles when the environmental challenges are not dealt with in a timely and effective manner, and the deteriorating environment increases tensions between humans. A virtuous circle demands synchronous efforts and progress in peacebuilding between humans and towards nature. Peacebuilding requires efforts made by different actors at different levels (local, regional and global) and in different sectors. Civil society plays a key role in motivating, mobilising and enabling citizens to build a better home at the local and national levels. Equally important are regional and global cooperation. This implies multiple loyalties to the nation, region and the world as a whole. Not only diplomats but also citizens should be aware of and understand the wider world and their place in it. They can take an active role in their communities to make their human and natural habitats more peaceful, sustainable and equitable.

We need to prepare for worst-case scenarios and negotiate new deals for global and local sustainable peace and development, not only because of the existential threat of climate change but also due to the high level of armed and unarmed violence between people. The global peace index of 2019 rated 45 countries as low and very low on peace. The five countries with the largest proportion of people who had experienced physical violence or knew someone who had are all in Sub-Saharan Africa. Namibia has the highest rate in the world, at 63%, followed by South Africa, Lesotho, Liberia and Zambia (IEP 2021). The pandemic revealed an absurd degree of inequality. Only 2% of the population of emerging economies are vaccinated, whereas European countries have nearly 80% vaccination scores. High levels of inequality and a lack of opportunities at domestic and international levels discourage human development, undermine trust and raise relative deprivation and tension. The globalisation of communication and information has raised expectations and the sense of relative deprivation of many more people. During my own lifetime, the world population tripled from 2.5 billion in 1950 to 7.8 billion in 2020. The population growth in Africa increased sixfold (1.341 billion). Sub-Saharan Africa has the highest projected population increase. In 2050, it is expected to reach 2.1 billion citizens.

The world has moved into a post-American era. The coercive diplomacy and regime change of the West, especially in the Middle East and North Africa, have been disastrous. In the first 20 years of this century, the West has committed the most external violence. The costs, including opportunity costs of the regime change in Afghanistan, have been immense. The Americans have spent $2.3 trillion. The costs for the Afghan people remain to be calculated; they are second- and third-class victims. A comprehensive definition of the costs of war comprehends not only the human, material and financial costs but also the political, psychological, ecological, cultural and spiritual costs. America and its closest allies, representing less than 10% of the world population, have lost the legitimacy to prescribe and run the future world order. Their coercive diplomacy and military force have not resolved conflicts. They have destroyed regimes and increased the suffering of citizens. Foreign-and-defense policymaking is the least transparent sector in the political decision-making process. It caters to specific internal and foreign interests. The post-American international system will rest on multiple pillars: the US and its close allies; China; the European Union; Russia; India; and the emerging regional communities or unions of Africa, Latin America and the Middle East. The strengthening of an African or sub-African pillar is of crucial importance to building sustainable peace. Africa could become more independent or positively interdependent to defend its own economic interests and increase its international bargaining power. Although colonialism ended in the 1960s, the decolonisation of history and the critical race theory started 60 years later.

Much research has already been conducted on the prevention and mitigation of human and natural violence. Sustainable peace is enhanced by a series of conditions that can be clustered into 5+1 groups or building blocks. The blocks are metaphors. Peace is not a Lego-block structure or temple to be built or assembled. All the building blocks are necessary; they are partial contributions to building peace or mitigating violence. Positive synergies within and between the building blocks raise the sustainability of peace. The first building block is an effective and reliable system of communication, consultation, negotiation and information. The second building block consists of (a) political structures with a high degree of good governance, high on effectiveness and legitimacy and low on corruption; (b) economic structures that generate sustainable development and positive expectations about the future and that eliminate gross vertical and horizontal inequalities; and (c) defense and security structures that provide objective and subjective security for people by effectively dealing with both internal and external threats. The third block is the software of peace. It involves the creation of an integrative social-political climate, characterised by hope, trust, reconciliation, we-ness feelings, respect for nature, etc. The fourth block consists of supportive institutions, such as an effective and legitimate judicial system, an empowering educational system, sustainable-peace-enhancing media, a health system accessible to all citizens, an administration promoting environmentally friendly behaviour and humanitarian aid in the immediate post-conflict phase. The fifth building block is a peace-enhancing regional and global political neighbourhood. The installation of the building blocks requires skills and a critical mass of peacebuilding leadership. In each of the building blocks, attention is given to the natural environment. The sustainable peace paradigm is comprehensive and not reductionist; all conditions are necessary for sustainable peace. Sustainable peace work entails transdisciplinary research and thinking about how to deal with the complex interactions and synergies between the building blocks. Sustainable peacebuilding is inclusive and emancipatory. All stakeholders must feel that they are being heard. Peace, or pacification imposed from a position of strength, tends to be unjust and dishonourable and, therefore, unsustainable.

The building of sustainable peace is seriously hampered by time, or the way decision makers deal with time (temperament). Although most people and policymakers value and respect time, a great deal of the temporal analysis and temporal behaviour in human and environmental conflicts shows gross deficiencies. Examples are the non-anticipation of the negative effects of the regime change interventions in Libya and Syria (civil war, ISIS, migration); the premature declarations of success in Afghanistan, Iraq and Libya; the missed opportunities and immense opportunity costs of regime changes; the prevalence of reactive, rather than proactive, violence prevention; fear and despair; the endless use of fast and slow violence vis-à-vis the Palestinians; the distortion of colonial history; the speed of trust; the denial of climate change; wasting the time of others; time running out to achieve a net-zero world regime; etc. The time to radically change the way we deal with time and develop a more adaptive temperament is now. Time is precious, irreversible and non-renewable. More systematic attention should be given to the role of time in conflicts and the impact of the temporal behaviour of stakeholders. This implies (a) knowledge of the multidimensional nature of time and temperament; (b) systematic analysis of the role of time and temperament in conflict situations; (c) an assessment of the costs and benefits; and (d) demanding accountability for temporal incompetence, misconduct and violence. Temporal misconduct is wrongful and involves improper conduct motivated by premeditated or intentional purposes or by obstinate indifference to the consequences of one's acts. Temporal violence refers to depreciating quantitative and qualitative life expectancies by protracting conflicts, long-term sanctions and structural violence, ecological deterioration or wasting the time of others.

Civil society in the peacebuilding in Sub-Saharan Africa in the Anthropocene significantly contributes to our understanding of peacebuilding in many ways. It acquaints the reader with peace research and peacebuilding from a Sub-Saharan perspective. It elicits the normative, theoretical and epistemic assumptions underlying the peacebuilding efforts in the Sub-Saharan region and highlights the complex linkages between ecological peace and peacebuilding. Suitable field research findings are presented from the Democratic Republic of Congo, South Africa, South Sudan and Zimbabwe. A great deal of the research focuses on the role of civil society (the grassroots, religious organisations, women groups and informal peace committees). The study shows the richness of African conflict transformation knowledge and know-how. I wonder why it is still termed *indigenous* or *traditional*. Interesting is the depiction of the impact of Western, Middle Eastern and Asian thinking about human and natural peace. The book also focuses closely on the relations with the Global North and China and emphasises the importance of research partnerships at local and international levels. Intellectual solidarity and the commitment of scholars to use research and education for advancing peace are very important in the pursuit of peace research and action in the midst of conflict, violence, tension, groupthink and intolerance. Critical peace work uses courage, collective responsibility and joy. This insightful book maps out the challenges of peacebuilding in Sub-Saharan Africa and shows how peace research can improve the prospects of avoiding war and reducing unarmed violence.

Leuven, Belgium
October 2021

Prof. em. Luc Reychler

Prof. em. Luc Reychler Ph.D. Harvard is an Emeritus Professor of International Relations at the Faculty of Social Sciences at the University of Leuven. He was previously Director of the *Centre for Peace Research and Strategic Studies* (CPRS) and *Secretary General of the International Peace Research Association* (IPRA, 2004–2008). He has published widely on sustainable peacebuilding architecture, the planning and evaluation of violence prevention and peacebuilding interventions, and multilateral negotiations. His books include *Le défi de la paix au Burundi* (Paris: L'Harmattan, 1999); *Democratic Peacebuilding and Conflict Prevention: the Devil is in the Transition* (Leuven: Leuven University Press, 1999); *Peacebuilding: A Field Guide* (Boulder, CO: Lynne Rienner, 2001); with Thania Paffenholz, *Aid for Peace: A Guide for Planning and Evaluation in Conflict Zones* (Baden-Baden: Nomos, 2007); *De volgende genocide* (The next genocide; Leuven: Leuven University Press, 2004); *Time for Peace: The Essential Role of Time in Conflict and Peace Processes* (Queensland: University of Queensland Press, 2015); with Arnim Langer, and *Luc Reychler: Pioneer in Sustainable Peacebuilding Architecture* (Cham: Springer Nature Switzerland, 2020).

Address: Prof. em. Luc Reychler, Kerkstraat 4, 321 Binkom, Belgium.

Email: luc.reychler@soc.kuleuven.be.

Websites: https://soc.kuleuven.be/lines/staff-1/staff/00005289;
https://www.uqp.com.au/authors/luc-reychler;
https://en.wikipedia.org/wiki/Luc_Reychler.

Acknowledgements

We would like to express our utmost gratitude to all the institutions and individuals who have contributed to bringing this book to fruition. It has been a long-term endeavour with roots in the exceptional relationships that bind Professor Geoff Harris and his former doctoral students, who are emerging as significant scholars in the field of peacebuilding. Their zeal and passion to develop a peaceful continent during the Anthropocene are key drivers of this contributed volume. That said, their efforts would have remained limited if they had not been supported by other Africa-based scholars who also contributed chapters to the book.

As we nurtured the thought of producing a multidisciplinary research volume, many colleagues came onboard. We are indebted to the Durban University of Technology (DUT), particularly to its Deputy Vice Chancellor of Research, Engagement and Innovation, Prof. Sibusiso Moyo; and Mr. Crispin Cole Hemson, Director of the International Centre of Nonviolence, for their unwavering support. We also acknowledge the generosity of the Divine Word Missionaries (SVD) of the Democratic Republic of Congo for providing material support and logistics during our research and editing of this volume, more importantly for the entire SVD Congregation; for inspiring this book via one of the characteristic dimensions of its mission and charism, notably justice and peace and integrity of creation.

We thank Springer Nature Switzerland, our chosen publishing house, for agreeing to publish this contributed volume. We are grateful to the reviewers who gave their time anonymously to assess the scientific value of the draft chapters submitted for this book. We thank the following language editors for their professionalism and commitment to the timely submission of drafts to the language editors: Ms. Elizabeth Norton-Amor, Ms. Vanessa Greatorex and Mr. Alan Bell.

Finally, we express our particular gratitude to Hans Günter Brauch. This book is only the latest of a range of cutting-edge contributions to global peace and security, peace ecology and environmental justice for which he has been responsible. We greatly appreciate his unselfish assistance in bringing this book to completion; and we are profoundly indebted to those who prepared and produced the book in Heidelberg and Chennai (Tamil Nadu, India).

Durban, South Africa
October 2021

Jean Chrysostome K. Kiyala
Geoff Thomas Harris

Contents

Part I Civil Society Organisations as Key Players in Peacebuilding in the Anthropocene

1 **General Introduction** 3
Jean Chrysostome K. Kiyala and Geoff Thomas Harris

2 **Peace Ecology in the Anthropocene for Africa** 17
Hans Günter Brauch

3 **Civil Society, Climate Change, Security Risks Management and Peacebuilding in the Anthropocene** 67
Jean Chrysostome K. Kiyala

4 **Faith-Based Organisations and Environmental Peace: Impact of *Laudato Si'* and Its Relevance in the Anthropocene** 95
Jean Chrysostome K. Kiyala

5 **Healing Nature and Creation in the Anthropocene: A Reflection on the Role of Religion** 127
Dumisani Maqeda Ngwenya

Part II Micro-level Case Studies

6 **Empowering Women to Build Peace: A Case Study of Grace to Heal Trust in Lupane District** 145
Ntombizakhe Moyo-Nyoni and Jean Chrysostome K. Kiyala

7 **Peacebuilding Through Indigenous Knowledge Systems: Lessons from Civil Society Organisations in Zimbabwe** 191
Ashton Murwira

8 **Civil Society and Peacebuilding in the Kivu Provinces of the Democratic Republic of Congo** 205
Georges Bomino Bosakaibo

9 **Zimbabwe's National Peace and Reconciliation Commission and Civil Society: Partners in Peacebuilding?** 235
Lawrence Mhandara

10 **Civil Society and Peacebuilding in the Democratic Republic of Congo: Opportunities, Challenges and Recommendations** 251
Jean Chrysostome K. Kiyala

11 **Promoting Democratisation in DR Congo: The Case Study of Act for Transparent and Peaceful Elections** 271
Jean Chrysostome K. Kiyala and Gerard Mpang'de Bisambu

12 **Civil Society Organisations and School Peace Clubs in South Africa: An Outcome Evaluation** 307
Dorothy Moyo

13 **The Challenges and Opportunities for Civil Society in Peacebuilding: Lessons from the Matabeleland Collective, Zimbabwe** 319
Dumisani Maqeda Ngwenya

14 **The Role of Civil Society Participation in Social Cohesion Processes in KwaZulu-Natal, South Africa** 337
Paul Kariuki

15 **Building the Capacity of Civil Society Organisations in Nonviolent Campaigning: A Case Study from South Sudan** 355
Moses Monday John

16 **How Informal Peace Committees Complement Macro-infrastructures for Peace in Zimbabwe** 383
Norman Chivasa

17 **Protecting and Nurturing Children Born from Rape in South Kivu Communities: A Challenge for Civil Society** 397
Josephine Kimanu Mauwa, Sylvia Blanche Kaye, and Denis Mukengere Mukwege

Part III Macro-level Case Studies

18 **Indigenous Peacebuilding Approaches and the Accountability of Former Child Soldiers: African Case Studies** 423
Jean Chrysostome K. Kiyala

19 **Navigating Channels for Civil Society Participation in Conflict Prevention and Peace-Making in the African Peace and Security Architecture** 465
Michael Aeby

Part IV Enhancing the Role of Civil Society in Peacebuilding

20 Centre-Periphery Relations Between Civil Society Organisations and External Funding Partners: A Case Study of South Kivu's Peace and Development Agenda ... 501
Philippe Mulume-Oderhwa Kaganda and Jean Chrysostome K. Kiyala

A Manifesto with Proposals for Action Towards a Peace Ecology and an Ecological Peace Policy: A Global Research Partnership Between Organisations in Sub-Saharan Africa and Countries in the Global North ... 527

On Durban University of Technology ... 543

On International Centre of Nonviolence (ICON) ... 545

On University of Bandundu ... 547

On the Editors ... 549

On the Contributors ... 553

Index ... 557

Acronyms and Abbreviations

14P	Infrastructures for peace
AAEC	Association of African Earthkeeping Churches
ACC	Amadiba Crisis Committee
ACCORD	African Centre for the Constructive Resolution of Disputes
ADEPAE	*Action pour le Développement et la Paix Endogènes* [the Action for Endogenous Development and Peace]
ADI-KIVU	Association for Integrated Development of Kivu
ADR	Alternative Dispute Resolution
AEFAD	*Action d'Encadrement de Famille pour le Développement Intégrale* [Family Support Action for Integral Development]
AETA	Act for Transparent and Peaceful Elections
AFD	*Agence Française de Dévelopement*
AFESD	Action for Children in Difficult Situations
AICs	African Initiated Churches
ANC	African National Congress
APC	Action for Peace and Concord
APC	Arms Procurement Commission
APSA	African Peace and Security Architecture
AR	Action Research
ASONET	African Solidarity Network
ATR	African traditional religions
AU	African Union
AUC	African Union Commission
AVP	Alternatives to Violence Project
AWCs	Association of Women's Clubs
AWG	Anthropocene Working Group
AZTREC	Association of Zimbabwean Traditional Ecologists
CACH	Cape for Change
CAN	Aid to the Church in Need
CASAC	Council for the Advancement of the South African Constitution
CBOs	Community-based Organisations

CCJP	Catholic Commission for Justice and Peace
CCL	Christian Council of Lesotho
CCMT	Centre for Conflict Management and Transformation
CDMC	*Comité de Médiation et de Conciliation* [Mediation and Conciliation Committee]
CEDA	Centre d'Etudes et de Documentation et d'Animation Civique
CEDAC	Documentation and Civic Animation
CENCO	*Conférence Épiscopale Nationale du Congo* [National Episcopal Conference of THE Congo]
CENI	*Commission Electorale Nationale Independence* [National Independent Electoral Commission]
CEWS	Continental Early Warning System
CFCs	Chlorofluorocarbons
CIAM	Centre for Information and Missionary Animation
CIDO	The Citizens and Diaspora Directorate
CLC	Lay Coordination Committee
CLOC	*Comité locale Communautaire* [Local Community Committee]
CNSA	National Agreement Monitoring Council
COMESA	Common Market for Eastern and Southern Africa
CoW	ECOWAS Council of the Wise
CPA	Comprehensive Peace Agreement
CS	Civil Society
CSAC	Superior Council of Audiovisual and Communication
CSI	Child Soldiers International
CSOs	Civil Society Organisations
CUG-LaS	Catholic University of Goma-La Sapientia
CW	Corruption Watch
DDP	Democracy Development Program
DR Congo	The Democratic Republic of the Congo
DRC	Democratic Republic of Congo
DUT	Durban University of Technology
ECLF	Ecumenical Church Leaders Forum
ECOSOCC	Economic Social and Culture Council
ECOWARN	ECOWAS Early Warning and Response Network
ECOWAS	Economic Community of West African States
ECPF	ECOWAS Conflict Prevention Framework
EFEAC	*École de Formation Électorale en Afrique Centrale* [The Electoral Training School in Central Africa]
EFZ	Evangelical Fellowship of Zimbabwe
ENGOs	Environmental non-governmental organisations
ES	Earth system
ESA	Earth Systems Analysis
ESAP	Economic Structural Adjustment Programme
ESS	Earth Systems Scienc
ESS	Ecumenical Support Services

FAO	Food and Agricultural Organisation of the United Nations
FAT	*Forum des Amis de la Terre* [the Friends of the Earth Forum]
FBOs	Faith-based organisations
FCC	Common Front for Congo
FDLR	Democratic Forces for the Liberation of Rwanda
FGDs	Focus Group Discussions
FIES	Food Insecurity Experience Scale
FNL	National Liberation Forces
FOCCISA	Fellowship of Christian Councils in Southern Africa
FOPAC	Federation of Agricultural Production Organisations of Congo
FPCs	formal peace committees
GBV	Gender-Based Violence
GEC	global environmental change
GHG	greenhouse gas
GIZ	German Corporation for Development Cooperation
GNU	Government of National Unity
GPA	Global political agreement
GTH	Grace to Heal
HCR-PT	High Council of the Republic-Parliament of Transition
HD	Centre for Humanitarian Dialogue
HDI	Human Development Index
HSCD	Human Security and Civil Society Division
ICC	International Criminal Court
ICD	Inter-Congolese Dialogue
ICJ	International Court of Justice
I-CSOs	International-Civil Society Organisations
IEC	Electoral Commission
IGOs	Intergovernmental Organisations
IJR	The Institute for Justice and Reconciliation
IKS	Indigenous Knowledge Systems
INGOs	International Non-Governmental Organisations
I-NGOs	International Non-Governmental Organisations
IPCC	Intergovernmental Panel on Climate Change
IPCs	Informal peace committees
IPSS	Institute for Peace and Security Studies
IPV	intimate partner violence
IRC	International Rescue Committee
ISS	Institute for Security Studies
ISSC	International Social Science Council
IVP	Life and Peace Institute
KZN	KwaZulu-Natal
KZNCC	KwaZulu-Natal Christian Council
KZNCSOC	KwaZulu-Natal Civil Society Organisations Coalition
KZNSCMRC	KwaZulu-Natal Social Cohesion, Moral Regeneration Council
LAV	Let Africa Live

LDH	Human Rights League
L-NGOs	Local Non-Governmental Organisations
LOMA	Law and Public Order Maintenance Act
LRA	Lord's Resistance Army
LSA	Learner Support Agents
MCC	The Mennonite Central Committee
MDC	Movement for Democratic Change
MDIs	multi-donor interventions
MDTFs	Multi-Donor Trust Funds
MFD	The Mediation Facilitation Division
MK	Umkonto We Sizwe
MMR	Mixed Methods Research
MONUSCO	United Nations Mission for the Stabilisation of the Congo
MoU	Memorandum of Understanding
MPR	*Mouvement Populaire de la Révolution* [Popular Revolutionary Movement]
MRG	Mediation Reference Group
MSC	Mediation and Security Council
MSU	Mediation Support Unit
NAYO	National Association of Youth Organisation
NCA	National Constitutional Assembly
NCD-NGOs	National Council of Development NGOs
NCP	National Congress Party
NFCCS	National Framework for consultations Civil Society
NGOs	Non-Governmental Organisations
NOEE	National Order of Electoral Experts
NPRC	National Peace and Reconciliation Commission
NTJWG	National Transitional Justice Working Group
OAU	Organisation of African Unity
OECD	Organisation for Economic Co-operation and Development
ONAD	Organisation for Nonviolence and Development
ONHRI	Organ on national healing, reconciliation and integration
OPAC	Convention on the Rights of the Child on the Involvement of Children in Armed Conflict
OSISA	Open Society Initiative for Southern Africa
PAC	Provincial Advisory Council
PBs	Planetary boundaries
POLAD	Political Actors Dialogue
PoW	Panel of the Wise
PSD	Peace and Security Department
QDA	Qualitative Data Analysis
RADHOSKI	Network of Human Rights Associations in South Kivu
RARCSS	Revitalised Agreement on Resolution of Conflict in the Republic of South
RCD-NGOs	Regional Council of Development NGOs

RECs	Regional Economic Communities
REWC SADC	Regional Early Warning Centre
RF	Rhodesian Front
RFDP	*Réseau des Femmes pour la Défense des Droits et la Paix* [Women's Network for the Defense of Rights and Peace]
RIO	Réseau d'innovation organizationalle/Organisational Innovation Network
RODHECIC	Network of Human Rights and Civic Education Organisations of Christian Inspiration
RSICC	Rome Statute of the International Criminal Court
SADC	Southern African Development Community
SAFCEI	Southern African Faith Communities Environment Institute
SAPS	South African Police Services
SAT	Southern Africa Trust
SATUCC	Southern African Trade Union Coordination Council
SIPRI	Stockholm International Peace Research Institute
SONAD	Sudanese Organisation for Nonviolence and Development
SPLM/A	Sudan People's Liberation Movement/Army
SPLM-IO	Sudan People's Liberation Movement—In Opposition
SPMC	*Solidarité des personnes marginalisées dans la communauté* [Solidarity of marginalised people in the community]
SRANC	Southern Rhodesia African National Congress
SSCC	South Sudan Council of Churches
STRN	Sustainability Transition Research Network Sudan
SYDIP	*Syndicat de Défense des Intérêts Paysans* [the Farmers' Defence Union]
TCG	Transnational Climate Governance
TfP	Training for Peace
TGoNU	Transitional Government of National Unity
UANC	United African National Council
UDF	United Democratic Front
UN	United Nations
UNDP	United Nations Development Programme
UNEP	United Nations' Environmental Programme
UNFCCC	United Nations Framework Convention on Climate Change
UNHCR	United Nations High Commission for Refugees
UNISA	University of South Africa
UNMISS	United Nations Mission in the Republic of South Sudan
UPEC	Civic Education Production Unit
US	United States
USIP	United States Institute of Peace
VICO	Guinea Pig Village
VSV	*La Voix de Sans Voix* [The Voice of the Voiceless]
WACSI	West African Civil Society Institute
WACSOF	West African Civil Society Forum

WANEP	West African Network for Peacebuilding
WARN	West African Early Warning and Response Network
WCC	World Council of Churches
WRI	War Resisters' International
ZANLA	Zimbabwe African National Liberation Army
ZANU	Zimbabwe African National Union
ZANU PF	Zimbabwe African National Union, Patriotic Front
ZAPU	Zimbabwe African People's Union
ZCA	Zimbabwe Christian Alliance
ZCBC	Zimbabwe Catholic Bishops' Conference
ZCC	Zimbabwe Council of Churches
ZCTU	Zimbabwe Congress of Trade Union
ZDI	Zimbabwe Democracy Institute
ZEC	Zimbabwe Election Commission
ZESN	Zimbabwe Election Support Network
ZHRC	Zimbabwe Human Rights Commission
ZIMCET	Zimbabwe civic Education Trust
ZINASU	Zimbabwe National Students' Union
ZIPA	Zimbabwe People's Army
ZIRRCON	Zimbabwean Institute of Religious Research and Ecological Conservation
ZNA	Zimbabwe National Army
ZPRA	Zimbabwe People's Revolutionary Army
ZPSP	Zimbabwe Peace and Security Programme
ZUM	Zimbabwe Unity Movement

Part I
Civil Society Organisations as Key Players in Peacebuilding in the Anthropocene

Chapter 1
General Introduction

Jean Chrysostome K. Kiyala and Geoff Thomas Harris

1.1 Introduction

Peacebuilding efforts over the past 3 decades have mitigated armed conflicts at the international and localised levels, but a new range of threats to human security and peace have emerged as a result of climate change. The planetary boundary framework (PBF)[1] assists in providing scientific data and its analysis concerning the risk that human activity poses to destabilising the Earth system. Planetary boundaries have been described as "scientifically based levels of human perturbation of the [Earth system] beyond which [its] functioning may be substantially altered" (Steffen et al. 2015, p. 736). Although this framework does not dictate the way society should develop, by "identifying a safe operating space for humanity on Earth, the PBF can make a valuable contribution to decision-makers in charting desirable courses for societal development" (Steffen et al. 2015, p. 736).

Being mindful of international and local patterns of conflicts and the violence that ensues, a holistic approach is needed to anticipate, prevent and tackle the various types of violence which are likely to be exacerbated by environmental conflicts. These types of violence, the ways of understanding them and effective means of tackling them are central issues in this book. Tsitsi Dangarembga, in her acceptance

[1] For a description of the planetary boundary framework, see https://www.stockholmresilience.org/research/planetary-boundaries.html and https://stockholmuniversity.app.box.com/s/avnyhh4xzshxb19j82hn5mf3hxyuvqj0.

Jean Chrysostome K. Kiyala is a Senior Lecturer, International Centre of Violence, Durban University of Technology, South Africa; Associate Professor and visiting lecturer the University of Bandundu, Catholic University of Bukavu and Evangelical University in Africa based in Bukavu (DR Congo).

Geoff Thomas Harris, Professor, International Centre of Nonviolence, Durban University of Technology.

J. C. K. Kiyala and G. T. Harris (eds.), *Civil Society and Peacebuilding in Sub-Saharan Africa in the Anthropocene*, The Anthropocene: Politik—Economics—Society—Science 34,
https://doi.org/10.1007/978-3-030-95179-5_1

Speech for the 2021 Peace Prize Awards of the German Book Trade, gave some hints about violence that humanity is grappling with:

> Obviously, peace cannot thrive under these conditions [physical, psychological, political, economic, metaphysical and genocidal violence] which are too often the order of the day in postcolonial countries. Only violence thrives in conditions of violence… It is a well-known fact that violence begets violence and we see this all over the world today, even in the various homes of empire. Imperial violence created conditions that caused many people to leave their homes and migrate to imperial countries. Citizens of the imperial countries resent this and mete out violence onto the bodies of the immigrants in various ways, including through institutional violence that is justified as an administrative necessity, a justification that was also given during the colonial era. At the same time, nationals of imperial nations who have a more developed sense of peace and justice take on their country mates who mete out violence to immigrants and conflict results in the imperial heartland. It is clearly a no-win situation. What are we to do, then, to foster peace? Clearly the world structure that ushered in the specific kinds of violence of our era cannot easily be undone… The more than seven billion of us human residents of this planet are today all connected to and embedded in that global system.

Dangarembga explains that the world is living in the context of structural violence, which encourages physical, psychological and genocidal violence, and the drivers of such violence are persistent political, economic, metaphysical and cultural structures. She links structural violence to the ecological crisis provoked by contemporary humans and emphasises that philosophy has helped develop erroneous conceptualisation of the knowing persons, particularly the cartesian philosophy, which has resulted in "multiple errors in our knowledge", including the damage that humans have inflicted on the ecosystem. She explains the need to think very differently:

> We must invent new thought, drag it out from where it is nascent in the folds of the universe to effect the paradigm shift in our ways of knowing and valuing and ascribing meaning that is necessary for our survival as we see oceans polluted, ozone depleted, climate changing, temperatures and shorelines rising, diseases ravaging in spite of science, hunger proliferating, and black bodies drowning in oceans on their way to join those who first sailed to join them, becoming this epoch's most enduring sacrifice to what it calls progress (Dangarembga 2021).

While stigmatising the knowledge system responsible for the multifaceted violence we experience in society, it is still the knowing person that is endowed with not only the scientific capacity but also ethical capability to pursue a new thinking and decision-making directions. It is imperative that our thinking and decision-making considers others – very broadly defined – and our natural environment. We must weigh the consequences of what we decide to do on ourselves, others and the word around us. This echoes Tsitsi Dangarembga's thinking; she avers:

> Our choices of thought content and process are ultimately a choice between violence producing and peace producing contents and narratives. This is true whether these contents and narratives are expressed only to ourselves - in thought - or whether we go on to express them to others around us. Both are generative (Dangarembga, 2021).

Against the backdrop of the multidimensional types and causes of violence and the consequences on humans and the environment, this book delves into eco-friendly conceptual frameworks that are crucial for the future of the ecosystem, global peace and security. These are the new paradigms of peace studies that have recently emerged, such *as environmental peacemaking* (Conca 1994, 2002), peace ecology (Kyrou 2007), and *environmental justice*.[2] Environmental peacemaking and peacebuilding emphasise the long-term advantages of "an environmental consciousness combined with a peace consciousness instead of an unguided effort at tracing the circumstantial, and amorphous 'peace revenues' from individual environmental projects" (Kyrou 2007, p. 79).

These paradigms seek to bring about a comprehensive approach to deal with conflicts in human communities, preserve ecosystems, and prevent further conflicts, whether these arise from economic, political, cultural, or social issues. Kyrou (2007) has argued that a new discourse – environmental peacemaking – has emerged, which works with the notions of "preventive diplomacy, peacemaking, peacekeeping and peace enforcement" (p. 600). The intricacies of each paradigm, its roles in promoting global peace, and its connections to conventional peacebuilding theory have been discussed by Barnett et al. (2007), Lederach (1997a). Emerging paradigms, such as peace ecology, have been surveyed by Brauch (2016), Brisman (2016), and Kyrou (2007), and environmental justice has been addressed by Schlosberg (2004, 2013) and Westra (2009).

This chapter discusses the aims and scope of the book, its central concepts – civil society, peacebuilding, and the Anthropocene – and their interconnections.

1.2 Aim and Scope

The book's overall aim is to examine interventions by civil society organisations (CSOs) in sub-Saharan Africa, with particular reference to those combatting climate change and tackling the anthropogenic causes of the Earth system's disintegration. These interventions resonate with African indigenous wisdom but also connect with the efforts of CSOs worldwide. The following issues are covered:

- Civil society and its peacebuilding commitment in the context of ecological changes and the pursuit of environmental peace in the Anthropocene
- Civil society and peacebuilding after large-scale violent conflicts, with particular emphasis on the challenges of accountability, reconciliation, healing, and nation-building
- Civil society and peacebuilding in the context of democratic processes and political transition
- The distinctive contributions from African social philosophies, such as Ubuntu (African humanism), Ujamaa (African socialism), and African Communitarianism, and traditional procedures and institutions, such as Baraza, Bushingantahe, Fambul Tok, Gacaca, and Mato oput

[2] For a brief introduction to the concepts on which there are extensive literature sources, see https://www.sciencedirect.com/topics/earth-and-planetary-sciences/environmental-justice.

- The partnership between local CSOs and their funding partners in the Global North
- The role of faith-based CSOs in efforts to care for and protect the planet

These themes are analysed by scholars from different fields who intend to awaken global conscience about how humanity should conduct its business to avoid a total collapse of the ecosystem. The section that follows justifies the significance of the contributions in this book.

1.3 Civil Society

The notion of civil society originates from the ancient Greek and Roman civilisations. Over the past 2 centuries, the concept has undergone historical and semantic development. In The Philosophy of Right (1820), Hegel asserted that the modern political community is composed of three levels: the family, civil society, and the state (Arato 1988). According to Hegel, civil society evolves as the place of economic development and labour to meet societal needs, while the state regulates everything by means of its legal and political organisation (Pelczynski 1984). Civil society, then, remains at the centre of the whole, independent from the family, the state, and business and represents "the pluralism of values, ideals and ways of life embodied in its institutions" (DeWiel 1997, p. 3). The state regulates the whole system because, without such rules and control, civil society would degrade into a state of nature which, in Hobbesian terms, represents the war of all against all. One definition speaks of civil society as follows:

> [Civil society] represent[s] a separate sphere of human relations and activity, differentiated from the state but neither public nor private or perhaps both at once, embodying not only a whole range of social interactions apart from the private sphere of the household and the public sphere of the state, but more specifically a network of distinctively economic relations, the sphere of the market-place, the arena of production, distribution and exchange (Wood 1990, p. 61).

A contemporary understanding of civil society's interaction with the state has been provided by the World Bank's 2007 report, acknowledging the legal and regulatory framework of the state which also supports civil society's activities with its subsidies. In turn, civil society operates as a liaison institution between the state and citizens – to promote values and accountability while also serving as the voice and conduit of information between the two (World Bank 2007, p. 4). According to Paffenholz and Spurk (2006, p. 2), civil society is the "sector of voluntary action" with institutional forms that are distinct from those of the state, family, and market. With this understanding, civil society exists to both resist the state when it follows undesired paths and promote positive interventions to bring about desired changes. Civil society has a strongly voluntary foundation and consists of "non-state actors and associations that are not purely driven by private or economic interests, are autonomously organised, and interact in the public sphere"; and finally, "it is independent from the state, but it is oriented toward and interacts closely with the state and the political sphere" (Paffenholz/Spurk 2006, pp. 2–3).

As suggested by the World Bank (2007), the domain of civil society's action includes the context of armed conflict. It responds to societal needs created by armed conflict when the state is paralysed, and business has been incapacitated. Armed conflict often destroys hospitals and health facilities, schools, markets, and other service delivery institutions. In times of civil war or other hostilities, shops are looted; families are displaced, and some of their members may even be killed; houses are destroyed; and the judicial structure is destroyed along with other infrastructure. In such contexts, Paffenholz/Spurk (2006) acknowledged the support that civil society receives. Such aid ought to be considerate of the dramatic impact of armed conflict on the lives of all people: it alters all levels, starting from "individual changes in attitudes and behaviour (trust and confidence) over economic and social change, to ultimate shifts of power relations in communities, regions and the society as a whole" (Paffenholz 2015, p. 11). Civil society stands as an intermediate sector that can facilitate social cohesion, rebuild social capital, and participate in community and national reconstruction.

In summary, CSOs comprise special interest groups, faith-based organisations, traditional and community groups, researchers and research institutions, humanitarian or development service delivery organisations, human rights and advocacy organisations, conflict resolution and peacebuilding non-governmental organisations (NGOs) and international NGOs, social and political movements, business associations, and various networks (Paffenholz 2015, pp. 108–109).

1.4 Functions of Civil Society

Civil society engages in a range of functions, which Paffenholz (2015) and Paffenholz/Spurk (2006) have categorised as follows:

- Local and international protection is assured by NGOs, such as Peace Brigades International, Amnesty International, Human Rights Watch, Coalition to Stop the Use of Child soldiers, Save the Children, etc.
- Local and international monitoring refers to activities that include monitoring the human rights situation, putting in place structures that implement, evaluate, and make recommendations to policymakers or human rights advocacy groups. This monitoring can be useful to individual government and armed groups incriminated in human rights violations and warn the international community (such as the UNOCHA [United Nations Office for the Coordination of Humanitarian Affairs], ECOWAS [Economic Community of West African States], and regional NGO peace networks). The International Crisis Group plays a dual role in monitoring, protection, and advocacy.
- Local and international advocacy entails putting in place an agenda of pressure action to ease negotiations and the implementation of negotiated settlements

to curb the recurrence of armed hostilities. Thus, advocacy can mean a public or non-public action. When it is public, advocacy involves signing petitions, leading demonstrations, sending press releases, making use of social media, and resorting to public campaigns; when it is non-public, it employs informal means of communication or dialogues and relations.

- Socialisation action "refers to in-group bonding that supports democratic behaviour and promotes tolerant and peaceful values within society" (Paffenholz 2015, p. 111).
- Intergroup social cohesion ensures that various groups, whose "social capital" has been degraded and destroyed during wars, learn to live together in peaceful cohesion and harmony.
- Facilitation and mediation are also evident. These describe civil society's role to facilitate and assist stakeholders in a conflict to come together in a transition process, at both the local and national levels.
- Service delivery is also critical because war destroys structures and the fabric of society and, with these, the flow of service delivery. In this environment, civil society may act to bring aid and social services to the needy; it is also able to help war-affected communities and support their efforts at reconstruction on a broader level.

The multiplicity of functions in which civil society engages reveals its importance and relevance in peacemaking and post-conflict reconstruction on all fronts, including monitoring peace processes and human rights conditions, democratisation, rebuilding social capital among antagonist groups, and the protection of war-affected people.

Of particular importance is the repair of social capital, that is, the resources available to people and entities due to their networks and the social relations that bind people together, and the shared values which stem from those networks. Figure 1.1 shows the range of issues covered by the concept of social capital and its interconnectedness.

When society and individuals cooperate and act on ethical principles, such as trust, morality, fairness, justice, and reciprocity, all communities enjoy an increase

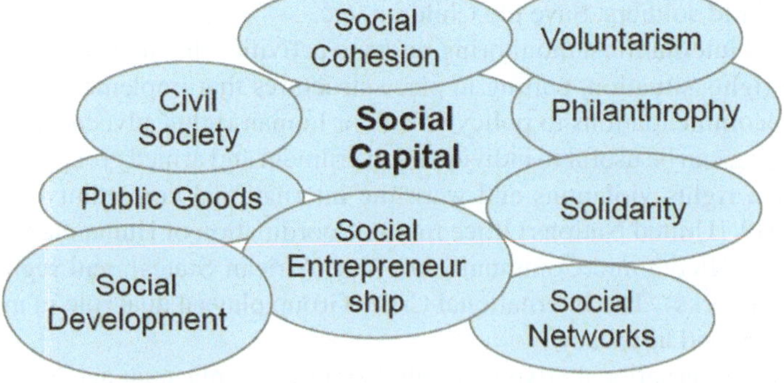

Fig. 1.1 Social capital: components and connections. *Source* Dill (2015, p. n.p.)

in economic development and the safety of citizen participation in social and political life. These elements that comprise the social fabric are damaged and may be destroyed by armed conflict. In such contexts, civil society can help repair the social fabric and restore wounded interpersonal relations.

1.5 Civil Society and Peacemaking

Kenneth Boulding (2002) defined peace as "a mosaic of identities, attitudes, values, beliefs, and patterns that lead people to live nurturantly with one another and the earth itself without the aid of structured power differentials, to deal creatively with their differences and share their resources" (p. 6, emphasis added). According to Ehrhart/Schnabel (2005), building peace implies the creation of political, economic, and social conditions to support sustainable justice, security, and development – the authors also emphasised that "In post-conflict situations, internal and external actors must cooperate in mutually reinforcing the socio-economic, governance, and security dimensions of a highly fragile environment" (p. 7).

Peacebuilding includes both the prevention of violence and the processes which help nations, communities, and individuals to recover from violence and set them on the road to sustainable peace. The United Nations Development Programme (UNDP 2015, 2020) assumed that peacebuilding and violence prevention are virtually synonymous and used the two concepts interchangeably (Barnett et al. 2007, p. 42). Similarly, the World Bank (2007) underlined that peacebuilding has multiple objectives – the prevention and management of armed conflict and sustaining of peace for at least a decade after violence has attenuated or ended, providing the conditions for rebuilding the economy, and accelerating development and democratisation. On this basis, peacebuilding operates in three time periods – "prevention prior to the outbreak of violence, conflict management during armed conflict, and post-conflict peacebuilding for up to 10 years after the conflict ends" (World Bank 2007, pp. 6–7).

Such a broadening of the concept of peacebuilding, however, means that almost every socioeconomic development initiative can fall under its agenda, making it difficult to determine where peacebuilding activities stop and "normal" development activities begin. Similarly, "implicit causal linkages are constructed between general development activities and peacebuilding objectives, but there is little evidence to substantiate these links … the political peacebuilding objectives seems to be tenuous and requires further exploration" (World Bank 2007, p. 9).

One of the major concerns of societies in transition from armed conflict is the creation of an environment that promotes durable peace. Civil society can play a central role in dialogues and negotiations aimed to manage, resolve, and transform conflicts. While Track I diplomacy is normally conducted by high-level government officials, civil society may be included at the negotiation table. If they are not, they may be members of official consultative forums that often run parallel to official negotiations. Moving down the scale of direct involvement, civil society may engage in less formal consultations, inclusive post-agreement mechanisms, high-level civil society initiatives, public participation, mass action, campaigns, demonstrations, street action, protests, and petitions (Paffenholz 2014a). These examples of CSO involvement are not applied individually because more than one may apply during a peace process (Paffenholz 2014a).

It is apparent that CSOs are involved at all levels. At level 1 (top leadership), religious leaders with high reputations may represent civil society; at level 2 (middle-range leadership), ethnic and religious leaders, academics or intellectuals, and leaders from humanitarian CSOs may take part in peace processes; and at level 3 (grassroots leadership), leaders feature prominently. The actions undertaken by or expected of these components can be summarised as follows: to unravel the drivers of conflict, deal with the implosion and explosion of conflicts that have turned violent and find realistic alternatives to contain these, and create infrastructure to maintain peace and build social cohesion.

The pre-eminence of the middle-range actors is explained by their positioning between the top range and grassroots levels, and middle-range leaders enjoy the trust of both top-level and grassroots actors. As Lederach (1997b, p. 94) stated, "They have more flexibility of thought and movement than top-level leaders and are far less vulnerable in terms of daily survival than those at the grassroots". The pyramid focuses on local CSOs, whose role is crucial and indispensable in initiating, building, and sustaining peace. Their actions may require financial aid, which normally comes from external partners, but their programmes for peace and reconciliation will emerge from within the community. Traditional or indigenous peacebuilding mechanisms may be particularly effective in achieving the agenda of peace.

Amidst several theories of peacebuilding, the works of Lederach (1997a), Paffenholz (2014b), and others have underlined the extent to which local initiatives can leverage peace efforts and accelerate social cohesion, which is vital for the transition towards sustainable peace. In post-conflict contexts, much of the reconciliation, healing, and caring for victims is carried out by various CSOs. Lederach's (1997a) pyramid of peacebuilding (shown in Fig. 1.2) depicts the actors involved in peacebuilding during this transition and the approaches that are adopted.

Beyond the model proposed by Lederach, civil society has involved also in peace diplomacy, especially when peace negotiations demand the application of Track II Diplomacy whereby facilitation initiatives typically involve non-government actors and may occur in parallel to official negotiations; they typically emphasise a problem-solving approach (Paffenholz 2014a). Such "interactive conflict resolution programs" (Chataway 1998, p. 270) are useful because they can create opportunities to consider new ideas, work across party lines, and possibly bring about changes in the political culture of each side (Chataway 1998; Paffenholz 2014a)

An important instrument of peacebuilding can be "community mobilisation", described by Erasmus (2001, p. 249) as "a means of tapping into the knowledge and resources of the local community and fostering a spirit of community ownership". He argued that this is one of the most effective means of reducing conflict at the grassroots level and emphasised that it needs to be sustained through dialogue, formally or informally, with all stakeholders (Erasmus 2001, p. 249). This approach should be

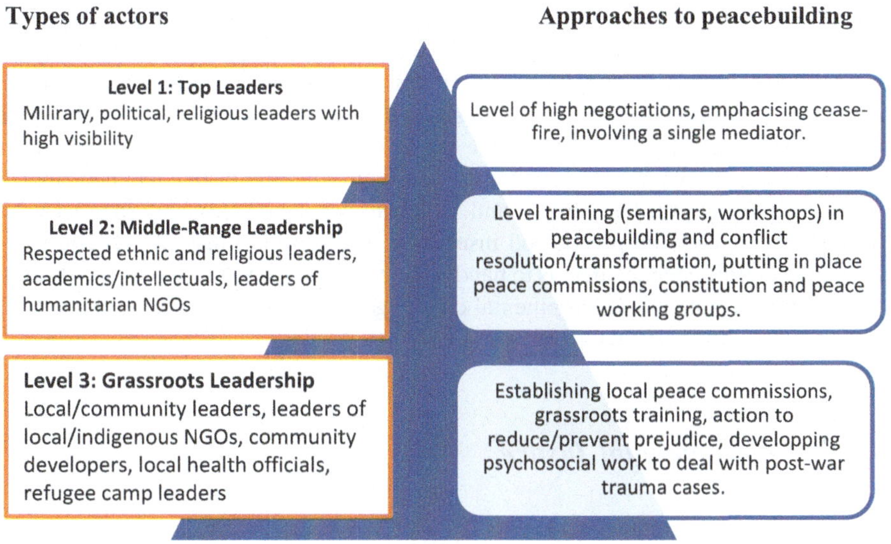

Fig. 1.2 Pyramid of peacebuilding actors and approaches. *Source* Authors' adaption of Lederach's peacebuilding pyramid (1997a, p. 39)

participatory in nature, in the sense that the community – possibly via a representative team – is involved in planning, implementing, and evaluating the process.

1.6 The Anthropocene

The Anthropocene concept was devised by Paul Crutzen around the turn of the century (Crutzen 2016). Hans Günter Brauch (in Chap. 2 of this book) explains it as a turning point in Earth's history in 1945; as a new context for human activities; as multiple challenges posed by anthropogenic global environmental and climate change; and as an opportunity for a peaceful transition towards sustainability resulting from transformations to society, the economy, and the political processes during the 21st century.

During the first phase of the Anthropocene (1945–2020), sub-Saharan Africa experienced a range of armed conflicts along with natural disasters in the form of floods, droughts, forest fires, locust plagues, and epidemics. The connection between armed conflict and these disasters was not established to any significant extent (Bronkhurst 2011). Due to the continent's projected population growth, the demand for drinking water, agricultural land, and food supplies will increase significantly, while there may be an even greater lack of food supplies and jobs for Africans during the Anthropocene's second phase (2021–2100). These climate-related environmental scarcities may result in new forms of violent climate conflicts. Tipping points in the climate system may well trigger increasing geopolitical conflicts in Africa.

The concept of peace ecology is intrinsically linked to peace studies (Kyrou 2007) as well as environmental justice (Schlosberg 2013; Westra 2009). The creation of refugees and internally displaced people as a result of armed conflict emanating from climate change poses massive challenges for environmental justice and contributes to the potential for further conflict and violence over land. This may interact with droughts to create widespread food insecurity. The essential task of peacebuilders during the Anthropocene, then, is to handle the complex combination of global security, environmental, and developmental challenges (Brauch 2016; Brauch/Scheffran 2012; Oswald Spring/Brauch 2021).

1.6.1 Environmental Justice

The theory of environmental justice has evolved horizontally, covering a larger scope of issues; vertically, it has focused on the comprehensive essence of environmental injustices, and conceptually, it has considered "the human relationship with the non-human world" (Schlosberg 2013, p. 37). The vertical perspective justifies the engagement of faith-based organisations (FBOs) and social movements to stand against politically and economically driven policies that subjugate and subdue the planet (Deen 1996; Deane-Drummond 2016; Eaton 2007; Gottlieb 2003). Schlosberg (2004) envisioned environmental justice beyond the assumptions of equity and the distributions of environmental goods and ills:

> The argument is that the justice demanded by global environmental justice is really threefold: equity in the distribution of environmental risk, recognition of the diversity of the participants and experiences in affected communities, and participation in the political processes which create and manage environmental policy (p. 517).

In the midst of the massive environmental challenges of the Anthropocene, environmental justice calls for international solidarity between humanity and the Earth; peace ecology is a means of environmental justice.

1.6.2 Peace Ecology

According to Kyrou (2007, p. 79), "Peace Ecology creates conceptual space for looking at the peacebuilding potential of environmental practices and projects regardless of whether they are driven by problem solving or by a worldview, whether they focus on some task at hand or on human consciousness". Kyrou (2007) described peace ecology as a new paradigm that is inspired by the concept of "environmental peacemaking" – understood as "a theoretical framework, broad and integrative enough to allow a full understanding, functionally as well as philosophically, of the inherent capacities of the environment to inform and sustain peace" (p. 73).

The well-being of the Earth system is inherently linked to peace and security (UNDP 2015, 2020). The threats posed by human action to ecosystems leave the Earth and its inhabitants vulnerable and insecure. To quote Dalby (2007, p. 155), "Recent innovations in earth system science have added compelling arguments for the integration of environmental matters into security". There has been a shift in security concerns from scarcity- and resource-related conflicts in the Global South to worries over climate change and other consequences of environmental degradation resulting from intense human activities. This reorientation implies the need to create an atmosphere of durable peace in regions that are the most affected by climate change and global warming. Both environmental security (Dalby 2007) and peace ecology can be enhanced by humankind's action and attitude towards its common home.

1.7 Structure of the Book

The book consists of 21 chapters arranged in four parts. The first – CSOs, FBOs, and peacebuilding in the Anthropocene – acts as a foundation in which the key concepts are explained and connected to each other. In parts two and three, a range of micro and macro case studies of civil society and peacebuilding are presented. In the final part, particular emphasis is given to the interactions of CSOs from the Global South and those from the Global North.

References

Arato, A. (1988). A Reconstruction of Hegel's Theory of Civil Society. *Cardozo Law Review*, 10, 301–321

Barnett, M., Kim, H., O'Donnell, M., & Sitea, L. (2007). Peacebuilding: What Is in a Name? *Global Governance*, 13(2007), 35–58

Boulding, E. (2002). Peace culture. In M. Afkhami (Ed.), *Toward a compassionate society* (pp. 8–15). Bethesda: Women's Learning Partnership

Brauch, H. G. (2016). Sustainable Peace in the Anthropocene: Towards Political Geoecology and Peace Ecology. In H. Brauch, Ú. Oswald Spring, J. Grin, & J. Scheffran (Eds.), *Handbook on Sustainability Transition and Sustainable Peace* (pp. 187–236): Springer. Hexagon Series on Human and Environmental Security and Peace book series (HSHES, volume 10). Cham: Springer International Publishing. https://doi.org/10.1007/978-3-319-43884-9_9

Brauch, H. G., & Scheffran, J. (2012). Introduction: Climate change, human security, and violent conflict in the Anthropocene. In J. Scheffran, M. Brzoska, H. Brauch, P. Link, & J. Schilling (Eds.), *Climate change, human security and violent conflict Hexagon Series on Human and*

Environmental Security and Peace, vol 8. (pp. 3–40). Berlin, Heidelberg: Springer. https://doi.org/10.1007/978-3-642-28626-1_1

Brisman, A. (2016). Peace ecology. *Contemporary Justice Review*, 13(3), 419–421 https://doi.org/10.1080/10282580.2016.1200886

Bronkhorst, S. (2011). *Climate change and conflict: lessons for conflict resolution from the southern Sahel of Sudan*. Umhlanga Rocks: ACCORD

Chataway, C. J. (1998). Track II diplomacy: From a Track I perspective. *Negotiation Journal*, 14(3), 269–287

Conca, K. (1994). In the Name of Sustainability: Peace Studies and Environmental Discourse. In J. Käkonen (Ed.), *Green Security or Militarized Environment* (pp. 7–24). Dartmouth: Aldershot

Conca, K. (2002). The Case for Environmental Peacemaking. In K. Conca & G. Dabelko (Eds.), *Environmental Peacemaking* (pp. 1–22). Baltimore: Johns Hopkins University Press

Crutzen, P. J. (2016). Geology of mankind. In C. PJ & B. HG (Eds.), *Paul J. Crutzen: A Pioneer on Atmospheric Chemistry and Climate Change in the Anthropocene* (Vol. 50, pp. 211–215). Cham: Springer Nature Switzerland AG

Dalby, S. (2007). Ecology, security, and change in the Anthropocene. *The Brown Journal of World Affairs*, 13(2), 155–164

Dangarembga, T. (2021). For those who are inside the whale: We need a new enlightenment. Acceptance Speech, 2021 Peace Prize Awards of the German Book Trade. https://www.youtube.com/watch?v=adQKTQrQUPE

Deane-Drummond, C. (2016). Pope Francis: Priest and Prophet in the Anthropocene. *Environmental Humanities*, 8(2), 256–262. doi:https://doi.org/10.1215/22011919-3664369

Deen, M. Y. I. (1996). Islamic environmental ethics, law, and society. In R. S. Gottlieb (Ed.), *This Sacred Earth. Religion, Nature, Environment* (2nd ed., pp. 164–173). New York: Routledge

DeWiel, B. (1997). A conceptual history of civil society: From Greek beginnings to the end of Marx. *Past Imperfect*, 6. https://doi.org/10.21971/P7MK5N

Dill, A. (2015). What is Social Capital 2015 – a review. http://commons.ch/wp-content/uploads/What_is_Social_Capital_2015.pdf Accessed 26 June 2021

Eaton, H. (2007). This sacred earth at the nexus of religion, ecology and politics. *European Journal of Science and Theology*, 3(4), 23–38

Ehrhart, H.-G., & Schnabel, A. (2005). *Security sector reform and post-conflict peacebuilding*. Tokyo: United Nations University Press

Erasmus, V. (2001). Community mobilization as a tool for peacebuilding. In L. Reychler & T. Paffenholz (Eds.), *Peace-building: a field guide* (pp. 246–258). Boulder: Lynne Rienner Publishers

Gottlieb, R. S. (2003). *This sacred earth: Religion, nature, environment*. London: Routledge

Kyrou, C. N. (2007). Peace ecology: An emerging paradigm in peace studies. *International Journal of Peace Studies*, 12(1), 73–92

Lederach, J. P. (1997a). *Building Peace: Sustainable Reconciliation in Divided Societies*. Washington, D.C.: United States Institute of Peace Studies

Lederach, J. P. (1997b). *Sustainable reconciliation in divided societies*. Washington, DC: USIP

Oswald Spring, Ú., & Brauch, H. G. (2021). *Decolonising Conflicts, Security, Peace, Gender, Environment and Development in the Anthropocene*. Cham: Springer Nature Switzerland AG

Paffenholz, T. (2014a). Civil society and peace negotiations: beyond the inclusion–exclusion dichotomy. *Negotiation Journal*, 30(1), 69–91

Paffenholz, T. (2014b). International peacebuilding goes local: analysing Lederach's conflict transformation theory and its ambivalent encounter with 20 years of practice. *Peacebuilding*, 2(1), 11–27

Paffenholz, T. (2015). Civil society and peacebuilding. *Development Dialogue*, 63, 108–118

Paffenholz, T., & Spurk, C. (2006). *Civil society, civic engagement, and peacebuilding. Social Development Papers: Conflict Prevention and Reconstruction*, 36, 10–55

Pelczynski, Z. (1984). *The State and Civil Society: Studies in Hegel's Political Philosophy*. Cambridge: Press Syndicate of the University of Cambridge

Schlosberg, D. (2004). Reconceiving environmental justice: global movements and political theories. *Environmental Politics*, 13(3), 517–540

Schlosberg, D. (2013). Theorising environmental justice: the expanding sphere of a discourse. *Environmental Politics*, 22(1), 37–55. https://doi.org/10.1080/09644016.2013.755387

Steffen, W., Richardson, K., Rockström, J., Cornell, S. E., Fetzer, I., Bennett, E. M., & De Wit, C. A. (2015). Planetary boundaries: Guiding human development on a changing planet. *Science*, 347(6223), 736–747. https://doi.org/10.1126/science.1259855

United Nations Development Programme. (2015). *Human Development Report 2015. Fighting Climate Change*. New York: UNDP

United Nations Development Programme. (2020). *Human Development Report 2020. The Next Frontier: Human Development and the Anthropocene*. New York: UNDP

Westra, L. (2009). *Environmental justice and the rights of ecological refugees (1st ed.)*. London: Routledge

Wood, E. M. (1990). *The uses and abuses of 'civil society'*. Socialist Register, 26

World Bank. (2007). *Civil Society and Peacebuilding: Potential, Limitations and Critical Factors*. Washington, D.C.: The International Bank for Reconstruction and Development/The World Bank

Chapter 2
Peace Ecology in the Anthropocene for Africa

Hans Günter Brauch

Abstract The term 'Anthropocene' was promulgated by Paul J. Crutzen in 2000. Anthropogenic threats to human survival posed by the atom bomb and global climate change began in the new geological epoch of the Anthropocene. They require a joint approach within the framework of *holistic peace ecology* and a transformative strategy towards an *ecological peace policy*. Since 1945 Sub-Saharan Africa has experienced many violent conflicts and human-induced disasters. Due to population growth, the demand for water, agricultural land and food supplies will increase, while there may be an even greater shortage of food supplies and jobs by 2100. These climate-related environmental scarcities may result in new forms of violent climate conflicts. Tipping points in the climate system may trigger geopolitical conflicts. This chapter summarises the key ecological challenges which Africa has faced since 1945, reviews the conflicts Africa has experienced, and assesses their implications for peace research and environmental studies in Africa and the limited work on the connections between the two fields of research. This text discusses the relevance of a peace ecology approach and the need for an ecological peace policy for Africa and reflects on the need to rethink and integrate research and action in Africa in the Anthropocene.

Keywords Anthropocene · Demographic projection · Climate change · Climate models · Ecological peace policy · Environmental peacemaking · Human development report · Peace ecology · Political geo-ecology · Global environmental change · Post-conflict peacebuilding · Tipping points

---

PD Dr. Hans Günter Brauch, chairman of AFES-PRESS; chairman of the board of the Hans Günter Brauch Foundation on Peace and Ecology in the Anthropocene (HGBS), and editor of this Anthropocene (APESS) book series.

J. C. K. Kiyala and G. T. Harris (eds.), *Civil Society and Peacebuilding in Sub-Saharan Africa in the Anthropocene*, The Anthropocene: Politik—Economics—Society—Science 34,
https://doi.org/10.1007/978-3-030-95179-5_2

2.1 The Anthropocene: A Turning Point, Context, Challenge and Opportunity

The Anthropocene is a new concept that gained ground on 23 February 2000 when the Nobel Prize laureate in chemistry (Crutzen 2002; Benner et al. 2022), used it during a meeting of the *International Geosphere Biosphere Programme* (IGBP) in Mexico, when he claimed that due to human interventions in nature and the Earth System humankind had entered a new geological epoch in the Earth's history: the 'Anthropocene'.[1] This concept triggered an intense debate in the natural and social sciences, humanities and law that in just 20 years resulted in more than 10 000 scientific publications, including over 5 000 peer-reviewed journal articles (Brauch 2022, 2023).[2] Three specific aspects of the Anthropocene have previously been identified (Brauch 2021b):

- It is a *turning point* in the Earth's, human and political history that, according to a majority proposal of the *Anthropocene Working Group* (AWG), began between the first test of an atom bomb in Alamogordo (USA) on 16 July 1945 and the Great Acceleration (McNeill/Engelke 2014) of both the Earth System's and socio-economic trends since 1950. To date, this is the only time that fundamental changes in the Earth's history (from the Holocene to the Anthropocene) have coincided with major changes in the international order as well as in political and human history.
- The *context* of both the Earth's history and human history has fundamentally changed since the end of World War II:
  - American dominance in the political, economic (WB, IMF, GATT) and security realm (military alliances [e.g. NATO] instead of the UN Charter's [1945] collective security system)[3];

[1] This text addresses two themes the author has been working on for some time and thus builds on several previous texts that were authored and co-authored on 'peace ecology' and the 'Anthropocene' (Brauch 2012, 2014, 2016, 2016a; Brauch et al. 2011, 2015; Brauch/Oswald Spring 2011, 2015; Brauch et al. (2016); Crutzen/Brauch 2016; Oswald Spring et al. 2009, 2014a). This chapter builds on Brauch (2021), the copyright of which the author retained. Additional new texts on these themes by this author are in preparation and will be published in the years to come.

[2] I appreciate the constructive comments on a first draft of 25 April 2021 received from Dr Hans Happes (Germany), former headmaster of a grammar school (Nikolaus-Kistner-Gymnasium), who has been working in Sumbawanga (Tanzania) since his retirement; the co-editors of this book, Dr Kiyala Jean Chrysostome (Democratic Republic of Congo) and Prof. Dr Geoff Thomas Harris (Australia), both of the Durban Institute of Technology, Durban, South Africa; Prof. Dr Jürgen Scheffran, Geography Department, Hamburg University and Head of CLISEC; Prof. Em. Dr Michael Brzoska, Hamburg University, former Director of the Institute for Peace Research and Security Policy, Hamburg University (ISFH); Prof. Em. Nils Petter Gleditsch, Peace Research Institute Oslo (PRIO), Norway; Prof. Em. Dr Peter Wallensteen, Uppsala University, Sweden.

[3] See Keal (1983), Kissinger (2014), Ikenberry (1999, 2011), Mazarr et al. (2016), Parmar (2018).

- The bipolar structure of a divided world during the Cold War that has also affected Africa, which became a battleground for several proxy conflicts and wars[4];
- For Africa the past 76 years have coincided with the process of decolonisation from colonial rule and the acquisition of political independence, while many colonial dependencies and conflicts have remained in the postcolonial order.[5]

- Multiple complex environmental and socio-economic and political *challenges*:
  - The 'Great Acceleration' was first observed in studies on socio-economic and Earth trends by the IGBP led by Will Steffen. These were first published in 2004 and updated in 2010. They detected a take-off since 1950 that has intensified since the end of the Cold War in 1990.
  - Anthropogenic global environmental and climate change has been discussed by natural scientists since the 1970s; it became a global political issue in 1988 and was discussed as a new security issue in the early 21st century.

But also – as I add here – the Anthropocene has brought new *opportunities* to launch countermeasures to adapt to and mitigate the physical effects of climate change and its socio-economic outcomes. These are increasingly taken into account in strategies to transition to sustainability through the decarbonisation of the economy and society in order to achieve a climate-neutral world between 2050 (European Commission 2019; European Union 2019), 2060 (Chen et al. 2020) and 2100 and to reduce CO_2 emissions by half by 2030, as US President Biden announced on Earth Day 2021 (Stone 2021), or by 65% by 2030, 88% by 2040 and 100% by 2045, as the German government announced in May 2021.[6]

The Anthropocene has been proposed as a new geological epoch for humankind in which two fundamentally different anthropogenic threats to the *survival* of humankind have emerged:

- *Atomic weapons* and the other weapons of mass destruction have been legitimated by the military, political and economic threats posed by rival superpowers (or neighbours in South Asia); these have been analysed as an object of peace and security studies.[7]

[4] See: Bourantonis/Wiener (1995), Volgy/Imwalle (1995), Westad (2007) Jackson/O'Malley (2018), Conca (2015); "List of proxy wars", at: https://en.wikipedia.org/wiki/List_of_proxy_wars; "Proxy Wars in Africa", at: https://www.academia.edu/Documents/in/Proxy_Wars_in_Africa; (Manyok 2008): "War of Proxy, Legacies of the Cold War on the Third World Countries: The Case of Congo and Angola", at: https://www.researchgate.net/publication/292368497_War_of_Proxy_Legaciesof_the_Cold_War_on_the_Third_World_Countries_The_Case_of_Congo_and_Angola (11 June 2021).

[5] Freund (1984), Birmingham (1995), Clapham (2020).

[6] See: Die Bundesregierung (2021): "Climate Change Act 2021: Intergenerational contract for the climate"; at: https://www.bundesregierung.de/breg-de/themen/klimaschutz/climate-change-act-2021-1913970 (8 June 2021).

[7] For an annual overview of global military expenditure, see: "SIPRI Military Expenditure Database", at: https://www.sipri.org/databases/milex. This database "contains consistent time series on the military spending of countries for the period 1949–2020".

- The threats posed by *anthropogenic global environmental changes* are the result of our way of life, economic strategies, production processes and consumption preferences. They cannot be addressed by Hobbesian logic or realist geopolitics[8] but only by a fundamental transformation based on global cooperation in which the only useful role of the military infrastructure is to help face, contain and cope with the consequences of human-triggered natural hazards and disasters. These issues have been primarily addressed by environmental studies and ecological considerations aimed at sustainable development that takes its planetary consequences (UNDP 2020) into account.

These challenges have been addressed in the social sciences by two distinct research programmes between which only very limited scientific cooperation, exchanges, controversies and interdisciplinary projects have occurred. Peace and security issues have been addressed from the two different perspectives and theoretical approaches of *security studies* – following a Hobbesian or realist logic – and *peace studies* inspired by idealistic values and often by Kantian policy goals of a world based on cooperation and peaceful social change.

In environmental and ecological studies the acquisition of knowledge about anthropogenic global environmental change was obtained through discussions and research in the natural sciences. Social science studies later joined in once these topics had become political issues that were handled in environmental regimes (on climate change, biodiversity, ozone, water and soil). To address these two approaches to global environmental change, Brauch et al. (2011) proposed a 'political geo-ecology for the Anthropocene'.

Anthropogenic threats to human survival posed by the atom bomb and by global environmental and climate change began in 1945 in the new geological epoch of the Anthropocene. They require both a joint analytical approach in the framework of *holistic peace ecology* and the integrated transformative strategy of an *ecological peace policy* that is capable of addressing the causes, interdependent processes and outcomes in both policy fields in the remaining decades of the 21st century.

This chapter addresses, primarily from a conceptual approach, the two interrelated concepts of peace ecology (Sect. 2.2) and ecological peace policy (Sect. 2.3) and reflects on the proposals developed by the UNDP in its Human Development Report 2015 on *Fighting Climate Change* (UNDP 2007–2008) and its 2020 publication on *Human Development and the Anthropocene* (Sect. 2.4).

In an empirical part this chapter summarises the key ecological challenges Africa has faced in the Anthropocene in the colonial and post-colonial context (Sect. 2.5), reviews the conflict types Africa has experienced since 1945 (Sect. 2.6), and tries to assess the implications of both for peace research and environmental studies in Africa and the limited work so far undertaken on the interconnections between the two research fields (Sect. 2.7).

[8] Thomas Hobbes (1588–1679) was a British political philosopher who, in International Relations, is considered a pioneer of modern realist thinking, which is influenced by the role of military and economic power.

In the concluding part this text discusses the relevance of a peace ecology approach in Africa (Sect. 2.8) and the need for an ecological peace policy for Africa south of the Sahara (Sect. 2.9), and the author reflects on the need to rethink and integrate research and action in Africa in the Anthropocene (Sect. 2.10).

With its conceptual, empirical and reflective parts and holistic perspective, this chapter aims at *consilience*, which the American biologist Wilson (1988) noted as a growing interlocking of causal explanations across disciplines, in which the "interfaces between disciplines become as important as the disciplines themselves" and "touch the borders of the social sciences and humanities". The key issue addressed in the peace ecology approach is the complex linkage between anthropogenic changes in the climate system and their societal outcome as multiple forms of conflicts that sometimes lead to violence or societal instability, and how the latter can be managed, prevented, or avoided. To this end, reactive or proactive political strategies, policies, and measures may deal with the cause by reducing *greenhouse gas* (GHG) emissions, and can address the impacts by political adaptation and mitigation measures to avoid escalation into violent conflicts.

In the Anthropocene a complex dual causal relationship exists between human beings, the Earth System and the social system. During the Anthropocene, humankind has for the first time directly intervened in the Earth System through the burning of fossil fuels, resulting in an anthropogenic increase in greenhouse gases, most particularly *carbon dioxide* (CO_2), in the atmosphere. Thus, we as human beings have for the first time directly interfered in the Earth System, triggering geophysical effects that directly affect social systems and livelihoods.

2.2 Peace Ecology in the Anthropocene Since 1945

The peace ecology concept was proposed by Kyrou (2007) inspired by Ken Conca (1994, 2002). It was developed further by Oswald Spring et al. (2014) and from a peace activist perspective by Amster (2014). Brauch (2021) framed peace ecology within the Anthropocene, arguing that since 1945 the separate approaches to studying peace and security issues on the one hand and, later, economic and ecological themes in the Anthropocene on the other hand should be replaced with a holistic peace ecology perspective as a scientific approach that integrates peace research and environmental studies.[9]

The link between peace and environmental issues has been addressed by only a few social scientists, among them the economist and peace activist Kenneth Boulding (1966, 1978) and his wife, the sociologist and peace educator Elise Boulding (1988, 1989, 1992, 2000). Elise Boulding linked peace to ecology as a result of practical

[9] This text addresses two themes the author has been working on for some time and thus builds on several previous texts that were authored and co-authored on 'peace ecology' and the 'Anthropocene' (Brauch 2014, 2016, 2017; Brauch et al. 2011, 2015, 2016; Brauch/Oswald Spring 2011, 2015, 2017; Crutzen/Brauch 2016; Oswald Spring et al. 2009, 2014, 2014a). Additional new texts by this author are in preparation and will be published in the years to come.

daily experience. She argued that there is no true peace without ecological links, such as respect for nature and human ecology (Morrison 2005; Boulding 2017, 2017a, 2017b, 2017c).

Since the 1990s, environmental security research has shifted from environmental scarcity, degradation and conflict to the dangers posed by global environmental and climate change (Sygna et al. 2013). With the direct impacts of humans upon ecosystems in the Anthropocene (Crutzen 2002, 2011) and the progressive securitisation of issues of global environmental change (GEC) since 2003, these anthropogenic changes are increasingly threatening human lives and livelihoods. Worldwide, the destruction of key ecosystem services, the pollution of air, water and soil, land use change and extreme weather events are creating new 'anthropogenic challenges' for humankind, although these do not pose a threat of violent conflict or war.

Peace ecology in the Anthropocene was conceptualised by Oswald Spring et al. (2014a: 18–19) within the framework of *six conceptual pillars of peace, security, equity, sustainability, culture and gender*, in which *negative peace* (non-war) is defined by the linkages between peace and security, while the concept of *positive peace* is defined by peace with social justice and global equity; for interactions between peace, gender and environment the concept of *cultural peace* was proposed, and for the relationships between peace, equity and gender the concept of *engendered peace* was suggested (Oswald Spring 2020: 19; Fig. 2.1).

These *five pillars of peace ecology* point to different conceptual features of peace. The classic relationship between 'international peace and security' in the UN Charter refers only to narrow *negative peace* without war or violent conflict. Its aim is the prevention, containment and resolution of conflicts and violence and the absence of 'direct violence' in wars and repression. In order to achieve peace with equity – known as *positive peace* – the absence of 'structural violence' (Galtung 1969) is necessary. This is accomplished by overcoming social inequality, discrimination, marginalisation and poverty where there is no access to adequate food, water, health or educational opportunities.

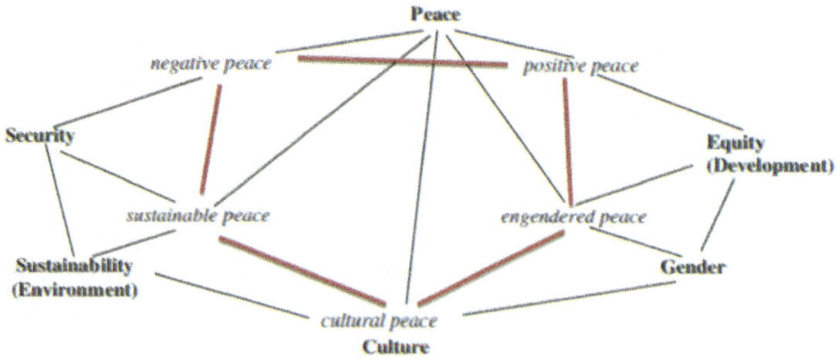

Fig. 2.1 Six conceptual pillars of peace, security, equity, sustainability, culture and gender; five pillars of peace ecology and their linkage concepts of negative, positive, sustainable, cultural and engendered peace. *Source* Oswald Spring et al. (2014: 19)

'Peace ecology' is defined as a holistic, multi- and interdisciplinary scientific perspective and approach that addresses both the peace and security focus of the UN Charter (1945) and the new development and environment studies that have gradually evolved during the decolonisation process since the 1950s, the emerging environment policies at national level since the 1960s and, in the UN context, since the Stockholm Conference (1972), the setting up of the *UN's Environment Programme* (UNEP), and the Rio Earth Summit (1992). The decision-makers who prepared and wrote the UN Charter in spring 1945 "showed no sign of thinking about the natural world. The charter makes no mention of the Earth, ecosystems, pollution, natural resources, or sustainability … and the theme was missing from the debate on the new organisation's purpose, structure, and rules" (Conca 2015: 33). The first secret nuclear test on 16 July 1945 in Alamogordo was the starting point of the nuclear era, while its radioactive isotopes where interpreted 70 years later by the *Anthropocene Working Group* (AWG) as evidence for the start of the Anthropocene as a new geological epoch that can be identified in sediments in the atmospheric testing sites of nuclear weapons powers (US, USSR, UK, France, China, India, Pakistan etc.).

The peace ecology approach combines the scientific, technological, peace, and security dimensions with environmental perspectives and methods by looking not only at these distinct features but also at the complex interactions and feedbacks of anthropogenic processes in the climate system that were triggered by humanity and may put the very survival of humankind at risk as a result of the mass use of nuclear weapons (in a nuclear war) or the exponential increase in the burning of hydrocarbon energy sources by individual human beings as part of their production and consumption processes and lifestyles (causing anthropogenic climate change). These two anthropogenic threats to the survival of humankind as a result of a nuclear war and/or climate change require combined but different instruments, strategies and policy outcomes.

Crutzen (2016) was a pioneer of investigations into the 'ozone layer depletion' caused by the human use of chlorofluorocarbons (CFCs) and halons in aerosol spray cans and refrigerants. He also conducted research into the potential effects of a 'nuclear winter' (Crutzen 1980; Badash 2009) which could result from the mass use of nuclear weapons with disastrous consequences for the environment and food security, and he suggested using the 'Anthropocene concept' as a new framework for a transformative science.

Human beings, as policy-makers (diplomats, international civil servants in the UN system), citizens and activists (in peace movements by creating societal awareness and putting political pressure on governments) can only avoid a nuclear winter through nuclear arms control and disarmament, while individual human beings can alter their behaviour to minimise the dangers of global environmental and climate change.

Containing and overcoming the ozone hole has become a unique success story of ozone diplomacy (Benedick 1998; The Vienna Convention for the Protection of the Ozone Layer, 1985; Montreal Protocol on Substances that Deplete the Ozone Layer 1989, and so far six amendments, most recently the Kigali Amendment in 2016) that became possible because of the development of substitutes for chlorofluorocarbons (CFCs) and halons, strong political will, and efficient leadership and diplomacy.

The peace ecology approach deals with the manifold links between peace, security and the environment, whereby humankind and the environment, as the two interdependent parts of Planet Earth, face the consequences of destruction, extraction and pollution (Oswald Spring 2008). The concept of sustainable peace also includes the processes of recovering from environmental destruction and reducing the human footprint in ecosystems through less carbon-intensive processes and, in the long term, possibly carbon-free and dematerialised production, so that future generations still have the opportunity to decide on their own resource and development strategies.

Policies aimed at 'sustainability transition' (Brauch et al. 2016; Brauch 2021b) are thus part of a positive strategy that addresses possible new causes of instability, crises, conflicts and, in the worst case, even war. These causes may be either the scarcity of fossil energy sources or the possible security consequences of anthropogenic global environmental and climate change, either of which may be triggered by linear trends as well as chaotic 'tipping points' (Lenton et al. 2008, 2019).

The relationship between peace, the environment and gender may result in the *cultural peace* that facilitates the creation of peace in the minds and actions of humankind. It socialises people so that religious and social discrimination can be overcome by establishing human rights granted equally to all people. This enables them to develop the ability to negotiate solutions to present and future conflicts peacefully and to share political, economic, social and cultural powers. The rights also respect different ecosystems by taking into account their vulnerability to human actions.

It is appropriate that the temporal context for this peace ecology research programme is the Anthropocene (Crutzen/Stoermer 2000; Crutzen 2002), since this new epoch in the Earth's and human history began when a turning point in the geological time of the Earth's history coincided with the long-term structural turning points in human history after the end of World War II, with the emerging nuclear era, the order of the Cold War (1947–1989), and the post-Cold War disorder (1990-present).

The environmental dimensions of this change could not be socially constructed until the late 1980s, when the exponential increase in greenhouse gases, particularly CO_2, since the Great Acceleration in the 1950s made the issues relating to global environmental change and climate change all too apparent. Whereas the CO_2 concentration in the atmosphere increased from 279 ppm in 1750 to 310 ppm in 1950 – i.e. by 31 ppm in 200 years – between 1950 and 1990 it leapt up by 44 ppm in just 40 years (from 310 ppm to 354 ppm), and in the 30 years between 1990 and May 2020 it increased by a further 62 ppm (from 354 ppm to 416 ppm – 137 ppm higher than the level in 1750).

Thus, with the exponential increase of *greenhouse gases* (GHG) in the atmosphere, scientific knowledge of the physical impacts of global climate change has increased since the 1980s, initially in the natural sciences. Several new research and training programmes and research institutes have been set up in the newly emerging areas of *Earth Systems Science* (ESS) and *Earth Systems Analysis* (ESA) (Steffen et al. 2020), and new degree courses in geo-ecology[10] are being offered in the sphere of physical geography. After the build-up of new scientific capabilities, the scientific knowledge of GEC has expanded rapidly. Its peer-reviewed research output was assessed by the Intergovernmental Panel on Climate Change in its first five assessment reports (IPCC 1990, 1996, 2001, 2007, 2013, 2014) (process of *scientisation*).[11] Among policy-makers in governments and parliaments political awareness of anthropogenic climate change emerged in 1988, and after only four years resulted in the first two framework agreements on climate change and biological diversity at the Rio Earth Summit (process of *politicisation*). In 1988 analysis of the security implications of climate change gradually began. In spring 2003 the national public debate and studies in think tanks in several OECD countries (UK, US, Germany) took off (process of *securitisation*), and after 2007 related research in the social sciences also began (Brauch 2002, 2009; Brauch/Scheffran 2012; Scheffran et al. 2012; Brzoska 2020; Brzoska/Scheffran 2020).

Le Billon/Duffy (2019) pointed to several impediments to building bridges between political ecology and peace and conflict studies:

> Conflict is at the core of many political ecology studies. Yet there has been limited engagement between political ecology and the field of peace and conflict studies. This lack of connection reflects in part the broader disciplinary context of these two fields. Whereas political ecology research mostly comes from disciplines that eschewed environmental determinism, such as human geography, much of peace and conflict studies is associated with political science using positivist approaches to determine the causal effects of environmental factors on conflicts. Yet greater connections are possible, notably in light of political ecology's renewed engagement with 'materialism', and peace and conflict studies' increasingly nuanced mixed-methods research on environment-related conflicts. Furthermore, political ecology's emphasis on uneven power relations and pursuit of environmental justice resonates with the structural violence approaches and social justice agenda of peace and conflict studies (Le Billon/Duffy 2019).

Both authors have remained within the mainstream debates on human geography, political science and international relations, and have so far not addressed the proposed concepts of 'political geoecology' (Brauch et al. 2011, 2015) and peace ecology (Brauch 2016a, 2021), nor have they contextualised their analysis in the Anthropocene.[12]

---

[10] The concept of 'geoecology' was introduced by Huggett (1995). In 2021, many geography departments, primarily in Central and Eastern Europe, are offering Bachelors and Masters degree courses in geoecology.

[11] The Sixth Assessment Report of the IPCC is scheduled to be released between October 2021 (WG 1) and October 2022 (WG II, WG III, Synthesis Report) and may be accessed here: https://www.ipcc.ch/assessment-report/ar6/ (8 June 2021).

[12] 'Geoecology' is a concept formulated by Huggett (1995) within physical geography. The proposal of 'political geoecology' is to combine approaches in physical and human geography or between the natural and the social sciences, while 'peace ecology' is an approach that proposes integrating ecological approaches with research programmes in the social sciences and peace studies.

In my opinion, the peace ecology approach should:

- overcome the overspecialisation in both the social and natural sciences;
- offer *holistic* approaches that address interlinkages between themes addressed by peace research and environmental studies;
- be contextualised in the Anthropocene Epoch of the Earth's history and address scientific and political linkages since the end of World War II and the Great Acceleration;
- be interdisciplinary; and
- methodologically, in the Anthropocene a peace ecology approach should be developed further from interdisciplinary and transdisciplinary approaches into a *transformative research programme.*

Between 2000, when Crutzen uttered the term 'Anthropocene', and 2020 a debate on linkage concepts between political science, international relations and peace research and environmental studies remained underdeveloped within the social sciences. The much-needed integration of knowledge derived from climate and Earth Systems Science and analysis in the natural sciences with analysis in the social sciences has remained rare despite the many pleas for interdisciplinary research.

Anthropogenically-induced global warming has triggered four major physical effects: (a) a global increase in the average temperature, (b) variations in precipitation, (c) low-onset sea-level rise, and (d) an increase in the probability and intensity of extreme weather events, which, to date, have been reviewed in five published assessment reports (IPCC 1990, 1995, 2001, 2007, 2013, 2014) in the forthcoming sixth assessment report (IPCC 2021, 2022, 2022a, 2022b), and in several special reports (IPCC 1997, 2011, 2012, 2018, 2019, 2019a, 2019b, 2020). The physical effects of anthropogenic climate change have had multiple societal impacts which may cause severe domestic or international crises, conflicts, and, in the very worst case, even violent wars (Brauch 2002, 2009).

The peace ecology approach or programme in the Anthropocene must combine and integrate the results of peace research in the social sciences with the research in Earth Systems science and analysis which has primarily been conducted from the perspective of natural sciences. The peace ecology approach must therefore cross the narrow disciplinary and research programme boundaries and move from multidisciplinary perspectives to interdisciplinary assessments. A peace ecology research programme that aims to contribute to an ecological peace policy should be transformative by including the transformation of the status quo into the research design.

So far, this debate is just emerging among a few social scientists in North America and Europe and has not yet been intensively discussed by the peace research and ecological research communities. Between 2012 and 2018 this debate partly took place in IPRA's Ecology and Peace Commission and is documented in five volumes.[13]

[13] See: Oswald Spring/Brauch/Tidball 2014; Brauch/Oswald Spring/Bennett/Serrano Oswald 2016; Oswald Spring/Brauch/Serrano Oswald/Bennett 2016; Brauch/Oswald Spring/Collins/Serrano Oswald 2018; Oswald Spring/Brauch 2021.

In countries of the Global South (in Africa, Asia and Latin America) this debate has hardly taken place, although in these regions environmental challenges, hazards and disasters have often caused internal, regional and intercontinental distress migration,[14] crises and violent conflicts, though rarely wars. In Africa there has been an increasing interest in peace research among African scholars, but their institutional funding, library resources and research equipment are extremely limited despite the urgency of the security and environmental challenges posed.

So far, eight Nobel Peace Prizes have been awarded to Africans south of the Sahara – three to four South Africans (Albert Lutuli, President of the ANC in 1960; Desmond Tutu in 1984; Nelson Mandela and F.W. de Klerk in 1993), one to former UN Secretary General Kofi Annan (Ghana 2001), two to three women (environmentalist Wangarĩ Maathai, Kenya, 2004; Ellen Johnson Sirleaf and Laymah Gbowee both from Liberia in 2011), one to Denis Mukwege (a gynaecologist and Pentecostal pastor from the Democratic Republic of the Congo in 2018), and one to Abij Ahmed, the Prime Minister of Ethiopia in 2019, who has been involved since November 2020 in a civil war with Tigray's People's Liberation Front.

All but Wangarĩ Maathai (2004) and Denis Mukwege (2018) were involved in classical issues of peace and security fighting against apartheid and for a peaceful transition in South Africa. Ellen Johnson Sirleaf (2011), as the first female President of Liberia (2006–2018), and Laymah Gbowee, who was leading a women's nonviolent peace movement, *Women of Liberia Mass Action for Peace*, both helped to bring an end to the Second Liberian Civil War in 2003. Wangarĩ Muta Maathai (1940–2011) was a Kenyan social, environmental, and political activist and the first African woman to win the Nobel Peace Prize in 2004. In 1977 she founded the *Green Belt Movement*, an environmental non-governmental organization focused on the planting of trees, environmental conservation, and women's rights. Their combined causes have addressed what activist peace ecology and ecological peace policy from below are about: struggling for political and human rights and gender equality.

2.3 Ecological Peace Policy in the Anthropocene

Peace ecology and ecological peace policy are two sides of the same coin. So far, the concept of 'ecological peace policy' does not exist in either peace research or environmental studies. However, several related concepts have been used in the literature, such as:

- "environment and peace" (IUCN)[15];

[14] See: Fachkommission Fluchtursachen (2021: 181–215): at: fk-fluchtursachen@bmz.bund.de.

[15] The Commission on Environment, Economic and Social Policy of the *International Union for Conservation of Nature* (IUCN) addresses "the integration of natural resource management in conflict prevention, mitigation, resolution, and recovery to build resilience in communities affected by conflict" that are "cross-cutting and relevant in all areas of conservation, sustainable development and security". This IUCN's commission works on these five areas: 1. social conflicts and peace,

- "environmental peacebuilding" (Ecopeace Middle East[16] and BioScience[17]);
- "Building Sustainable Peace: Understanding the Linkages between Social, Political, and Ecological Processes in Post-War Countries" (Krampe 2016)[18];
- "Building Peace Through Environmental Conservation" (Notaras at UNU 2010)[19];
- "Environment of Peace" (SIPRI)[20];
- "Ecological Threats to Peace" (USIP);[21]
- "Making Peace with Nature" (UNEP Report);[22]

2. security and peace, 3. ecological conflicts and peace, 4. peacebuilding, and 5. conflict resolution. See at: https://www.iucn.org/commissions/commission-environmental-economic-and-social-policy/our-work/environment-and-peace (11 April 2021).

[16] For an overview with many sources, see: https://en.wikipedia.org/wiki/Environmental_peacebuilding (11 April 2021). See also the websites on environmental peacebuilding, at: https://www.epmooc.org/ (11 April 2021).

[17] For a summary of the emerging debate by Lesley Evans Ogden on "Environmental peacebuilding", see: *BioScience*, 68,3 (March 2018): 157–163.

[18] See Florian Krampe, summary of PhD thesis at Uppsala University, at: http://uu.diva-portal.org/smash/record.jsf?pid=diva2%3A945418&dswid=-1351 (11 April 2021).

[19] See Mark Notaras (UNU), at: https://ourworld.unu.edu/en/building-peace-through-environmental-conservation (11 April 2021).

[20] In SIPRI's research programme "climate change and environmental degradation are already impacting peace and security in diverse ways. At the same time, the change needed to transition to lower-carbon, greener economies is fraught with risks, but also offers many opportunities to contribute to more peaceful, sustainable societies. Environment of Peace will synthesize the best available evidence on environmental change and its societal impacts. It will present new insights on the risks, challenges and promising solutions. And it will illuminate pathways for policy and action. The initiative will release a major report in 2022, marking 50 years since the landmark United Nations Conference on the Human Environment (the Stockholm Conference). See at: https://www.sipri.org/research/peace-and-development/environment-peace (11 April 2021).

[21] On 22 September 2020 the US Institute of Peace presented: "The new Ecological Threat Register (ETR), produced by the Institute for Economics and Peace, [that] synthesizes and visualizes data on environmental indicators to estimate which countries, regions, and areas are most vulnerable to environment-induced conflict. In particular, the ETR underscores that 141 countries are vulnerable to ecological threats, and that approximately 1.2 billion people could be displaced globally by ecological disasters in the next 30 years. On September 22, USIP and the Institute for Economics and Peace examined the inaugural Ecological Threat Register, as experts explored the nexus between conflict and climate change and considered strategies for boosting resilience to climate-induced insecurity." See at: https://www.usip.org/events/ecological-threats-peace (11 April 2021).

[22] See: "Making Peace with Nature", at: https://www.dw.com/en/making-peace-with-nature/a-56615328 (11 April 2021). This UNEP (2019) report *Global Environmental Outlook 2019 – Healthy Planet – Healthy People* (Cambridge: Cambridge University Press) says "A new UN blueprint offers an integrated 'peace plan' to tackle three interlinked environmental emergencies – the climate crisis, biodiversity loss and pollution – that cannot be solved in isolation." See UNEP's GEO-6 (Nairobi 2019) report https://www.unep.org/resources/global-environment-outlook-6, which suggests a holistic approach:

Fifth, environmental policy is necessary but inadequate by itself to address systemic ecological problems, solutions to which require a more holistic approach. Current (inter)national policies are not on track to address the key environmental challenges effectively and equitably, in line with the aspirations of the SDGs. Environmental considerations need to be integrated into all policy areas,

- "War and peace in an age of ecological conflicts" (Latour 2014).[23]

The tasks of an 'ecological peace policy' are to project the probable ecological consequences of the physical effects of global environmental change; to launch early preventive actions to avoid environmental crises, conflicts or wars; and to solve these environmental consequences peacefully without any outbreaks of violence. The theme of an ecological peace policy has already been addressed by the Brundtland Commission (1987) in *Our Common Future*, which linked both themes from a policy perspective by discussing the linkage between "Peace, Security, Development, and the Environment" with regard to "environmental stress as a source of conflict", "conflict as a cause of unsustainable development", and "steps towards security and sustainable development". This political agenda-setting addressed the following four linkage problems:

1. Among the dangers facing the environment, the possibility of nuclear war, or military conflict of a lesser scale involving weapons of mass destruction, is undoubtedly the gravest. Certain aspects of the issues of peace and security bear directly upon the concept of sustainable development.
2. Environmental stress is both a cause and an effect of political tension and military conflict Nations have often fought to assert or resist control over raw materials, energy supplies, land, river basins, sea passages, and other key environmental resources. Such conflicts are likely to increase as these resources become scarcer and competition for them increases.
3. The environmental consequences of armed conflict would be most devastating in the case of thermonuclear war. But there are damaging effects too from conventional, biological, and chemical weapons, as well as from the disruption of economic production and social organisation in the wake of warfare and mass migration of refugees. But even where war is prevented, and where conflict is contained, a state of 'peace' might well entail the diversion into armament production of vast resources that could, at least in part, be used to promote sustainable forms of development.
4. A number of factors affect the connection between environmental stress, poverty, and security, such as inadequate development policies, adverse trends in the international economy, inequities in multi-racial and multi-ethnic societies, and pressures of population growth. These linkages among environment, development, and conflict are complex and, in many cases, poorly understood. But a

---

such that the potential and actual implications for natural resources and the environment are robustly included in policies for economic growth, technological development and urban design, so that there is effective long-term decoupling between economic growth, resource use and environmental degradation. Climate mitigation needs to be accompanied by policy for the equitable adaptation to committed climate change. Policies will only be effective if they are well designed, involving clear goals and flexible mixes of policy, including monitoring, instruments aimed at achieving them [...] and when access to judicial remedies are available [...]. Such a holistic approach need not require additional economic costs.

[23] Latour (2014); for an English translation, see: https://www.cairn.info/revue-revue-juridique-de-l-environnement-2014-1-page-51.htm (11 April 2021).

> comprehensive approach to international and national security must transcend the traditional emphasis on military power and armed competition. The real sources of insecurity also encompass unsustainable development, and its effects can become intertwined with traditional forms of conflict in a manner that can extend and deepen the latter (Brundtland Commission 1987).

The concept of an 'ecological peace policy' is a normative concept that still needs to be systematically developed. It must address the causes and consequences of anthropogenic global change as well as policy responses at local, national and international level. This proposed 'ecological peace policy' is not limited at *international level* to global and regional foreign, development and environment policy. It starts at the individual level with the values, attitudes and behaviour of individual citizens, e.g. with their food and consumptive behaviour, their preferred transportation system and their individual and family ecological footprint. It thus becomes an issue of education by parents and in school from kindergarten, elementary and senior school level to professional training, teaching and research at institutions of higher education.

Many ecological decisions are made by families, the local village, town or city councils, state governments, and national or federal governments. In ecological peace policy the threats that are often used to legitimate political decisions have shifted fundamentally from 'the other' (neighbour, other ethnic or religious group, country or military alliance). We ourselves have become the major cause or threat with our behaviour and consumptive decisions. Here the motto 'think globally but act locally' matters. Thus a major issue area starts with education from the kindergarten to the high school in teaching the local practice of achieving sustainable development goals.

At state and national political level an ecological peace policy does not just address the classical environmental policy areas of pollution of the soil, water and air, which also directly affects the quality of our food and our health. Major areas of analysis, debate and innovation are economic, energy, transportation, agriculture and health policies. Implementing the *Sustainable Development Goals* (SDGs) and strategies for *sustainable transition* affects all levels of policy-making and is of relevance for all societal, economic, political and scientific actors.

At international level an ecological peace policy requires the holistic approach suggested by the UNEP in its GEO-6 Report in 2019, which necessitates an integrated approach between foreign (diplomacy), security and defence policies and environment and development concerns. The European Commission's goal to achieve a European Green Deal and climate neutrality by 2050 affects multiple policy fields at the level of the European Union and its 27 member states.

With regard to ecological peace policy, two phases of human and world history can be identified which offer current researchers a reasonably long-term perspective of the Anthropocene:

(a) the *past* eighty years of world history since the start of World War II in 1939 and the Japanese attack on Pearl Harbor (1941) up to the year 2020; and
(b) the *next* eighty years of world history from the present to the end of the twenty-first century (2021–2100).

In 2022, we are midway between these two phases. We can analyse and interpret the first eight decades of the Anthropocene and project trends for the next eight decades – i.e. until 2050 and 2100, which is the horizon of most climate models assessed by the IPCC. The outcome for the planet – and humankind – depends on the political strategies and programmes that are launched now or later or not at all.

It is impossible to be sure how countries which have previously failed to implement their obligations under climate change agreements will behave in the future. However, deductions can be made about societal and political outcomes by using climate models which correlate projections of future population levels with the production and consumption levels associated with different lifestyles.

An ecological peace policy should aim to provide a preventive strategy so that (1) the projected physical effects of anthropogenic climate change can be contained, resolved and structurally prevented through multilateral diplomacy and cooperation, and (2) domestic and international violent crises, conflicts and wars can be avoided. Jointly and proactively addressing resource scarcity, resource pollution and stress (on water, soil, air) will make it possible to enhance the health and security of the affected people.

2.4 Rethinking Human Development in the Anthropocene

Since 1990 the *United Nations Development Programme* (UNDP) has released an annual *Human Development Report* (HDR) that addresses the following themes which are conceptually relevant for the proposed peace ecology approach:

- *New Dimensions of Human Security* (UNDP 1994);
- *International Cooperation at a Crossroads: Aid, Trade and Security in an Unequal World* (UNDP 2005);
- *Beyond Scarcity: Power, Poverty and the Global Water Crisis* (UNDP 2006);
- *Fighting Climate Change: Human Solidarity* in a Divided World (UNDP 2007/2008);
- *Sustaining Human Progress: Reducing Vulnerabilities and Building Resilience* (UNDP 2014);
- *The Next Frontier: Human Development and the Anthropocene* (UNDP 2020);

Of these six HDRs the Report on *Human Security* (UNDP 1994) had a significant influence on triggering a debate in peace research and in parts of the liberal security studies community (Brauch et al. 2009). From an ecological perspective, four reports are of relevance: (a) *Power, Poverty and the Global Water Crisis* (UNDP 2006); (b) *Fighting Climate Change: Human Solidarity in a Divided World* (UNDP 2007/2008); (c) *Sustaining Human Progress: Reducing Vulnerabilities and Building Resilience* (UNDP 2014); *and* (d) *The Next Frontier: Human Development and the Anthropocene* (UNDP 2020). For the purpose of this chapter only the most recent report on Human Development and the Anthropocene will be reviewed, specifically with regard to Africa. UNDP's Administrator, Achim Steiner, argued in his foreword:

> The pressures we exert on the planet have become so great that scientists are considering whether the Earth has entered an entirely new geological epoch: the Anthropocene, or the age of humans. It means that we are the first people to live in an age defined by human choice, in which the dominant risk to our survival is ourselves. Advancing human development while erasing such planetary pressures is the next frontier for human development, and its exploration lies at the heart of this 30th anniversary edition of UNDP's Human Development Report. To survive and thrive in this new age, we must redesign a path to progress that respects the intertwined fate of people and planet and recognizes that the carbon and material footprint of the people who have more is choking the opportunities of the people who have less (UNDP 2020, p. iii).

Looking ahead, Achim Steiner outlined alternative routes of action:

> If people have the power to create an entirely new geological epoch, then people also have the power to choose to change. We are not the last generation of the Anthropocene; we are the first to recognize it. We are the explorers, the innovators who get to decide what this – the first generation of the Anthropocene – will be remembered for. Will we be remembered by the fossils we leave behind...? Or will we leave a much more valuable imprint: balance between people and planet, a future that is fair and just? *The Next Frontier: Human Development and the Anthropocene* sets out this choice, offering a thought-provoking, necessary alternative to paralysis in the face of rising poverty and inequalities alongside alarming planetary change. With its new, experimental planetary pressures and adjusted Human Development Index, we hope to open a new conversation on the path ahead for each country – a path yet unexplored (UNDP 2020, p. iii).

The HDR 2020 addresses three themes in order to expand human development "easing planetary pressures" by (i) "renewing human development for the Anthropocene"; (ii) discussing "mechanism of change to catalyse action"; and (c) "exploring new metrics" with a special focus on human development in Africa (UNDP 2020).

In part I – on renewing human development for the Anthropocene – Chap. 1 discusses human development in the context of the Anthropocene, while Chap. 2 deals with "unprecedented planetary and social imbalances and their interactions", and Chap. 3 argues that "working together in the pursuit of equity, innovation and planet stewardship can steer actions towards the transformational changes required to advance human development in the Anthropocene" (UNDP 2020: 17). Chapter 1 concludes that

> the Anthropocene brings new evidence and concepts to inform public debate about the changes – normative, economic, technological, behavioural – needed to ease the unprecedented pressures we are putting on the planet. There can be no doubt that only people can effect these changes, but the Anthropocene and its planetary imbalances are superimposed on social imbalances and tensions (UNDP 2020, p. 43).

Chapter 2 argues that human activity has driven dangerous planetary change with an exponential increase since the Great Acceleration began around 1950. The report argues that the Anthropocene "implies enormous uncertainty for people and societies" (UNDP 2020, p. 56), where the risks are greater and different and where an unprecedented planetary change may trigger "unprecedented shocks on human development". Among the drivers climate change has weakened economic progress, increased inequality and hunger and the impacts of natural hazards (UNDP 2020, pp. 60–61) that have increasingly resulted in displacements to a minor part by geophysical hazards (e.g. earthquakes) and to a major part by floods that have affected the continents of the Global South differently (Fig. 2.2), primarily due to floods.

According to Fig. 2.3, temperatures are projected to rise "outside the range of survivability – more over the next 50 years than in the past 6,000 years", especially in Africa, South America, South and South East Asia, China and in northern Australia, which may increase the inequity of the poor and powerless. In Chap. 3, the HDR 2020 asks how human beings may be empowered "for equity, innovation and stewardship of nature" (UNDP 2020, p. 70). The chapter concludes that sustainable development "will require more than adaptations and gradual changes. It will require transformations that break current locked-in systems of unsustainability" (UNDP 2020, p. 98).

Chapter 3 concludes that:

> We need to aim for transformative changes in how societies relate to the biosphere, focus on distributive approaches, and ensure extraction and emission rates align with the rates at which resources are produced and waste and emissions can be absorbed by the environment. Outcomes, such as biodiversity conservation and climate stabilization, can be measured as single variables, but the goals of sustainable human development must be rooted in integrated, transdisciplinary understandings of the connections of societies in the biosphere. Development pathways and goals will vary over time and space, as they are met or redefined. This requires adaptive management, the ability to better understand, learn and act accordingly in an endless, iterative process. ... Sustainable human development is ... a dynamic and continued process, and ample research, human will and political power – as well as urgency – exist to actively engage in that process (UNDP 2020, p. 98).

Part II of the HDR 2020:

> explores mechanisms of change that can mobilize action by individuals, communities, governments, civil society and businesses. .... Three specific mechanisms of change are considered. *First*, social norms, which frame socially permissible – or forbidden – behaviours. ... *Second*, incentives for change [that] determine in part what consumers choose to buy, what

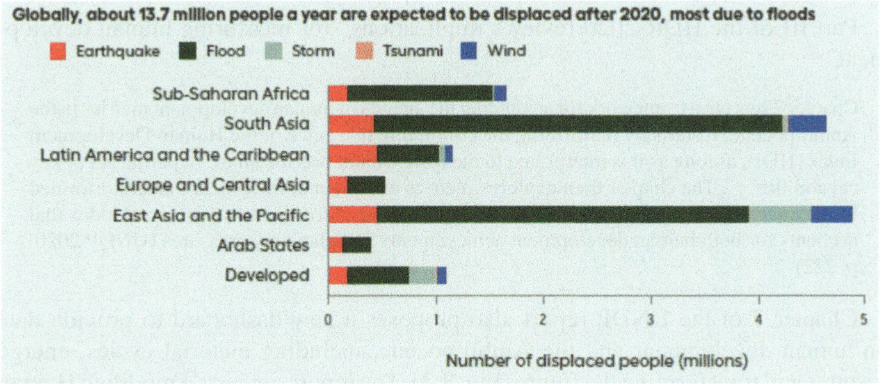

Fig. 2.2 Major hazards and displacements by world regions. *Source* UNDP (2020: 61) based on IDMC (2002a, 2020b). The figure is copyrighted under the Creative Commons Attribution 3.0 IGO license. Permission granted

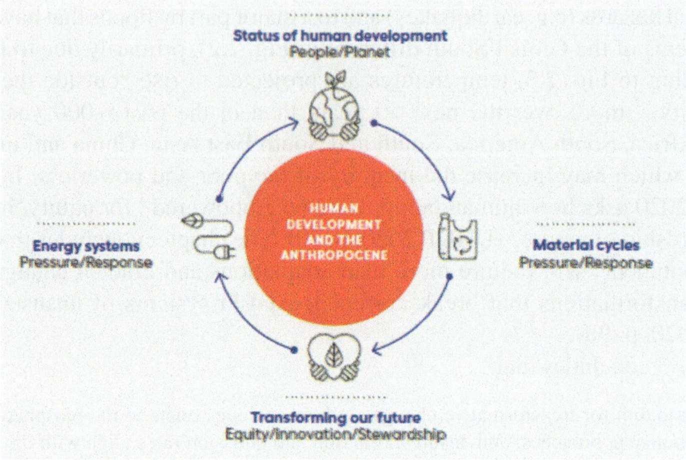

Fig. 2.3 Human development and the Anthropocene. *Source* UNDP (2020, p. 228). The figure is copyrighted under the Creative Commons Attribution 3.0 IGO license. Permission granted

> firms produce and trade, where investors put their money and how governments cooperate. Incentives and social norms interact with one another, but incentives are also crucial in their own right: Even if people do not change their minds, they may still respond to incentives based on what they can afford and where they see opportunities to meet their aspirations. … It also explores how these incentives could evolve in ways that would ease planetary pressures and move societies towards the transformative changes required for human development in the Anthropocene. It considers three domains shaped by considerations related to incentives: finance, prices and international collective action. Third, just as social norms and incentives can be harnessed for transformational change, so can a new generation of nature-based solutions. They can protect, sustainably manage and restore ecosystems, simultaneously promoting wellbeing and mitigating biosphere integrity loss. They embrace equity, innovation and stewardship of nature (UNDP 2020, p. 129).

Part III of the HDR 2020 reviews implications "for measuring human development":

> Chapter 7 sets out a framework for advancing the agenda of human development metrics in the Anthropocene. It starts by reaffirming the continuing relevance of the Human Development Index (HDI), as long as it is interpreted to measure what it was meant to – a partial set of key capabilities. … The chapter then explores metrics of human development that are informed by the analysis in this Report. It concludes with a proposal for a new experimental index that accounts for both human development achievements and planetary pressures (UNDP 2020, p. 222).

Chapter 7 of the UNDP report also proposes a new dashboard to provide data on human development and the Anthropocene, including material cycles, energy systems and transforming the future (Fig. 2.3). The report suggests a modified Human Development Index that is adjusted to planetary pressures (UNDP 2020, p. 236).

No African country is featured among the 66 states with very high human development. Four North African countries, South Africa and Gabon are in the group with high human development, and 11 countries are in the medium group of human development, while 31 countries are in the low development group. The material footprint[24] of Sub-Saharan African countries is 2.5 tonnes per capita and thus far below the global average of 12.3, or about a tenth of the average footprint of people in the OECD countries.

In 2019, the *Human Development Index* (HDI) value of Sub-Saharan Africa was 0.547 – slightly above the average of the least developed countries (0.538) but significantly below the global average of 0.737 and the OECD countries' average of 0.900. The value of the *planetary-pressures adjusted HDI* (PHDI)[25] for Sub-Saharan Africa was 0.539 in 2019 – slightly above 0.533 for the least developed countries but below the global average of 0.683 and significantly below the average of the OECD countries (0.766).

2.5 Demographic and Ecological Challenges Facing Africa in the Anthropocene

Estimations concerning the extent of the ecological challenges facing Africa by end of the 21st century are based on projections of population growth between 2019 and 2100 – indicating the likely demand for water, soil, food, and housing – and on the economic models (growth rates) and climate models which predict the impact these changes will have on temperature increase, precipitation changes, sea-level rise and the increase in the number and intensity of extreme weather events.

These projected challenges point to potential hotspots where both population growth rates and the impacts of global environmental change precipitate major changes. Sub-Saharan Africa is the region with the highest projected population increase – from 1,011 billion in 2019 to 1,400 billion in 2030, 2,118 in 2050 and 3,775 billion in 2100 – while the population in Europe and North America is projected to remain relatively stable, at between 1,114 billion in 2019 and 1,120 in 2100.

The ranking of the world's ten most populous countries, based on these official UN figures in 1990 and 2019 and on the medium-variant projections for 2050 and 2100, takes into account the following changes in these 110 years:

24 'Material footprint' refers to the amount of raw materials extracted from Planet Earth to meet human needs. See: https://unstats.un.org/sdgs/report/2017/goal-12/.

25 The UNDP HDR (2020: 244) defined the *planetary-pressures adjusted HDI* (PHDI) as a: "Planetary pressures-adjusted HDI (PHDI): HDI value adjusted by the level of carbon dioxide emissions and material footprint per capita to account for excessive human pressures on the planet. It should be seen as an incentive for transformation. See technical note at: http://hdr.undp.org/sites/default/files/phdi_tn.pdf for details on how the PHDI is calculated."

In 1990, with 95 million people, Nigeria was the 10th most populous country. In 2019, Nigeria had become the 7th most populous nation, with its population more than doubling to 201 million. Ethiopia had become the 12th most populous country, with 112 million people, and the Democratic Republic of the Congo was the 16th most populous nation, with 67 million people.

By 2050, Nigeria is projected to reach 401 million people, with the third-largest population; Ethiopia will be number 8 with 205 million people; the Democratic Republic of the Congo will be number 9 with 194 million people; Egypt number 11 with 160 million; and the United Republic of Tanzania number 15 with a population of 129 million (Table 2.1).

By 2100 Nigeria's population is projected to reach 733 million, placing it at number 3 after India and China; the Democratic Republic of the Congo will be number 6 with 362 million people; Ethiopia is projected to be number 8 with 294 million people; the United Republic of Tanzania will be number 9 with 286 million; and Egypt will be number 10 with 225 million people. Thus, by the end of this century half of the 10 most populous countries will be in Africa (Fig. 2.4).

The projections of global environmental change and climate change in Sub-Saharan Africa by 2050 and 2100 based on various climate models will pose major challenges for any ecological peace policy in the decades ahead, and are currently

Table 2.1 Population of the world, SDG regions and selected groups of countries in 2019, 2030, 2050 and 2100, according to the medium-variant projection. *Source* UNPD (2019: 6). The figure is copyrighted under the Creative Commons Attribution 3.0 IGO license. Permission granted at: http://hdr.undp.org/en/content/

Region	Population (millions)			
	2019	2030	2050	2100
World	7713	8 548	9735	10 875
Sub-Saharan Africa	1 066	1 400	2 113	3 775
Northern Africa and Western Asia	517	609	754	924
Central and Southern Asia	1 991	2227	2 496	2 334
Eastern and South-Eastern Asia	2 335	2 427	2411	1 967
Latin America and the Caribbean	648	706	762	630
Australia/New Zealand	30	33	33	49
Oceania*	12	15	19	26
Europe and Northern America	1 114	1 132	1 136	1 120
Least developed countries	1 033	1 314	1 877	3 047
Land-locked Developing Countries	521	659	926	1 406
Small Island Developing States	71	73	37	33

Data source United Nations, Department of Economic and Social Affaire, Population Division (2019). World Population Prospects 2019. *excluding Australia and New Zealand

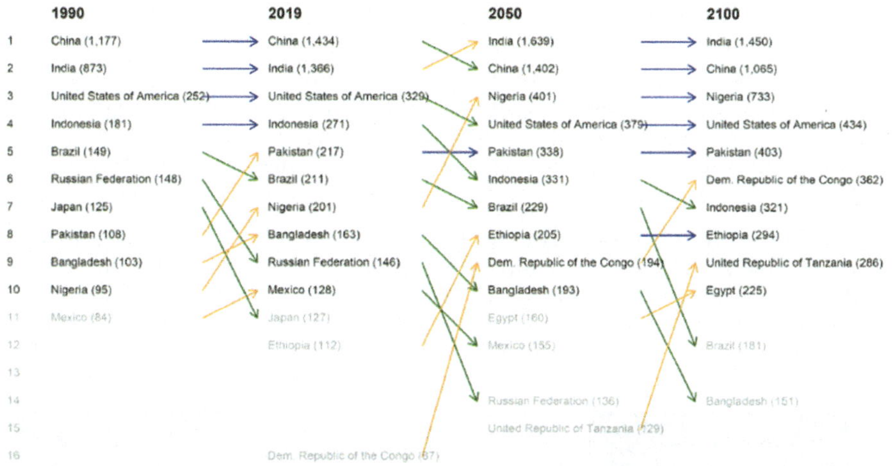

Fig. 2.4 Rankings of the world's 10 most populous countries in 1990 and 2019, and the medium-variant projections for 2050 and 2100. *Source* UNPD (2019): 14. The figure is copyrighted under the Creative Commons Attribution 3.0 IGO license. Permission granted

based on the fifth assessment report (AR5) of Working Group II of the IPCC (Niang/Ruppel/Abdrabo 2014). The most recent data are due to be published in the IPCC's Sixth Assessment Report (IPCC AR6 2021, 2022, 2022a, 2022b).

The task of Working Group II of the IPCC is to review and assess the following problems:

- Co-benefits, risks and co-costs of mitigation and adaptation, including interactions and trade-offs, technological and financial challenges and options.
- Ethics and equity: climate change, sustainable development, gender, poverty eradication, livelihoods, and food security.
- Perception of risks and benefits of climate change, adaptation and mitigation options, and societal responses, including psychological and sociological aspects.
- Climate engineering, greenhouse gas removal, and associated feedbacks and impacts.
- Regional and sectorial climate information.
- Epistemology and different forms of climate-related knowledge and data, including indigenous and practice-based knowledge.

Box 2.1 IPCC's Fifth Assessment Report (2014), WG II, Executive Summary for Africa. *Source* IPCC (2014: 1202–1204). The IPCC's (2022) Sixth Assessment Report will contain the most recent information on Africa in Chap. 9, and will available to download from: https://www.ipcc.ch/report/sixth-assessment-report-working-group-ii/ (Geneva: IPCC, 28 February 2022).

IPCC, AR 4, WG II, chapter on Africa (IPCC 2005: 433–467. *Source* https://www.ipcc.ch/report/ar4/wg2/	IPCC, AR 5, WG II, chapter on Africa (IPCC 2014: 1,202–1,204). *Source* https://www.ipcc.ch/site/assets/uploads/2018/02/WGIIAR5-Chap22_FINAL.pdf	IPCC, AR 6, WG II, chapter on Africa (IPCC 2022: i.p.). *Source* https://www.ipcc.ch/site/assets/uploads/2018/02/WGIIAR5-Chap22_FINAL.pdf

IPCC: *Climate Change 2007 – Impacts, Adaptation and Vulnerability: Contribution of Working Group II to the Fourth Assessment Report of the IPCC* (Geneva: IPCC; Cambridge: Cambridge University Press).	IPCC: *Climate Change 2014 – Impacts, Adaptation and Vulnerability: Contribution of Working Group II to the Fifth Assessment Report of the IPCC* (Geneva: IPCC; Cambridge: Cambridge University Press).	IPCC: *Climate Change 2022 – Impacts, Adaptation and Vulnerability: Contribution of Working Group II to the Fifth Assessment Report of the IPCC* (Geneva: IPCC; Cambridge: Cambridge University Press).

Excerpts of the Executive Summary on Africa, IPCC, WG 2, Chapter 22 (2014: 1,202–1,204):

> The mean annual temperature rise over Africa, relative to the late 20th century mean annual temperature, is likely to exceed 2°C in the Special Report on Emissions Scenarios (SRES) A1B and A2 scenarios by the end of this century (medium confidence). …
>
> A reduction in precipitation is likely over Northern Africa and the southwestern parts of South Africa by the end of the 21st century under the SRES A1B and A2 scenarios (medium to high confidence). …
>
> African ecosystems are already being affected by climate change, and future impacts are expected to be substantial (high confidence). …

Climate change will amplify existing stress on water availability in Africa (high confidence). Water resources are subjected to high hydro-climatic variability over space and time, and are a key constraint on the continent's continued economic development. …

Climate change will interact with non-climate drivers and stressors to exacerbate vulnerability of agricultural systems, particularly in semi-arid areas (high confidence). …

Progress has been achieved on managing risks to food production from current climate variability and near-term climate change but these will not be sufficient to address long-term impacts of climate change (high confidence). …

Climate change may increase the burden of a range of climate-relevant health outcomes (medium confidence). Climate change is a multiplier of existing health vulnerabilities (high confidence), including insufficient access to safe water and improved sanitation, food insecurity, and limited access to health care and education.

Executive Summary, Africa, IPCC, WG 2, Chapter 22 (2014: 1,202–1,204) on adaption policies:

In all regions of the continent, national governments are initiating governance systems for adaptation and responding to climate change, but evolving institutional frameworks cannot yet effectively coordinate the range of adaptation initiatives being implemented (high confidence). … Disaster risk reduction, social protection, technological and infrastructural adaptation, ecosystem-based approaches, and livelihood diversification are reducing vulnerability, but largely in isolated initiatives. … Despite implementation limitations, Africa's adaptation experiences nonetheless highlight valuable lessons for enhancing and scaling up the adaptation response, including principles for good practice and integrated approaches to adaptation (high confidence). … Ecosystem-based approaches and pro-poor integrated adaptation-mitigation initiatives hold promise for a more sustainable and system-oriented approach to adaptation, as does promoting equity goals, key for future resilience, through emphasizing gender aspects and highly vulnerable groups such as children. …

Given multiple uncertainties in the African context, successful adaptation will depend on building resilience. … Growing understanding of the multiple interlinked constraints on increasing adaptive capacity is beginning to indicate potential limits to adaptation in Africa (medium confidence). Climate change combined with other external changes (environmental, social, political, technological) may overwhelm the ability of people to cope and adapt, especially if the root causes of poverty and vulnerability are not addressed. Evidence is growing for the effectiveness of flexible and diverse development systems that are designed to reduce vulnerability, spread risk, and build adaptive capacity. These points indicate the benefits of new development trajectories that place climate resilience, ecosystem stability, equity, and justice at the center of development efforts. … There is increased evidence of the significant financial resources, technological support, and investment in institutional and capacity development needed to address climate risk, build adaptive capacity, and implement robust adaptation strategies (high confidence).

Executive Summary, WG 2, Chapter 22: Africa (IPCC 2014: 1,204), on climate change and conflict:

Strengthening institutional capacities and governance mechanisms to enhance the ability of national governments and scientific institutions in Africa to absorb and effectively manage large amounts of funds allocated for adaptation will help to ensure the effectiveness of adaptation initiatives (medium confidence). Climate change and climate variability have the potential to exacerbate or multiply existing threats to human security including food, health, and economic insecurity, all being of particular concern for Africa (medium confidence). ... Many of these threats are known drivers of conflict (high confidence).

Causality between climate change and violent conflict is difficult to establish owing to the presence of these and other interconnected causes, including country-specific socio-political, economic, and cultural factors. For example, the degradation of natural resources as a result of both overexploitation and climate change will contribute to increased conflicts over the distribution of these resources. ... Many of the interacting social, demographic, and economic drivers of observed urbanization and migration in Africa are sensitive to climate change impacts. ... Of nine climate-related key regional risks identified for Africa, eight pose medium or higher risk even with highly adapted systems, while only one key risk assessed can be potentially reduced with high adaptation to below a medium risk level, for the end of the 21st century under 2 °C global mean temperature increase above preindustrial levels (medium confidence). Key regional risks relating to shifts in biome distribution, loss of coral reefs, reduced crop productivity, adverse effects on livestock, vector- and water-borne diseases, undernutrition, and migration are assessed as either medium or high for the present under current adaptation, reflecting Africa's existing adaptation deficit. ... The assessment of significant residual impacts in a 2°C world at the end of the 21st century suggests that, even under high levels of adaptation, there could be very high levels of risk for Africa. At a global mean temperature increase of 4°C, risks for Africa's food security (see key risks on livestock and crop production) are assessed as very high, with limited potential for risk reduction through adaptation.

Neither the population projections of the UN's Populations Division (UNPD 2019, 2021) nor the climate change assessments of the IPCC's WG II (2014, 2022) are reliable predictions, but they refer to important trends that have to be taken into account in any framework for an ecological peace policy for Africa during the second phase of the Anthropocene (2021–2100).

From a peace research perspective, empirical research on conflicts in Sub-Saharan Africa is of vital importance, especially since independence (1950s–1990s). The WBGU's analysis in its 2008 report on *Security Risk Climate Change* and the IPCC's fifth assessments (2014) in the chapter on human security and violent conflicts may offer the best available scientific information based on the assumptions and results of existing knowledge. However, such data may be overturned by new trends emerging from exponential developments triggered by tipping points in the chaotic climate system.

In addition to its sixth Report on the *Global Environmental Outlook*, the UN's Environmental Programme (UNEP 2016) published a series of regional assessments for the global regions. Its *Regional Assessment for Africa* is of specific relevance to this chapter. According to UNEP (2016a), based on Africa's Agenda 2063:

> Africa aims to establish a prosperous region characterized by sustainable inclusive growth, peace and good governance. The region's growth path shall be led by increased agricultural productivity, industrialization, investment in infrastructure development and renewable energy, conservation of biodiversity, sustainable and fair and equitable use of its genetic resources, clean air and water, and better adaptive capacity to climate change.

The UNEP report also noted that

> Africa faces a great challenge of sustaining rapid economic growth as its population is expected to double to approximately 2.5 billion by 2050, while safeguarding the life-support system provided by its rich natural capital, which underpins the realization of its long-term vision. It is therefore imperative that such growth must consider the region's relatively weak environmental governance and a paucity of accurate and up-to-date environmental and socio-economic data for evidence-based decision-making.

In its key findings the UNEP's regional report emphasised:

> The GEO-6 Regional Assessment for Africa affirms the importance of both Agenda 2063 and Agenda 2030 as defined by the Sustainable Development Goals (SDGs). Both contain common elements for a development trajectory that will provide Africa with a healthy living environment while ensuring good health and quality of life for her people. The two are also critical to preserving and valuing Africa's natural capital for the benefit of its citizens and their livelihoods. In order to realize these visions, Africa's public institutions are called upon to build flexible and adaptive governance structures.

The report focused on all the major environmental problems: (1) indoor air pollution, (2) outdoor pollution, (3) renewable forms of energy, (4) access to potable water, (5) groundwater management, (6) Africa's fisheries, (7) land resources, (8) urbanisation, (9) food production, (10) land tenure, (11) biodiversity. The report recommended

> that Member States actively include a system of factoring biodiversity and ecosystem services into national accounting systems. Africa should also ensure that the African Union strategy on illegal trade in wildlife is translated into action, fully implemented and regularly monitored. Africa faces both enormous challenges in relation to environmental management, and equally huge opportunities for 'doing things better'. The goal to build an integrated, prosperous and peaceful region that is resilient to future shocks can only be reached with the understanding that clean and healthy air, water, land and biodiversity are necessary to support this transformation. All efforts must thus be taken to ensure the protection and integrity of these resources that are critical life-support systems for sustained human wellbeing. Whilst the inherent uncertainty and diversity in potential futures makes it tenuous for a set of prescriptive policies to be established, policy decisions should aim to minimize environmental and developmental trade-offs, and maximize Africa's ability to safeguard its natural capital effectively. Emphasis should be placed on improving protection of the environment, addressing critical data gaps, and developing the human and technical capacities required for a sustainable future. The assessment concludes that low-carbon, climate-resilient choices in infrastructure, energy and food production coupled with effective and sustainable natural resource governance are key to protecting the continent's ecological assets that underpin a healthy society.

Projections of Africa's rapid population rise during the second phase of the Anthropocene (2020–2100) indicate different global and regional trends in the working age populations during both phases of the Anthropocene (1950–2100). While for the rest of the World (without Africa) the working age population has been rising since 1950, reached its height in 2010, has declined since and is projected to continue declining until 2100, it is estimated that in Africa the highest number of employed people will be reached in 2060.[26]

2.6 Conflicts Threatening Africa in the Anthropocene

At least three peace research centres or institutes and the largest institute in strategic studies (IISS) offer regular reviews and assessments of conflicts globally and in Africa:

- *The Uppsala Conflict Data Program* (UCDP) maintained by the Department of Peace and Conflict Research at Uppsala University "offers a number of datasets on organised violence and peacemaking"[27];
- The *Stockholm International Peace Research Institute* (SIPRI) publishes annual reviews and assessments of "global developments in armed conflicts, peace processes and peace operations" in its SIPRI yearbook (Davis 2020);
- The Conflict Barometer of the *Heidelberg Institute for International Conflict Research* (HIIC, 2020) publishes an annual review and assessment of "disputes, non-violent crises, violent crises, limited wars and wars".
- The *International Institute for Strategic Studies* (IISS 2020, 2020a) annually issues (a) *The Armed Conflict Survey* containing a "worldwide review of political, military and humanitarian trends in current conflicts" and (b) a *Strategic Survey* containing "assessments of geopolitics".

In Fig. 2.5 the Uppsala Conflict Data Program provides a regional survey of armed conflicts during the first (past) phase of the Anthropocene (1946–2019), while Fig. 2.6 shows battle-related death by regions over a shorter timespan of three recent decades (1989–2019). There are few systematic long-term analyses of possible causal relationships between environmental and climate-related violent events in Africa.

In his analysis in the *SIPRI Yearbook 2020* of the armed conflicts in Africa during 2019, Ian Davis briefly mentioned that "while Africa is responsible for only 4 per cent of global carbon dioxide emissions, it is particularly vulnerable to the double burden of climate-related factors and political fragility" (Davis 2020, p. 178). He referred to the *African Peace and Security Architecture Roadmap 2016–2020*, which highlighted "climate change as one of the cross-cutting issues affecting peace and security" (African Union Commission 2015).

---

[26] For information on changes in the working age of the population, 1950–2100, see: UNPD (2019): 20.

[27] See at: https://ucdp.uu.se/downloads/charts/ (15 April 2021).

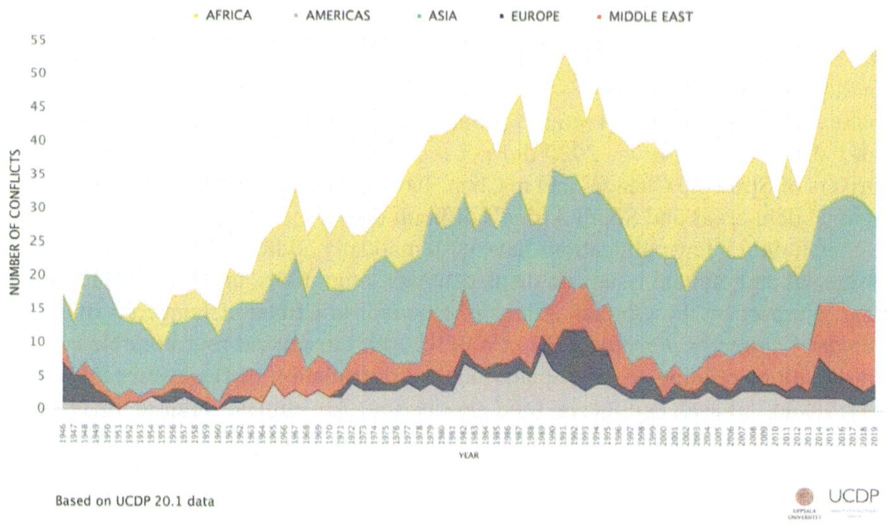

Fig. 2.5 Armed conflict by region (1946–2019). *Source* Based on UCDP, 20.1 data are available free of charge; at: https://www.pcr.uu.se/research/ucdp/charts-graphs-and-maps/

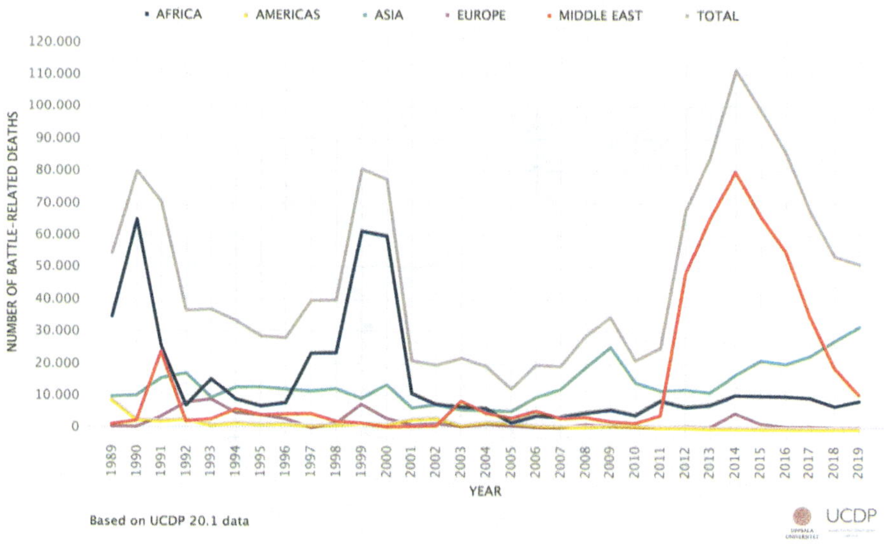

Fig. 2.6 Battle-related death by region (1989–2019). *Source* Based on UCDP data are availabe free of charge; at: https://www.pcr.uu.se/research/ucdp/charts-graphs-and-maps/

Since 1992 the *Conflict Barometer* has offered detailed country-specific conflict data on the changes in conflict densities, the frequencies of conflict intensities and conflict type (HIIK 2020: 66–111). The IISS's *Armed Conflict Survey* was also country-specific, but did not refer to climate change or the environment in the subject index. Its *Strategic Survey 2020* did not include environmental factors among the 'Drivers of Strategic Change' and in Chap. 10 on sub-Saharan Africa focused on the regional debt crisis and South Africa's difficult choices.

So far, only a few publications have systematically addressed the impact of environmental and climate issues on displacements, migrations, crises and conflicts. In 2007, a report by the German Advisory Council on Climate Change (WBGU) on *Climate Change as a Security Risk* (WBGU 2008) discussed four possible conflict constellations relating to (a) climate-induced degradation of freshwater resources, (b) climate-induced decline in food production, (c) climate-induced increase in storm and flood disasters, and (d) environmentally-induced migration. It further analysed 10 regions as hotspots of climate change, among them two in Africa: (i) the Mediterranean (Southern Europe and North Africa) and (ii) Sub-Saharan Africa (Fig. 2.7).

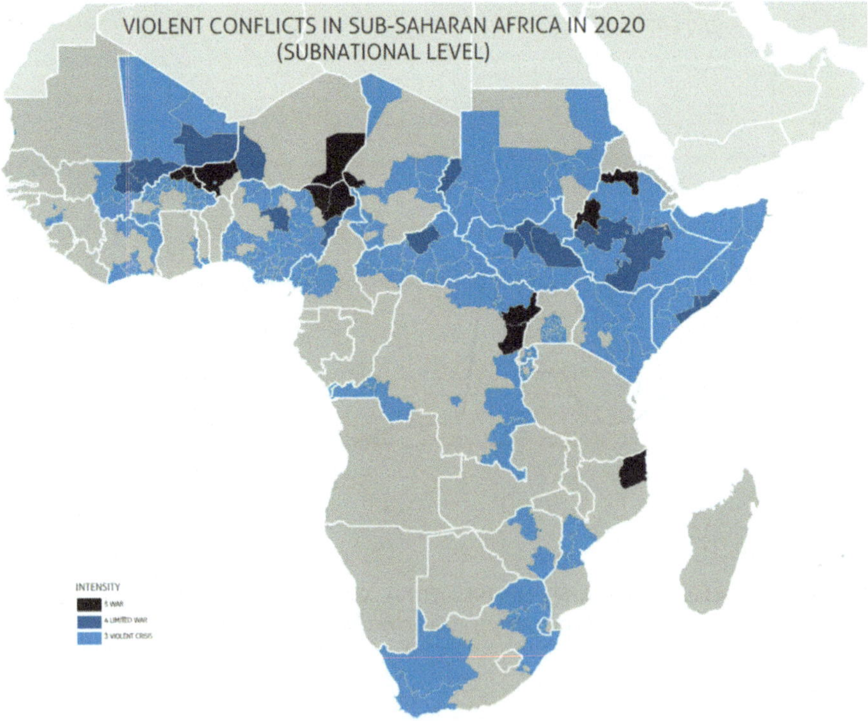

Fig. 2.7 Violent sub-national conflicts in Sub-Saharan Africa in 2020. *Source* HIIC (2021): 70. Reprinted with permission

The WBGU Report proposed nine initiatives "for the mitigation of destabilization and conflict risks associated with climate change" within three sections: (a) Fostering a cooperative setting for a multipolar world, (b) Climate policy as security policy I: Preventing conflict by avoiding dangerous climate change, and (c) Climate policy as security policy II: Preventing conflict by implementing adaptation strategies.

Although the WBGU report was picked up by the European Council in March 2008 (EU, 2008), in a report by the UN Secretary General in September 2009 at the request of the General Assembly, and in several discussions in the UN Security Council since 2007 during the presidency of the UK and Germany and during a co-chaired presidency of France and Germany, the systematic and long-term knowledge that exists so far on Africa, which has experienced the most regional conflicts, has remained very limited.[28]

These issues have primarily been addressed by the policy briefs of consultants and by only a few systematic assessments, among them a chapter of the IPCC's 5AR (2014) on human security, climate change and conflicts. While the global environmental and climate change linkages with national, international and human security have been put on the international agenda, systematic studies and concrete actions on adaptation, mitigation and resilience-building specifically on and in Africa have been scarce.

2.7 Impact on Peace Research and Ecology in Africa

In his report on *Climate change and its possible security implications* of 11 September 2009, the Secretary-General of the United Nations reviewed the linkages between climate change and security, and interpreted climate change as a 'threat multiplier'. Sustainable development includes the provision of 'coping tools' whereby climate change adaptation, mitigation and conflict prevention (Fig. 2.9) become 'threat minimizers'. This figure illustrates what 'peace ecology', as a combined scientific approach in the social sciences, and an 'ecological peace policy' are supposed to be about.

---

[28] See: Joint paper by the Commission and the Secretary-General/High Representative concerning "Climate change and international security" presented to the European Council, Brussels, 3 March 2008; at: https://www.consilium.europa.eu/uedocs/cms_data/docs/pressdata/en/reports/99387.pdf; Press Conference by Security Council President, 4 April 2007; at: https://www.un.org/press/en/2007/070404_Parry.doc.htm; UN Security Council, SC/9000, 5663rd meeting, 17 April 2007: "Security Council holds first-ever debate on impact of climate change on peace, security, hearing 50 speakers"; at: https://www.un.org/press/en/2007/sc9000.doc.htm; UN, 2007: "Security Council Holds First-Ever Debate on Impact of Climate Change on Peace, Security, Hearing over 50 Speakers, UN Security Council, 5663rd Meeting, 17 April 2007"; at: https://www.un.org/press/en/2007/sc9000.doc.htm; UN (2009): "Climate change and its possible security implications". Resolution adopted by the General Assembly, A/RES/63/281 (New York: United Nations General Assembly, 11 June); UN (2009a), (2010), (2011): "Informal Thematic Debate on Human Security"; at: http://www.un.org/en/ga/president/65/initiatives/HumanSecurity.html

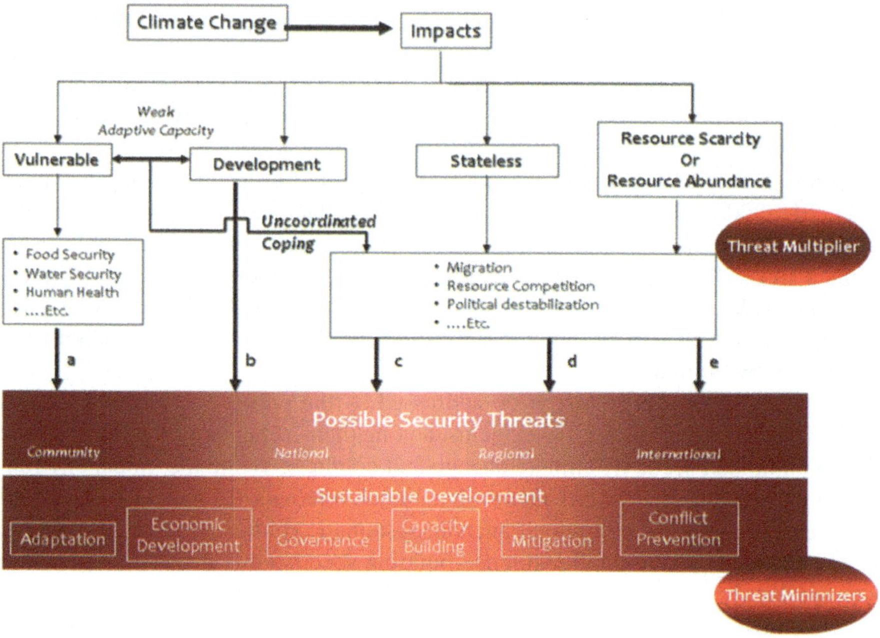

Fig. 2.8 Threat multipliers and threat minimizers: the five channels. *Source* UN SG (2009): 6

Most research on environmental challenges in Africa and on displacement, migration and violent conflicts is still taking place in isolation without any conceptual or empirical integration. A few articles have reviewed the state of research on "Climate as risk factor for conflict", arguing (Fig. 2.8):

> Research findings on the relationship between climate and conflict are diverse and contested. Here we assess the current understanding of the relationship between climate and conflict, based on the structured judgments of experts from diverse disciplines. These experts agree that climate has affected organized armed conflict within countries. However, other drivers, such as low socio-economic development and low capabilities of the state, are judged to be substantially more influential, and the mechanisms of climate–conflict linkages remain a key uncertainty. Intensifying climate change is estimated to increase future risks of conflict (Mach/Kraan/Adger et al. 2019, pp. 193-197).

These authors argue that "the lack of clarity on current knowledge limits informed management of the risks of conflict to states and human security, and of the risks of continuing greenhouse gas emissions." The experts and authors ranked four drivers as influential conflict risks; (1): low socio-economic development, (2) low capabilities of the state, (3) intergroup inequality, and (4) recent history of violent conflict. They conclude that there

> is agreement that climate variability and change shape the risk of organized armed conflict within countries. In conflicts to date, however, the role of climate is judged to be small compared to other drivers of conflict, and the mechanisms by which climate affects conflict

are uncertain. As risks grow under future climate change, many more potential climate–conflict linkages become relevant and extend beyond historical experiences.

The usefulness of this approach is emphasised by these authors:

For those scholars and policy-makers who are focused on climate, a synoptic understanding of the climate–conflict relationship is important, even if the role of climate is relatively minor compared to other drivers of conflict. Given that conflict has pervasive detrimental human, economic and environmental consequences, climate–conflict linkages – even if small – would markedly influence the social costs of carbon and decisions to limit future climate change.

For those scholars and policy-makers focused on conflict, the assessment has pointed to the different ways in which climate may interact with the major drivers of conflict risk. Effectively managing such interactions will require mainstream and holistic, rather than myopic, considerations of the role of the climate across diverse settings and attention to uncertainties that will persist.

And finally, appreciation of the future role of climate change and its security effects can help to prioritize societal responses, which could include enhanced global aid and cooperation.[29]

Nina von Uexkull/Halvard Buhaug (2021) have summarised the security implications of climate change after 10 years of scientific progress:

The study of security implications of climate change has developed rapidly … into an important and thriving research field that traverses epistemological and disciplinary boundaries. Here, we take stock of scientific progress by benchmarking the latest decade of empirical research against seven core research priorities collectively emphasized in 35 recent literature reviews. ... Overall, we find that the research community has made important strides in specifying and evaluating plausible indirect causal pathways between climatic conditions and a wide set of conflict-related outcomes and the scope conditions that shape this relationship.

Both articles rely on a review of the primarily quantitative literature, ignoring qualitative debate and conceptual discussion on the linkage between peace research and ecology in the Anthropocene, and they make no specific reference to Africa. They are consequently of little relevance when formulating policies to address the linkages between environmental and climatic challenges that Africa has faced during the Anthropocene between 1945 and 2020 and since independence.[30]

The policy-orientated and empirical literature on this dual policy linkage between climate and conflict pertaining to Africa and by African authors is still very limited. Scheffran et al. (2019)offer a comprehensive overview of the state of knowledge on ‘Climate and Conflict in Africa’, referring to climate as a “risk multiplier” in “fragile regions and hotspots” in Africa:

where poverty, violence, injustice, and social insecurity are prevalent. The linkages have been most extensively studied for the African continent, which is affected by both climate change and violent conflict. Together with other drivers, climate change can undermine human security and livelihoods of vulnerable communities in Africa through different pathways. These

[29] See the figure: Scope of the expert assessment – Climate as a risk factor for armed conflict. Source Mach et al. (2019: 193–197).

[30] Collective thematic scope of special issue contributions. *Source* von Uexkull und Buhaug (2021), 58,1: 3–17.

> include variability in temperature and precipitation; weather extremes and natural disasters, such as floods and droughts; resource problems through water scarcity, land degradation, and food insecurity; forced migration and farmer-herder conflict; and infrastructure for transport, water, and energy supply. Through these channels, climate change may contribute to humanitarian crises and conflict, subject to local conditions for the different regions of Africa. While a number of statistical studies find no significant link between reduced precipitation and violent conflict in Africa, several studies do detect such a link, mostly in interaction with other issues. The effects of climate change on resource conflicts are often indirect, complex, and linked to political, economic, and social conflict factors, including social inequalities, low economic development, and ineffective institutions.

Regions dependent on rain-fed agriculture are more sensitive to civil conflict following droughts. Scheffran et al. (2019) argue that

> [r]ising food prices can contribute to food insecurity and violence. Water scarcity and competition in river basins are partly associated with low-level conflicts, depending on socio-economic variables and management practices. Another conflict factor in sub-Saharan Africa is shifting migration routes of herders who need grazing land to avoid livestock losses, while farmers depend on land for growing their harvest. Empirical findings reach no consensus on how climate vulnerability and violence interact with environmental migration, which also could be seen as an adaptation measure strengthening community resilience. Countries with a low human development index (HDI) are particularly vulnerable to the double exposure to natural disasters and armed conflict. Road and water infrastructures influence the social and political consequences of climate stress. The high vulnerabilities and low adaptive capacities of many African countries may increase the probability of violent conflicts related to climate change impacts.

This text, published in the *Oxford Research Encyclopaedia of Climate Science* (2019), offers a comprehensive overview of the empirical, theoretical and conceptual literature on and in Africa, and includes a broad bibliography focusing on (i) the climate-conflict nexus, (ii) climate change vulnerability, human security and violent conflict in Africa, (iii) factors of security risks and conflict potentials in Africa, (iv) North Africa, (v) East Africa, (vi) West Africa, vii) Central Africa, (viii) Southern Africa, (ix) reflections on the climate-conflict nexus, (x) emphasis on the specific conditions and outcomes in African sub-regions:

> No scientific consensus has emerged on whether and how climate change triggers or contributes to violent conflict in Africa. This is because different factors matter in different regions of Africa and the overall link between climate change and violent conflict in Africa is likely to be indirect, complex, and related to multiple political, economic, and social factors. Several studies indicate that climate-induced natural disasters can destabilize societies with weak economies, mixed political regimes, and pre-existing conflicts. Other studies find resource conflicts arising from unequal distribution of aid, changing power relationships, and opportunities for warlords. Droughts tend to affect civil conflict in regions dependent on rain-fed agriculture. Qualitative regional case studies and conceptual frameworks can help to capture the complex linkages between climate change, vulnerability, and violent conflict. Conflict escalation cannot be explained by a single factor such as climate change. However, in 'hotspots', multiple drivers, actors, and actions can interact in complex vicious circles that undermine state authority and overstretch problem-solving capabilities. The interaction between vulnerability to climate change and conflict is subject to regionally specific conditions across Africa.

Scheffran et al. (2019)summarise the different drivers for North, East, West, Central and South Africa.

> In North Africa, water allocation (particularly in the Nile River Basin), food, land use, and agriculture are most likely related to violent conflicts. Rising food prices may have contributed to cascading events that destabilized the region in the wake of the Arab Spring. In East Africa, environmental change combines with social and political issues, such as marginalization and exclusion, food problems, population pressure, refugee movements, and political instability, likely contributing to conflicts between pastoralists and other groups, which remain controversial with regard to their significance. West Africa experiences vulnerability to climate change and land use together with violent conflict, for example, between farmers and herders. Less systematic research on the linkages between climate change and conflict is available for Central Africa and South Africa, which are both shaped by adverse environmental impacts, migration, and major conflicts related to natural resource exploitation and scarcity. A political ecology perspective not only challenges simplistic linkages of farmer-herder conflicts with resource scarcity, and environmental and climate change, but also considers political power relations, complex actor networks, and social processes as well as citizenship and land use rights. (…) The conflict situation affects the conditions for resource management. New integrative strategies assess how different lifestyles are affected by ecosystem services provided by natural resources to local people. Vulnerabilities and adaptation measures to altered climate conditions are considerably different for farming or pastoralist livelihood strategies (…), encompassing local knowledge and the capacity building of institutions and governance structures that can support rural communities affected by climate change (…). All geographical levels (local, regional, and federal) and all types of society (government, private sector, and civil society) are to be included.

Scheffran et al. (2019)also distinguish between different outcomes:

> Cooperative solutions for sustainable peace and environmental peace may emerge when communities see environmental change as a common security issue that needs to be jointly addressed by collaboration to use and share resources in a sustainable and efficient way (…). In crises, cooperation often prevails over conflict, and the number of international water agreements and cooperation have been increasing (…). Policies and institutions can reduce conflicts linked to climate change, using a range of policy levers from mitigation and adaptation to development. Measures to reduce agro-pastoral conflicts include the joint management of common pool resources, the protection of resources from degradation against harsh climate conditions, and soil and water conservation techniques, as well as land fertilization by cattle manure. With regard to the societal boundary conditions, the revision of land tenure and pastoral legislation, notably on access to resources, the strengthening of local structures for conflict settlement and prevention with participation of traditional leaders, representatives of the local administration, and civil society organizations, as well as sensitization and information sharing, could be useful strategies to reduce conflict risk (Cabot 2017).

In Africa, scientific and political discussions on the links between climate change and conflicts have been taking place for about a decade within both security studies and peace research. In May 2012, the Institute for Security Studies in Pretoria, Addis Ababa, Dakar and Nairobi published a paper by Roban Adano Wario and Faruma Daudi (2012) from Kenya which aimed

> to review and assess the existing evidence on the security threat of climate change, with particular reference to Africa. The paper addresses the question to what extent climate change poses a threat to security and conflict in Africa. It further seeks to identify manifestations of climate change, the sectors and regions most likely to suffer from the adverse impacts of climate change, and the associated incidence of conflict.

In November 2011 Pius Yanda; Salomé Bronkhorst (2011) published a *Policy and Practice Brief* on "Climate change and conflict: Conflict-sensitive climate change adaptation in Africa" for the *African Centre for the Constructive Resolution of Disputes* (ACCORD) in Durban, based on a

> two-day expert seminar to identify issues and recommendations for ensuring that adaptation to climate change is conflict-sensitive. Practitioners and scientists presented new research on the linkages between climate change and conflict in Africa, and analysed various tools, policies and approaches to ensure that conflicts arising from climate change are addressed and climate change adaptation measures are conflict-sensitive. (...) The first section reports on the scientific evidence for the linkages between climate change and conflict. The second section discusses what it means to have conflict-sensitive climate change adaptation – hereafter referred as 'conflict-sensitive adaptation' – drawing on the research and discussion outcomes from the seminar. The third section outlines key considerations for conflict-sensitive adaptation. The final section includes recommendations for continental, regional and national bodies on conflict-sensitive adaptation (Bob/Bronkhorst 2014).

Since this conference in preparation for COP 17 of the UNFCCC in Durban, policy reports (von Soest 2020), scientific articles[31] and several books[32] have been published on the linkages between a specific environmental theme (climate change) and possible policy outcome (conflict) or impact (security and peace). However, none of these publications were framed in the context of peace ecology or an ecological peace policy in the Anthropocene.

2.8 Towards Peace Ecology for Africa in the Anthropocene

The themes of global environmental and climate change as a cause, and displacements, migration, crises and violent conflicts, including as a worst case civil and international wars, as possible outcomes and multiple impacts of policies aiming at peace and security are concerns which peace ecology and an ecological peace policy need to address in Africa in the social sciences and in practice during the Anthropocene epoch.

Brauch (2021) recommends analysis, reflection and action on the following problem areas between peace and security issues on the one hand, and environment and sustainable development problems on the other, in which various relationships and linkages have been addressed in social science literature since the 1980s:

[31] Scheffran (2020: 19-48); Balbo, Rothe, Scheffran (2020: 287–296); van Baalen, Mobjörk (2018: 547–575); von Uexkull, Pettersson (2018: 953–968); Kamta, Schilling, Scheffran (2021: 27); Kogoui Kamta, Schilling/Scheffran (2020: 6830); Kamta et al. (2020: 95–104); Marie, Yirga, Haile, et al. (2021); Schilling, Hertig, Tramblay, Scheffran (2020: 15); Shaaban, Scheffran, Böhner, Elsobki (2019: 4); Noagah Bukhari, Sow, Scheffran (2019: 161-185); Schilling, Locham, Scheffran (2018: 571-600); Solomon, Birhane Gordon, et al. (2018: 284–290); Evadzi, Scheffran, Zorita, Hünicke (2018: 183-197).

[32] Swain, Bali, Swain Anders, Krampe (2011); Cabot (2017).

- the early conceptual debate on the linkages between peace and the environment[33];
- studies on the impacts of weapons and wars on the environment[34];
- the conceptual debate on environmental and ecological security (Brauch 2021, pp. 115–118);
- the impact of environmental degradation and stress on environmental conflict (Brauch 2021, pp. 118–122);
- environmental peace-making and the role of the environment in post-conflict peace-building (Brauch 2021, pp. 122–125);
- the emerging discourse and policy debate on climate change and conflicts (Brauch 2021, pp. 125–132);
- the early approaches to peace ecology and their shortcomings (Brauch 2020, pp. 132–141).

However, these multiple research projects and studies have not yet resulted in a joint comprehensive peace ecology research programme within the framework of the Anthropocene epoch in Earth and human history. Since the 1990s, environmental security research has shifted from environmental scarcity, degradation and conflict to the dangers posed by global environmental and climate change (Sygna et al. 2013). With the direct impacts of humans upon ecosystems in the Anthropocene (Crutzen 2002, 2011) and with the progressive securitisation of GEC issues since 2003, these anthropogenic changes are increasingly threatening human lives and livelihoods. Worldwide, the destruction of key ecosystem services, the pollution of air, water and soil, land use change and extreme events are creating new 'anthropogenic challenges' for humankind, although these do not pose a threat of violent conflict or war.

Peace ecology in the Anthropocene was conceptualised by Oswald Spring et al. (2014, pp. 18–19) within the framework of peace, security, equity, sustainability, culture and gender (Fig. 2.1). In order to develop a research framework and programme that can conceptually and methodologically encompass the many research projects that have emerged in the social sciences since the end of the Cold War, these authors suggest a wider peace ecology concept than that proposed by Kyrou (2007)/Amster (2014).

It is appropriate that the temporal context for this peace ecology research programme is the Anthropocene (Crutzen/Stoermer 2000; Crutzen 2002), since this epoch of Earth and human history began when a turning point in the geological time of the Earth's history coincided with the long-term structural turning points in human history after the end of World War II, the emerging nuclear era, the Cold War (1947–1989) and the post-Cold War disorder (1990-present).

The environmental dimensions of this change could not be socially constructed until the late 1980s, when the exponential increase in greenhouse gases, particularly CO_2, since the *Great Acceleration* in the 1950s made the issues relating to global environmental change and climate change all too apparent. With the exponential

[33] Brauch (2021): 111–114, Boulding/Kenneth (1978, 1989), Boulding/Elise (1988, 1992, 2000), Oswald Spring (2008a).

[34] Brauch (2021): 114–115; Westing (1976, 2003, 2013, 2013a).

increase of *greenhouse gases* (GHG) in the atmosphere since 1950, scientific knowledge of the physical impacts of global climate change has increased since the 1980s, initially in the natural sciences. Several new research and training programmes and research institutes have been set up in the newly emerging areas of *Earth Systems Science* (ESS) and *Earth System Analysis* (ESA). After the build-up of new scientific capabilities, knowledge of GEC has expanded rapidly.

In *Anthropocene Geopolitics* Dalby (2020, pp. 169–187, 184) briefly discusses the previously proposed interdisciplinary research programme of 'political geoecology in the Anthropocene' (Brauch et al. 2011), which brings natural science perspectives into primarily social-science-orientated research.

In my opinion, a peace ecology approach should:

- overcome the overspecialisation in both the social and the natural sciences;
- offer *holistic* approaches that address interlinkages between themes addressed by peace research and environmental studies;
- be contextualised in the Anthropocene epoch of the Earth's history and address scientific and political linkages since the end of World War II and the Great Acceleration:
- be interdisciplinary; and in the Anthropocene it should be further developed from interdisciplinary and transdisciplinary approaches into a *transformative research programme.*

Each discipline has its specific epistemology, premises and methods of generating new knowledge.[35] As the problems and issues that need to be examined scientifically become more complex, *multidisciplinarity* offers a first step towards analysing complex problems from different disciplinary perspectives. These multidisciplinary studies rely on the methodologies of their respective disciplines.

Given the complexity of the Anthropocene, global environmental change and resource scarcity, several research centres and think tanks have proposed *transdisciplinarity* as a new scientific approach to overcome the narrow disciplinary boundaries of specialised subfields and epistemic schools of knowledge creation. In short, *transdisciplinarity* refers to a research strategy that establishes a common research objective that crosses disciplinary boundaries.

Schneidewind, Singer-Brodowski/Augenstein (2016) proposed moving from a 'transdisciplinary' approach to 'transformative science', while Swilling (2016) suggested 'anticipatory science'. The concept of 'transformative research' or 'science' has been used since the 2000s for a new approach that cuts across the dominant *scientific paradigms.* The US National Science Board (2007) adopted the following working definition of 'transformative research':

> [it] *involves ideas, discoveries, or tools that radically change our understanding of an important existing scientific or engineering concept or educational practice or lead to the creation of a new paradigm or field of science, engineering, or education. Such research challenges current understanding or provides pathways to new frontiers.*

[35] This section relies on Oswald Spring et al. (2016: Chapter 43).

Building on this approach, in *World in Transition – A Social Contract for Sustainability*, the WBGU (2011, pp. 21–23, 321–356) referred to "four transformative pillars of the knowledge society": transformation research and transformation education, as well as transformative research and transformative education. It asserted (WBGU 2011, p. 21) that transformation research "specifically addresses the future challenge of transformation realisation" by exploring "transitory processes in order to come to conclusions on the factors and causal relations of transformation processes" and should "draw conclusions for the transformation to sustainability based on an understanding of the decisive dynamics of such processes, their conditions and interdependencies." Transformative research supports transformation processes with specific innovations in the relevant sectors and should encompass, for example, "new business models such as the shared use of resource-intensive infrastructures, and research for technological innovations like efficiency technologies" by aiming at a "wider transformative impact". Schneidewind/Singer-Brodowski (2013) and Göpel (2017)have developed this transformative approach further for climate policy and research on sustainability transition.

In its report on the *Transformative Cornerstones of Social Science Research for Global Change*, the *International Social Science Council* (ISSC 2012: 21–22) identified six cornerstones: (1) historical and contextual complexities; (2) consequences; (3) conditions and visions for change; (4) interpretation and subjective sense-making; (5) responsibilities; and (6) governance and decision-making. The report concluded that

> the transformative cornerstones framework speaks to the full spectrum of social science disciplines, interests and approaches – theoretical and empirical, basic and applied, quantitative and qualitative. By not fashioning a global change research agenda around a substantive focus on concrete topics – water, food, energy, migration, development, and the like – the cornerstones are not only inclusive of many social science voices but, perhaps most importantly, show that climate change and broader processes of global environmental change are organic to the social sciences, integral to social science preoccupations, domains par excellence of social science disciplines. ... The transformative cornerstones of social science function not only as a framework for understanding what the social sciences can and must contribute to global change research. They function as a charter for the social sciences, a common understanding of what it is that the social sciences can and must do to take the lead in developing a new integrated, transformative science of global change.

Various initiatives by the US *National Science Board* (2007), the ISCC (2012), and the *Sustainability Transition Research Network* (STRN 2016) have called for a new scientific paradigm for research into both global environmental change and sustainability transitions. The policy dimension should be included in the research design by moving from knowledge creation to action, policy initiatives, development and implementation.

These efforts are still highly dependent on the top-down efforts of governments and multinational enterprises. A transformative research programme implies moving "from knowledge to action" by addressing the challenges for peace and sustainable development arising from the impacts of global environmental and climate change in the Anthropocene as a result of anthropogenic interventions in multiple Earth

Systems processes. Peace ecology – as introduced above – is action research, in which the transformative action is already reflected in the research design.

2.9 Ecological Peace Policy for Africa in the Anthropocene

The two suggested concepts of peace ecology and ecological peace policy in the Anthropocene are closely related. Peace ecology applies to integrated scientific analyses of peace, security, development and environmental issues since the end of World War II, when the Anthropocene began, triggered by the development, test and employment of nuclear weapons (1945) and their atmospheric testing (until 1963) and by the Great Acceleration (since 1950) that resulted in an exponential increase in *greenhouse gases* (GHG), especially *carbon dioxide* (CO_2).

Peace ecology differs from other more narrow concepts of environmental cooperation, environmental peacemaking, peacebuilding and post-conflict peacebuilding (Ide 2017, 2018, 2018a, 2019, 2019a, 2020; Ide/Detges 2018; Ide et al. 2021). The proposed peace ecology approach in the Anthropocene offers a specific contextualisation and a broader thematic focus on peace, security and ecological linkages than the more narrow specialised concepts in the context of environmental diplomacy.

For industrialised countries, ecological peace policy comprises a normative approach to politics that reflects ecological concerns, challenges posed by global environmental and climate change, sustainable development issues and sustainability transition designed to yield a social, business and overall policy capable of achieving the *Sustainable Development Goals* (SDGs)[36] and the UN's Agenda 2030 for Sustainable Development[37] through a process of decarbonisation.

As Africa is a continent that has already been seriously affected by the physical and societal effects of climate change, it is crucial to include environmental assessments and ecological concerns in the framework of domestic and external policy fields designed to achieve the African Union's Agenda 2063, the Framework Document and First Ten-Year Implementation Plan of which was adopted in June 2015 in South Africa.[38]

Africa is now also living in the Anthropocene. Although so far and probably for many years to come its GHG emissions are relatively low, the environmental impact of the global economic model and behaviour has already been significant, and in 2019 Africa had a 23% share of global fatalities from natural disasters, following

---

[36] SDGs; at: https://sdgs.un.org/goals (15 June 2021).

[37] UN's Agenda 2030 for Sustainable Development; at: https://sdgs.un.org/2030agenda.

[38] See: "01 Background Note: Agenda 2063: The Africa We Want"; at: https://au.int/sites/default/files/documents/33126-doc-01_background_note.pdf .

Table 2.2 African Region Profile for Natural Disasters from 1980 to 2008. Source: https://www.preventionweb.net/ (18 April 2021)

Affected People				Killed People			
Country	Disaster	Date	Total affected	Country	Disaster	Date	Killed
Kenya	Drought	1999	23,000,000	Ethiopia	Drought	1983	300,000
South Africa	Drought	2004	15,000,000	Sudan	Drought	1983	150,000
Ethiopia	Drought	2003	12,600,000	Mozambique	Drought	1981	100,000
Ghana	Drought	1983	12,500,000	Ethiopia	Epidemic	1988	7,385
Sudan	Drought	1991	8,600,000	Nigeria	Epidemic	1991	7,289
Sudan	Drought	1983	8,400,000	Nigeria	Epidemic	1996	4,346
Ethiopia	Drought	1983	7,750,000	Burkina Faso	Epidemic	1996	4,071
Ethiopia	Drought	1987	7,000,000	Niger	Epidemic	1995	3,022
Malawi	Drought	1992	7,000,000	Sudan	Epidemic	1988	2,770
Ethiopia	Drought	1989	6,500,000	Algeria	Earthquake*	1980	2,633

Asia (45%) and Europe (23.4%), with a higher percentage than the Americas (8%) and Oceania (0.6%).[39]

Based on older EMDAT data,[40] the Prevention Web[41] offered this summary of the disaster statistics for Africa (1980-2008), according to which between 1980 and 2008 most people in Africa were affected by drought and most people were killed as a result of droughts and epidemics primarily in the highly vulnerable Sahel zone (Table 2.2).

According to the World Bank's indicators, CO_2 emissions in Africa increased from 126,045.1 kt of CO_2 in 1960 to 438,422.2 kt in 1990 and 853,107.1 kt in 2016.[42] By 2020 the CO_2 emissions of 47 Sub-Saharan African countries represented about 1 billion people and constituted about 3–4 per cent of the global emissions. Based on a report on the *State of the Climate in Africa in 2019* (WMO, 2020), Africa was severely affected by increased temperatures, the unpredictability of precipitation, and extreme weather events, resulting in severe droughts and food shortages.

The tasks of an ecological peace policy are to project the probable ecological consequences of the physical effects of global environmental change; to launch early preventive actions to avoid environmental crises, conflicts or wars; and to solve these environmental consequences peacefully without any outbreaks of violence (Brauch 2002).

[39] See: Share of fatalities from natural disasters by continent 2019. *Source:* M. Szmigiera, *Statistica* (30 March 2021), at: http://www.statista.com/statistics/273890/countries-with-the-most-fatalities-from-natural-disasters/ (18 April 2021).

[40] See EMDAT, at: https://www.emdat.be/ (20 April 2021).

[41] See: PreventionWeb; at: https://www.preventionweb.net/english/ (20 April 2021).

[42] See: "CO_2 emissions (kt) – Sub-Saharan Africa". *Source*: Carbon Dioxide Information Analysis Center, Environmental Sciences Division, Oak Ridge National Laboratory, Tennessee, United States; at: https://data.worldbank.org/indicator/EN.ATM.CO2E.KT?locations=ZG (14 June 2020).

It is impossible to be sure how countries which have previously failed to implement their obligations under climate change agreements will behave in the future. However, deductions can be made about societal and political outcomes by using climate models which correlate projections of future population with the production and consumption levels associated with different lifestyles.

An ecological peace policy should aim to provide a preventive strategy so that (1) the projected physical effects of anthropogenic climate change can be contained, resolved and structurally prevented through multilateral diplomacy and cooperation, and (2) domestic and international violent crises, conflicts and wars can be avoided. Jointly and proactively addressing resource scarcity, resource pollution and stress (of water, soil, air) will make it possible to enhance the health and security of the affected people.

2.10 Outlook: Rethinking and Integrating Research and Action in the Anthropocene

Whether these conceptual and theoretical considerations for a new research framework (peace ecology) and an integrated policy approach (an ecological peace policy) matter for Sub-Saharan African countries is for African researchers and policy-makers and their advisers to decide.

Although several environmental linkages between peace and security issues have increasingly been addressed in the social sciences – political science, international relations, strategic studies and peace research – and a peace ecology approach has slowly emerged since 2007 (Kyrou 2007; Oswald et al. 2014; Amster 2014), an integrated comprehensive ecological peace policy has so far not been developed and discussed.

Although between 2000, when Crutzen uttered the term 'Anthropocene', and 2022 nearly a thousand books and several thousand scholarly articles were published (Brauch 2021a), a debate on linkage concepts between political science, international relations, peace research and environmental studies has remained underdeveloped within the social sciences, and the much-needed integration of knowledge derived from climate and Earth Systems Science and Earth Systems Analysis in the natural sciences with analysis in the social sciences has remained rare, despite many pleas for interdisciplinary research.

In the Anthropocene there is a complex dual causal relationship between human beings, the Earth System and the social system. During the Anthropocene, for the first time humankind has directly intervened in the Earth System through the burning of fossil fuels, resulting in an anthropogenic increase in greenhouse gases, most particularly *carbon dioxide* (CO_2), in the atmosphere. This global warming has triggered four major physical effects: (a) an increase in global average temperature, (b) variations in precipitation, (c) low-onset sea-level rise, and (d) an increase in the probability and intensity of extreme weather events. These phenomena have been assessed so far by five assessment reports of the IPCC and several special reports. The physical effects of anthropogenic climate change have had multiple societal impacts that may cause severe domestic or international crises, conflicts, and, in the very worst case, even violent wars (Brauch 2009).

A peace ecology approach or programme in the Anthropocene for Africa must combine and integrate the results of peace research in the social sciences with the research in Earth Systems Science (ESS) and Earth Systems Analysis (ESA) which has primarily been conducted in the natural sciences. A peace ecology approach must thus cross the narrow disciplinary and research programme boundaries and move from multidisciplinary perspectives to interdisciplinary assessments. A peace ecology research programme that aims to contribute to an ecological peace policy should be transformative by including the transformation of the status quo in the research design.

References

Adano Wario, Roban; Daudi, Faruma, 2012: *Links between climate change, conflict and governance in Africa*, Institute for Security Studies Paper (Pretoria: ISS).

African Union Commission, 2015: *African Peace and Security Architecture. APSA Roadmap 2016-2020* (Addis Abbaba: African Union Commission, Peace and Security Department): 20–21.

Amster, Randall, 2014: *Peace Ecology* (Boulder, CO: Paradigm).

Badash, Lawrence, 2009: *A Nuclear Winter's Tale. Science and Politics in the 1980s* (Cambridge, MA: MIT Press).

Balbo, Andrea L.; Rothe, Delf; Scheffran, Jürgen, 2020: "The Anthropocene: an opportunity for transdisciplinary and inclusive science?", in: Brzoska, Michael; Scheffran, Jürgen (Eds.): *Climate Change, Security Risks, and Violent Conflicts. Essays from Integrated Climate Research in Hamburg* (Hamburg: Hamburg University Press): 287–296.

Benedick, Richard Elliot, 1998: *Ozone Diplomacy - New Directions in Safeguarding the Planet*, Enlarged Edition (Cambridge, Mass.: Harvard University Press).

Benner, Susanne; Lax, Gregor; Crutzen, Paul J.; Lelieveld, Jos; Pöschl, Ulrich: Brauch, Hans Günter (Eds.), 2022: *Paul J. Crutzen and the Anthropocene – A New Earth's Epoch* (Cham: Springer International Publishing).

Birmingham, David, 1995: *The Decolonization of Africa* (London: Routledge).

Bob, Urmilla; Bronkhorst, Salome (Eds.), 2014: *Conflict Sensitive adaptation to climate change in Africa* (Berlin: Berliner Wissenschafts-Verlag).

Boulding, Elise, 1988: *Building a Global Civic Culture: Education for an Interdependent World* (New York, NY: Teachers College Press).

Boulding, Elise (Ed.), 1992: *New Agendas for Peace Research: Conflict and Security Re-examined* (Boulder, Co: Lynne Rienner).

Boulding, Elise, 2000: *Cultures of Peace: The Hidden Side of History* (New York: Syracuse University Press).

Boulding, J. Russell (Ed.), 2017: *Elise Boulding: A Pioneer in Peace Research, Peacemaking, Feminism and the Family: From a Quaker Perspective*, with a Foreword by Betty Reardon and a Preface by Úrsula Oswald Spring (Cham – Heidelberg – New York – Dordrecht – London: Springer International Publishing).

Boulding, J. Russell (Ed.), 2017a: *Elise Boulding: Writings on Peace Research, Peacemaking and the Future* (Cham – Heidelberg – New York – Dordrecht – London: Springer International Publishing).

Boulding, J. Russell (Ed.), 2017b: *Elise Boulding: Writings on Feminism, the Family and Quakerism* (Cham – Heidelberg – New York – Dordrecht – London: Springer International Publishing).

Boulding, J. Russell (Ed.), 2017c: *Elise Boulding: Autobiographical Writings and Selections from Unpublished Journals and Letters* (Cham Boulding, Kenneth E. 1966: "The Economics of the Coming Spaceship Earth", in: Jarrett, H. (Ed.): *Environmental Quality in a Growing Economy* (Baltimore: Johns Hopkins Press).

Boulding, Kenneth E., 1978: *Ecodynamics* (Beverly Hills, CA: Sage).

Boulding. Elise. 1989: *One Small Plot of Heaven: Reflections on Family Life by a Quaker Sociologist* (Philadelphia, PA: Pendle Hill Press, 1989).

Bourantonis, Dimitris; Wiener, Jarrod (Eds.), 1995: *The United Nations in the New World Order: The World Organization at Fifty* (London: Macmillan)

Brauch, Hans Günter, 2002: "Climate Change, Environmental Stress and Conflict: AFES-PRESS Report for the Federal Ministry for the Environment, Nature Conservation and Nuclear Safety", in: Federal Ministry for the Environment, Nature Conservation and Nuclear Safety (Ed.): *Climate Change and Conflict: Can Climate Change Impacts Increase Conflict Potentials? What is the Relevance of this Issue for the International Process on Climate Change?* (Berlin: Federal Ministry for the Environment, Nature Conservation and Nuclear Safety, 2002): 9–112; at: http://www.afes-press.de/pdf/Brauch_ClimateChange_BMU.pdf.

Brauch, Hans Günter, 2009: "Securitizing Global Environmental Change", in: Brauch, Hans Günter; Oswald Spring, Ursula; Grin, John; Mesjasz, Czeslaw; Kameri-Mbote, Patricia; Behera, Navnita Chadha; Chourou, Béchir; Krummenacher, Heinz (Eds.), 2009: *Facing Global Environmental Change: Environmental, Human, Energy, Food, Health and Water Security Concepts* (Berlin – Heidelberg – New York: Springer-Verlag): 65–102.

Brauch, Hans Günter, 2011: "Global Climate Change Impacts for the Mediterranean in the 21st Century: Challenges for Human and Environmental Security", in: Hans Günter Brauch, Úrsula Oswald Spring, Czeslaw Mesjasz, John Grin, Patricia Kameri-Mbote, Béchir Chourou, Pal Dunay, Jörn Birkmann (Eds.): *Coping with Global Environmental Change, Disasters and Security – Threats, Challenges, Vulnerabilities and Risks* (Berlin – Heidelberg – New York: Springer-Verlag, 2011): 485–524.

Brauch, Hans Günter, 2012: "Environmental Security", in: Juergensmeyer, Mark; Anheier, Helmut K. (Eds.): *Encyclopedia of Global Studies* (London: Sage, 2012).

Brauch, Hans Günter, 2014: "From Climate Change and Security Impacts to Sustainability Transition: Two Policy Debates and Scientific Discourses", in: Oswald Spring, Úrsula; Brauch, Hans Günter; Tidball, Keith G. (Eds): *Expanding Peace Ecology: Peace, Security, Sustainability, Equity and Gender* (Cham: Springer International Publishing): 33-62.

Brauch, Hans Günter, 2016: "Historical Times and Turning Points in a Turbulent Century: 1914, 1945, 1989 and 2014?", in: Brauch, Hans Günter; Oswald Spring, Úrsula; Bennett, Juliet; Serrano Oswald, Serena Eréndira (Eds.): *Addressing Global Environmental Challenges from a Peace Ecology Perspective* (Cham – Heidelberg – New York – Dordrecht – London: Springer International Publishers): 11–54.

Brauch, Hans Günter, 2016a: "Building Sustainable Peace by Moving Toward Sustainability Transition", in: Brauch, Hans Günter; Oswald Spring, Úrsula; Bennett, Juliet; Serrano Oswald, Serena Eréndira (Eds.): *Addressing Global Environmental Challenges from a Peace Ecology Perspective* (Cham – Heidelberg – New York – Dordrecht – London: Springer International Publishers): 145–180.

Brauch, Hans Günter, 2016b: "Conceptualizing Sustainable Peace in the Anthropocene: A Challenge and Task for an Emerging Political Geoecology and Peace Ecology", in: Brauch, Hans Günter; Oswald Spring, Úrsula; Grin, John; Scheffran; Jürgen (Eds.): *Handbook on Sustainability Transition and Sustainable Peace* (Cham – Heidelberg – New York – Dordrecht – London: Springer): 187–236.

Brauch, Hans Günter, 2019: "Sustainable Peace Through Sustainability Transition as Transformative Science: A Peace Ecology Perspective in the Anthropocene", Brauch, Hans Günter;

Oswald Spring, Ursula (Eds.): *Climate Change, Disasters, Sustainability Transition and Peace in the Anthropocene* (Cham: Springer International Publishing): 175-234.

Brauch, Hans Günter, 2021: "Peace Ecology in the Anthropocene", in: Oswald Spring, Úrsula; Brauch, Hans Günter (Eds.): *Decolonising Conflicts, Security, Peace, Gender, Environment and Development in the Anthropocene* (Cham: Springer Nature Switzerland).

Brauch, Hans Günter, 2022: "The Anthropocene Concept in the Natural and Social Sciences, the Humanities and Law – A Bibliometric Analysis and a Qualitative Interpretation (2000-2020)", in: Benner, Susanne; Lax, Gregor; Crutzen, Paul J.; Lelieveld, Jos; Pöschl, Ulrich; Brauch, Hans Günter (Eds.): *Paul J. Crutzen and the Anthropocene – A New Earth's Epoch* (Cham: Springer International Publishing).

Brauch, Hans Günter, 2023: "The Anthropocene in the Social Sciences: A Bibliometric Analysis of Political Science, International Relations, Political Theory, Environment, Ecology Security, Gender Studies and Peace Research", in: Brauch, Hans Günter (Ed.): *Rethinking Politics, Policy and Polity in the Anthropocene* (Cham: Springer Nature Switzerland).

Brauch, Hans Günter; Dalby, Simon; Oswald Spring, Úrsula, 2011: "Political Geoecology for the Anthropocene", in: Brauch, Hans Günter; Oswald Spring, Úrsula; Mesjasz, Czeslaw; Grin, John; Kameri-Mbote, Patricia; Chourou, Béchir; Dunay, Pal; Birkmann, Jörn (Eds.): *Coping with Global Environmental Change, Disasters and Security – Threats, Challenges, Vulnerabilities and Risks* (Berlin – Heidelberg – New York: Springer-Verlag): 1453-1486.

Brauch, Hans Günter; Dalby, Simon; Oswald Spring, Úrsula, 2015: "Political Geoecology for the Anthropocene" [in Chinese], in: Brauch, Hans Günter; Oswald Spring, Úrsula; Mesjasz, Czeslaw; Grin, John; Cheng, Liu (Eds.): *Coping with Global Environmental Change, Disasters and Security – Threats, Challenges, Vulnerabilities and Risks* [in Chinese] (Nanjing: PR China: Nanjing Press Company).

Brauch, Hans Günter; Oswald Spring, Úrsula, 2011: "Introduction: Coping with Global Environmental Change in the Anthropocene", in: Brauch, Hans Günter; Oswald Spring, Úrsula; Mesjasz, Czeslaw; Grin, John; Kameri-Mbote, Patricia; Chourou, Béchir; Dunay, Pal; Birkmann, Jörn (Eds.): *Coping with Global Environmental Change, Disasters and Security – Threats, Challenges, Vulnerabilities and Risks* (Berlin Heidelberg – New York: Springer-Verlag, 2011): 31-60.

Brauch, Hans Günter; Oswald Spring, Úrsula, 2016: "Sustainability Transition and Sustainable Peace: Scientific and Policy Context, Scientific Concepts and Dimensions", in: Brauch, Hans Günter; Oswald Spring, Úrsula; Grin, John; Scheffran; Jürgen (Eds.): *Handbook on Sustainability Transition and Sustainable Peace* (Cham: Springer International Publishing): 3-66.

Brauch, Hans Günter; Oswald Spring, Úrsula, 2018: *Climate Change, Disasters, Sustainability Transition and Peace in the Anthropocene* (Cham: Springer Nature Switzerland)

Brauch, Hans Günter; Oswald Spring, Úrsula; Bennett, Juliet; Serrano Oswald, Serena Eréndira (Eds.): *Addressing Global Environmental Challenges from a Peace Ecology Perspective* (Cham – Heidelberg – New York – Dordrecht – London: Springer International Publishers).

Brauch, Hans Günter; Oswald Spring, Ursula; Grin, John; Mesjasz, Czeslaw; Kameri-Mbote, Patricia; Behera, Navnita Chadha; Chourou, Béchir; Krummenacher, Heinz (Eds.), 2009: *Facing Global Environmental Change: Environmental, Human, Energy, Food, Health and Water Security Concepts* (Berlin – Heidelberg – New York: Springer-Verlag).

Brauch, Hans Günter; Oswald Spring, Úrsula; Grin, John; Scheffran, Jürgen, (Eds.), 2016: *Handbook of Sustainability Transition and Sustainable Peace* (Heidelberg – New York – Dordrecht – London: Springer-Verlag).

Brauch, Hans Günter; Oswald Spring, Úrsula; Mesjasz, Czeslaw; Grin, John; Dunay, Pal; Behera, Navnita Chadha; Chourou, Béchir; Kameri-Mbote, Patricia; Liotta, P.H. (Eds.), 2008: *Globalization and Environmental Challenges: Reconceptualizing Security in the 21st Century* (Berlin – Heidelberg – New York: Springer-Verlag).

Brauch, Hans Günter; Scheffran, Jürgen, 2012: "Introduction: *Climate Change, Human Security, and Violent Conflict in the Anthropocene", in:* Scheffran, Jürgen; Brzoska, Michael; Brauch, Hans Günter: Link, Michael; Schilling, Janpeter (Eds.), 2012: *Climate Change, Human Security and Violent Conflict: Challenges for Societal Stability* (Berlin—Heidelberg: Springer-Verlag): 3-40.

Brundtland Commission (World Commission on Environment and Development), 1987, 1988: *Our Common Future. The World Commission on Environment and Development* (Oxford – New York: Oxford University Press).

Brzoska, Michael, 2020: " Klimawandel und Gewaltkonflikte aus sozialwissenschaftlicher Perspektive", in: *S+F: Sicherheit und Frieden - Security and Peace*, 38,4: 200-205.

Brzoska, Michael; Scheffran, Jürgen (Eds.), 2020: *Climate Change, Security Risks, and Violent Conflicts. Essays from Integrated Climate Research in Hamburg* (Hamburg: Hamburg University Press).

Cabot, Charlene, 2017: *Climate Change, Security Risks and Conflict Reduction in Africa: A Case Study of Farmer-Herder Conflicts over Natural Resources in Côte d'Ivoire, Ghana and Burkina Faso* (Berlin – Heidelberg – New York: Springer, 2016).

Chen, Baiping; Fæste, Lars; Jacobsen, Rune; Teck Kong, Ming; Lu, Dylan: Palme, Thomas, 2020: "How China Can Achieve Carbon Neutrality by 2060" (14 December); at: https://www.bcg.com/publications/2020/how-china-can-achieve-carbon-neutrality-by-2060 (23 April 2021).

Clapham, Christopher, 2020: "Decolonising African Studies?", in: *Journal of Modern African Studies*, 58,1: 137-153

Conca, Ken, 1994: "In the Name of Sustainability: Peace Studies and Environmental Discourse", in: Käkonen, Jyrki (Ed.): *Green Security or Militarized Environment* (Dartmouth: Aldershot): 7-24.

Conca, Ken, 2002: "The Case for Environmental Peacemaking", in: Conca, Ken; Dabelko, Geoffrey (Eds.): *Environmental Peacemaking* (Baltimore: Johns Hopkins University Press): 1-22.

Conca, Ken, 2015: *An Unfinished Foundation: The United Nations and Global Environmental Governance* (Oxford: Oxford University Press).

Crutzen, Paul J., 2002: "Geology of Mankind", in: *Nature*, 415,3 (January): 23.

Crutzen, Paul J., 2011: "The Anthropocene: A Geology of Mankind", in: Brauch, Hans Günter; Oswald Spring, Úrsula; Mesjasz, Czeslaw; Grin, John; Kameri-Mbote, Patricia; Chourou, Béchir; Dunay, Pal; Birkmann, Jörn (Eds.), 2011: *Coping with Global Environmental Change, Disasters and Security – Threats, Challenges, Vulnerabilities and Risks* (Berlin – Heidelberg – New York: Springer-Verlag): 3-4.

Crutzen, Paul J.; Birks, John W., 1982: "The Atmosphere After a Nuclear War: Twilight at Noon", in: *Ambio: A Journal of the Human Environment*, 11 (January): 114-125.

Crutzen, Paul J.; Brauch, Hans Günter (Eds.), 2016: *Paul J. Crutzen: A Pioneer on Atmospheric Chemistry and Climate Change in the Anthropocene* (Cham: Springer International Publishing).

Crutzen, Paul J.; Stoermer, Eugene F., 2000: "The Anthropocene", in: *IGBP Newsletter*, 41-42.

Dalby, Simon, 2020: *Anthropocene Geopolitics: Globalization, Security, Sustainability Politics and Public Policy* (Ottawa: University of Ottawa Press).

Davis, Ian, 2020: "7. Armed conflicts and peace-processes in Sub-Saharan Africa", in: SIPRI (ed.): *SIPRI Yearbook 2020 – Armaments, Disarmament and International Security* (Stockholm: SIPRI – Oxford, New York: Oxford University Press): 171-216.

EU [European Commission; Council], 2008: *Climate Change and International Security*. Doc 7249/08 (Brussels: European Commission, 14 March).

European Commission, 2019: "2050 long-term strategy"; at: https://ec.europa.eu/clima/policies/strategies/2050_en (23 April 2020). With links to relevant EU documents.

European Union, 2019: Going Climate Neutral by 2050 (Luxembourg, Publications Office of the EU).

Evadzi, P.I.K.; Scheffran, J.; Zorita, E.; Hünicke, B. 2018: "Awareness of sea-level response under climate change on the coast of Ghana", in: *Journal of Coastal Conservation*, 22,1: 183–197.

Evans Ogden, Lesley, 2018: "Environmental peacebuilding", in: *BioScience*, 68,3 (March): 157-163.

Fachkommission Fluchtursachen, 2021: *Krisen vorbeugen, Perspektiven schaffen, Menschen schützen: Bericht der Fachkommission Fluchtursachen der Bundesregierung* (Bonn: Bundesministerium für wirtschaftliche Zusammenarbeit und Entwicklung).

Freund, Bill, 1984: "The Decolonisation of Africa, 1940–1960", in: Freund, Bill: *The Making of Contemporary Africa* (London: Palgrave): 191-233.

Galtung, Johan, 1969: "Violence, peace, and peace research", in: *Journal of Peace Research*, 6,3: 167-191.

Göpel, Maja, 2017: *The Great Mindshift: Why We Need a New Economic Paradigm for Sustainability Transformations* (Cham – New York – Heidelberg – Dordrecht – London: Springer).

HIIC [Heidelberg Institute on International Conflict Research], 2021: *Conflict Barometer* 2020: *Disputes, non-violent crises, violent crises, limited wars and wars* (Heidelberg: HIIK).

Huggett, Richard John, 1995: *Geoecology. An Evolutionary Approach* (London – New York: Routledge).

IDMC (Internal Displacement Monitoring Centre), 2020: *Internal Displacement 2020: Mid-Year Update* (Geneva: IDMC).

IDMC (Internal Displacement Monitoring Centre), 2020a. "Global Internal Displacement Database." Geneva; at: https://www.internal-displacement.org/database (25 November 2020).

Ide, Tobias, 2017: "Space, Discourse and Environmental Peacebuilding", in: *Third World Quarterly*, 38,3: 544-562.

Ide, Tobias, 2018, 2019: "The Impact of Environmental Cooperation on Peacemaking: Definitions, Mechanisms, and Empirical Evidence", in: *International Studies Review*, 21,3 (September): 327–346.

Ide, Tobias, 2018a: "Does Environmental Peacemaking between States Work? Insights on Cooperative Environmental Agreements and Reconciliation in International Rivalries", in: *Journal of Peace Research*, 55,3: 351-365.

Ide, Tobias, 2019: *Environmental Peacemaking and Environmental Peacebuilding in International Politics*, CLISEC Working Paper 32.

Ide, Tobias, 2019a: "The Impact of Environmental Cooperation on Peacemaking: Definitions, Mechanisms and Empirical Evidence", in: *International Studies Review*, 21,3: 327-346.

Ide, Tobias, 2020: "Environmental Peacebuilding", in: *Oxford Bibliographies in International Relations*; doi: https://doi.org/10.1093/obo/9780199743292-0290.

Ide, Tobias; Bruch, Carl; Carius, Alexander; Conca, Ken; Dabelko, Geoffrey D.; Matthew, Richard; Weinthal, Erika, 2021: "The Past and Future(s) of Environmental Peacebuilding", in: *International Affairs*, 97,1: 1-16.

Ide, Tobias; Detges, Adrien, 2018: "International Water Cooperation and Environmental Peacemaking", in: *Global Environmental Politics*, 18,4: 63-84.

Ide, Tobias; Sümer, Vakur; Aldehoff, Larissa M., 2018: Environmental Peacebuilding in the Middle East, in: Swain, Ashok; Öjendal, Joakim (Eds.): *Routledge Handbook on Environmental Conflict and Peacebuilding* (London: Routledge): 175-187.

Ikenberry, G. John, 1999: "Liberal Hegemony and the Future of American Postwar Order", in: Paul, T. V.; Hall, John A. (Eds.): *International Order and the Future of World Politics*, (Cambridge, UK: Cambridge University Press).

Ikenberry, G. John, 2011: *Liberal Leviathan: The Origins, Crisis, and Transformation of the American World Order* (Princeton, N.J.: Princeton University Press).

Intergovernmental Panel on Climate Change, 1990: *Climate Change: The IPCC Impacts Assessment* (Geneva: WMO; UNEP; IPCC).

Intergovernmental Panel on Climate Change, 1996: *Climate Change 1995: The Science of Climate Change: Contributions of Working Group I to the Second Assessment Report of the Intergovernmental Panel on Climate Change* (Cambridge: Cambridge University Press).

Intergovernmental Panel on Climate Change, 1997: *The Regional Impacts of Climate Change: An Assessment of Vulnerability* Cambridge: Cambridge University Press).

Intergovernmental Panel on Climate Change, 2001: *Climate Change 2001: The Scientific Basis* (Cambridge – New York: Cambridge University Press).

Intergovernmental Panel on Climate Change, 2007: *Climate Change 2007: The Physical Science Basis: Working Group I Contribution to the Fourth Assessment Report of the IPCC* (Cambridge – New York: Cambridge University Press); also at: https://www.ipcc.ch/report/ar4/wg1/.

Intergovernmental Panel on Climate Change, 2011: *Special Report: Renewable Energy Sources and Climate Change Mitigation* (Cambridge: Cambridge University Press, April).
Intergovernmental Panel on Climate Change, 2012: *Managing the Risks of Extreme Events and Disasters to Advance Climate Change Adaptation* (Cambridge: Cambridge University Press, April).
Intergovernmental Panel on Climate Change, 2013: *Climate Change 2013: The Physical Science Basis: Working Group I Contribution to the Fifth Assessment Report of the Intergovernmental Panel on Climate Change* (Cambridge – New York: Cambridge University Press); at: http://www.ipcc.ch/report/ar5/wg1/.
Intergovernmental Panel on Climate Change, 2014: *Climate Change 2014: Synthesis Report: Summary for Policymakers* (Geneva: IPCC); at: https://www.ipcc.ch/report/ar5/syr/.
Intergovernmental Panel on Climate Change, 2018: *Global Warming of 1.5°C* (Cambridge: Cambridge University Press, October).
Intergovernmental Panel on Climate Change, 2019: *The Ocean and Cryosphere in a Changing Climate* (Cambridge: Cambridge University Press, September).
Intergovernmental Panel on Climate Change, 2019a: *Climate Change and Land* (Cambridge: Cambridge University Press, August).
Intergovernmental Panel on Climate Change, 2019b: *Methodology Report: 2019 Refinement to the 2006 IPCC Guidelines for National Greenhouse Gas Inventories* (Cambridge: Cambridge University Press, May).
Intergovernmental Panel on Climate Change, 2020: *Methodology Report on Short-lived Climate Forcers* Cambridge: Cambridge University Press, June).
Intergovernmental Panel on Climate Change, 2021: *AR6 Climate Change 2021: The Physical Science Basis, The Working Group I contribution to the Sixth Assessment Report* (Cambridge – New York: Cambridge University Press, July); at: https://www.ipcc.ch/report/ar6/wg1/.
Intergovernmental Panel on Climate Change, 2022: *AR6 Climate Change 2022: Impacts, Adaptation and Vulnerability The Working Group II contribution to the Sixth Assessment Report* (Cambridge – New York: Cambridge University Press, February); at: https://www.ipcc.ch/report/sixth-assessment-report-working-group-ii/.
Intergovernmental Panel on Climate Change, 2022a: *AR6 Climate Change 2022: Mitigation of Climate Change. The Working Group III contribution to the Sixth Assessment Report* (Cambridge – New York: Cambridge University Press).
Intergovernmental Panel on Climate Change, 2022b: *AR6 Synthesis Report: Climate Change 2022* (Cambridge – New York: Cambridge University Press, October).
IISS (International Institute on Strategic Studies), 2020: *The Armed Conflict Survey. The* worldwide *review ofpolitical, military and humanitarian trends in current conflicts* (London: IISS, published by Routledge).
IISS (International Institute on Strategic Studies), 2020a: *Strategic Survey. The Annual Assessment of Geopolitics* (London: IISS, published by Routledge).
ISSC (International Social Science Council), 2012: *Transformative Cornerstones of Social Science Research for Global Change* (Paris: International Social Science Council).
Jackson, Simon; O'Malley, Alanna, 2018: *The Institution of International Order: From the League of Nations to the United Nations* (London: Routledge).
Kamta, Frederic; Azadi, Hossein; Scheffran, Jürgen, 2020: "The Root Causes of the Crisis in North-east Nigeria: Historical, Socioeconomic and Environmental Dimensions", in: *Mediterranean Journal of Social Sciences,*11,3: 95-104.
Kamta, Frederic Noel; Schilling, Janpeter; Scheffran, Jürgen, 2021: "Water Resources, Forced Migration and Tensions with Host Communities in the Nigerian Part of the Lake Chad Basin", in: *Resources*, 10,4: 27.
Keal, Paul, 1983: "World War II and Plans for Post-War Order. In: Unspoken Rules and Superpower Dominance", in: Keal, Paul: *Unspoken Rules And Superpower Dominance* (London: Palgrave Macmillan).
Kissinger, Henry, 2014: *World Order* (New York: Penguin).

Kogoui Kamta, Frederic Noel; Schilling, Janpeter; Scheffran, Jürgen, 2020: "Insecurity, Resource Scarcity, and Migration to Camps of Internally Displaced Persons in Northeast Nigeria", in: *Sustainability*. 12,17: 6830.

Kyrou, Christos, 2007: "Peace Ecology: An Emerging Paradigm in Peace Studies", in: *The International Journal of Peace Studies*, 12,2.

Latour, Bruno, 2014: "War and peace in an age of ecological conflicts", in: *Revue juridique de l'environnement*, 2014/1, 39: 51-63.

Le Billon, Philippe; Duffy, Rosaleen, 2018, 2019: "Conflict Ecologies: Connecting Political Ecology and Peace and Conflict Studies", in: *Journal of Political Ecology*, 24: 239-260.

Lenton, Timothy; Held, Hermann; Kriegler, Elmar; Hall, Jim W.; Lucht, Wolfgang; Rahmstorf, Stefan; Schellnhuber, Hans Joachim, 2008: "Tipping Elements in the Earth's Climate System", in: *Proceedings of the National Academy of Science* (PNAS), 105,6 (12 February): 1,786-1,793.

Lenton, Timothy; Rockström, Johan; Gaffney, Owen; Rahmstorf, Stefan; Richardson, Katherine; Steffen, Will; Schellnhuber, Hans Joachim, 2019: "Climate tipping points — too risky to bet against", in: *Nature*, 575 (28 November): 592-595.

Mach, Katharine J.; Kraan, Carolien M.; Adger, W. Neil; Buhaug, Halvard; Burke, Marshall; Fearon, James D.; Field, Christopher B.; Hendrix, Cullen S.; Maystadt, Jean-Francois; O'Loughlin, John; Roessler, Philip; Scheffran, Jürgen; Schultz, Kenneth A.; von Uexkull, Nina, 2019: "Climate as risk factor for conflict", in: *Nature, 571, 7764:* 193-197.

Marie, M.; Yirga, F.; Haile, M.; Ehteshammajd, S.; Azadi, J.; Scheffran, J., 2021: "Time-Series Trend Analysis and Farmer Perceptions of Rainfall and Temperature in Northwestern Ethiopia", in: *Environment Development and Sustainability*.

Mazarr, Michael J.; Priebe, Miranda; Radin, Andrew; Stuth Cevallos, Astrid, 2016: *Understanding the Current International Order* (Santa Monica, CA: RAND Corporation).

McNeill, John R.; Engelke, Peter, 2014: *The Great Acceleration; An Environmental History of the Anthropocene since 1945* (Cambridge, MA – London: The Belknap Press).

Morrison, Mary Lee, 2005: *Elise Boulding: A Life in the Cause of Peace* (Jefferson, NC: McFarland).

Niang, I.; Ruppel, O.C.; Abdrabo, M.A.; Essel, A.; Lennard, C.; Padgham, J.; Urquhart, P., 2014: "Africa", in: *Climate Change 2014: Impacts, Adaptation, and Vulnerability. Part B: Regional Aspects. Contribution of Working Group II to the Fifth Assessment Report of the Intergovernmental Panel on Climate Change* [Barros, V.R., C.B. Field, D.J. Dokken, M.D. Mastrandrea, K.J. Mach, T.E. Bilir, M. Chatterjee, K.L. Ebi, Y.O. Estrada, R.C. Genova, B. Girma, E.S. Kissel, A.N. Levy, S. MacCracken, P.R. Mastrandrea, and L.L. White (eds.)] (Cambridge – New York: Cambridge University Press): 1199-1265.

Noagah Bukari, Kaderi; Sow, Papa; Scheffran, Jürgen, 2019: "Real or Hyped? Linkages Between Environmental / Climate Change and Conflicts – The Case of Farmers and Fulani Pastoralists in Ghana", in: Behnassi, Mohamed; Gupta, Himangana; Pollmann, Olaf (Eds.): *Human and Environmental Security in the Era of Global Risks: Perspectives from Africa, Asia and the Pacific Islands* (Cham: Springer International Publishing): 161-185.

Oswald Spring, Úrsula 2020: *Earth at Risk in the 21st Century: Rethinking Peace, Environment, Gender, and Human, Water, Health, Food, Energy Security, and Migration* (Cham: Springer International Publishing).

Oswald Spring, Úrsula, 2008: "Sustainable Development", in: De Rivera, Joe (Ed.): *Culture of Peace* (New York: Springer-Verlag): 211-227.

Oswald Spring, Úrsula, 2008a: "Peace and Environment: Towards a Sustainable Peace as Seen from the South", in: Brauch, Hans Günter; Oswald Spring, Úrsula; Mesjasz, Czeslaw; Grin, John; Dunay, Pal; Behera, Navnita Chadha; Chourou, Béchir; Kameri-Mbote, Patricia; Liotta, P.H. (Eds.): *Globalization and Environmental Challenges: Reconceptualizing Security in the 21st Century* (Berlin – Heidelberg – New York: Springer-Verlag): 113–126.

Oswald Spring, Úrsula; Brauch, Hans Günter; Serrano Oswald, S.E.; Bennett, Juliet (Eds.), 2016: *Regional Ecological Challenges for Peace in Africa, the Middle East, Latin America and Asia Pacific Perspective* (Cham: Springer International Publishers).

Oswald Spring, Úrsula; Brauch, Hans Günter (Eds.), 2021: *Decolonising Conflicts, Security, Peace, Gender, Environment and Development in the Anthropocene* (Cham: Springer Nature Switzerland).

Oswald Spring, Úrsula; Brauch, Hans Günter; Dalby, Simon, 2009: "Linking the Anthropocene, HUGE and HESP in a Fourth Phase of Human, Environmental and Gender Security and Peace Research", in: Brauch, Hans Günter; Oswald Spring, Úrsula; Grin, John; Mesjasz, Czeslaw; Kameri-Mbote, Patricia; Behera, Navnita Chadha; Chourou, Béchir; Krummenacher, Heinz (Eds.): *Facing Global Environmental Change: Environmental, Human, Energy, Food, Health and Water Security Concepts* (Berlin – Heidelberg – New York: Springer-Verlag): 1277–1294.

Oswald Spring, Úrsula; Brauch, Hans Günter; Scheffran, Jürgen, 2016: "Sustainability Transition with Sustainable Peace: Key Messages and Scientific Outlook", in: Hans Günter Brauch, Úrsula Oswald Spring, John Grin, Jürgen Scheffran (Eds.): *Handbook of Sustainability Transition and Sustainable Peace* (Heidelberg – New York – Dordrecht – London: Springer-Verlag): 887–928.

Oswald Spring, Úrsula; Brauch, Hans Günter; Tidball, Keith G. (Eds.), 2014: *Expanding Peace Ecology: Peace, Security, Sustainability, Equity and Gender: Perspectives of IPRA's Ecology and Peace Commission* (Heidelberg – Dordrecht – London – New York: Springer-Verlag).

Oswald Spring, Úrsula; Brauch, Hans Günter; Tidball, Keith G., 2014a: "Expanding Peace Ecology: Peace, Security, Sustainability, Equity and Gender", in: Oswald Spring, Úrsula; Brauch, Hans Günter; Tidball, Keith G. (Eds.): *Expanding Peace Ecology: Peace, Security, Sustainability, Equity and Gender: Perspectives of IPRA's Ecology and Peace Commission* (Heidelberg – Dordrecht – London – New York: Springer-Verlag): 1–30.

Parmar, Inderjeet, 2018: "The US-led liberal order: imperialism by another name?", in: *International Affairs*, 94,1: 151–172.

Scheffran, Jürgen, 2020: "Climate change and weather extremes as risk multipliers: Tipping points, cascading events, and societal instability", in: Brzoska, Michael; Scheffran, Jürgen (Eds.): *Climate Change, Security Risks, and Violent Conflicts. Essays from Integrated Climate Research in Hamburg* (Hamburg: Hamburg University Press).

Scheffran, Jürgen; Brzoska, Michael; Brauch, Hans Günter: Link, Michael; Schilling, Janpeter (Eds.), 2012: *Climate Change, Human Security and Violent Conflict: Challenges for Societal Stability* (Berlin—Heidelberg: Springer-Verlag).

Scheffran, Jürgen; Link, Peter Michael; Schilling, Janpeter, 2019: "Climate and Conflict in Africa", in: *Oxford Research Encyclopedia of Climate Science* (Oxford: Oxford University Press): Online publication.

Schilling, J.; Hertig, E.; Tramblay, Y.; Scheffran, J., 2020: "Climate change vulnerability, water resources and social implications in North Africa", in: *Regional Environmental Change* 20,1: 15.

Schilling, J.; Locham, R.; Scheffran, J., 2018: "A local to global perspective on oil and wind exploitation, resource governance and conflict in Northern Kenya", in: *Conflict, Security & Development*, 18,6: 571-600.

Schneidewind, Uwe; Singer-Brodowski, Mandy; Augenstein, Karoline, 2016: "Transformative Science for Sustainability Transitions", in: Brauch, Hans Günter; Oswald Spring, Úrsula; Grin, John; Scheffran, Jürgen (Eds.): *Handbook of Sustainability Transition and Sustainable Peace* (Heidelberg – New York – Dordrecht – London: Springer-Verlag): 123-136.

Shaaban, M.; Scheffran, J.; Böhner, J.; Elsobki, M.S.,2019: "A dynamic sustainability analysis of energy landscapes in Egypt", in: *Journal of Artificial Societies and Social Simulation*, 22, 1: 4.

Solomon, N.; Birhane E.; Gordon, C.; Haile, M.; Taheri, F.; Azadi, H.; Scheffran, J.,2018: "Environmental impacts and causes of conflict in the Horn of Africa: A review", in: *Earth-Science Reviews*,177: 284–290.

Steffen, W.; Richardson, K.; Rockström, J. *et al.*, 2020: "The emergence and evolution of Earth System Science", in: *Nat Rev Earth Environ*, 1, 54–63; at: https://doi.org/10.1038/s43017-019-0005-6.

Stone, Madeleine, 2021: "Biden wants to cut U.S. climate pollution in half—here's how" (22 April); at: https://www.nationalgeographic.com/environment/article/biden-wants-to-cut-us-climate-pollution-in-half-heres-how (23 April)

STRN, 2016: "Exploring Transition Research as Transformative Science - IST Conference 2016 in Wuppertal", at: https://wupperinst.org/en/a/wi/a/s/ad/3253.

Swain, Ashok; Bali, Ranjula; Swain Anders, Themnér; Krampe, Florian, 2011: *Climate Change and the Risk of Violent Conflicts in Southern Africa* (Pretoria: Global Crisis Solutions).

Swilling, Mark, 2016: "Preparing for Global Transition: Implications of the Work of the International Resource Panel", in: Brauch, Hans Günter; Oswald Spring, Úrsula; Grin, John; Scheffran, Jürgen (Eds.): *Handbook of Sustainability Transition and Sustainable Peace* (Heidelberg – New York – Dordrecht – London: Springer-Verlag): 391-418.

Sygna, Linda; O'Brien, Karen; Wolf, Johanna (Eds.), 2013: *A Changing Environment for Human Security. Transformative Approaches to Research, Policy and Action* (London – New York: Routledge).

UN, 2009a: *Climate change and its possible security implications. Report of the Secretary-General.* A/64/350 of 11 September 2009 (New York: United Nations).

UN, 2010: *Human Security - Report of the Secretary-General.* A/64/701 of 8 May 2010 (New York: United Nations).

UNDP, 1994: *Human Development Report 1994: New Dimensions of Human Security* (New York: UNDP).

UNDP, 2005: *Human Development Report 2005: International cooperation at a crossroads: Aid, trade and security in an unequal world* (New York: UNDP).

UNDP, 2006: *Human Development Report 2006: Beyond scarcity: Power, poverty and the global water crisis* (New York: UNDP).

UNDP, 2007/2008: *Human Development Report 2007: Fighting climate change: Human solidarity in a divided world* (New York: UNDP).

UNDP, 2014: *Human Development Report 2014: Sustaining Human Progress: Reducing Vulnerabilities and Building Resilience* (New York: UNDP).

UNDP, 2020: *Human Development Report 2020: The Next Frontier: Human Development and the Anthropocene* (New York: UNDP).

UNEP (United Nations Environmental Programme), 2016: GEO-6 *Global Environmental Outlook* (Nairobi: United Nations Environment Programme).

UNEP (United Nations Environmental Programme), 2016a: *Regional Assessment for Africa* (Nairobi: UNEP).

UNEP (United Nations Environmental Programme), 2019: *Global Environmental Outlook (GEO 6) - Healthy Planet – Healthy People* (Cambridge: Cambridge University Press).

UNPD (United Nations, Department of Economic and Social Affairs, Populations Division), 2019: *World Populations Prospects 2019* (New York: United Nations).

UNPD (United Nations, Department of Economic and Social Affairs, Populations Division), 2021: *World Populations Prospects 2021* (New York: United Nations).

US National Science Board, 2007: *Enhancing Support of Transformative Research at the National Science Foundation* (Washington, D.C.: National Science Foundation, 7 May).

van Baalen, Sebastian: Mobjörk, Malin., 2018: "Climate Change and Violent Conflict in East Africa: Integrating Qualitative and Quantitative Research to Probe the Mechanisms", in: *International Studies Review*, 20, 4 (December): 547–575.

Volgy, Thomas J.; Imwalle, Lawrence E., 1995: "Hegemonic and Bipolar Perspectives on the New World Order", in: *American Journal of Political Science,* 39,4 (Nov.): 819–834.

Von Soest, Christian, 2020: "A Heated Debate: Climate Change and Conflict in Africa", in: *GIGA Focus*, No. 2 (April):

Von Uexkull, Nina: Pettersson, Therese, 2018: "Issues and Actors in African Nonstate Conflicts: A New Data Set." *International Interactions*, 44.5: 953–968.

Von Uexkull, Nina; Buhaug, Halvard, 2021: "Security Implications of Climate Change: A Decade of Scientific Progress", in: *Journal of Peace Research,* 58,1: 3–17.

WBGU (German Advisory Council on Climate Change), 2007, 2008: *Climate Change as a Security Risk* (Berlin: WBGU).

WBGU (German Advisory Council on Climate Change), 2011: *World in Transition – A Social Contract for Sustainability* (Berlin: WBGU, 2011: 21–23, 321–356).
Westad, Odd Arne, 2007: *The Global Cold War* (Cambridge: Cambridge University Press).
Westing, Arthur H., 1976: *Ecological Consequences of the Second Indochina War* (Stockholm: Almqvist & Wiksell).
Westing, Arthur H., 2003: "Environmental Dimension of the Gulf War", in: Brauch, Hans Günter; Liotta, P.H.; Marquina, Antonio; Rogers, Paul F.; Selim, Mohammad El-Sayed (Eds.): *Security and Environment in the Mediterranean: Conceptualising Security and Environmental Conflicts* (Berlin – Heidelberg – New York – Hong Kong – London – Milan – Paris – Tokyo: Springer-Verlag): 523–534.
Westing, Arthur H., 2013: *Arthur H. Westing: Pioneer on the Environmental Impact of War.* (Heidelberg – Dordrecht – London – New York: Springer-Verlag).
Westing, Arthur H., 2013a: *Texts on Environmental and Comprehensive Security* (Heidelberg – New York – Dordrecht – London: Springer-Verlag).
Wilson, Edward O., 1988: *Consilience* (New York: Knopf).
WMO (World Meteorological Organisation), 2020: *Climate in Africa in 2019* (Geneva: WMO).
Yanda, Pius; Bronkhorst, Salome, 2011: *Climate change and conflict: Conflict-sensitive climate change adaptation in Africa*, Policy & Practice Brief (Durban: ACCORD).

Chapter 3
Civil Society, Climate Change, Security Risks Management and Peacebuilding in the Anthropocene

Jean Chrysostome K. Kiyala

Abstract The field of peace studies is seeing an increased interest in environmental peace-making or peace ecology. This chapter explores the security threats and conflicts induced by climate change on humanity while advocating available tools aimed to attenuating environmental instability in several regions of the globe and maintaining the sustainability of biodiversity. The pursuit of peace in climate change context requires a pluridisciplinary approach that encompasses a better understanding of environmental conflicts, environmental justice, peace ecology, ecoeducation, ecoethics, and developing climate-sensitive adaptation and conflict-sensitive mechanisms to alleviate the effects of conflicts induced by climate change. The chapter argues that a synergic cooperation between civil society, business, corporations and political actors has the potential to lead a global and concerted implementation of healthy ecological policies. Finally, the interplay of various dimensions of human agency to protect the ecosystem are held as the pathways to mitigating the global environmental crisis we are confronted with in our time, and to achieving ecological sustainability.

Keywords Anthropocene · Climate change · Ecojustice · Environmental conflicts · Human agency · Environmental peace · Peacebuilding · Security risk management

3.1 Introduction

The last three decades have revealed growing concerns over the future of the Earth as the Earth system appears to be deteriorating, creating a new landscape of conflict

[1] "The prevalence of moderate or severe food insecurity in the population, based on the Food Insecurity Experience Scale (FIES) is an estimate of the percentage of a country's population that faces difficulties in accessing enough safe and nutritious food for normal growth and development and an active and healthy life." (*Hunger and food insecurity.* Food and Agricultural Organization of the United Nations FAO 2021). http://www.fao.org/hunger/en/. Accessed 28 July 2021.

Jean Chrysostome K. Kiyala is a Senior Lecturer, International Centre of Violence, Durban University of Technology, South Africa; Associate Professor and visiting lecturer the University of Bandundu, Catholic University of Bukavu and Evangelical University in Africa based in Bukavu (DR Congo).

J. C. K. Kiyala and G. T. Harris (eds.), *Civil Society and Peacebuilding in Sub-Saharan Africa in the Anthropocene*, The Anthropocene: Politik—Economics—Society—Science 34,
https://doi.org/10.1007/978-3-030-95179-5_3

resulting from climate change (Bob et al. 2014; Bronkhorst 2011; Willms/Werner 2009). Thanks to research and the conceptual development of planetary boundaries (Rockström et al. 2009; Steffen et al. 2015), we are provided with scientific data and evidence of the deterioration of our planet which calls for new ways of pursuing economic development without destroying the Earth and the future of humanity.

The negative consequences of climate change in Africa are experienced in Somalia (Eklöw/Krampe 2019), in the Horn of Africa (Krampe et al.2020), in the Sahel Region (Bronkhorst 2011), in Mozambique (Artur/Hilhorst 2012; Osbahr et al. 2008), and other parts of Africa (Bob/Bronkhorst 2014; Lisk 2009). This situation requires a host of approaches and questions intended to mitigate the negative anthropogenic causes of climate change, find adequate means of climate change adaptation (Bob et al. 2014; Pius Yanda/Bronkhorst 2011), and take preemptive measures to deal with climate change-induced conflicts (Pihkala 2018), or develop conservatory preventive approaches to protect the environment, all of which are theorised as peace ecology (Brauch 2016; Brisman 2016).

In the same perspectives, new paradigms aimed at preserving the natural beauty of the environment have emerged, namely environmental peacebuilding (Hardt/Scheffran 2019), environmental justice (Schlosberg 2004, 2013; Sze/London, 2008), the whole range of activism against climate change by civil society organisations (Böhmelt 2013; Caniglia et al. 2015; Newell 2008), and individual agency such as the leading role being played by "teen green" Greta Thunberg (Kühne 2019; Thunberg 2019b, 2020).

Climate change has been acknowledged as an issue of paramount international disquietude at the UN level (Wilson 2020, p. 33). Nonetheless, finding a consensus between foreign and domestic policy priorities among powerful states have been very slow and continue to inhibit more proactive responses on the part of the Security Council, despite elevating climate change to the stage of international concern (Wilson 2020). The survival of humanity (McNeely 2020) is at stake and it is urgent to save our "common home" (Pope Francis 2015). This is a moral obligation: firstly because the atmospheric conditions become hostile to humanity; secondly because of the risks of further direct violence due to the lack of tools to handle environmental disputes related to natural resources; and lastly because climate change has become a large-scale instrument of destruction of infrastructure and the conditions for decent livelihoods.

This chapter is justified by the imperative to mitigate the risks associated with failure to deal with climate change and its consequences that may create insecurity, and hinder local, regional, transregional peace (Krampe 2017; Krampe/Mobjörk 2018). The science is crystal clear but progress towards mitigating climate change has been very slow. Humanity remains vulnerable without a real action plan or determination to prevent the worst from happening, which echoes the frustration of Rowlatt (2021), the BBC Chief Environment Correspondent who notes:

> In Paris, the world agreed to avoid the worst impacts of climate change by trying to limit global temperature increases to 2°C above pre-industrial levels by the end of the century. The aim was to keep the rise to 1.5°C if at all possible (p. n.p.).

The International Panel on Climate Change (IPCC) suggested the following direction to mitigate climate change: "Pathways consistent with 1.5°C of warming above pre-industrial levels can be identified under a range of assumptions about economic growth, technology developments and lifestyles" (Fifita et al. 2018). However, there is no substantial improvement. Rowlatt (2021)further remarks:

> We are way off track. On current plans the world is expected to breach the 1.5°C ceiling within 12 years or less and to hit 3°C of warming by the end of the century. Under the terms of the Paris deal, countries promised to come back every five years and raise their carbon-cutting ambitions (p. n.p.).

If this is the trend for the future, coupled with denialism from world leaders of the threat posed by climate change, our world and the future of humanity is in jeopardy, as pointed out by Rowlatt (2020) who noted that the Antarctic is melting and shows the devastating consequences of climate change and *the journey to the doomsday glaciers.*

For Africa, climate disruption is already having extreme consequences, according to the UNEP Global Environment Outlook 2000 report (UNEP 1999) by Clarke (1999):

> Poverty is a major cause and consequence of the environmental degradation and resource depletion that threaten the region. Major environmental challenges include deforestation, soil degradation and desertification, declining biodiversity and marine resources, water scarcity, and deteriorating water and air quality. Urbanisation is an emerging issue, bringing with it the range of human health and environmental problems well known in urban areas throughout the world. Growing 'environmental debts' in many countries are a major concern because the cost of remedial action will be far greater than preventive action (p. 6).

The main objective of this essay is to raise international awareness and urge all strata of society to become peace actors in the fight against climate change ensured from the degrading condition of the Earth System; this perspective requires a great sense of urgency and active participation of the entire human family.

The argument put forward here is drawn from literary research utilising scholarly online resources (internet web-browsing devices) which are considered as some of the many tools and techniques of literary research (Towheed 2009, p. 11). These materials were analysed and interpreted according to content analytical methods which are essentially interpretive (Hsieh/Shannon 2005; Krippendorff 2004).

This essay is structured as follows: (1) the introduction that substantiates the existing ecological and vital challenges humanity is faced with in the Anthropocene; (2) a brief conceptualisation of the Anthropocene Epoch; (3) a brief discussion of planetary boundaries framework to monitor and act against climate change and its effects on the ecosystem; (4) an overview of security issues triggered by climate change, and the need of environmental peacebuilding; (5) an exploration of innovational mechanisms of climate sensitive adaptation to prevent climate change-induced conflicts; (6) a discussion about environmental conflict and ecojustice; (7) an exploration of environmental peacemaking or peace ecology; (8) a discussion on the relation between ecological education and ecological ethics, and their relevance to action against climate change; (9) an investigation of the role played by civil society to fight climate change; (10) a discussion on the impact of human agency in alleviating the burden of human activities on the atmospheric condition; and finally, (11) some concluding remarks are drawn from the analysis of civil society's engagement in climate action and environmental peacemaking.

3.2 The Anthropocene Epoch

The Anthropocene is a new geological epoch in Earth's history that is discernible in the set of anthropogenic perturbations on atmospheric conditions resulting from human activities, and the extent to which that effects the well-being of the ecosystem. Rafferty (n.d.) explains the Anthropocene Epoch as deriving from the Greek and means the "recent age of man". It is an unauthenticated interval of geologic time that makes up the third worldwide division of the Quaternary Period (2.6 million years ago to the present) (Rafferty n.d.) and:

> [It is identified as] the time in which the collective activities of human beings (Homo sapiens) began to substantially alter Earth's surface, atmosphere, oceans, and systems of nutrient cycling. A growing group of scientists argue that the Anthropocene Epoch should follow the Holocene Epoch (11,700 years ago to the present) and begin in the year 1950 (Rafferty n.d., n.p.).

While it is known that the American biologist Eugene Stoermer coined the concept of Anthropocene in the late 1980s, it is the Dutch chemist and Nobel Prize Laureate Paul Josef Crutzen who is predominantly credited with vulgarising this concept in 2000 (Rafferty n.d.). Later, in 2008, British geologist Jan Zalasiewicz and his colleagues advanced the first proposal to recognise the Anthropocene Epoch as a formal geological interval (Rafferty, n.d.). Finally, in the Anthropocene Working Group of the International Union of Geologic Sciences (IUGS) that took place in 2016 a vote passed recommending the Anthropocene to be acknowledged as a formal geologic epoch; that was at the 35th International Geological Congress (Rafferty n.d.).

Prof. Jos Lelieveld, the Director of Max Planck Institute for Chemistry (Otto Hahn Institute), notes that geologists have traditionally called the most recent 12,000 years the Holocene, Paul J. Crutzen contends, "in the past centuries the impact of humanity on the Earth's surface is so large, and unique, that a renaming of the geological timescale is justified" (Jos Lelieveld cited in Crutzen/Brauch 2016, p. xii).

The starting date of the great impact of anthropogenic effects on the environment is debatable. Thus, Stephen et al. (2011) propounds the following about the term Anthropocene:

> (i) that the Earth is now moving out of its current geological epoch, called the Holocene and (ii) that human activity is largely responsible for this exit from the Holocene, that is, that humankind has become a global geological force in its own right (p. 843).

Crutzen (2016) places the start of Anthropocene in the late eighteenth century; he elucidates, "when analyses of air trapped in polar ice showed the beginning of growing global concentrations of carbon dioxide and methane" (p. 211). He expounds that further stating:

> It seems appropriate to assign the term 'Anthropocene' to the present, in many ways human-dominated, geological epoch, supplementing the Holocene—the warm period of the past 10–12 millennia. The Anthropocene could be said to have started in the latter part of the eighteenth century, when analyses of air trapped in polar ice showed the beginning of growing global concentrations of carbon dioxide and methane. This date also happens to coincide with James Watt's design of the steam engine in 1784 (Crutzen 2016, p. 211).

According to Zalasiewicz et al. (2010, p. 2231), "The Anthropocene represents a new phase in the history of both humankind and of the Earth, when natural forces and human forces became intertwined, so that the fate of one determines the fate of the other". Human agency is responsible for dramatic atmospheric transformations as pointed out by Oswald Spring /Brauch (2021):

> Direct human interventions into the Earth System through the accumulation of greenhouse gases and carbon dioxide (CO_2) in the atmosphere have caused multiple societal impacts resulting in rapid increases in production, consumption, urbanisation, pollution, migration, crises and conflicts (p. 32).

According to the industrial view, most effects occurred in the early industrialisation period around 1850, whereas the early anthropogenic opinion reckons that significant impacts date to thousands of years before (Ruddiman 2013). There are three indicators that Ruddiman discusses to define the Anthropocene; these are mainly "global-scale human influence: forest clearance (and related land use), emissions of greenhouse gases (CO_2 and CH_4), and effects on global temperature" (2013, p. 45). He contends that "anthropogenic impacts on Earth's environments" were manifest in the past 150 years, and not before 1850, since during the preindustrial era when world populations were lower than in the early industrial view, there were "very small early anthropogenic effects on land use and greenhouse-gas emissions" (Ruddiman 2013). In contrast, it is argued:

> [H]istorical and archeological data reveal much larger forest clearance in preindustrial times because early farmers used much more land per capita than those in recent preindustrial centuries [...]. This early deforestation, along with other effects of early agricultural activities, resulted in large greenhouse-gas emissions [...] consistent with the anomalous CO_2 and CH_4 increases seen when compared with decreases during previous interglaciations (Ruddiman 2013, p. 64).

Ruddiman (2013) considers two periods of the Anthropocene, based on the historical and archeological evidence, showing that deforestation in preindustrial times was much more sizeable that it appears during the industrial era; also, the emissions of greenhouse gases during the preindustrial epoch were smaller, though substantial; and finally, he remarks that "the net anthropogenic effect on global temperature was probably larger in preindustrial than industrial times" (p. 65). He proposes a two-phase Anthropocene to resolve the timing (start) of this new geological epoch. They are:

- An early phase with anthropogenic effects that began at a very small level thousands of years ago but slowly grew to considerable size by the end of preindustrial times.
- A later explosive phase of wide-ranging anthropogenic impacts during the industrial era.

It is plausible to assert that agricultural activities, besides other human searches for better living conditions prompted by complex human existential needs, have caused the degradation of the Earth system along with world population increase, the development of civilisations, the onset of the industrial revolution, and the looming super interglacial climate, thus making the informal use of the term Anthropocene more meaningful (Ruddiman 2013). According to Steffen et al. (2011), the beginning of the Anthropocene Era coincides with advent of the industrial revolution around 1800.

Conceding that humans' actions are at the origin of the Anthropocene Epoch, it is also straight-thinking to reckon that the pernicious after-effects of this geological period can be controlled by humanity to slow further ecological deterioration. To deal with the anthropogenic repercussions of atmospheric perturbation, Crutzen has already appealed to scientists and engineers to lead society through this new geological time with adequate information and behavioural adaptation to the new epoch; he writes:

> A daunting task lies ahead for scientists and engineers to guide society towards environmentally sustainable management during the era of the Anthropocene. This will require appropriate human behaviour at all scales, and may well involve internationally accepted, large-scale geo-engineering pro-jects, for instance to 'optimize' climate. At this stage, however, we are still largely treading on terra incognita (Crutzen 2016, p. 212).

In the Anthropocene, we are navigating in "terra incognita" as Paul Crutzen expresses it. Africa is no exception because it is already highly vulnerable. The northern Mauritania is a case in point as people experience first-hand the impact of global warming; this is reported by BBC (2021): "The rising temperatures and desertification are wiping out communities and as the Life at 50C series has discovered, many are being forced to leave their ancestral homes in search of a better life" (BBC News 2021). That calls for an increase in research, and engaging actors at various levels of society; political leaders, business and corporations, and civil society (non-governmental organisations, religious leaders, grassroots leadership, think tanks, academics, intellectuals, etc.). Building a global partnership and cooperation among all these components of society will assist in reading the signs of the time, studying and analysing the development of incidences associated with past, present and future atmospheric conditions in order to maintain peace and harmony within the ecosystem, more practically on the Earth. One agent this chapter is concerned with is civil society; it has been participating in climate action and it can achieve more to alleviate the burden of human activities on the Earth system.

Thus, it is necessary to follow scientific data that research puts at our disposal via the evidence of planetary boundaries that is explored in the next section. The planetary boundaries framework is useful to assess the impact of human action on the Earth system and to envisage the series of interventions required to lessen the risks of complete disintegration of our planet and eventually to avert the disastrous consequences of anthropogenic climate change on peace, security and livelihoods of humanity. The framework that provides a scientific ground and proven evidence of how human activities, since the dawn of the Industrial Revolution, have become the major cause of global environmental change is known as planetary boundaries (PBs; Steffen et al. 2015).

3.3 Planetary Boundaries and the Survival of Humanity

The concept of planetary boundaries was introduced by Johan Rockström and colleagues; it refers to:

> The approach recognizes the severe risks associated with trying to deliberately manipulate the Earth system to counteract deleterious human influences, given the lack of knowledge of the functioning of the Earth system and the possibility of abrupt and/or irreversible changes, some of them very difficult to anticipate, when complex systems are perturbed. The planetary boundaries approach is thus explicitly based on returning the Earth system to the Holocene domain, the environmental envelope within which contemporary civilisation has developed and thrived (Steffen et al. 2011, p. 860).

The planetary boundaries framework indicates the 'safe operating space' for humanity, with care for the Earth system; they are founded on a handful of subsystems or processes, the multitude of which display abrupt change performance when critical thresholds are crossed (Steffen et al. 2011, p. 860). It follows then that need to create viable and satisfactory living conditions which depends on development gets scientist entangled in a dilemma of how to use Earth's resources to pursue humanity's sustainability and well-being without endangering the ecosystem and prompting its further disintegration (Steffen et al. 2011, p. 860) (Fig. 3.1).

Steffen et al. (2015) define planetary boundaries as follows:

> [They are] scientifically based levels of human perturbation of the ES beyond which ES functioning may be substantially altered. Transgression of the PBs thus creates substantial risk of destabilising the Holocene state of the ES in which modern societies have evolved (p. 736).

Humanity needs natural resources such as forest, bush, water, air space, etc., which are made available through technological innovations; how to restrain technological impulse to protect and present the environment? (Steffen et al. 2011, p. 860). The lack or insufficiency of those natural resources are becoming the driver to insecurity, conflicts and violence.

With the UN 2030 Sustainable Development Goals (SDG) Agenda, we are faced with the growing challenge to operate without destabilising further the natural regulation of climate. The dilemma we are faced with is providing for the needs of the whole world, and simultaneously attenuate the devastating effects of climate change (Morton et al. 2019). It is important to argue that the UN SDGs constitutes a project

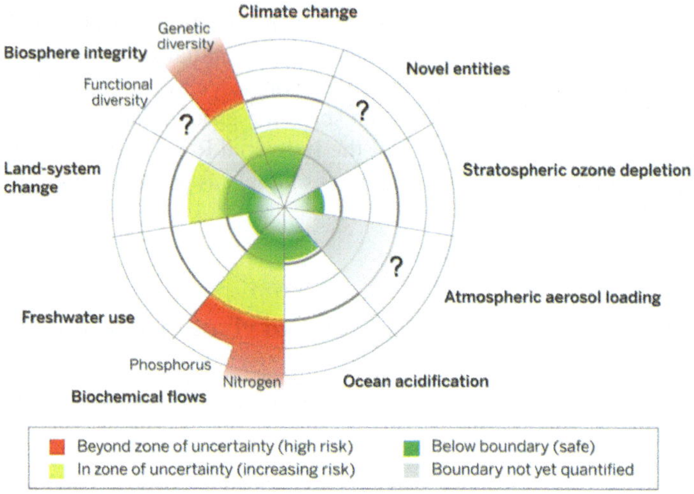

Fig. 3.1 Planetary boundaries. *Source* Steffen et al. (2015, p. 6)

of global development that puts in place the mechanisms to ensure the goals poverty eradication, protection of the environment and social justice are attained without hurting our planet; this agenda intends to make the Earth a habitable home, and make sure that all people relish peace and prosperity (Morton et al. 2019). Thus:

> The SDGs integrate the three dimensions of sustainable development (economic, environmental and social), they apply to high-income countries as well as developing countries and there are mechanisms to hold countries to account (Morton et al. 2019, p. 65).

The real existential and vital conundrum humanity is confronted with in the Anthropocene is to attain a greater quality of life and satisfy various human needs such as water, electricity and nutrition for the overall world population (estimated at 9 billion) without infringing planetary ethical standards or destabilising planetary processes (O'Neill et al. 2018). These scholars examined the possibility of having a "safe and just" development space; they evaluated the quantified resource use that is susceptible of meeting basic human needs; and they compared this to downscaled planetary boundaries of an estimated 150 nations. They came up with the following findings: "no country meets basic needs for its citizens at a globally sustainable level of resource use. Physical needs such as nutrition, sanitation, access to electricity, and the elimination of extreme poverty could likely be met for all people without transgressing planetary boundaries" (O'Neill et al. 2018, p. 1). The challenge remains significant in any strategy to ameliorate "physical and social provisioning systems" in order to move nations towards sustainability (O'Neill et al. 2018).

Human activities injure the sacred Erath system space as the ozone layer is wounded. The invasive activity of humans in the environment has damaged the ozone layer; the consequence of this renders humans and non-human creations vulnerable. To illustrate this point, Steffen et al. (2011) remarked:

> Following the discovery of the ozone hole over Antarctica, with its undeniably anthropogenic cause, the realization that the emission of large quantities of a colourless, odourless gas such as carbon dioxide (CO_2) can affect the energy balance at the Earth's surface has reinforced the concern that human activity can adversely affect the broad range of ecosystem services that support human (and other) life [1,2] and could eventually lead to a 'crisis in the biosphere' ([3], cited in Grinevald [4]) (pp. 842–843).

It appears that the invasion of the atmosphere by substances that toxify the ozone is a multiplier factor that occasions global warming and the rise in sea levels and threatens the survival of all oxygen-dependent creatures (humans, flora, fauna, and all vegetal and animal species). The Earth's degradation is far, far more concerning than the destruction resulting warfare. If such violence is avoidable and controllable (as has been shown with the notion of "democratic peace", antimilitarism and other forms of pacifism, conflict mediation, and diplomacy, in the face of ecological violence such as "tsunami" wild fires, global warming, and the rise in sea levels), the sole rational defence left for humanity remains a concerted engagement to which Pope Francis and his predecessors have called the world: a radical change in behaviour and use of the resources that nature provides (Pope Francis 2015). This requires a conscientious use of technological innovations and the recognition of the sacredness of the ozone that regulates climate and atmospheric conditions. All these matters not only concern human security but moreover entail environmental security (Kyrou 2007); to some extent, the lack of food security is also a cause of concern in the Anthropocene (Steffen et al. 2011). Thus, global warming will significantly affect humans (Zalasiewicz et al. 2010). Therefore, we now briefly examine global-warming-induced climate change, the extent to which peace and security are fragilised, and how peacebuilding initiatives can alleviate the burden of atmospheric perturbation on both humans and non-human beings and their natural environment.

3.4 Climate Change, Security and Peacebuilding

From the perspectives of Bob et al. (2014), borrowed from Houghton and the World Meteorological Organization (2002), the concept of climate is described as "the average state of the atmosphere for a given time scale (hour, day, month, season, year, decades and so forth) and generally for a specified geographical zone" (Bob et al. 2014, p. 27). This atmospheric condition has caused conflicts and is likely to generate more disputes in the future, which will require some adaptation and risk management that would call for the participation of several international and regional actors and partners. This view is raised by Pius Yanda/Bronkhorst (2011) who remark:

> Climate change and climate-related conflict are at once a challenge to livelihoods, for natural resource management and for peacebuilding. Conflict-sensitive climate change adaptation is and should remain at the core of existing and future work in the fields of sustainable [development], the environment and peace. Given that the linkages between climate change and conflict are complex, and operate at different scales across time and space, there is a need

> for different scientific disciplines to work together on research, including local knowledge from communities who have already shown resilience (p. n.p.).

Civil wars, insurrections, and regional conflicts have fallen in number and intensity, but climate change may re-ignite such wars. A new front of conflicts is opening up, namely environment-related conflicts, where sustainable development and peace are at stake because of climate change.

Thus research and reflection are needed to frame a more comprehensive agenda to manage natural resources in the aftermath of conflicts, namely an inclusive environmental peacebuilding scheme that embraces environmental cooperation and resource risk management and the necessity to mitigate resource-induced instability by carrying through environmental cooperation projects (Krampe 2017).

It is worth noting, for instance, that such cooperation has grown between the UN Peacebuilding Commission and the UN Environment Programme (UNEP) to generate a synergic research project aimed at overseeing resource risks, and which incorporates the Environmental Law Institute, the University of Tokyo, and McGill University (Krampe 2017, p. 4). It is argued, finally:

> Future research needs to acknowledge the complexity of the post-conflict landscape and advance environmental peacebuilding research to realize the potential and the risks of natural resource management. This is urgently needed, because this complexity lies at the core of the SDGs that will guide UN policies in the coming decades and is instrumental to building a sustainable peace (Krampe 2007, pp. 6–7).

Due to the fact that "Climate-related security risks are increasingly transforming the security landscape in which multilateral peacebuilding efforts are taking place" (Eklöw/Krampe 2019, p. vii), environmental peacebuilding research is growing as in important area of inquiry in the Anthropocene. The work carried out by Krampe/Mobjörk (2018)demonstrates the significance of responding to climate-related security risks in four regional intergovernmental organisations (IGOs) in Asia and Africa—ASEAN (South East Asia), SAARC (South Asia), ECOWAS (West Africa), and IGAD (East Africa). The ability of the world to deal with the complexities of peace and security at various levels of society and prevent the vulnerability that is being prompted by climate change that is likely to affect livelihood conditions and development too, relies not only on civil society organisations, but also on IGOs (Krampe/Mobjörk 2018). There is an increasing awareness among IGOs about security risks related to climate which need to be addressed via international cooperation because of the transnational dimension of climate-related security risks and the emerging challenges the world has to face; this requires further studies on the effective implementation of climate-management risk policy frameworks that are relevant in regional adaptation processes (Krampe/Mobjörk 2018) in order to prevent and resolve environmental conflicts nonviolently. To assuage the effects of global warming that imperils peace, and destabilises people's peaceful settlement in their natural environment, is important to consider climate conflict-sensitive adaptation theory and praxis; that is examined to the next section.

3.5 Climate Conflict-Sensitive Adaptation

The repercussions of climate disruptions include their effects on the *ecogeomorphology of coastal wetlands* (Day et al. 2008), *warming water*, *coral bleaching*, *rising sea levels* and their repercussions on different regions of the Earth (McClanahan/Cinner 2012), etc. However, new perspectives have surfaced, naturally the "climate sensitive-adaptation" along "conflict-sensitive" concept (Bob/Bronkhorst 2014; Bob et al. 2014; Pius Yanda/Bronkhorst 2011). The thinking behind this model of response to climate crisis is to acknowledge the established fact about the irreversibility of Earth systems to their original healthy status. A number of researchers support conflict-sensitivity in the pursuit of climate change sensitive-adaptation, because it firstly climate change can engender scarcities of resources; secondly, because it can heighten existing threats to peace and security (Pius Yanda/Bronkhorst 2011). The UN Secretary-General in 2009 described climate change not only as a 'threat-multiplier' but also emphasised that adaptation to climate change can be utilised as a 'threat-minimiser' (UN Secretary-General 2009, Document A/64/350 cited in Pius Yanda/Bronkhorst 2011, p. 3).

The reality of the Anthropocene can neither be changed nor reverted by humans and technological impulses. However, following Crutzen's (2016) ecological recommendations, scientific and engineering innovations should lead humanity to a more responsible way of handling our behaviours and conducting ourselves vis-à-vis natural resources provided by the Earth.

Climate Sensitive-Adaptation has evolved as another response to environmental conflicts; it entails developing approached to sustainable development built on positive impacts of climate change. Pius Yanda/Bronkhorst (2011, p. 3) remark:

> [...] conflict-sensitive adaptation is concerned with optimising the positive impacts of climate change and of adaptation, not only avoiding or mitigating the negatives. Positive impacts of climate change may include more rain in certain places, while in adaptation, positive impacts and objectives – besides conflict prevention – could be peacebuilding and sustainable development.

Bob et al. (2014) ascertain the existence of projects aimed to building mechanisms of adaptation to climate change in Africa; they are currently foremost among policymakers, donors, NGOs and researchers in view of tackling climate change issues and follows through the December 2011 COP 17 (17th Conference of the Parties to the UNFCCC), during which funding instruments for climate adaptation were sanctioned. Because of the linkages between climate change and conflicts (Pius Yanda/Bronkhorst 2011), it has become imperative to study adaptation prospects to prevent and deal with such conflicts when they arise (Bob/Bronkhorst 2014; Bob et al. 2014; Pius Yanda/Bronkhorst 2011). Several scholars maintain that climate change adaptation goes along conflict-sensitive approaches; that is essential for global peace (Babcicky 2013; Tänzler et al. 2010). Countries that are not prepared to arise above atmospheric disturbances due to insufficient capacity of adaptation could experience social tensions and potential conflict (Tänzler et al. 2013), which can be an impediment to peace, sustainability stability in the Anthropocene. Climate sensitive-adaptation should also be conflict-sensitive (Pius Yanda/Bronkhorst 2011).

Therefore, it is mandatory that all countries in all regions of the world embark on the journey of climate-sensitive adaptation and develop the tools to address environmental conflicts when they arise. To foster environmental peacebuilding, the discipline of Peace Studies has seen the emergence of new concepts to assist its role in mitigating climate change and its consequences. These include environmental justice and peace ecology or environmental peacemaking; these concepts are examined in the section that follows.

3.6 Environmental Conflicts and Environmental Justice

Environmental conflict is the new major challenge for Africa and can further deepen her vulnerability in the decades to come. For instance, shortages of water and fertile soils and the impacts of extreme weather events could trigger more violent conflicts; these are already manifest in the southern Sahel of Sudan (Bronkhorst 2011), and in the Horn of Africa (Molvaer 1991), and they can be exacerbated by scarcity of energy resources (Ijumba/Kaya 2016). Schlosberg (2004) envisions environmental justice beyond the assumptions of equity and the distributions of environmental goods and ills, expanding it further as contended here:

> The argument is that the justice demanded by global environmental justice is really threefold: equity in the distribution of environmental risk, recognition of the diversity of the participants and experiences in affected communities, and participation in the political processes which create and manage environmental policy (p. 517).

As humanity finds itself in the midst of anthropogenic causes of environmental disasters and their consequences that affect both the Earth, our home—as it is hit by drought, wildfires, flooding, etc. and ourselves, its inhabitants, through forced displacements of population ultimately causing ecological refugees, Westra (2009) warns that this reality creates a new political and social condition to be dealt with in the future. The future of world peace is threatened, and this requires an urgent and global response, which takes into account ecology-related paradigms, analyse them, and advocate environmental justice.

The theory of environmental justice has evolved horizontally, covering a large scope of issues; vertically, it focuses on the comprehensive essence of environmental injustices; and conceptually, it looks at "the human relationship with the non-human world" (Schlosberg 2013, p. 37). This theory relates to *socio-environmental conflicts* or *ecological distribution conflicts* that have been spelled out by Temper et al. (2018); they argued from an unperceivable angle that socio-environmental conflicts can also become the catalyst of social, economic, and ecological sustainability, as asserted here:

> We contend that protests are not disruptions to smooth governance that need to be managed and resolved, but that they express grievances as well as aspirations and demands and in this

way may serve as potent forces that can lead to the transformation towards sustainability of our economies, societies and ecologies (Temper et al. 2018, p. 573).

In view of this assertion, this chapter holds that environmental justice can serve as an instrument of stability, sustainability, and mitigation of violent conflicts that may erupt in various geographical spaces. It should also underline that it is one of the drivers of social justice, a view held by Schlosberg (2013) as well who affirmed the relevance of both the vertical and conceptual aspects of environmental justice, linking it to the understanding of it being one of the factors that "are understood to create the conditions for social justice" (p. 37). These latter perspectives justify the engagement of civil society faith-based organisations and social movements in standing against politically and economically driven policies that subjugate and subdue our planet. One of the components of environmental justice is ecojustice philosophy which "merges social and environmental justice theories by emphasising physical, spiritual, and emotional connections between an environment and the residing social group" (Tippins/Britton 2015). Unfortunately, human freedom has resulted in a disastrous intrusion into nature, thus corrupting the Earth system. Therefore, as noted by Steffen et al. (2015, p. 736), "There is an urgent need for a new paradigm that integrates the continued development of human societies and the maintenance of the Earth system in a resilient and accommodating state". Achieving this prospect demands ethical engagement with nature, responsible human agency, ethical stewardship, and developing pragmatic strategies to implement peace ecology which provides objective means of environmental justice; that is explored in the next section.

3.7 Peace Ecology

Before discussing the concept of peace ecology, it is helpful to understand peace culture, which informs it. Boulding (2002, p. 6) defined peace as "a mosaic of identities, attitudes, values, beliefs, and patterns that lead people to live nurturantly with one another and the earth itself without the aid of structured power differentials, to deal creatively with their differences and share their resources". The reference to the Earth is not accidental, but rather essential, because interpersonal relationships require an environment where these are built, without which there would be no society.

Peace ecology has emerged as a new worldview in efforts to understand the concept of peace within the context of environmental studies, environmental security, and peace studies. Kyrou (2007) described peace ecology as a new paradigm that is inspired by the concept of "environmental peacemaking"—comprehended as "a theoretical framework, broad and integrative enough to allow a full understanding, functionally as well as philosophically, of the inherent capacities of the environment to inform and sustain peace" (Kyrou 2007, p. 73). More importantly, this worldview places emphasis on the contingency of the impact of ecological degradation on human vulnerability because of the interaction between humanity and nature and the inherent impact of environmental deterioration on the survival of humankind.

The well-being of the Earth system is inherently linked to peace and security. The threats posed by human action on the ecosystem leave the Earth and its inhabitants vulnerable and insecure. To borrow from Dalby (2007, p. 155), "Recent innovations in earth system science have added compelling arguments for the integration of environmental matters into security". There has been a shift in security concerns from scarcity- and resource-related conflicts in the Global South to worries over climate change and other consequences of environmental degradation resulting from intense human activities (Dalby 2007). This scholar emphasised the significance of reducing "the total throughput of materials and energy in the biosphere to limit disruptions while simultaneously building resilience and habits of international cooperation into human societies to better cope when disaster strikes" (Dalby 2007, p. 155). This security issue also implies the need to create an atmosphere of durable peace in regions that are worst hit by climate change and global warming. Thus, *environmental security* (Dalby 2007) and *peace ecology* can be enhanced by humankind's action and attitude towards "our common home".

Kyrou (2007) framed the interaction and interdependency of human and nature in terms of an "ecological web":

> We breathe from its air, drink from its water, are subject to and depend on its climate. We extract from it materials for utility and basic survival. There we find our spiritual resources, draw our inspiration for art, and even the raw materials of which our dreams and myths are made (pp. 79–80).

This essay is constructed on the assumption that, generally, violence has detrimental effects on humanity as well as nature, as argued here:

> [I]n the case of direct or physical violence, victims include people and the environment […]. The impacts of war on the environment do not end with a cease-fire; they persist for decades due to demolished infrastructure, movement of refugees and internally displaced people, the remaining risks from hazards such as mines and depleted uranium, and the political shortcomings of reconstruction (Kyrou 2007, p. 80).

The need for *environmental peacemaking*—building peace with nature—is sustained by the existing *ecological violence* that "is directly related to environmental studies and refers to the direct injury to the environment through pollution, degradation, overexploitation, and other forms of injury, especially in cases of severe or irreversible damage" (Kyrou 2007, p. 81). Such violence results from environmental conflicts that can be dealt with and prevented. For instance, Bronkhorst conducted a study that helped to determine to what extent environment-related conflicts can be addressed, and her investigation can be applied in areas affected by climate change, discernible in "drought, desertification, water scarcity, and competition over grazing and pasture which can contribute to conflict" (Bronkhorst 2011, p. 5). Her findings focused on the southern Sahel, stretching across Africa to the Sudan, and especially across the state of Southern Kordofan and its neighbours. With three points, she underlined how human security can be built to prevent environmental conflicts (Bronkhorst 2011, pp. 47–48):

- By putting in place legitimate mechanisms for dealing with vulnerabilities resulting from deprivations (poverty, lack of livelihood, etc.), with the involvement of NGOs to get affected communities reconciled
- By developing mechanisms susceptible of addressing environmental threats and climate changes such as "water scarcity, drought and a lack of land for farming and grazing" (Bronkhorst 2011, p. 48) that are detrimental to human security
- By creating the conditions that can assist in responding to deprivations and exclusions and other environment-related drivers of conflicts; NGOs are key partners for the success of these strategies

She argued that these elements, and other underlying factors of human insecurity, originate from unattended environmental crises that include, but which are not limited to, drought, food insecurity, poverty, a lack of livelihoods, and structural factors that create the condition of vulnerability to climate change—mainly exclusions and deprivations. These are potential causes of environmental conflicts that can lead to personal violence, and they are embedded structures that need to be transformed through development activities.

3.8 Ecoeducation, Eco-Ethics and Environmental Ethics

Ecological education is emerging as new trend in reconceptualising environmental security concerns and developing global consciousness about human agency to respond to ecological crisis in the Anthropocene. As with the conventional form of peace education that aims to prevent conflicts and violence (UNESCO 2013), ecoeducation should aim to instil individual and collective responsibility in people—to care for the ecosystem in order to prevent and deal with environment-induced conflicts nonviolently.Ecoeducation has a pluridisciplinary tendency which embraces environmental studies, philosophy, ethics and philosophy of nature and life; it aims to shape human attitudes and behaviours towards their natural environment for a sustainable living (Iovan 2014). Such as innovative pedagogy is needed today; it should be inspired by ecoethics, environmental ethics, and ecojustice philosophy.

To render these theories more intelligible, it is necessary to clarify their meanings and interconnectedness. The distinction between eco-ethics and environmental ethics can be explained by the fact that the former entails "the essential foundation for sustainable use of the planet. Such a foundation must consist of a series of value judgments to which humanity is committed" (Cairns Jr. 2002, p. 79); the latter falls under philosophical disciplines; it "studies the moral relationship of human beings to, and also the value and moral status of, the environment and its non-human contents" (Brennan/Lo 2002, p.1). Further, it incorporates environmental affairs and controversies in various ways, such as promoting equity in relation to non-Western cultures, fighting the abuse of indigenous groups that occur through land exploitation processes; scrutinising economic prosperity when land use is concerned, and conscientising people about the danger of modifying the environment to suit certain human lifestyles (Tippins/Britton 2015).

The dawn of eco-ethics and environmental ethics can be traced in the search for a moral stance, contrasting moral consequentialism in decision-making and environmental policy-making frameworks. It is important to nuance the concept of eco-ethics and environmental ethics, only for the sake of a better understanding of the two, while in fact and pragmatically, both serves the deontological purpose of care and protection of the ecosystem. What concerns us most is the open-ended eco-ethics concept, based on the distinction made by (Skolimowski 1984):

> Environmental ethics must be distinguished from eco-logical ethics. While the former concerns itself with the appropriate management of natural resources and is often guided by cost-benefit analysis, the latter (ecological ethics) is much broader as it spells out the relationships between man and nature; and also analyses those attributes of man which can make him an ecological animal. Eco-values are based on the recognition of intrinsic values of which reverence for life is one, and perhaps the most important one (Skolimowski 1984, p. 45).

Despite the different nuances denoted by these two forms of ethics, they are anchored in similar moral values and principles which apply to either interpersonal relationships, or to human-environment relations. Thus, moral responsibility vis-à-vis the environment leads to *echo-deontology*, which can be framed as a set of morally sound normative actions towards the natural world; they universally conform to the Kantian categorical imperative, which bear on nature and environment wellness. Such moral obligations by which humans are bound is not limited to inter-human relationships but it goes beyond humanity to include moral standards regarding the ways and means by which humankind handles non-human creatures. So, ecofriendly stewardship is an imperative of interhuman and human-nature interactions.

Eco-theology also features along the same leading edge educational line or environmental ethics because it revives the natural, spiritual and ontological bonds that bind humanity together with the whole creation and foster essential interconnectedness of humans and nature; that should give rise to responsible and ethical stewardship over the Earth. In this process, it is imperative to reconnect education science and ecojustice with youth activism, cultures and natural system, environmentalism, sustainability, marketing that is concerned with the ecosystem; thus, such pedagogical innovations are susceptible to enhance ecojustice philosophy (Mueller/Tippins 2015). Ongoing youth activism or environmentalism shown by Gretta Thumberg and other youths is encouraged, and their generation should continue to put across the message on environmental justice.

To end this section, the urgency to tackle the challenges set off by climate change opens up to new teaching paradigms, namely eco-pedagogy. Ecoethics remains one of the essential approaches that can help reduce ecological vulnerability, subsequently minimise the risks of natural disasters (Etkin/Stefanovic 2005). Here comes the role of education as the chief vehicle for imparting knowledge about biosphere and geosphere; ecoeducation is a useful practice for the dissemination of eco-friendly values that underscore respect in the interactions between humans and their nature environment. The role of civil society in environmental protection and the relief of climate change threats to peace and security is discussed in the following section.

3.9 Civil Society and Climate Change

While Environmental Non-Governmental Organisations (ENGOs) seek access to climate change-related negotiations as delegates alongside corporations and business lobbying groups at the United Nations Framework Convention on Climate Change (UNFCCC) with a view to affecting official delegates' policies, it has been found that "only business groups are likely to exert a causal influence on states' climate delegations. However, contrary to expectations, these groups appear to have enhanced states' efforts towards environmentally friendly policies" (Böhmelt 2013, p. 698). In contrast to the finding according to which business lobbying groups are more likely to leverage healthy environmental policies than ENGOs, the perspective of Ginanjar/Mubarrok (2020) provides a different perspective. They note, "civil society can provide the opportunities for the people to contribute to global governance related to climate change. Civil society here can be interpreted as a political space, where associations of people can work towards the formation of social and legal rules through non-formal political channels" (Ginanjar/Mubarrok 2020, p. 41).

The reason for the authors' optimistic view is based on the fact that "Extinction Rebellion's involvement in global governance is an indirect involvement with resistance as their mode of participation, which is indicated by the emergence of mass protest in various cities around the world" (Ginanjar/Mubarrok 2020, p. 41). Furthermore, some people, particularly at the grassroots level, are apprehensive of the inaction of global governance when it comes to climate change as perceived today; governmental stances are not adequate and effective to yield changes, and "because the nature of the global climate regime is somehow voluntary and non-binding" (Ginanjar/Mubarrok 2020, p. 41).

The IPCC (2018) acknowledges the role of civil society in climate change action; it states in Section D on *Strengthening the Global Response in the Context of Sustainable Development and Efforts to Eradicate Poverty*, D.7.:

> Strengthening the capacities for climate action of national and sub-national authorities, civil society, the private sector, indigenous peoples and local communities can support the implementation of ambitious actions implied by limiting global warming to 1.5°C (high confidence). International cooperation can provide an enabling environment for this to be achieved in all countries and for all people, in the context of sustainable development. International cooperation is a critical enabler for developing countries and vulnerable regions (high confidence) (IPCC 2018, p. 25).

The IPCC, in D.7.1 and D.7.2 of the same summary, views the partnerships of non-state public and private actors, including institutional investors, the banking system, civil society and scientific institutions as the instruments that would facilitate adequate actions and responses consistent with the policy of limiting global warming (IPCC 2018, p. 25). In D.7.2, it observes:

> Cooperation on strengthened accountable multilevel governance that includes non-state actors such as industry, civil society and scientific institutions, coordinated sectoral and cross-sectoral policies at various governance levels, gender sensitive policies, finance including innovative financing, and cooperation on technology development and transfer can ensure participation, transparency, capacity building and learning among different players (high confidence) (IPCC 2018, p. 25).

Despite cautions against the role of civil society in climate action, civil society organisations (CSOs) have remained consistent and vocal in awakening human conscience, and in confronting businesses/markets, corporations, and governments about their denial or lack of concern over the collapsing Earth system.

At the 23 September 2019 United Nations Climate Action Summit where Greta Thunberg accused world leaders of denialism, robbing her of her dreams and childhood, and failing the youth in her poignant opening speech: "I shouldn't be up here. I should be back in school on the other side of the ocean. Yet, you all come to us young people for hope. How dare you!" (Thunberg 2019a). Thunberg's activism epitomises what individual agency should look like, along with Pope Francis' exhortation to urgently work to save the Earth. He writes, "The urgent challenge to protect our common home includes a concern to bring the whole human family together to seek a sustainable and integral development, for we know that things can change" (Francis 2015). Since climate change is caused by human activities, it is also believed that human agency is indispensable in all efforts to reduce or halt the worsening atmospheric condition.

3.10 Human Agency

The new geological epoch into which the Earth has entered is an established fact, with ongoing research and scientific endeavours to uncover more implications of the Anthropocene and reimagine humanity's agency and stewardship in handling the ES. This section explores the interplay of factors between agents, actors, and stewardship of the Anthropocene that could advance innovative economic progress while safeguarding the ES and taking the edge off climate change. In the Anthropocene, individual and collective agency are required to mitigate and slow the deterioration in the ES.

The concept of human agency, in a broader sense, can be interpreted as "the capacity of individual and collective actors to change the course of events or the outcome of processes" (Pattberg/Stripple 2008cited in Otto 2020, p. 2); this is reflected in the "everyday agency of individual human agents" (Otto et al. 2020, p. 9), which is responsible for the anthropogenic cause of the atmospheric transformation of the ES with climate change as one of its consequences (Crutzen/Brauch 2016; Steffen et al. 2015; Grinevald et al. 2011).

The notion of *agency* is distinct from that of *actors*, which bears a collective connotation, such as governments, civil society, businesses, scientific communities, non-state actors, cities and their affiliated political interests, and actions in the field

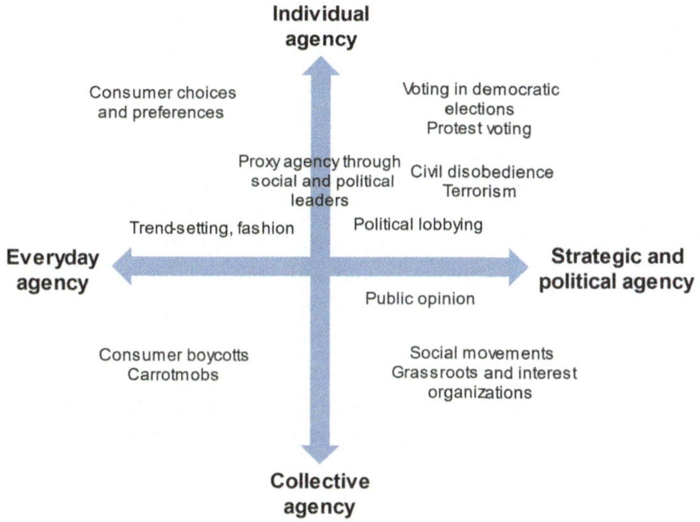

Fig. 3.2 Agency dimensions. *Source:* Otto et al. (2020, p. 9). Adapted from Lister (2004) and Coulthard (2012) with empirical examples of social phenomena

of climate governance (Pattberg/Stripple 2008). To these institutions, we can add corporations and families to form the category of actors. Unfortunately, the important role that business and non-state actors play in lessening the impact of climate change is seldom represented in international negotiations (Pattberg/Stripple 2008).

We are in a critical time where individual and collective agency, and the inclusion and participation of all in Transnational Climate Governance (TCG), are imperative to rescue the Earth from, if not to limit, the scale of atmospheric disruptions and, subsequently, to minimise climate change and its detrimental consequences for the entire creation.

The multifaceted representation of agency in Figure 3.2 shows how the degradation of the Earth system is precipitated by several agents. A remedial global action demands the involvement of all stakeholders to rebuild hope and peace; this does not entail doing away with the root of agency in Earth sciences. For instance, Latour (2014) noted:

> Geo-physiology as well as geo-morphology, geo-physics, geo-graphy, geo-politics should not eliminate any of the sources of agency—including those generated by former humans, those I call Earthbound—if they want to converge toward a common geostory (p. 14).

This proposition features in the path to a solution that is presented by Pope Francis: an all-inclusive strategy that involves actors at various degrees of agency—consumers, politicians, the business class, and civil society (social movements, grassroots and interest groups and associations, etc.). The interactions of these institutions as platforms of climate governance suggest the imperative of concerted efforts in the struggle to alleviate the burden of anthropogenic atmospheric disruptions. In this regard, Pattberg/Stripple (2008) acknowledged that varied institutions interact to forge policies that aim to protect the environment; these include "the related norms and rules and the resulting roles and responsibilities of actors within the field of climate change as a transnational arena of climate governance" (p. 372).

The apparent agitation and lack of peace in the world can be regarded as the consequence of the crumbling ES that becomes manifest in climate change and its corollaries. This is a clear signal that the peoples of the world must rise to save Earth, although many industrialised nations have not reached the stage of trepidation as atmospheric conditions continue to deteriorate. The growing sense of disquietude revealed by increased interest in and activism towards peace ecology in our time is justified by the fact that "the earth is quaking"—causing worries, though some politicians remain in denial, as noted by Latour (2014). The anxiety echoed in "Save the Earth" advocacy is not only limited to protecting "our house on fire" (Thunberg 2019b, n.p.) and the generational conflict stemming from the failure of adults to safeguard the environment for future generations (Kühne 2019; Thunberg 2020) but also relates to the fear of violent conflict for human communities. This is well explained by Bronkhorst (2011), who provided insights into how to mitigate environment-related disputes between pastoralist and agriculturalist communities via interventions by local and international NGOs. Her research is relevant in the sense that it discovers and anticipates the problem of environmental security amidst potential environmental conflicts and, at the same time, opens the pathway to resolving such conflicts. This responsibility is incumbent upon humankind. The role of humans in healing the Earth is evident because their activities have led to the new geological epoch of the Anthropocene (Otto et al. 2020). Similarly, their agency is needed to redress the perilous turn that the Earth is facing.

Individual and collective agency are necessary and should be applied simultaneously in global efforts to save "our common home" and to ensure that humanity's response as radical commitment to redeem or rescue the disintegrating ES is more effective. The Earth system shows the nature of actions that individuals can undertake, namely, the choices made by and preferences of consumers; and, at the collective level, social movements and grassroots and interest groups can leverage boycott campaigns against certain products to protect the environment (Otto et al. 2020). Individuals are potential agents who can contribute to mitigating the effects of anthropogenic climate change; civil society and its constituent FBOs fall under "collective actors" in transnational climate governance, as highlighted by Pattberg/Stripple (2008):

> Collective actors derived from civil society, the market and various communities become effectively public with a potential to govern people and issues. As we will see in a moment, this is an accord that harmonizes with recent writings on the public and private in world politics (p. 371).

Individual agents and collective actors are the major players in the fight against anthropogenic disruptions of the Earth systems. I should point out two aspects of stewardship, namely, that which is based on advocacy and the change of lifestyles, and that which involves governments, corporations, and multinational organisations. The latter requires an international consciousness and determination to control the factors that contribute to global warming; the following causes of rising emissions were listed by the European Commission:

> Burning coal, oil and gas produces carbon dioxide and nitrous oxide; cutting down forests (deforestation); increasing livestock farming; cows and sheep produce large amounts of methane when they digest their food; Fertilisers containing nitrogen produce nitrous oxide emissions; Fluorinated gases are emitted from equipment and products that use these gases. Such emissions have a very strong warming effect, up to 23 000 times greater than CO_2 (European Commission 2020, p. n.p.).

It is granted that trees are helpful in regulating the climate via absorbing CO_2 from the atmosphere; thus, when they are cut down, the said beneficial results are lost as the carbon stored in the trees is ultimately released into the atmosphere, with the consequence of adding to the greenhouse effect (European Commission 2020, p. n.p.). Thus, deforestation negatively affects the natural adjustment and balancing of CO_2 that protects the ecosystem. To reduce growing risks of global warming, national agency is paramount when we consider the findings of studies conducted by Matthews et al. (2014). According to the new estimate concerning national contributions to climate warming, such as CO_2 emissions originating from fossil fuels, land use change, and other greenhouse gases ("methane, nitrous oxide and sulfate aerosol emissions"), "many countries have dominant contributions from land-use CO_2 and non-CO_2 greenhouse gas emissions, emphasising the importance of both deforestation and agriculture as components of a country's contribution to climate warming" (Matthews et al. 2014, p. 1). For instance, agriculture itself "is also a significant driver of many of the PBs [planetary boundaries] still in the safe zone" (Campbell et al. 2017, p. n.p.).

This begs answers to a twofold question:

- How to overcome this conundrum while it is reported that the number of hungry people in the world approximated 690 million in 2020, and the population of undernourished people was 678.1 million in 2019, which is expected to increase to 841.4 million by 2030[1] (Food and Agricultural Organization of the United Nations (FAO 2021)
- How to achieve Goal 2 of the United Nations 2030 Sustainable Development Agenda that calls for global support for the livelihoods of small-scale food producers, ameliorating the resilience of food production systems and enhancing the sustainable use of natural resources that are deemed necessary to fulfilling the mandate of SDG2: to make the world hunger-free, and halt food insecurity and malnutrition (FAO 2021; United Nations 2015)

To respond to the ways in which human activity perturbs the ES, Pope Francis invites the world to an environmental global dialogue to save the Earth. His appeal reflects the concerns shown by faith-based organisations (FBOs) in building both societal peace and environmental peace. From a similar perspective, FBOs, as a component of civil society, have been in the forefront of the struggle for environmental justice, protection of human natural environment, livelihood and food security.

In addition, sulfate aerosol emissions, despite their short lifetime, contribute largely to climate warming, and across most developed countries, "per-capita contributions are not currently consistent with attempts to restrict global temperature change to less than 2°C above pre-industrial temperatures" (Matthews et al. 2014, p. 1). National climate policies need to improve, and this is incumbent upon each country's agency towards climate change.

Individual, communal, national, and international consciousness that is anchored in the faith-oriented understanding and practice of stewardship are required today in order to save our ecosystem. Because climate governance is not the sole responsibility of state authorities, private sectors and civil society are also partners in protecting the environment. To achieve this, responsible stewardship is needed at all levels of society.

3.11 Conclusion

This chapter sought to substantiate existing various climate-change-induced security threats and conflicts we experience in the Anthropocene. It started with presenting the overall landscape of present atmospheric transformations and the extent to which Earth systems are damaged and affect humanity and non-human beings. It was demonstrated also that climate change has become the most severe threat to the survival of humanity and the ecosystem, to global peace, world economic and social security. That is admitted as historical truth supported by science.

The need for urgent solution to climate warming, locally, regionally and internationally becomes obvious and undeniable. Action is needed to prevent total disintegration of Earth systems. Thanks to scientific advancement which is credited for developing planet boundaries framework to monitor both humans' attempts to revert the course of geosocial epochs, and to presenting new findings about atmospheric disturbances that continue to unfold in the Anthropocene.

Various interventions are underway to alleviate the burden of human activities on Earth systems, despite some cases of denialism shown by those who paradoxically decry the threats posed by climate change while simultaneously support activities that contribute to the deterioration of the ecosystem. To sustain concerns associated with climate change hypocritical and denialists behaviour patterns, how many nations admit scientific truth according to which production and use of coal, burning fossil fuels, deforestation, unregulated fishing and unlawful waste disposal, unlawful exploitation, and commercialisation of wood; deregulated farming and unfriendly use of methods to keep and maintain livestock are the predominant contributors the global worming? How many have significantly developed policy frameworks to move to renewable energy? New sources of renewable energy include wind and sunlight: they are the Earth's natural resources; they are not finite or exhaustible; they are renewable energy that need to be developed and supplied to phase out fossil fuels and their harmful effects on the environment.

Because of the gravity of ecological turmoil, it was noted that advocacy to fighting climate change has gained impetus. Various actors, including civil society are engaged in environmentalist or ecological projects. Social movements, research and think tanks have all increasing developing innovative theoretical paradigms and pragmatic solutions to save Earth planet; a case in point is the repair of the ozone hole "after the phasing out of ozone-depleting substances" (Steffen et al. 2015, p. 347). This gives hope for resilience of science, and the needed openness to turn crisis into innovative opportunity (Rockström 2019).

The planetary boundaries framework was an essential scientific innovation in terms of leading humanity to resilience and sustainable development without transgressing the boundary paradigms; because such attempt could lead to irreversible crisis and unprecedented planetary catastrophe (Rockström 2019; Steffen et al. 2015). In terms of meeting the risks of global-warming-driven security issues and conflicts, "climate change-sensitive-adaptation" and "conflicts-sensitive" concepts, approaches, praxis or philosophy are indispensable in framing ecofriendly policy frameworks which is susceptible to mitigate environmental instability, insecurity and global peace disruption. To get there, a synergic cooperation between civil society, business, corporations, political actors, science and think tanks are urged to be adopted as thc pathway to leading an international and concerted implementation of healthy ecological policies. Such an enterprise cannot be successful without theoretical paradigms developed in peace studies that we discussed throughout this chapter, namely: (1) environmental justice, (2) environmental peace-making/peace ecology, and (3) ecological education which embraces ecoethics and environmental ethics.

Finally, the interplay of various dimensions of human agency to act against climate change is maintained as another essential track to follow that it would be possible to reduce the global environmental crisis of our time, to achieve ecological sustainability. All human, social, technological and scientific interventions remain a significant determinant to recreating an ecofriendly agenda of sustainable development, global peace and fostering humanity climate change-resilience for sustainable biodiverse. In all these

References

Artur, L., & Hilhorst, D. (2012). Everyday realities of climate change adaptation in Mozambique. *Global Environmental Change, 22*(2), 529–536

Babcicky, P. (2013). A conflict-sensitive approach to climate change adaptation. *Peace Review, 25*(4), 480–488

BBC News (Producer). (2021). Life at 50C: Fleeing Sahara's shifting sands. https://www.bbc.com/news/av/science-environment-58925616. Accessed 22 Oct 2021

Bob, U., & Bronkhorst, S. (2014). *Conflict-sensitive adaptation to climate change in Africa*. Berlin: Berliner Wissenschafts-Verlag

Bob, U., Bronkhorst, S., & Sala, S. (2014). Climate change and conflict: Conflict-sensitive climate change adaptation in Africa. In U. Bob & S. Bronkhorst (Eds.), *Conflict-sensitive adaptation to climate change in Africa* (pp. 27–54). Berlin: Berliner Wissenschafts-Verlag

Böhmelt, T. (2013). Civil society lobbying and countries' climate change policies: a matching approach. *Climate policy,* 13(6), 698–717

Boulding, E. (2002). Peace culture. In M. Afkhami (Ed.), *Toward a compassionate society* (pp. 8–15). Bethesda: Women's Learning Partnership

Brauch, H. G. (2016). Sustainable Peace in the Anthropocene: Towards Political Geoecology and Peace Ecology. In H. Brauch, Ú. Oswald Spring, J. Grin, & J. Scheffran (Eds.), *Handbook on Sustainability Transition and Sustainable Peace* (pp. 187–236). Cham: Springer

Brennan, A., & Lo, Y.-S. (2002). Environmental ethics. In. online: Stanford Encyclopedia of Philosophy. https://plato.stanford.edu/entries/ethics-environmental/. Accessed 23 Oct 2021

Brisman, A. (2016). Peace ecology. *Contemporary Justice Review,* 13(3), 419–421. https://doi.org/10.1080/10282580.2016.1200886

Bronkhorst, S. (2011). *Climate change and conflict: lessons for conflict resolution from the southern Sahel of Sudan.* Umhlanga Rocks: ACCORD

Cairns Jr, J. (2002). A declaration of eco-ethics. *Ethics in Science and Environmental Politics, 18,* 79–81. https://doi.org/10.3354/esep002079

Campbell, B. M., Beare, D. J., Bennett, E. M., Hall-Spencer, J. M., Ingram, J. S., Jaramillo, F., Ortiz, R, Ramankutty, N, Sayer, J A, Shindell, D. (2017). Agriculture production as a major driver of the Earth system exceeding planetary boundaries. *Ecology and Society,* 22(4). https://doi.org/10.5751/ES-09595-220408

Caniglia, B. S., Brulle, R. J., & Szasz, A. (2015). Civil society, social movements, and climate change. *Climate change and society: Sociological perspectives,* 1, 235–268

Clarke, R. (1999). *Global Environment Outlook 2000 report (UNEP 1999).* Nairobi: United Nations Environment Programme

Crutzen, P. J. (2016). Geology of mankind. In P.J. Crutzen & H.G. Brauch (Eds.), *Paul J. Crutzen: A Pioneer on Atmospheric Chemistry and Climate Change in the Anthropocene* (Vol. 50, pp. 211–215). Cham: Springer Nature Switzerland AG

Crutzen, P. J., & Brauch, H. G. (2016). *Paul J. Crutzen: a pioneer on atmospheric chemistry and climate change in the Anthropocene* (Vol. 50). Cham: Springer Nature Switzerland AG

Dalby, S. (2007). Ecology, security, and change in the Anthropocene. *The Brown Journal of World Affairs,* 13(2), 155–164

Day, J. W., Christian, R. R., Boesch, D. M., Yáñez-Arancibia, A., Morris, J., Twilley, R. R., . . . Schaffner, L. (2008). Consequences of climate change on the ecogeomorphology of coastal wetlands. *Estuaries and Coasts, 31*(3), 477–491. https://doi.org/10.1007/s12237-008-9047-6

Eklöw, K., & Krampe, F. (2019). *Climate-related security risks and peacebuilding in Somalia.* Stockholm: Stockholm International Peace Research Institute (SIPRI)

Etkin, D., & Stefanovic, I. L. (2005). Mitigating natural disasters: The role of eco-ethics. In *Mitigation of natural hazards and disasters: International perspectives* (pp. 135–158). online: Springer

European Commission. (2020). *Climate change action.* Retrieved from https://ec.europa.eu/clima/climate-change/causes-climate-change_en. Accessed 10 Oct 2021

European Union (EU). (2008). *Climate Change and International Security.* Paper from the High Representative and the European Commission to the European Council, S113/08. 14 March 2008 European Union (EU). online. https://www.consilium.europa.eu/uedocs/cms_data/docs/pressdata/en/reports/99387.pdf. Accessed 10 Oct 2021

Fifita, S., Forster, P., Ginzburg, V., Handa, C., Kheshgi, H., Kobayashi, S., . . . Vilariño, M. V. (2018). Mitigation Pathways Compatible with 1.5°C in the Context of Sustainable Development. https://www.ipcc.ch/site/assets/uploads/sites/2/2019/05/SR15_Chapter2_High_Res.pdf. Accessed 6 Oct 2021

Food and Agricultural Organization of the United Nations (FAO) (2021). Hunger and food insecurity. http://www.fao.org/hunger/en/. Accessed 15 July 2021

Ginanjar, W. R., & Mubarrok, A. Z. (2020). Civil Society and Global Governance: The Indirect Participation of Extinction Rebellion in Global Governance on Climate Change. *Journal of Contemporary Governance and Public Policy, 1*(1), 41–52

Hardt, J. N., & Scheffran, J. (2019). Environmental peacebuilding and climate change: peace and conflict studies at the edge of transformation. *Policy Brief,* 68(December 2019), 1–20

Houghton, D. D., & World Meteorological Organization. (2002). *Introduction to climate change: Lecture notes for meteorologists.* Geneva: Secretariat of the World Meteorological Organization Geneva, Switzerland

Hsieh, H.-F., & Shannon, S. E. (2005). Three approaches to qualitative content analysis. *Qualitative Health Research,* 15(9), 1277–1288

Ijumba, N., & Kaya, H. (2016). Energy Resources and Environmental Conflicts in Africa: Implications on Indigenous Knowledge Systems and Environmental Policy. *Journal of Human Ecology,* 53(2), 96–100. https://doi.org/10.1080/09709274.2016.11906960

Iovan, M. (2014). Eco-education: A required element of public policies for sustainable social and economic development. *Academicus International Scientific Journal*(09), 14–28

IPCC. (2018). Summary for Policymakers. In V. Masson-Delmotte, P. Zhai, H.-O. Pörtner, D. Roberts, J. Skea, P. R. Shukla, A. Pirani, W. Moufouma-Okia, C. Péan, R. Pidcock, S. Connors, J. B. R. Matthews, Y. Chen, X. Zhou, M. I. Gomis, E. Lonnoy, T. Maycock, M. Tignor, & T. Waterfield (Eds.), *Global Warming of 1.5°C. An IPCC Special Report on the impacts of global warming of 1.5°C above pre-industrial levels and related global greenhouse gas emission pathways, in the context of strengthening the global response to the threat of climate change, sustainable evelopment, and efforts to eradicate poverty.* Geneva: World Meteorological Organization

Krampe, F. (2017). Toward sustainable peace: A new research agenda for post-conflict natural resource management. *Global Environmental Politics,* 17(4), 1–8. https://doi.org/10.1162/GLEP_a_00431

Krampe, F., & Mobjörk, M. (2018). Responding to climate-related security risks: reviewing regional organizations in Asia and Africa. *Current Climate Change Reports,* 4(4), 330–337. https://doi.org/https://doi.org/10.1007/s40641-018-0118-x

Krampe, F., Van De Goor, L., Barnhoorn, A., Smith, E., & Smith, D. (2020). Water Security and Governance in the Horn of Africa. *Stockholm International Peace Research Institute SIPRI, SIPRI Policy (Paper No. 54).* https://www.diva-portal.org/smash/get/diva2:1424710/FULLTEXT01.pdf. Accessed 7 Oct 2021

Krippendorff, K. (2004). Content analysis: An introduction to its methodology. Thousand Oaks Calif.: Sage

Kühne R.W. (2019, September 2). Climate change: the Science behind Greta Thunberg and Fridays for Future. https://doi.org/10.31219/osf.io/2n6kj

Kyrou, C.N. (2007). Peace ecology: An emerging paradigm in peace studies. *International Journal of Peace Studies,* 12(1), 73–92. http://www.jstor.org/stable/41852955

Latour, B. (2014). Agency at the time of the Anthropocene. *New Literary History,* 45(1), 1–18. https://doi.org/10.1353/nlh.2014.0003

Lisk, F. (2009). Overview: The Current Climate Change Situation in Africa. In H. Besada & N. Sewankambo (Eds.), *Climate Change in Africa: Adaptation, Mitigation and Governance Challenges* (pp. 8–15). Waterloo: The Centre for International Governance Innovation (CIGI)

Matthews, H. D., Graham, T. L., Keverian, S., Lamontagne, C., Seto, D., & Smith, T. J. (2014). National contributions to observed global warming. *Environmental Research Letters,* 9(1), 014010. https://doi.org/10.1088/1748-9326/9/1/014010

McClanahan, T. R., & Cinner, J. (2012). *Adapting to a changing environment: confronting the consequences of climate change*: OUP USA

McNeely, J. A. (2020). Today's protected areas: supporting a more sustainable future for humanity. *Integrative zoology,* 15(6), 603–616

Molvaer, R. K. (1991). Environmentally induced conflicts? A discussion based on studies from the Horn of Africa. *Bulletin of Peace Proposals,* 22(2), 175–188

Morton, S., Pencheon, D., & Bickler, G. (2019). The sustainable development goals provide an important framework for addressing dangerous climate change and achieving wider public health benefits. *Public Health, 174*, 65–68. https://doi.org/10.1016/j.puhe.2019.05.018

Mueller, M. P., & Tippins, D. J. (2015). *Ecojustice, citizen science and youth activism.* Cham: Springer International Publishing. https://doi.org/https://doi.org/10.1007/978-3-319-11608-2

Newell, P. (2008). Civil society, corporate accountability and the politics of climate change. *Global Environmental Politics,* 8(3), 122–153

O'Neill, D. W., Fanning, A. L., Lamb, W. F., & Steinberger, J. K. (2018). A good life for all within planetary boundaries. *Nature Sustainability,* 1(2), 88–95. https://doi.org/10.1038/s41893-018-0021-4

Osbahr, H., Twyman, C., Adger, W. N., & Thomas, D. S. (2008). Effective livelihood adaptation to climate change disturbance: scale dimensions of practice in Mozambique. *Geoforum,* 39(6), 1951–1964

Otto, I. M., Wiedermann, M., Cremades, R., Donges, J. F., Auer, C., & Lucht, W. (2020). Human agency in the Anthropocene. *Ecological Economics,* 167, 1–31. https://doi.org/10.1016/j.ecolecon.2019.106463

Pattberg, P., & Stripple, J. (2008). Beyond the public and private divide: remapping transnational climate governance in the 21st century. *International Environmental Agreements: Politics, Law and Economics,* 8(4), 367–388. https://doi.org/10.1007/s10784-008-9085-3

Pihkala, P. (2018). Eco-anxiety, tragedy, and hope: psychological and spiritual dimensions of climate change, *Zygon®,* 53(2), 545–569

Pius Yanda, P., & Bronkhorst, S. (2011). Climate change and conflict: Conflict-sensitive Climate Change Adaptation in Africa. *Policy & Practice Brief,* 014 (November 2011). https://media.africaportal.org/documents/policy_practice14.pdf Accessed 5 Oct 2021

Pope Francis (2015). *Encyclical Letter Laudato Si' of The Holy Father Francis on Care for Our Common Home.* Rome: Libreria Editrice Vaticana

Rafferty, John P. (n.d.). "Anthropocene Epoch". In *Encyclopedia Britannica,* https://www.britannica.com/science/Anthropocene-Epoch. Accessed 7 Oct 2021

Rockström, J., Steffen, W., Noone, K., Persson, Å., Chapin III, F. S., Lambin, E., . . . Schellnhuber, H. J. (2009). Planetary boundaries: exploring the safe operating space for humanity. *Ecology and society, 14*(2). https://doi.org/10.1126/science.1259855originally published online January 15, 2015

Rockström, J. (2019). Safe Future for Humanity on Earth. *This is a Session at Republica19. The Thirteen Edition of the Conference (some call it a festival) took place on May 6, 7 and at Station Berlin.* https://www.youtube.com/watch?v=wJtfWZGxnGI&feature=emb_rel_end. Accessed 10 Oct 2021

Rowlatt, J. (2021). Why 2021 could be turning point for tackling climate change. *BBC News, 1 January.* https://www.bbc.com/news/science-environment-55498657. Accessed 6 Oct 2021

Ruddiman, W. F. (2013). The anthropocene. *Annual Review of Earth and Planetary Sciences,* 41, 45–68; https://doi.org/10.1146/annurev-earth-050212-123944

Schlosberg, D. (2004). Reconceiving environmental justice: global movements and political theories. *Environmental Politics,* 13(3), 517–540 (2004). https://doi.org/10.1080/0964401042000229025

Schlosberg, D. (2013). Theorising environmental justice: the expanding sphere of a discourse. *Environmental Politics,* 22(1), 37–55. https://doi.org/10.1080/09644016.2013.755387

Skolimowski, H. (1984). Eco-ethics as the foundation of conservation. *Environmentalist, 4*(7), 45–51

Spring, Ú. O., & Brauch, H. G. (2021). *Decolonising Conflicts, Security, Peace, Gender, Environment and Development in the Anthropocene.* Cham: Springer Nature Switzerland AG

Steffen, W., Grinevald, J., Crutzen, P., & McNeill, J. (2011). The Anthropocene: conceptual and historical perspectives. *Philosophical Transactions of the Royal Society A: Mathematical, Physical and Engineering Sciences,* 369(1938), 842–867

Steffen, W., Richardson, K., Rockström, J., Cornell, S. E., Fetzer, I., Bennett, E. M., Biggs, R., Carpenter, S.R., De Vries, W., . . . Sörlin, S. (2015). Planetary boundaries: Guiding human development on a changing planet. *Science,* 347(6223), 736–747. https://doi.org/10.1126/science.1259855

Sze, J., & London, J. K. (2008). Environmental justice at the crossroads. *Sociology Compass,* 2(4), 1331–1354. https://doi.org/10.1111/j.1751-9020.2008.00131.x

Tänzler, D., Maas, A., & Carius, A. (2010). Climate change adaptation and peace. *Wiley Interdisciplinary Reviews: Climate Change, 1*(5), 741–750

Tänzler, D., Carius, A., & Maas, A. (2013). The need for conflict-sensitive adaptation to climate change. *Environmental Change and Security Program Report, 14*(2), 5

Temper, L., Demaria, F., Scheidel, A., Del Bene, D., & Martinez-Alier, J. (2018). The Global Environmental Justice Atlas (EJAtlas): ecological distribution conflicts as forces for sustainability. *Sustainability Science, 13*(3), 573–584. https://doi.org/10.1007/s11625-018-0563-4

Thunberg, G. (2019a). Greta Thunberg tells world leaders 'you are failing us', as nations announce fresh climate action. https://news.un.org/en/story/2019/09/1047052. Accessed 26 July 2021

Thunberg, G. (2019b). Our house is on fire. *The Guardian,* 20. https://www.theguardian.com/books/2020/mar/04/our-house-is-on-fire-review. Accessed 29 July 2021

Thunberg, G. (2020). Greta Thunberg: 'My generation will not give up without a fight'. *Green Left Weekly,* 01 January 2020(1250), 13. https://doi.org/10.3316/informit.934151085409524. Accessed 29 July 2021

Tippins, D., & Britton, S. A. (2015). Ecojustice pedagogy. In R. Gunstone (Ed.), *Encyclopedia of Science Education. Springer, Dordrecht.* online. https://doi.org/10.1007/978-94-007-2150-0_297

Towheed, S. (2009). Tools and techniques for literary research: Using online and printed sources. In D. d. S. Correa & W. R. Owens (Eds.), *The Handbook to Literary Research* (2nd ed., pp. 9–35). London: Routledge

UNESCO. (2013). Promoting a culture of peace and non-violence in Africa through education for peace and conflict prevention Phase 1: Mapping – Final Report. https://unesdoc.unesco.org/ark:/48223/pf0000234682. Accessed 18 Jan 2020

United Nations (2015). Transforming our World: the 2030 Agenda for Sustainable Development. Resolution adopted by the General Assembly on 25 September 2015 [without reference to a Main Committee] (A/70/L.1)] 70/1. https://www.un.org/en/development/desa/population/migration/generalassembly/docs/globalcompact/A_RES_70_1_E.pdf. Accessed 25 July 2021

United Nations. (2021). UN General Assembly and Climate Week NYC to Accelerate Climate Action. https://unfccc.int/news/un-general-assembly-and-climate-week-nyc-to-accelerate-climate-action. Accessed 6 Oct 2021

Westra, L. (2009). *Environmental Justice and the Rights of Ecological Refugees* (1st Ed.). London: Routledge. https://doi.org/10.4324/9781849770088

Willms, D. G., & Werner, K. (2009). Climate change and conflict in Africa. In H. Besada & N. K. Sewankambo (Eds.), *Climate Change and Conflict in Africa: Adaptation, Mitigation and Governance Challenges* (pp. 26–30). Waterloo: The Centre for International Governance Innovation

Wilson, G. (2020). Climate Change as a" Threat to the Peace": Responding to Climate Change through the UN Collective Security System. *International Journal of Climate Change: Impacts & Responses, 12*(3)

Zalasiewicz. J., Williams, M., Steffen, W., & Crutzen, P. (2010). The new world of the Anthropocene. *Environmental Science & Technology,* 44(7), 2228–223

Chapter 4
Faith-Based Organisations and Environmental Peace: Impact of *Laudato Si'* and Its Relevance in the Anthropocene

Jean Chrysostome K. Kiyala

Abstract This chapter looks into the extent to which faith-based civil society can contribute to environmental peace and justice in the Anthropocene. It focuses on Pope Francis' Encyclical Letter *Laudato Si'* ["Praise be to you, my Lord"], "On Care for Our Common Home" which is used here as an exploratory case study. It argues that humanity can withstand the effects of industrial revolution, the advancement of science and technology and governments' unfair policies that are deemed anthropogenic causes of the disintegrating Earth system (ES). The potential of faith-based civil society to lead a promising transformative action to save our planet is situated at both a theological/spiritual level (world regions' population, fundamentals, beliefs, and teaching on care for the environment) and an experiential/pragmatic level (human agency, resilience and stewardship). Based on the conceptual and practical analysis of Earth sciences and how these apply to the Anthropocene, this chapter put forth four strategies as pathways to mitigate the crumbling of the ES and its corollaries, mainly climate change, as well as drought, famine, loss of lands and displacement of populations. These strategies are (1) promoting ecological peace education, (2) popularising environmental justice, (3) encouraging a change in the belief that satisfying individual needs is the basic purpose of life, and lastly (4) emboldening humanity's conscientious stewardship over creation and its natural environment.

Keywords Anthropocene · Catholicism · Civil society · Faith-based organisations · Environmental justice · Laudato Si' · Peace ecology · Peacebuilding

---

Jean Chrysostome K. Kiyala is a Senior Lecturer, International Centre of Violence, Durban University of Technology, South Africa; Associate Professor and visiting lecturer the University of Bandundu, Catholic University of Bukavu and Evangelical University in Africa based in Bukavu (DR Congo).

J. C. K. Kiyala and G. T. Harris (eds.), *Civil Society and Peacebuilding in Sub-Saharan Africa in the Anthropocene*, The Anthropocene: Politik—Economics—Society—Science 34,
https://doi.org/10.1007/978-3-030-95179-5_4

4.1 Introduction

The challenges that both humanity and non-human creation face in the age of environmental crisis keep amplifying, putting the future of the planet at high risk of total disintegration as the consequences of human activities are reaching alarming proportions. This is perceptible in global climate/atmospheric change, toxic waste, loss of land, loss of species, loss of wilderness, devastation of indigenous peoples, human patterns and quantities of consumption, and genetic engineering (Gottlieb 2003). In the same vein, Patriarch Bartholomew called humanity to repent for mishandling creation; he warned against the ongoing destruction of our planet and the risk that future generations face as Earth system (ES) continues to disintegrate; he said:

> All of us [--] scientists, as well as religious and political leaders, indeed all people [--] are witnessing a rise in the atmosphere's temperature, extreme weather conditions, the pollution of ecosystems both on land and in the sea, and an overall disturbance [--] sometimes to the point of utter destruction -- of the potential for life in some regions of the world, [...] (Catholic News Service 2012, p. n.p.).

The effects of this crisis are already experienced in Africa, where the impacts of climate change are already seen as a reality (Lisk 2009; Willms/Werner 2009), and these are ingredients and triggers for violent conflicts (Ahmed 2010; Brauch/Scheffran 2012) where violence is already endemic and multidimensional, partly due to ethnicity and religious factors.

Amidst the anxiety over the crumbling ES, faith-based organisations (FBOs) have taken a stance to navigate against the currents of political and economic pressure and the development strategies and schemes that have affected our planet. World religious leaders and ecotheologians offer a range of reflections, thoughts, philosophies and faith-based ethics that are necessary to change the course of human history in this time of ecological crisis.

To address this quandary, Gottlieb (2003) leads the discussion that involves *eco-theological* perspectives in the age of environmental crisis, and his argument is grounded in the views held by traditional religions on nature, and the overall trends on *religious practice for a sacred earth, ecology, religion, and society.* These concepts would need to be further explored because they are capable of swaying human attitudes and behaviours towards nature and the use of its resources. Unfortunately, a full debate on the strategies of global world religions for environmental justice and peace ecology is beyond the purview of this chapter, which focuses on Catholicism, particularly the impact that Pope Francis' Encyclical Letter *"Laudato si', mi' Signore"* – "Praise be to you, my Lord" – has had on ecological peace discourse since its publication.

Laudato Si' is the Encyclical Letter written by the Roman pontiff Pope Francis amidst the ecological crisis affecting the world in the Anthropocene Epoch. "The Anthropocene" is a term that was popularised by the Nobel laureate Paul J. Crutzen to describe "the current geological era, acknowledging the enduring influence of humankind on planet Earth. This concept conceives humans to be a geologic factor, influencing the evolution of our globe and the living beings populating it"

(Crutzen/Brauch 2016, p. 247). This Epoch is so defined as following the Holocene – the most recent 12,000 years as described by geologists. Crutzen contends that the effects of human activities on Earth's surface have been so significant and impactful over the past centuries that renaming or conceptualising such a "geological timescale is justified" (Crutzen/Brauch 2016, p. vii). In a similar vein, the amplitude of humanity's marks on the environment was observed by Steffen et al. (2011, p. 842) who remark: "The human imprint on the global environment has now become so large and active that it rivals some of the great forces of Nature in its impact on the functioning of the Earth system".

The purpose of this chapter is to explore the role that faith-based organisations (FBOs) can play as actors in the fight against further deterioration of atmospheric conditions and the disintegration of the ES; it uses Pope Francis' Encyclical Letter *Laudato Si'* as a multidisciplinary exploratory case study that consists of (1) investigating the potential of FBOs to lead their members to actions susceptible of rescuing the earth, starting from individual change of lifestyle and attitudes towards the use of the resources that the Earth puts at our disposal; (2) adopting literary research mode of inquiry that uses documentary analysis enriched by a multidisciplinary analytical approach that encompasses theological reflections and historical analysis within the interpretive paradigm.

Against the backdrop of this introductory statement, this chapter consists of eight parts and is structured as follows: following the introduction (1), which states the problem and lays down the perspectives of the argument put forward by Pope Francis' invitation to a global transformational dialogue to save our environment. The subsequent sections consist of (2) some key concepts; (3) an exploration of FBOs, particularly Catholicism, from which vantage point *Laudato Si'* is discussed and its impact on environmental peace assessed; followed (4) the analysis of core arguments based on advocacy and the potential of Christians and world religions to drive transformative activities that are susceptible of responding to the ecological challenges of our time; and (5) conclusions.

To better understand the connections between *Laudato Si'*, civil society, peacebuilding, environmental justice, and the Anthropocene, it is important to succinctly define these concepts.

4.2 Key Concepts

4.2.1 Civil Society

The concept of "civil society" is a multifaceted and complex one which has gone through several semantic developments since its inception in Ancient Greek ethics and philosophy. The origin of the concept of "civil society" dates to the 4th century BC, precisely to the Aristotelian notion of *koinônia politikè* (literally translated as "political community"), and in its many Latin translations (*societas civilis, communitas civililis, communicatio, communio et coetus*), which we owe to Cicero, and with the help of which the ancient world defined the political unity of the City (*polis, civitas*) (Laudani 2012). In Plato's *De Republica* ("The Republic") (Book I, XXXII), in the broader sense, which is a philosophical interpretation, "the civic society is a mode of organising and governing public life where the individual citizen is the core value, the main functioning subject and final end with his/her interests, requirements and rights" (Hovhannes 2016, p. 98). But the narrower meaning of civil society foregrounds the role of NGOs in the activities of "organising and governing public life" as Hovhannes explains: "the third sector is meant by civil society; that is non-governmental, non-profit organizations, foundations, associations of legal persons, artistic unions, charity organizations, civil movements and other forms of activisms" (p. 98).

According to Paffenholz/Spurk (2006), the notion of civil society refers to

> the sector of voluntary action within institutional forms that are distinct from those of the state, family and market... It comprises non-state actors and associations that are not purely driven by private or economic interests, are autonomously organized, and interact in the public sphere; and... it is independent from the state, but it is oriented toward and interacts closely with the state and the political sphere (pp. 2–3).

Civil society represents both a sector of action in interaction with family, business and state; and an intermediate sphere between family, state, and business (Paffenholz/Spurk 2006, p. 3). It is more effective when it maintains its independence. FBOs such as churches and Islamic associations are important components of civil society (Paffenholz 2015). In its contemporary understanding, civil society has become an important factor in peacebuilding (Belloni 2001; Last 2010; Paffenholz 2015). The next section briefly discusses peacebuilding theory in the context of interpersonal relationships and the interaction between humanity and nature.

4.2.2 Peacebuilding

Prominent peacebuilding and restorative justice scholars envision peace beyond the interaction of humans and aver that peacebuilding has a spiritual meaning. Noll (2003) notes that peace-making stems from the ancient Hebrew word Shalom, and this concept embraces a broader meaning that defines human interrelations whose essence and purpose is peace. As far as restorative justice is concerned, Zehr (2002) links this too to Shalom which means "all-rightness" (p. 19), not only with each other, but with the creator and with the environment. Zehr explains this interconnectedness from the concept of Shalom which represents "the vision of 'all-rightness' with each, with the creator, and with the environment" (2002, p. 17). The spiritual dimension in religious peacebuilding can create a sense of engagement and a commitment both to peace and to transforming a relationship of a mission.

World religious groups such as Catholics, the Quakers, the Baha'i Faith, and Protestant Christians have been committed to peace. The contribution of Pope Francis in civil society and peacebuilding goes beyond social, political, and theological discourse. It embraces a critical issue of our time, the anthropogenic perturbations of the ES in the Anthropocene. Peacebuilding in this context intersects with the conceptualisation of "shalom" according to Zehr (2002, p. 17). It entails creating harmonious living conditions between humans and their environment and among humans themselves. Shalom can be further understood as "all-rightness" with the whole creation: that is, a matter of justice and fairness with the whole of creation. Unfortunately, as the next section discusses it, the relation between humanity and creation has been distorted in the Anthropocene by human actions. That calls for environmental justice.

4.2.3 The Anthropocene

There exist a wide range of studies inspired by scientific innovations to explore and define the Anthropocene Geological Epoch (Crutzen 2016; Crutzen/Brauch 2016; Steffen et al. 2011; Zalasiewicz et al. 2008); simultaneously, there emerges a global search to examine the impact of the Anthropocene on global peace and security (Brauch/Scheffran 2012; Dalby 2007). Some pathways are looked into to mitigate the overall consequences of the Anthropocene (such as climate change and environmental conflicts) and the extent to which this affects global peace (Bronkhorst 2011; Hardt/Scheffran 2019; Willms/Werner 2009).

The Anthropocene is recognised as "a new epoch in Earth history, arguing that the advent of the Industrial Revolution around 1800 provides a logical start date for the new epoch" (Steffen et al. 2011, p. 842). The concept of the Anthropocene was developed by Paul Crutzen, the Nobel Prize–winning chemist, in 2002 to suggest that Earth has left the Holocene and has entered this "new Epoch—'the Anthropocene'—because of the global environmental effects of increased human population and economic development" (Zalasiewicz et al. 2008, p. 4). The term entails "the contemporary global environment dominated by human activity" (Zalasiewicz et al. 2008, p. 4).

Characterising the start of this New Epoch, Crutzen (2016, p. 211) explains: "The Anthropocene could be said to have started in the late eighteenth century, when analyses of air trapped in polar ice showed the beginning of growing global concentrations of carbon dioxide and methane". To give a closer estimate of this new epoch in Earth history, Steffen et al. (2011) argue that "the advent of the Industrial Revolution around 1800 provides a logical start date for the new epoch" (p. 842). They contend: "Climate change has brought into sharp focus the capability of contemporary human civilization to influence the environment at the scale of the Earth as a single, evolving planetary system" (Steffen et al. 2011, p. 842). The manifestation of climate change in the opening of the ozone hole above Antarctica and its Anthropocene-related causes raised concerns that human activities can have adverse effects on the ecosystem and could subsequently lead to a "biosphere crisis" (Steffen et al. 2011, p. 843).

It is plausible to deduce from these ecological perspectives the relevance of Pope Francis' invitation to enter into a global dialogue through which humanity could start healing the wounds sustained by the Earth and try to reverse the dangerous path taken by technological advancement in many sectors of nations' development (arms/weaponry, food, gas industries, clothing, footwear, electronic devices, pharmaceutical, chemical, aeronautic and automobile industries). In referring to his invitation in *Laudato Si'* as a dialogue, the conversational approach Pope Francis adopts demands an examination of human agency and stewardship over the resources entrusted to them by the Creator. That begs the question: What to do? The answer to this question is partially provided by exploring the role that civil society can play amidst this ecological crisis, particularly FBOs.

4.3 Faith-Based Organisations and Peacebuilding

The concept of FBOs refers to beliefs across various world religions that espouse the notion of peace as a vital need of human community and see peace as a transcendental reality. Institutions that approach peace and conflict resolution from the perspective of their cultural and religious beliefs are the Catholic tradition, the Islamic Tradition, the Baha'i´ Faith, the Mennonite Tradition, the St Egidio Community, the Interreligious Peacebuilding Project, and the Quaker Christian religious movement. Bercovitch/Kadayifci-Orellana (2009) define faith-based actors as "organizations, institutions and individuals who are motivated and inspired by their spiritual and religious traditions, principles, and values to undertake peace work" (p. 185).

The role of FBOs has not been limited to resolving conflict; they are great contributors to social service, charitable activities, and in providing community aid, humanitarian assistance and relief services (Bercovitch/Kadayifci-Orellana 2009, p. 177). The commitment of these organisations to peacebuilding has a late start in research; moreover, the role played by religion in peacemaking has not received ample attention, as Bercovitch/Kadayifci-Orellana (2009) note:

> One reason for this belated interest in religion is that secular, rational problem-solving approaches and methodological, epistemological perspectives developed by conflict resolution scholars viewed religion either as an instigator of conflict or ignored it altogether because religious issues involved in conflicts cannot be addressed from an empirical or positivist perspective (p. 177).

It is becoming clear that "the role of religion and religious networks" in peacebuilding is gaining scholarly theoretical, spiritual and pragmatic interest in public life and is becoming a field of interest in contemporary social sciences as a result of strong religious beliefs in non-Western society, mainly in Islam and Christianity (Whetho/Uzodike 2008). The late surge of interest in the study of what religion and faith-based networks could contribute to peacebuilding has been triggered by what Bercovitch/Kadayifci-Orellana (2009) describe as the inevitable need to conduct research and analysis on the relations between religions against the backdrop of increasing ethno-religious conflicts following the end of the Cold War.

Abu-Nimer (2001) explained that religion plays an important role in the causes of conflict, its dynamic and its settlement. Religious extremism, fundamentalism, violence, and terrorism are the main sources and consequences of conflict which occurs in the name of religious beliefs. However, the fact that there have been global efforts towards conflict resolution and peacebuilding by FBOs is overlooked (Abu-Nimer 2001). World religions can influence peace processes, and among the religions that show the potential to settle conflict and attain peace are Hinduism, Sikhism, Judaism and Islam (Abu-Nimer 2001, p. 686). Furthermore, Abu-Nimer asserts: "Religion can also bring social, moral, and spiritual resources to the peacebuilding Process" (Abu-Nimer 2001, p. 686). Bercovitch and Kadayifci-Orellana note: "Faith-based actors are increasingly becoming involved in ethno-religious and other conflicts as mediators, and not without success" (Bercovitch/Kadayifci-Orellana 2009, p. 176). There have been other faith-based initiatives for peacebuilding such as the Quaker Peacebuilding Projects in Northern Ireland between 1969 and 2007 (Kennedy 2015).

Leaders from different religions and faith practices have shown concerns about the disintegration of the ES and invited believers to conversion to save the natural environmental. Pope Francis (2015) contends: "the ecological crisis is also a summons to profound interior conversion"[1] (p. 158). All Christians and other believers are called to enact these new attitudes that lead to "ecological conversion" (p. 156). In addition, Pope Francis refers to Patriarch Bartholomew, who has spoken of the obligation for each person to repent of the ways that each of us has harmed the planet through generating small-scale ecological damage; and who urges us to take responsibility for our personal contribution, at differing degrees, in disfiguring and destroying creation.[2] The message from religious leaders could have significant impacts on believers in dealing with climate change. On this note, Hoffman (2015, p. 2) contends: "When people hear the message to address climate change and protect the environment from the church, mosque, synagogue or temple, it will have far more power to motivate action than a regulatory or economic message ever will".

Pope Francis taps into the long tradition of the Catholic Church in which is found the theology of Earth developed by the Apostle Paul of Tarsus, and into the social teaching of the Catholic Church reflected in his predecessors' apostolic exhortations to maintain peace on earth and peace *with* earth. The looming danger of anthropogenic impacts on ES could be more devastating than any warfare, because of the planetary scale of the destruction resulting from global warming, rises in sea level, wildfires and so forth.

In the context of international networking for climate, CSOs are partners with the Transnational Climate Governance (TCG) which pursues the plan of "climate change mitigation", as Pattberg and Stripple (2008) point out:

---

[1] *Laudati Si'* # 217.

[2] Patriarch Bartholomew's Message for the Day of Prayer for the Protection of Creation (1 September 2012) cited by Pope Francis (2015), p. 8. This Ecumenical Patriarch is recognized for his critical role in environmental advocacy; he also proclaims the significance of spiritual values in efforts to determine environmental ethics and actions.

> Public–private partnerships, that is networks of different societal actors, including governments, international agencies, corporations, research institutions and civil society organizations, have become a cornerstone of the current global environmental order, both in discursive and material terms (p. 380).

It becomes evident that FBOs stand as a strong partner of TCG both in terms of population and in loyalty to leaders and the organisation; thus, they are not left out whenever the health of Creation is under threat. For instance, the Pew Research Center projects a substantive increase of world religions' population from 6,895,850,00 in 2010 to 9,307,190,000 in 2050 (Pew Research Center 2021). These estimates represent the membership of eight main religions, namely: Christians, Muslims, Unaffiliated, Hindus, Buddhists, folk religions, other religions, and Jews. Should all animist and traditional religions be incorporated into these statistics, the likelihood of surpassing the numbers shown here increases.

The prospects of FBOs leading a successful pro-ecosystem campaign and influencing human behaviour change—from attitudes and actions that are harmful to nature, to embracing much more caring habits for the environment—are sustainable by two facts:

- The assent of believers to the directives and messages of their leaders
- The world religions' population is projected approximate 9 307 190 000 in 2050; that offers the potential for a significant sway away from wrongful habits and lifestyles (consumerism, arms races, unbridled economic gains, the drive for human comfort such as the excessive use of private means of transportation, etc.).

FBOs make a large part of actors in terms of environmental peace, and they constitute an important component of civil society organisations; they are considered as actors in the *global climate governance* alongside "governments, civil society, science, business, and public non-state actors such as cities, and their interlinked political activities in this field" (Pattberg/Stripple 2008, p. 368). The section that follows explores the extent to which FBOs are involved in environmental peacemaking.

4.3.1 Faith-Based Organisations and Environmental Peacemaking

Religions and myths have been part of the most consistent explanations of the origins and destiny of human and non-human existence. Gottlieb (2003) underlined the problem that humanity faces amidst ecological crises and attributed it to humanity having devastated the natural world. He wrote:

> Threats to the environment are so often threats to our own lives and the people—or parts of nature—that we love, threats about which we can often do virtually nothing. And the hazards involved are so enormous, so potentially irreversible, that it may seem easier to hide from the information than to take it in (Gottlieb 2003, p. 2).

To achieve a reform aimed at the protection of ecosystems, religion and religious movements have been instrumental in advocacy work and the conscientisation of believers about their duty to care for the environment. The question is then, what is meant by religion?

Borrowing from Gottlieb (2003), "religion [entails] those systems of belief, ritual, institutional life, spiritual aspiration, and ethical orientation which are premised on an understanding of human beings as other or more than simply their purely social or physical identities" (p. 7). The voice of the mainline religions regarding environmental peace is loud in the works of Gottlieb (2003), Eaton (2007), and others. For instance, Eaton (2007) acknowledged the interconnectedness of religion, ecology, and politics and asserted that the three are "theoretically illuminated through liberation, political, contextual and critical theologies" (p. 23). The notion of liberation linked to creation is also found in Christianity, expressed in eschatological categories in Pauline environmental theology:

> [20] For the creation was subjected to frustration, not by its own choice, but by the will of the one who subjected it, in hope [21] that the creation itself will be liberated from its bondage to decay and brought into the freedom and glory of the children of God. [22] We know that the whole creation has been groaning as in the pains of childbirth right up to the present time. [23] Not only so, but we ourselves, who have the firstfruits of the Spirit, groan inwardly as we wait eagerly for our adoption to sonship, the redemption of our bodies (Rom. 8:20–23).

The prophetic message of the Apostle Paul foretold the environmental crises that have been experienced as a result of humanity's failure to acknowledge the origins and destiny of creation—its sacredness, its actual disintegration, and the extent to which the survival of humanity and non-human creation is threatened. The freedom evoked by Paul mainly concerns the conversion of the heart, embracing responsible stewardship, and parting ways with sinful habits characterised by consumerism, the race for the accumulation of wealth at the expense of the earth's well-being, greed, and the lack of solidarity between humanity and non-human creation. A more meaningful and contextual interpretation of Romans 8:20–23 is expounded by Pope Francis (2015) when he stated:

> The violence present in our hearts, wounded by sin, is also reflected in the symptoms of sickness evident in the soil, in the water, in the air and in all forms of life. This is why the earth herself, burdened and laid waste, is among the most abandoned and maltreated of our poor; she "groans in travail" (Rom. 8:22).[3]

It is only ethical for humanity to show gratitude to its environment, which provides for its subsistence: "our very bodies are made up of her elements, we breathe her air and we receive life and refreshment from her waters".[4] The same message is echoed by the ecological patriarch Bartholomew, who is cited here by Pope Francis:

> [Patriarch Bartholomew] has repeatedly stated this firmly and persuasively, challenging us to acknowledge our sins against creation: "For human beings… to destroy the biological diversity of God's creation; for human beings to degrade the integrity of the earth by causing

---

[3] *Laudato Si'* # 2

[4] Ibid.

> changes in its climate, by stripping the earth of its natural forests or destroying its wetlands; for human beings to contaminate the earth's waters, its land, its air, and its life – these are sins".[5]

Both religious leaders, Pope Francis and the Apostle Bartholomew, sent the same message of conversion from sinful habits that show neither care nor respect for the ecosystem. The dialogue in which Pope Francis invites humanity to engage concerns a change of heart:

> I urgently appeal, then, for a new dialogue about how we are shaping the future of our planet. We need a conversation which includes everyone, since the environmental challenge we are undergoing, and its human roots, concern and affect us all. The worldwide ecological movement has already made considerable progress and led to the establishment of numerous organizations committed to raising awareness of these challenges.[6]

This note on faith-based organisations (FBOs) and the dialogical message of conversion put forward by world religious leaders prompted us to look at one aspect of Catholicism as unravelled in *Laudato Si'*. It is, thus, important to clarify the concepts of civil society at the outset. In doing so, we bore in mind that *Laudato Si'* was also developed within civil society, particularly the Roman Catholic Church, which is led by Pope Francis, the *Laudato Si'* Encyclical Letter's author.

Several scholars have described the Earth as a sacred element of creation (Eaton 2007; Gottlieb 2003). Eaton (2007) sees the interconnectedness of religion and politics as a driving force in that world religions generally link spirituality (faith/beliefs), ethics and creation. In this interaction, ethics and spirituality connect human creation and non-human creation and guide human behavior and attitudes vis-à-vis their natural environment (flora and fauna). For instance, Renugadevi (2012) asserts:

> The messages of environmental conservation contained in the Vedic and Puranic literatures, in the Upanishads, Arthashastra, Charak Samhita, Ramayana and Mahabharata are all based on Hindu religious philosophy. "Non-violence" that is, non-injury to both the living as well the non-living creations of nature such as plants, animals, air, water, land (earth), hill and forest is the core of Hindu religious philosophy which extended up to Jainism and Buddhism (p. 1).

In a similar vein, care for the environmental has animated the reforms undertaken by Christian faith-based efforts in Appalachia to "advance a framework for policy change based on the view that the roots of the contemporary environmental crisis are moral and spiritual in nature" (Moseley/Feldman 2003, p. 227). The said reforms touch on "a transformation of personal values, attitudes, and conduct in support of an environmental ethic of care" (Moseley/Feldman 2003, p. 227). The authors underline the importance of education in the pursuit of Appalachian initiatives and strategies to enhance personal and eventually societal transformation (Moseley/Feldman 2003).

The earth is the locus of life, interrelations, existential and ontological connections between human creation and non-human creation. Human survival depends on the

[5] The address of Patriarch Bartholomew in Santa Barbara, California (8 November 1997); cf. John Chryssavgis, On Earth as in Heaven: Ecological Vision and Initiatives of Ecumenical Patriarch Bartholomew, Bronx, New York, 2012.

[6] *Laudato Si'* # 14.

conditions that nature provides water, food, climate, soil, air, and other resources that are indispensable for their subsistence. In animist and African traditional religions too, nature and earth are protected through rituals performed in this regard and call for the conservation of the environment. For example, Daneel (2006) wrote that two different religious movements, namely the Association of Zimbabwean Traditionalist Ecologists and the Association of African Earthkeeping Churches came together to fight towards the liberation of creation, especially for the rehabilitation of the degrading environment and overcrowded lands in Zimbabwe. This strategy, known the *chimurenga*, [struggle to free creation] was sponsored by the Zimbabwean Institute of Religious Research and Ecological Conservation, and the campaign was led under the banner *War of the Trees* (Daneel 2006).

As far as Islam is concerned, there are varied opinions about Muslims' practice of Islam environmental ethics (Saniotis 2012). However, Deen (1996) utilised legal and ethical categories to explain the guiding principle of Islamic environmental tenets:

> In Islam, the conservation of the environment is based on the principle that all the individual components of the environment were created by God, and that all living things were created with different functions, functions carefully measured and meticulously balanced by the Almighty Creator (p. 142)

While some Muslims show no concern for environment, those in the West and in Muslim majority countries keep ethical environmental principles inspired by the religion (Saniotis 2012).

The analysis of faith-based organisations' capacity to build environmental peace in the Anthropocene, as conceived in this chapter, is limited to the socio-theological level and qualitative reflections on the existing literature in this field. It does not cover quantitative data review and analysis of *sophisticated information-compression techniques including simulation modelling* that are part of studies conducted in Earth sciences (Schellnhuber 1999), nor does it come up with new scientific evidence emanating from geochemistry of the atmosphere, atmospheric chemistry or geological studies to sustain the thesis according to which the scale of human activities has significantly interfered with the natural processes of the ES; and humans have caused "ecological disruptions and vulnerabilities" (Dalby 2007).

The Christian tradition considers the Earth as a sacred space (Eaton 2007) that needs care and protection. The success of this commitment depends on a sensitive inter-religious dialogue that highlights and seeks shared values, principles, beliefs, and a common understanding of the purpose and destiny of creation particularly the Earth—with which humanity has ontological and natural bonds that should define and guide humankind's conduct vis-à-vis our planet. Thus, human comportment ought to be ethical and reverential. That is the view shared by world religions and religious movements across the globe. But for this chapter, now we turn to the way environmental peace is conceived, perceived and practiced in Catholicismone of the main Christian constituents.

4.3.2 Catholicism and Peacebuilding

The global agenda of environmental peace pursued by FBOs, a component of transnational civil society, is discussed from the perspectives of Catholicism which makes a case study for this chapter. Considering the importance of non-state institutions in the Transnational Climate Governance (TCG), social science scholars and theologians explain the interconnectedness of the concepts of civil society, faith-based organisations, and peacebuilding. As a component of civil society, FBOs are actively involved in peacebuilding (Bercovitch/Kadayifci-Orellana 2009; Paffenholz 2015; Paffenholz/Spurk 2006). The section that precedes explored the notions of FBOs and their relation to peacebuilding, civil society and *Laudato Si'*. Framing an argument of *Laudato Si'* requires putting this document in its context, in which a FBO, in this case the Roman Catholic Church, through the invitation to dialogue in view of redeeming our "common home", is open to all religions and religious movements.

The agenda of peacebuilding in the Catholic Church was mainly inspired by the gospels of Jesus and Saint Augustine's *The City of God*, which was developed by St Pope John XXIII (born Angelo Giuseppe Roncalli) who envisioned a world where peace prevailed as a heavenly city (Hittinger 2013, p. 38). During the Second World War, Pope John XXIII reflected on the problem of nationalism, and imagined bishops to be "an image of the supranationalism of the Church…Roncalli resolved to read again afresh St. Augustine's *City of God*, and that gave him new insight leading to the inception of the new encyclical letter entitled *Pacem in terris (*Peace on Earth)" (Hittinger 2013, p. 38). His preliminary thoughts were set out in his book *Journey of the Soul*, translated by Dorothy White in 1999 (Hittinger 2013). He wrote:

> Peace is tranquility in the order of things, ordered obedience in fidelity to the eternal law. Order is giving each thing its place. The Peace of mankind is ordered harmony in the home, in the city, in man. Wretched, therefore, is the people that is alienated from God (John XXIII, *Journey of a Soul*, 1999 cited in Hittinger 2013, p. 38).

The concept of "peace as tranquility of order is a paradigm of singular importance for his encyclical, especially its teaching on human rights" (Hittinger 2013, p. 39). The full title of John XXIII's encyclical letter is "*Pacem in terris*: On Establishing Peace in Truth, Justice, Charity, and Liberty". *Laudato Si'* falls under this "both old and new" discourse of peace, as put forward by Rolston (2015):

> This is an encyclical about [the] environment and, equally, a fundamental socio-economic critique. There is remarkable wisdom here, ancient and contemporary, and many of us who have been saying these things for decades can rejoice in a new and powerful voice for saving the Earth (p. 57).

Scholars have found important linkages between peace ecology and *Laudato Si'* to the extent that Pope Francis is called "the ecological Pope" (Rolston 2015), and the impact he has made on the world through his Encyclical Letter *Laudato Si'* is evident (Franchi 2016; Hoffman, 2015); others like Deane-Drummond (2016) portray him as "Priest and Prophet in the Anthropocene". In the sphere of international relations, *Laudato Si'* is relevant on theoretical and pragmatic levels because it has some *planetary implications* (Ferrara 2019). On this note, establishing humanity's responsibility towards our wounded environmental systems is essential to mitigating further ecological damages. Reading the signs of the times and responding to the dialogue, the world engages with Pope Francis to save "Our Common Home" as the Holy Father writes: "The urgent challenge to protect our common home includes a concern to bring the whole human family together to seek a sustainable and integral development, for we know that things can change"[7] (Pope Francis 2015, p. 12). Pope Francis' voice arises from his social and religious identity that remains closely connected to civil society, because FBOs are constituents of civil society.

In the middle of this crisis in the Anthropocene, religion stands as the light to dispel the darkness of sin and bring redress that passes through the acknowledgement of the wrongs committed against nature and the resolve to experience metanoia and adopt a more caring attitude.

In anthropomorphic categories inspired by Francis of Assisi, Pope Francis launched a universal call for dialogue that would transform humankind's attitudes to and relations with the environment, which the Pope has described as "Our Common Home" that shelters and sustains people's existence: "our common home is like a sister with whom we share our life and a beautiful mother who opens her arms to embrace us" (Francis 2015, p. 3). In the Catholic tradition, many saints have related with nature in a friendly and humane manner. For instance, St. Francis of Assisi's influence on Pope Francis is evident in *Laudato Si'* through which the Holy Father laid out his argument to protect and save nature, which has been wounded by human activities. This echoes the canticle of Francis of Assisi: "Praise be to you, my Lord, through our *Sister* or *Mother* Earth, who sustains and governs us, and who produces various fruit with coloured flowers and herbs" (Francis of Assisi 2021 cited in Pope Francis 2015, p. n.p.). In similar terms, St. Anthony of Padua, who is known as one of the most famous disciples of St. Francis of Assisi, in his many sermons also preached to the fish, and the Catholic tradition recognises several documented reports of his popular and miraculous encounter and interaction with the fish. It has been reported:

> At last St Anthony, inspired by God, went down to the sea-shore, where the river runs into the sea, and having placed himself on a bank between the river and the sea, he began to speak to the fishes as if the Lord had sent him to preach to them, and said: "Listen to the word of God, O ye fishes of the sea and of the river, seeing that the faithless heretics refuse to do so." No sooner had he spoken these words than suddenly so great a multitude of fishes, both small and great, approached the bank on which he stood, that never before had so many been seen in the sea or the river. All kept their heads out of the water and seemed to be looking attentively on St Anthony's face; all were ranged in perfect order and most peacefully, the smaller ones in front near the bank, after them came those a little bigger, and last of all, where the water was deeper, the largest (St. Anthony of Padua, n.d., p. n.p.).

The same story is narrated by Staudt (2014) who saw strong bonds between humanity and creation and noted the responsibility that humans have to love, care for, and protect the earth and all that the natural environment houses, both humans and non-human creation.

---

[7] *Laudato Si'* # 13.

Laudato Si' echoes the principles of Catholicism that include the dignity of the human person, peace, social justice, subsidiarity, stewardship of creation, etc. (United States Conference of Catholic Bishops 2004).

4.3.3 Precedents of Laudato Si'

The dialogue Pope Francis invites the world to, namely the rethinking of humanity's attitudes to and handling of "Our Common Home", has been identified by several scholars as an ecological conversation and exhortation that appeals for a change in behavior to safeguard our environment. Direct links are established between *Laudato Si'* and "the Anthropocene" as the former becomes a response to the latter. For instance, Rolston (2015) describes Pope Francis as "an Ecological Pope" who challenges "the Anthropocene Epoch"; he writes:

> I greatly welcome the recent encyclical: *Laudato si' On Care for Our Common Home*. One of the world's great leaders, and a popular one, insists that the human relationship to nature can and ought to involve love and appreciation, gratitude and care (p. 52).

While acknowledging the need for humanity to use the resources that nature provides (water, air, space, flora and fauna, etc.), their usage requires greater ethical stewardship because humanity's survival depends largely on these resources. It is both an individual and a collective moral responsibility towards the ES. Rolston (2015, p. 53) notes that Pope Francis "links underdeveloped environmental ethics with an overdeveloped economy". It is further understood that the fate of Earth in the current ecological crisis cannot be interpreted outside the worlds of geoeconomics and geopolitics which echo the perspectives of Spring and Brauch's (2021) economic and political decisions.

Ferrara (2019) points out that *Laudato Si'* embraces the realm of international relations and world politics which have leverage over climate change, as he observes: "The Encyclical seems to suggest that practicing sustainable international relations means exiting the logic of power or hegemony, while simultaneously operationalising the concept of care" (p. 1). The Pope has earned significant recognition for *Laudato Si'* from scientists as well as theologians. This document is relevant in many fields of environmental studies because it has an interdisciplinary characteristic and bridges theoretical knowledge with pragmatic existential reality.

In the first lines of *Laudato Si'*, Pope Francis evokes the anthropomorphic allegory of the Earth presented as "our sister" and "our mother"; both concepts reveal love, care, and the gift of life, deduced from what the ES offers to humankind. Pope Francis writes: "our common home is like a sister with whom we share our life and a beautiful mother who opens her arms to embrace us" (Pope Francis 2015, p. 3). In the words of St Francis of Assisi (2021), this metaphor is expressed as follows in the "Canticles of the Creatures": "Praised be You, my Lord, through our Sister Mother Earth, who sustains and governs us, and who produces various fruit with colored flowers and herbs". The "groaning" of creation results from mankind's irresponsible use and abuse of the resources that Mother Earth provides, and this is a manifestation of the violence that is present in humans that has made air, water and earth a wasteful space (Pope Francis 2015).

Following John XXIII's Encyclical *Pacem in Terris,* which warned against the nuclear crisis, Pope Paul VI referred to ecological concerns as "a tragic consequence" of "unchecked human activity" resulting from irresponsible exploitation and misuse of nature (Hittinger 2013, p. n.p.). Mankind's activity is destroying the natural environment in the name of freedom to subdue non-human creation. Alluding to Pope Paul VI's Apostolic Letter *Octogesima Adveniens,* Pope Francis (2015) remarks: "scientific advances, the most amazing technical abilities, the most astonishing economic growth, unless they are accompanied by authentic social and moral progress, will definitively turn against man".[8] Pope Paul VI warned of the need for radical change in humans' attitude to and stewardship of nature in the context of extraordinary technological advancement (Pope Paul VI's Address to FAO on the 25th Anniversary of its Institution, 16 November 1970, cited in Pope Francis 2015). While humanity admires technological innovations in all sectors of life, unrestrained progress without global consciousness of the fate of the environment leaves the ecosystem vulnerable. Pope Paul VI warned the world of this as he expressed his concerns, stating that there was "the urgent need for a radical change in the conduct of humanity" because of the "ecological catastrophe under the effective explosion of industrial civilization".[9]

The increasing impact of human activities on the environment was further predicted by Saint John Paul II, then the Roman Pontiff, in his Encyclical Letter *Redemptor Hominis.* He expressed concerns regarding humans' relation to their immediate environment which was limited to finding its meaning in instantaneous use and consumption. Thus, the Holy Father called for a "global ecological conversion" (John Paul II, *Redemptor Hominis*, 15: AAS 71 [1979], 287 cited in Pope Francis 2015, p. 5). Pope Francis avers: "Authentic human development has a moral character. It presumes full respect for the human person, but it must also be concerned for the world around us"[10] (p. 6). Referring to St John Paul II's Encyclical Letter *Sollicitudo Rei Socialis*, Pope France cites: "take into account the nature of each being and of its mutual connection in an ordered system".[11]

The imperative moral responsibility of humanity to protect the ES stems from the fact that humanity itself is so bound to its ecology, without which there is no prospect of humanity's survival. It comes down to humans making a radical change in their perception of nature and how it can be treated, because this care reflects care for humans themselves. Pope John Paul II puts it in the following terms: "lifestyles, models of production and consumption, and the established structures of power which

[8] Pope Francis (2015) alludes to Pope Paul VI's Apostolic Letter *Octogesima Adveniens* (14 May 1971) in his Address to the Food and Agriculture Organization of the United Nations (FAO) on the 25th Anniversary of its Institution (16 November 1970), and AAS 62 (1970), p. 833.

[9] Ibid.

[10] *Laudado Si'* # 5.

[11] John Paul II, Encyclical Letter *Sollicitudo Rei Socialis* (30 December 1987), 34: AAS 80 (1988), 559.

today govern societies" (John Paul, 1979's Encyclical Letter *Centesimus Annus* [1 May 1991], 38: AAS 83 [1991], 841 cited in Pope Francis 2015, pp. 5–6).

Also concerned with humans' responsibility for the care of environment was Pope Benedict XVI, Pope Francis' predecessor, who addressed diplomats accredited to the Holy See, inviting them and their respective governments to eliminate the structural causes of a dysfunctional world economy and amend the models of growth which were ineffective to ensure respect for the environment (Pope Benedict XVI's address to the Diplomatic Corps accredited to the Holy See [8 January 2007]: AAS 99 [2007], 73, cited in Pope Francis 2015). Pope Francis' invitation to ecological dialogue opens up a global advocacy for radical change in the way humanity consumes the products of unbridled industrialisation which wound the ES. The relevance of *Laudato Si'* for ethics in the Anthropocene is evident because of the linkages between conflict and the environment which prompt a reflection on the role that civil society, through the reflective writing of Pope Francis, can play in awakening humanity's conscience and awareness of the conflict that humankind has engaged with in its own house—nature.

The stance taken by world religions (Christianity, Islam, Judaism, Buddhism, Hinduism, etc.), and religious movements such as Quakers, reflects the commitment of these collective actors to lead the transnational campaign to protect and ease the anthropogenic burden on atmospheric conditions. It is important to close this section with the invitation of Pope Francis: "Let ours be a time remembered for the awakening of a new reverence for life, the firm resolve to achieve sustainability, the quickening of the struggle for justice and peace, and the joyful celebration of life" (Earth Charter, The Hague, 29 June 2000, cited in Pope Francis 2015). This appeal echoes the role of stewardship that is discussed next.

4.3.4 Conscientious Stewardship

The concept of stewardship is utilised in its theological or biblical sense. Eaton (2007) defines it thus:

> Stewardship is a both a biblical motif as well as an easily acceptable ecological paradigm for many Christians. Here the ecological crisis is understood mostly in its physical manifestations: pollution, species extinction, global warming, biodiversity losses, severe water and soil declines and changing weather patterns. It is predominantly a problem of resources. Good stewardship and resource management are appropriate responses (p. 24).

Its theological meaning suggests care for God's creation; it is a process through which humanity supports God as co-creator, and this prompts a joint effort to care for creation (Eaton 2007). Furthermore, "stewardship is a light green paradigm, meaning that it maintains anthropocentrism, and no intrinsic or sacred value is attached to the natural world" (Eaton 2007, p. 25). From the perspective of University of Notre Dame Center for Social Concerns (2015) the concept of stewardship is associated with inter-creatures' solidarity as expressed here:

> There is a solidarity among all creatures arising from the fact that all have the same Creator and are ordered to glory and worship of the Creator. Humanity's dominion over inanimate and other living beings is not absolute; it is limited by concern for the quality of life of his or her neighbor, including generations to come. As such, the steward is a manager, not an owner. Accordingly, use of the mineral, vegetable and animal resources of the universe cannot be divorced from our moral responsibilities. Our stewardship of the earth enables us to be participants in God's act of creating and sustaining the world (p. n.p.).

Pope Francis remarks, "the world cannot be analyzed by isolating only one of its aspects, since 'the book of nature is one and indivisible' and includes the environment, life, sexuality, the family, social relations, and so forth" (Pope Francis 2015, p. 6).

A conscientious stewardship is imperative to save the social and natural environment that sustains humanity. This reflects the comments made by Staudt (2014)in his sermons on the feast day of St Anthony of Padua. He preached:

> […] we need to recognize our link to God's physical creation through our bodies. That connection can be very enriching, as we appreciate the beauty of that creation and shape it with our hands. But the connection is not just physical. We are meant to be the stewards of the world and to free it from bondage (2014, p. n.p.).

He continued, saying that the fish showed reverence to the recognition they received from Anthony of Padua, which led the people of Rimini to conversion, and that pleased Anthony who responded: "Blessed be the eternal God; for the fishes of the sea honor him more than men without faith, and animals without reason listen to his word with greater attention than sinful heretics" (St Anthony of Padua, cited in Staudt 2014, p. n.p.).

That great sign experienced by Anthony and the people Rimini can inspire our world today to love creation and be converted from our wrongful deeds that continue to harm God's creation, in which humanity and all non-human beings live, because humanity and the rest of creation, our natural environment, all that it houses, are one and intrinsically and ontologically interconnected.

Thus, "Pope Benedict asked us to recognize that the natural environment has been gravely damaged by our irresponsible behavior. The social environment has also suffered damage" (Pope Francis 2015, pp. 6–7). There is no reason to believe in limitless human freedom as the atmospheric conditions continue to deteriorate. It is all-important to seek the truth that can shepherd humanity and accompany it to the path of peoples' progress and prosperity. Pope Francis (2015, p. 6) observes with concern "the notion that there are no indisputable truths to guide our lives, and hence human freedom is limitless"; thus, the world remains in the darkness of an unbridled search for development that will engulf all that exists. Creation is destroyed by human ego, an evil that Pope Benedict XVI explains as follows: "The misuse of creation begins when we no longer recognize any higher instance than ourselves, when we see nothing else but ourselves".[12]

The 'Green Patriarch' Bartholomew identified the root causes of the deterioration of ES in humanity, not solely in technology: "[God] asks us to replace consumption

---

[12] Pope Benedict XVI's Address to the Bundestag, Berlin (22 September 2011): AAS 103 (2011), 664, cited in Pope Francis, 2015, p. 7.

with sacrifice, greed with generosity, wastefulness with a spirit of sharing, and with ascetism".[13] Conversion entails a harmonious living with others and the environment. A model to be followed is St Francis of Assisi, who maintained the existence of a natural bond between his humanness and other beings such as birds, flowers, and creation at large (Pope Francis 2015).

There is hope that humanity's response, inspired by FBOs' engagement in environmental issues, can lessen the negative anthropogenically-generated atmospheric perturbations. The likelihood of this happening comes from the fact that Christians and other world religions, as members of FBOs, are also concerned about caring for the environment (Eaton 2007; Gottlieb 2003; Moseley/Feldman 2003; Moyer 2013). Furthermore, Scriptures reveal how the apostle Paul of Tarsus was versed in environmental theology[14] whereby he appeals to Christians to engage in alleviating the plight of the non-human part of creation, as Bauckham (2011) observes: "While [Roman 8:22] may not exactly mandate environmental activity by contemporary Christians, it may be able to give us much needed orientation within the multiple ecological crises that we unavoidably face today" (p. 91). St Paul's prophetic alert on the plight of creation compels believers to be actively involved in a transformative mission to save the whole creation. The risk posed by human activity to Earth's systems does not only endanger non-human creation, but the very existence and survival of humanity depends on resources provided by nature. To this effect, Eaton (2007) suggests that *ecojustice* is needed to tackle the questions of "equitable access to and distribution of the Earth's resources" (p. 25).

Stewardship in the Anthropocene calls for *ecojustice* and ecological conversion, which entails the rehabilitation and restoration of the ecosystem that has been wounded by humans' uncaring activities. That requires a conscientious stewardship of the whole creation which is "groaning" as the Scriptures remind us: "We know that the whole creation has been groaning as in the pains of childbirth right up to the present time" (Romans 8:22). St Paul the Apostle does not exclude humanity from the groaning of creation, because humans are intrinsically and indivisibly part of creation. This scriptural passage is followed by a message of hope that is well explained by Hahne (2010) in theological terms; he asserts: "the present corruption of nature is a result of the Fall of Adam and Eve, and the eschatological redemption of nature will free it from corruption and transform it to freedom and glory" (p. 20).

4.3.5 Ecological Conversion for Pollution and Failed Stewardship

Patriarch Bartholomew grieved over the destruction of our natural environment; he expresses thus: "God made the world for human beings, he said, but gave them

[13] Patriarch Bartholomew's lecture at the Monastery of Utstein, Norway (23 June 2003) cited in Pope Francis, 2015, p. 9.

[14] Letter of St Paul to the Romains (Rm 8:22).

responsibility to care for it and make it fruitful, which means using it carefully and not exploiting it 'out of a sense of greed'" (Catholic News Service 2012, p. n.p.). In addition, the Patriarch invites humanity to repent for pollution and failure of stewardship over creation. His message was recorded by the National Catholic Reporter in the following terms:

> We are ultimately imploring God to change the mindset of the powerful in the world, enlightening them not to destroy the planet's ecosystem for reasons of financial profit" and selfishness […] At the same time, each individual must ask God's forgiveness because "we all generate small ecological damage in our individual capacity and ignorance. Therefore, in praying for the natural environment, we are praying for personal repentance for our contribution -- smaller or greater -- to the disfigurement and destruction of creation [.]" (Patriarch Bartholomew cited in Catholic News Service 2012, p. n.p.).

The need for repentance, in a common understanding, demands recognition of harm done to another person or group of persons, making reparation, and making the commitment to refrain from re-offending. Today such repentance is needed along reconciliation between humans and non-human creation because humanity in its entirety is part of creation and connected to the Earth. The demand of restoring harmony in the whole creation is justified by anthropomorphic recognition of the Earth as "mother" or "sister" by Francis of Assisi and Pope Francis, and humans as dwellers of that natural home, our planet. For this reason, Pope Francis' reflection underlines what follows: "the ecological crisis is also a summons to profound interior conversion" (Pope Francis 2015, p. 158). All Christians and other believers are called to enact these new attitudes that lead to "ecological conversion" (p. 156). Pope Francis message of change of heart resonates with that of Ecumenical Patriarch Bartholomew reported by the Catholic News Service (2012).

4.4 Analysis and Argument

4.4.1 Pope Francis' Path to Rebuild Earth System Wellness

The basic argument put forward on the potential of the agency of world religions that will follow after this section is drawn from Chapter Six of *Laudato Si'*: "Ecological Education and Spirituality". From the outset Pope Francis acknowledges that the scale of damage caused by humans' selfish attitudes and unethical means of economic development and social growth through technological innovations is so huge that the ES' original condition cannot be regained by science and technology. It rather demands both spiritual and human concerted commitment, the former consisting of returning to the truth (here characterising the origin of all natural resources) and the latter referring to humans' material agency (instruments of transformation) and spiritual agency (awareness of humankind's common origin and interconnectedness with nature that prompts human hearts to moderation and ethical use of the resources they share with other beings and non-human creation). Pope Francis calls humankind to *metanoia*—an adoption of normative and ethical conduct that is different from the past and yet better for creation. He asserts:

> Many things have to change course, but it is we human beings above all who need to change. We lack an awareness of our common origin, of our mutual belonging, and of a future to be shared with everyone. This basic awareness would enable the development of new convictions, attitudes and forms of life (Pope Francis 2015, p. 149).

When humans put themselves at the center and imagine the satisfaction of their needs being the ultimate destination of creation, nature is destroyed, deforestation will continue, climate change will remain a threat to the future of human existence, and global warming too will burden the world as other non-human creatures depend solely on their original and natural homes (forests, seas, soil, etc.). Pope Francis speaks against selfishness and humankind's *self-centeredness* and *self-enclosedness* that engender and increase greed, and he argues: "The emptier a person's heart is, the more he or she needs things to buy, own and consume. It becomes almost impossible to accept the limits imposed by reality" (Pope Francis 2015, p. 150).

To *consumerism*, Pope Francis adds "obsession" that is discernible in "a consumerist lifestyle"; moreover, when few people can maintain it, such conduct can push towards "violence and mutual destruction" (Pope Francis 2015, p. 151). As we are confronted with the deterioration of our environment, changes of behavior and lifestyle could help to ease the burden on ES when industrial production declines because of the more responsible consumption of goods.

Change of behavior includes combating consumerism. Pope Francis points out: "Compulsive consumerism is one example of how the techno-economic paradigm affects individuals" (p. 149), and people become enslaved in a frenzy of buying and spending needlessly. There is a need for postmodern humanity to develop a "new self-awareness capable of offering guidance and direction, and this lack of identity is a source of anxiety" (Pope Francis 2015, p. 149).

Universal awareness is what is much needed in our time. But how to achieve a global awareness of the danger we are faced with and ensure future generations enjoy a natural environment when some world leaders remain in denial about the ongoing deterioration of the earth? Pope Francis refers to the principles set forth in the Earth Charter launched at The Hague on 29 June 2000.[15] The Earth Charter puts forward a plan of action that embraces *Respect and Care for the Community of Life, Ecological Integrity, Social and Economic Justice, Democracy, Nonviolence, and Peace* (Earth Charter International 2000). The prospects of saving the Earth lie in a variety of plans that underpin different levels of agency which can address the five key Earth Charter areas mentioned above.

It is worth noting:

[15] The Earth Charter is a declaration of fundamental ethical principles for building a just, sustainable, and peaceful global society in the 21st century. It seeks to inspire in all people a new sense of global interdependence and shared responsibility for the well-being of the whole human family, the greater community of life, and future generations. It is a vision of hope and a call to action. https://static1.squarespace.com/static/5838ae02d1758efcf2a1256d/t/58b340812e69cf75a414acee/1488142466387/Earth+Charter.pdf (accessed 14 June 2021).

> The Encyclical's contribution to public discourse is significant from three different perspectives. First, theologically speaking, Pope Francis' approach helps believers to translate their faith convictions into active environmental concerns.… My preliminary argument is that the Encyclical deals less with environmental issues than with a political philosophy of world politics. More than this, it is a cosmological appraisal of the human condition on the planet. Pope Francis refers explicitly to 'a strategy for real change' that requires 'rethinking processes in their entirety, for it is not enough to include a few superficial ecological considerations while failing to question the logic which underlies present-day culture' (LS, §197) (Ferrara 2019, pp. 2–3).

Pope Francis also addresses individualism as one of the evils embodied in attitudes that are detrimental to our natural environment, and he invites us to the moral imperative of evaluating the effects of every action and decision we take, whether at a personal or collective level, and how this affects the world around us. He ascertains: "If we can overcome individualism, we will truly be able to develop a different lifestyle and bring about significant changes in society" (Pope Francis 2015, p. 153).

4.4.2 World Religions' Impact on Climate Change

The stance taken by Pope Francis aims to embark all the faithful on a consequential dialogue that creates a great hope of change of attitude, behavior, and lifestyle. The increasing number of believers, which is expected to grow by 0.2% to reach 2 411 340 000 in 2050, gives reason to trust that if there is a collective commitment by over 2 billion believers who can heed Pope Francis' message, then the anthropogenic effects of human mishandling of the environment can be lessened.

The dialoguing ecological appeal by Pope Francis requires a global commitment. It would sound vague to discuss a change of conduct and attitudes to save the Earth without presenting concrete examples of such commitment. *Laudato Si'*, offers, as an example of material ways of transformation and metanoia from individualism, obsession, and consumerism, the following: "A person who could afford to spend and consume more but regularly uses less heating and wears warmer clothes, shows the kind of convictions and attitudes which help to protect the environment" (Pope Francis 2015, p. 154).

Figures 4.1 and 4.2 and Tables 4.1 and 4.2 show that the traditional audience of Pope Francis' message and other world religions can also leverage a paradigm shift in the way believers transcend their existential pursuit of a better life that violates ethical use of earth's resources. The message of pastoral care imbued with environmental education could reach over 2.3 billion Christians, which is the largest population of believers (Fig. 4.1), and over 4 billion other believers. The hope is alive that if 6 billion believers can heed the message of their spiritual leaders, things can change for the better.

As concerns grow over the future of planet Earth, the world population grows as well, all these two realities need to be considered in ecological education. The increase in world population (Fig. 4.2) means the need for a concomitant increase

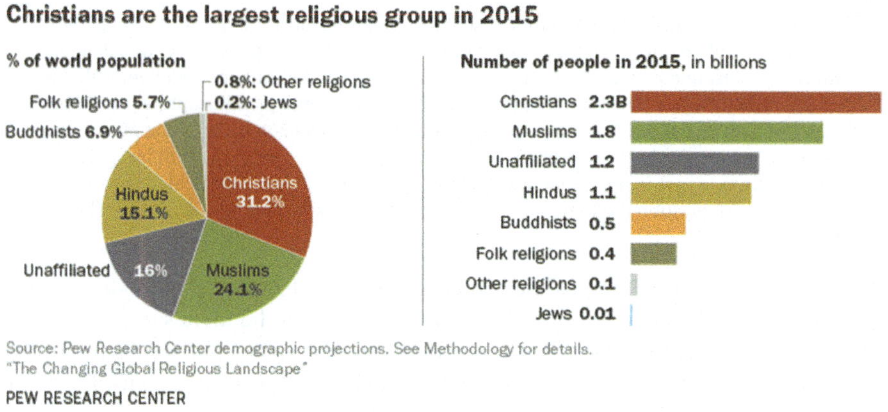

Fig. 4.1 By Conrad Hackett and David McClendon. *Source* Pew Research Center, https://www.pewresearch.org/fact-tank/2017/04/05/christians-remain-worlds-largest-religious-group-but-they-are-declining-in-europe/

in food security and other means of subsistence to satisfy this population. Pope Francis' message becomes even more relevant in the face of these two parameters. Here I allude to *Laudato Si*'s call for more solidarity, responsibility, moral attitudes and good conduct as far as consumerism, individualism, and obsession are concerned.

With the projected growth of major religious groups estimated to reach 9,307,190,000 by 2050 (Table 4.1), proportionate to a world population increase to approximately 9,735,033,990 by 2050 (Table 4.2), ecological education advocated by the main world religions can have a tremendous impact in curbing the trajectory of the degradation of ES.

Faith-based civil society has a tremendous role to play in protecting and saving our natural environment. Amongst many pioneers in this context are Pope Francis (the "ecological Pope") and his predecessors, Patriarch Bartholomew (the "Green Patriarch") and others who are making their appeal heard. There is hope that future generations could live in a green environment if we all commit ourselves to this cause by raising continuous awareness.

4.4.3 Ecological Education

The previous sections laid the foundation for a final analysis and core argument that this chapter will turn to defend now, essentially the impact that Pope Francis' Laudato Si', world religions' voice on global society, and the potential of peace education transform humanity's attitudes and conduct towards the Earth, our *common home.*

Pope Francis makes a twofold suggestion to respond to ecological challenges of our time in order to keep "our common home" for future generations:

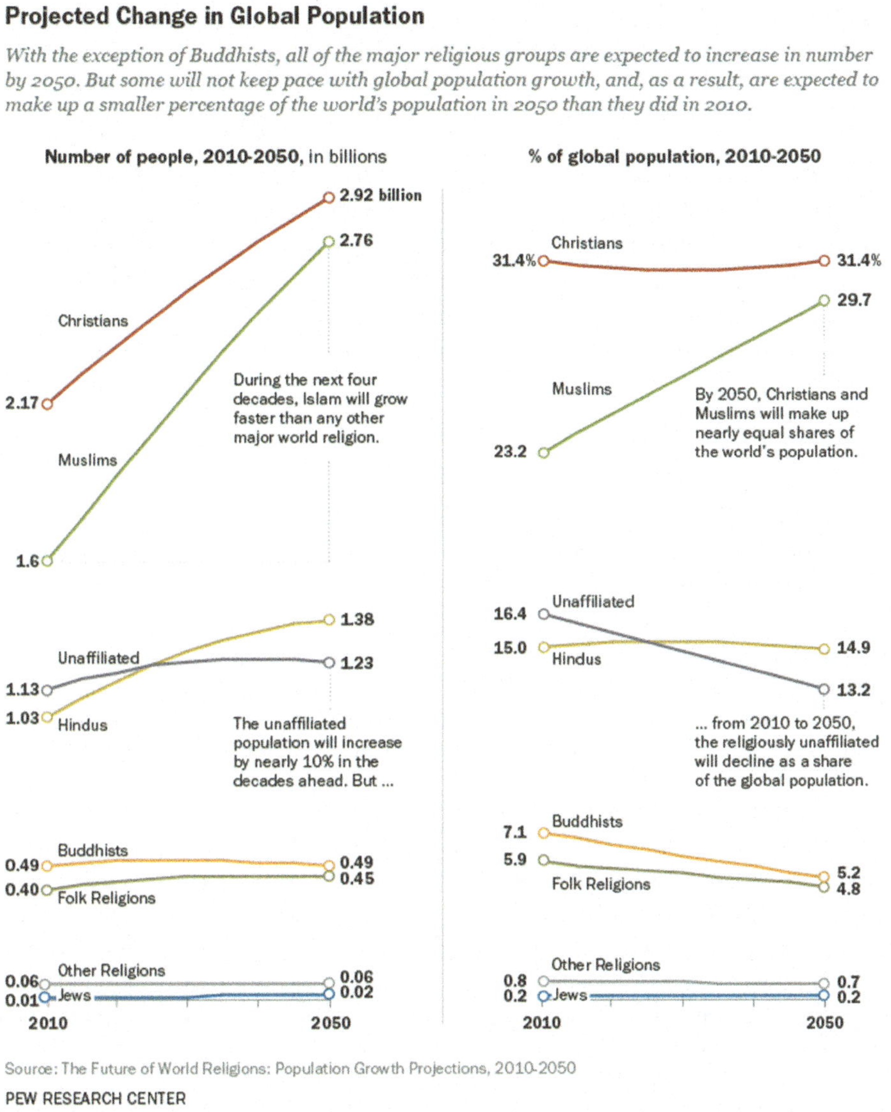

Fig. 4.2 World Religion's Projected Population Growth (Pew Research Center 2021). *Source* https://www.pewforum.org/2015/04/02/religious-projections-2010–2050/

- Change the belief that satisfying individual needs is the basic purpose of life.
- Adopt conscious stewardship through environmental peace education.

The Holy Father remains consistent in the marks he is making to save "our common home". Believers of world religions, including Christians, who represent 31.4% of world religious population, are followers of their leaders to whom they generally

Table 4.1 Size and projected growth of major religious groups. *Source* https://www.pewforum.org/2015/04/02/religious-projections-2010-2050/ (24 Jul 2021)

	2010 population	% of world population in 2010	Projected 2050 population	% of world population in 2050	Population growth 2010-2050
Christians	2,168,330,000	31.4	2,918,070,000	314	749,740,000
Muslims	1,599,700,000	23.2	2,761480,000	29,7	1,161,780,000
Unaffiliated	1,131,150,000	16,4	1,230,340,000	13,2	99,190,000
Hindus	1,032,210,000	15.0	1,384,360,000	14,9	352,140,000
Buddhists	487,760,000	7.1	486,270,000	5,2	−1,490,000
Folk Religions	404,690,000	5.9	449,140,000	4,8	44,450,000
Other Religions	58,150,000	0.8	61450,000	0.7	3,300,000
Jews	13,860,000	0.2	16,090,000	0.2	2,230,000
World total	6,895,850,000	100.0	9,307,190,000	100.0	2,411,340,000

Source The Future of World Religions: Population Growth Projections, 2010–2050,PEW RESEARCH CENTER

Table 4.2 World population forecast (2020–2050). *Source* Worldometer (2019). https://www.worldometers.info/world-population/ (24 Jul 2021)

Year (July 1)	Population	Yearly % Change	Yearly Change	Median Age	Fertility Rate	Density (P/Km^2)	Urban Pop %	Urban Population
2020	7,794,798,739	1.10	83,000,320	31	2.47	52	56.2	4,378,993,944
2025	8,184,437,460	0.98	77,927,744	32	2.54	55	58.3	4,774,646,303
2030	8,548,487,400	0.87	72,809,988	33	2.62	57	60.4	5,167,257,546
2035	8,887,524,213	0.78	67,807,363	34	2.70	60	62.5	5,555,833,477
2040	9,198,847,240	0.69	62,264,605	35	2.77	62	64.6	5,938,249,026
2045	9,481,803,274	0.61	56,591,207	35	2.85	64	66.6	6,312,544,819
2050	9,735,033,990	0.53	50,646,143	36	2.95	65	68.6	6,679,756,162

remain loyal as well as to their religions. The hope that they heed the message of Pope Francis, Patriarch Bartholomew and their successors, and other religious leaders is likely. This implies that the message of environmental peace—peacebuilding with nature and care for the environment—should be continuously proclaimed in churches, mosques, temples, synagogues, pagodas, places of worship, sanctuaries, shrines, etc. The message of peace with the environment should resound beyond these platforms: it should be covered in written and broadcast spaces through which the message of peace permeates to reach all bona fide people who care for earth, "our Common home".

Pope Francis states:

> Education in environmental responsibility can encourage ways of acting which directly and significantly affect the world around us, such as avoiding the use of plastic and paper, reducing water consumption, separating refuse, cooking only what can reasonably be consumed, showing care for other living beings, using public transport or car-pooling, planting trees, turning off unnecessary lights, or any number of other practices. All of these reflect a generous and worthy creativity which brings out the best in human beings. Reusing something instead of immediately discarding it, when done for the right reasons, can be an act of love which expresses our own dignity (Pope Francis 2015, p. 155).

On a similar note, Patriarch Bartholomew underlines the significance of environmental information and education to mitigate the adverse effects of anthropogenic causes of climate change; he observes:

> We need to have constant vigilance, information and education in order to understand clearly the relationship between today's ecological crisis and our human passions of greed, materialism, self-centeredness, and rapacity, which result in and lead to the current crisis that we face. Therefore, the only way out of this impasse is our return to the original beauty of order and economy, of frugality and ascesis, which can guide us toward a more careful management of the natural environment (Patriarch Batholomew 2021, p. n.p.).

The pathways proposed by the Pope Francis and Patriarch Bartholomew to restore wellness to creation incudes adequate flow of information and education on the original beauty of creation; that presupposes:

- embracing selfless stewardship over our planet and opting for moderation instead of the compulsive consumerism the world has become accustomed to;
- living out a global solidarity instead of individualism and obsession;
- adopting a lifestyle that upholds the Earth Charter;
- creating a universal awareness of the looming threat to the natural environment;
- promoting ecological education in various settings such as schools, families, the media, catechesis and elsewhere;
- repentance for pollution and failed stewardship;
- and instilling good education to plant seeds in our youth, which will bear lasting fruit;

Furthermore, the global campaign to protect and save the ecosystem lies beyond the boundaries of religions, beliefs, religious practices, spiritual aspirations, philosophical, metaphysical and ontological explanations of the interaction and bonds that exist between humanity and non-human creation. World religious leaders have been deeply concerned about what kind of environment we shall leave to future generations. That echoes the Message of His All Holiness, Patriarch Bartholomew, for the day of prayer for the protection of the Environment (01/09/2016) to *the Plenitude of the Church,* as he noted:

> We are conscious of the Pan-Orthodox declaration about "our greatest responsibility to hand down a viable natural environment to future generations and to use it according to divine will and blessing" (Encyclical of the Holy and Great Council) and "that not only present, but also future generations have a right to enjoy the natural goods granted to us by the Creator" (Decision of the Holy and Great Council on "The Mission of the Church") […] (Patriarch Bartholomew 2021, p. n.p).

Acknowledging the great importance of the family, Pope Francis describes it as: "the place in which life – the gift of God – can be properly welcomed and protected against the many attacks to which it is exposed and can develop in accordance with what constitutes authentic human growth".[16]

Ecological education is gaining ground and is seen as an additional front in the fight to save the Earth. Pope Francis notes that environmental education embraces a broader spectrum of goals that now, beside focusing on providing scientific information, awakening consciousness and preventing environmental risks, opens up to critiquing "the myths of a modernity grounded in a utilitarian mindset (individualism, unlimited progress, competition, consumerism, the unregulated market)" (Pope Francis 2015, p. 154). Furthermore, "it seeks to restore the diverse levels of ecological equilibrium, establishing harmony within ourselves, with others, with nature and other living creatures, and with God" (p. 154). Finally, Pope Francis argues: "Environmental education should facilitate making the leap towards the transcendent which gives ecological ethics its deepest meaning" (p. 154).

4.5 Conclusion

This chapter aimed to explore the role faith-based civil society organisations play in addressing the environmental crisis—by awakening worldwide consciousness to the perilous risks that creation runs in the Anthropocene Epoch. It was evidenced that human activities have damaged the environment, and this assertion was substantiated by the existing literature in peace ecology and the concept of the Anthropocene. On the one hand, governments' unbridled pursuit of wealth and denialism regarding the ongoing atmospheric challenges and the deterioration of the ozone layer were part of the causes of this crisis; on the other hand, it results from humanity's greed and wrongful conception of stewardship which is mistaken for unconscious dominion over creation at the expense of the environment's well-being.

Against these careless patterns of individual and government conduct, civil society and its FBO component were found to be in the forefront of the battle to save "our common home". At this level of intervention, Pope Francis' *Laudato Si'* invitation showed how it was relevant in the context of *ecotransformation* along with other Catholic Social Teaching and the fundamentals and beliefs of other world religions. These FBOs hold the principles of the dignity of the human person, care for creation, human stewardship of creation, etc. In addition, world religions maintain the teaching on change of heart and moderation in satisfying individual needs. This is contrasted with the modern mentality associated with the belief according to which the purpose of life entails the ultimate satisfaction of individual needs. It emerged from the discussion that more consideration towards the Earth is urgent. Pope Francis and Francis of Assisi describe the relation between humans and non-human creation and the Earth in anthropomorphic categories as they refer to Earth as a sister: "our

[16] *Laudato Si'* # 213.

common home is like a sister with whom we share our life and a beautiful mother who opens her arms to embrace us"[17] (Pope Francis 2015, p. 3). The interaction of humanity with nature demands compassion, more in this time than ever before, because our planet is wounded.

It is important to underline once again that the effectiveness of an intervention to save Mother Earth necessitates a global commitment by governments, FBOs, individuals, families, businesses/markets, corporations, and all categories of civil society who have the capacity to act now, to do so. Moreover, awakening international responsiveness about individual and collective agency remains an important factor that could sway global market policies and the meaning and goals of development, and the means to achieve that—which should be ethical. We all need to exercise reasonable and conscious stewardship in all our activities. This should be cognizant of the ongoing ecological distress that preoccupies the world. The path that Pope Francis suggests touches on the core of the crisis: the human heart that generates needs and whims that need to be controlled by reason, ethics and morality. These three elements reflect the transformational and anthropocentric perspectives on the dialogic engagement that *Laudato Si'* and the fundamentals of world religions invite humanity to adopt to prevent further deterioration of the atmospheric conditions and the looming risks of the ES' total collapse.

The conceptual interconnectedness between environmental justice, environmental peace, and peace ecology in the Anthropocene shows the necessity of taking the functions of FBOs beyond the contemporary theological discourse and seeing how a pluridisciplinary approach to tackling climate change and other effects of human activities on the ES becomes relevant within Earth sciences. Incorporating spirituality as a novel tool of environmental justice and environmental peace emerged in this chapter and should be taken further in international discourse and policy debates to save "our common home" in the Anthropocene. Four parameters are essential to attain this vision; these are: (1) promoting ecological peace education, (2) popularising environmental justice, (3) encouraging change of the belief that satisfying individual needs is the basic purpose of life, and lastly (4) emboldening humanity's conscious stewardship over creation and its natural resources.

In a final analysis, faith-based civil society is relevant in sub-Saharan Africa, where environmental conflicts are already occurring. The need to prevent further disruptions of the ecosystem is urgent in order to keep harmony between humans and other non-human creation. Relying on FBOs' actual and increasing population, it is plausible to nurture the hope that world leaders and world religious authorities can utilise their faith-based audience to advocate a universal awareness of the current state of the ES and encourage them, through ecological peace education, the teaching of individual and collective conversion, and leveraging transformative attitudes toward natural resources to turnaway from selfish use of the potentials of Earth and adopt a more generous, moderate lifestyle creating global solidarity over the goods that that our planet puts at our disposal, and enhance compassion towards "Mother Earth". Conversation and living up to the principles of the Earth Charter are

[17] *Laudato Si'* #1

key ways to change the course of ecological history in the Anthropocene. *Laudato Si'* would be more impactful if concerted strategies were put in place by FBOs through an ecumenical dialogue, and world religions renewed their commitment to environmental peace education and environmental justice in families, churches, schools and all professional milieus.

References

Abu-Nimer, M. (2001). Conflict resolution, culture, and religion: Toward a training model of interreligious peacebuilding. *Journal of Peace Research,* 38(6), 685–704

Ahmed, F. (2010). Approaches to and tools for managing environmental conflicts in coastal zones in Africa: Challenges and prospects in relation to Integrated Coastal Zone Management (ICZM). *African Journal on Conflict Resolution,* 10(2), 31–47

Bauckham, R. (2011). The Story of the Earth According to Paul: Romans 8: 18–23. *Review & Expositor,* 108(1), 91–97

Belloni, R. (2001). Civil society and peacebuilding in Bosnia and Herzegovina. *Journal of Peace Research,* 38(2), 163–180

Bercovitch, J., & Kadayifci-Orellana, A.S. (2009). Religion and mediation: The role of faith-based actors in international conflict resolution. *International Negotiation,* 14(1), 175–204. https://doi.org/10.1163/157180609X406562

Brauch, H. G., & Scheffran, J. (2012). Introduction: Climate change, human security, and violent conflict in the Anthropocene. In J. Scheffran, M. Brzoska, H.G. Brauch, P.M. Link, J. Schilling, (Eds.) *Climate change, human security and violent conflict* (pp. 3–40). Berlin, Heidelberg: Springer. https://doi.org/10.1007/978-3-642-28626-1_1

Bronkhorst, S. (2011). *Climate change and conflict: lessons for conflict resolution from the southern Sahel of Sudan.* Umhlanga Rocks: ACCORD

Catholic News Service. (2012). Patriarch says people need to repent for pollution. Catholic News Service, Aug 28, 2012. https://www.ncronline.org/news/patriarch-says-people-need-repent-pollution (Accessed 4 Oct 2021)

Crutzen, P.J. (2016). Geology of mankind. In J.P. Crutzen & H.B. Brauch (Eds.), *Paul J. Crutzen: A Pioneer on Atmospheric Chemistry and Climate Change in the Anthropocene* (Vol. 50, pp. 211–215). Cham: Springer Nature Switzerland AG. https://doi.org/10.1007/978-3-319-27460-7

Crutzen, P. J., & Brauch, H. G. (2016). *Paul J. Crutzen: a pioneer on atmospheric chemistry and climate change in the Anthropocene* (Vol. 50). Cham: Springer Nature Switzerland. https://doi.org/10.1007/978-3-319-27460-7

Dalby, S. (2007). Ecology, security, and change in the Anthropocene. *The Brown Journal of World Affairs,* 13(2), 155–164

Daneel, M. L. (2006). African initiated churches as vehicles of Earth-care in Africa. In *The Oxford handbook of religion and ecology.* Online Publication Date: Sept 2009. https://doi.org/10.1093/oxfordhb/9780195178722.003.0024

Deane-Drummond, C. (2016). Pope Francis: Priest and Prophet in the Anthropocene. *Environmental Humanities,* 8(2), 256–262. https://doi.org/10.1215/22011919-3664369

Deen, M. Y. I. (1996). Islamic environmental ethics, law, and society. In R. S. Gottlieb (Ed.), *This Sacred Earth. Religion, Nature, Environment* (2nd ed., pp. 164–173). New York and London: Routledge

Earth Charter International (2000). The Earth Charter: A declaration of fundamental principles for building a just, sustainable, and peaceful global society in the 21st century. https://static1.squarespace.com/static/5838ae02d1758efcf2a1256d/t/58b340812e69cf75a414acee/1488142466387/Earth+Charter.pdf. Accessed 26 June 2021

Eaton, H. (2007). This sacred Earth at the nexus of religion, ecology and politics. *European Journal of Science and Theology*, 3(4), 23–38
Eklöw, K., & Krampe, F. (2019). *Climate-related security risks and peacebuilding in Somalia*. Stockholm: Stockholm International Peace Research Institute (SIPRI)
Ferrara, P. (2019). Sustainable International Relations. Pope Francis' Encyclical *Laudato Si'* and the Planetary Implications of "Integral Ecology". *Religions*, 10(466), 1–20. https://doi.org/10.3390/rel10080466
Franchi, L. (2016). Laudato Si' and ecological education: implications for Catholic education. *Pensamiento Educativo: Journal of Latin American Educational Research*, 53(2), 1–13
Francis of Assisi (2021). Canticle of the Creatures: Early Documents, Vol 1, The Saint – page 113–114. http://franciscanseculars.com/the-canticle-of-the-creatures/. Accessed 26 June 2021
Gottlieb, R. S. (2003). *This sacred earth: Religion, nature, environment*. New York and London: Routledge
Hahne, H.A. (2010). The Whole Creation Has Been Groaning. Christian Reflection (Winter 2010), 19–28. https://www.baylor.edu/content/services/document.php/106707.pdf. Accessed 25 July 2021
Hardt, J. N., & Scheffran, J. (2019). Environmental peacebuilding and climate change: peace and conflict studies at the edge of transformation. *Policy Brief, 68* (December 2019), 1–20
Hittinger, R. (2013). Quinquagesimo ante: Reflections on Pacem in Terris fifty years later. The global quest for Tranquillitas Ordinis: Pacem in Terris, Fifty Years Later: The proceedings of the 18th plenary session (pp.38–60). Vatican City: Pontificia Accademia Scientiarum Socialium
Hoffman, A.J. (2015). Laudato Si and the Role of Religion in Shaping Humanity's Response to Climate Change. *Solutions*, 6(5), 40–47
Hovhannes, H. (2016). Idea and models of civil society: development tendencies, arguments pro and con. *Wisdom*, 2(7), 95–102
Ijumba, N., & Kaya, H. (2016). Energy Resources and Environmental Conflicts in Africa: Implications on Indigenous Knowledge Systems and Environmental Policy. *Journal of Human Ecology*, 53(2), 96–100. https://doi.org/10.1080/09709274.2016.11906960
Kennedy, M. (2015). Le Mare & McCartney's Coming from the Silence: Quaker Peacebuilding Initiatives in Northern Ireland 1969–2007 - Book Review. *Quaker Studies*, 15(2), 258–260
Krampe, F. (2017). Toward sustainable peace: A new research agenda for post-conflict natural resource management. *Global Environmental Politics, 17*(4), 1–8. https://doi.org/10.1162/GLEP_a_00431
Krampe, F., & Mobjörk, M. (2018). Responding to climate-related security risks: reviewing regional organizations in Asia and Africa. *Current Climate Change Reports, 4*(4), 330–337. https://doi.org/10.1007/s40641-018-0118-x
Last, D. M. (2010). Civil Society and Peacebuilding. In *Routledge Encyclopedia of Peace*. http://dx.doi.org/https://doi.org/10.2139/ssrn.2459768
Laudani, R. (2012). Une démocratie a reconstruire: aux origines de la société civile. *Le Monde diplomatique*, September 2012. https://www.monde-diplomatique.fr/2012/09/LAUDANI/48145. Accessed 29 May 2021
Lisk, F. (2009). Overview: The Current Climate Change Situation in Africa. In H. Besada & N. Sewankambo (Eds.), *Climate Change in Africa: Adaptation, Mitigation and Governance Challenges* (pp. 8–15). Waterloo: The Centre for International Governance Innovation (CIGI)
Molvaer, R. K. (1991). Environmentally induced conflicts? A discussion based on studies from the Horn of Africa. *Bulletin of Peace Proposals*, 22(2), 175–188

Moseley, L., & Feldman, D. L. (2003). Faith-based environmental initiatives in Appalachia: Connecting faith, environmental concern and reform. *Worldviews: Global Religions, Culture, and Ecology,* 7(3), 227–252

Moyer, J. (2013). Faith-Based Environmental Work in Canada: A Profile. *Western Geography,* 23, 60–85

Noll, D. (2003). *Peacemaking: Practicing at the Intersection of Law and Human Conflict*. Telford: Cascadia Publishing House

Oswald Spring, Ú., Brauch, H.G. (2021). *Decolonising Conflicts, Security, Peace, Gender, Environment and Development in the Anthropocene.* Heidelberg: Springer. https://doi.org/10.1007/978-3-030-62316-6

Paffenholz, T. (2015). Civil society and peacebuilding. *Development Dialogue,* 63, 108–118

Paffenholz, T., & Spurk, C. (2006). Civil society, civic engagement, and peacebuilding. *Social Development Papers: Conflict Prevention and Reconstruction,* 36, 10–55

Pattberg, P., & Stripple, J. (2008). Beyond the public and private divide: remapping transnational climate governance in the 21st century. *International environmental agreements: Politics, law and economics*, 8(4), 367–388. https://doi.org/10.1007/s10784-008-9085-3

Patriarch Bartholomew. (2021). Message of His All Holiness, Patriarch Bartholomew, for the day of prayer for the protection of the Environment (01/09/2016). Laudato Si Movement 202. https://www.laudatosi.org/ecumenical/day-of-creation/message-patriarch-bartholomew/ Accessed 4 Oct 2021

Pew Research Center. (2021). The Future of World Religions: Population Growth Projections, 2010–2050. https://www.pewforum.org/2015/04/02/religious-projections-2010-2050/pf_15-04-02_projectionsoverview_projectedchange640px/. Accessed 23 July 2021

Pope Francis (2015). Encyclical Letter Laudato Si' of The Holy Father Francis on Care for Our Common Home. Rome: Libreria Editrice Vaticana

Renugadevi, R. (2012). Environmental ethics in the Hindu Vedas and Puranas in India. *African Journal of History and Culture*, 4(1), 1–3. https://doi.org/10.5897/AJHC11.042

Rolston, H. (2015). Ecological Pope challenges the Anthropocene Epoch. In J B Cobb, Jr., and I Castuera (Eds), For Our Common Home: Process-Relational Responses to Laudato Si' (pp. 52-57). Anoka, MN: Process Century Press

Saniotis, A. (2012). Muslims and ecology: fostering Islamic environmental ethics. *Contemporary Islam*, 6(2), 155–171. https://doi.org/10.1007/s11562-011-0173-8

Schellnhuber, H. J. (1999). 'Earth system'analysis and the second Copernican revolution. *Nature*, 402(6761), C19–C23

St. Anthony of Padua. (n.d.). Preaching to the fish: The 49th chapter of the 14th century "Little Flowers of St. Francis". https://sites.google.com/site/paduastanthony/preaching-to-the-fish Accessed 24 April 2021

Staudt, R.J. (2014). St. Anthony's Fish and the Mission to Creation. Catholic Exchange, 13 June 2014. https://catholicexchange.com/st-anthonys-sermon-fish-bringing-gospel-creation. Accessed 17 July 2021

United States Conference of Catholic Bishops. (2004). *Compendium of the Social Doctrine of the Church*: Catholic Church Pontificium Consilium de Justitia et Pace Pontifical Council for Justice and Peace

Whetho, A., & Uzodike, U. O. (2008). Religious networks in post-conflict Democratic Republic of the Congo: A prognosis. *African Journal on Conflict Resolution,* 8(3), 57–84

Willms, D. G., & Werner, K. (2009). Climate change and conflict in Africa. In H. Besada & N. K. Sewankambo (Eds.), *Climate Change and Conflict in Africa: Adaptation, Mitigation and Governance Challenges* (pp. 26–30). Waterloo: The Centre for International Governance Innovation

Worldometer (www.Worldometers.info). (2019). Elaboration of data by United Nations, Department of Economic and Social Affairs, Population Division. World Population Prospects: The 2019 Revision (Medium-fertility variant). https://www.worldometers.info/world-population/ (24 Jul 2021)

Zalasiewicz, J., Williams, M., Smith, A., Barry, T. L., Coe, A. L., Bown, P. R., Benchley, P, Cantrill, D, Gale, AGibbard, P., Gregory, F J, Hounslow, M W, Kerr, A C, Pearson, P, Knox, R, Powell, J, Walters, C, Marshall, J, Oates, M, Rawson, P, Stone, P. (2008). Are we now living in the Anthropocene? GSA *Today,* 18(2), 4–8. https://doi.org/10.1130/GSAT01802A.1

Zehr, H. (2002). *The Little Book of Restorative Justice.* Intercourse PA: Good Books

Chapter 5
Healing Nature and Creation in the Anthropocene: A Reflection on the Role of Religion

Dumisani Maqeda Ngwenya

> *When creation is threatened by climate change we are called to speak out and act as an expression of our commitment to life, justice and love* (World Council of Churches, September 2007).

Abstract While religion is one of the greatest causes of conflict around the world, it has also been recognised as an effective instrument in peacebuilding. The influence religious leaders and the teachings of the various religious scriptures—on peace and good neighbourliness—have over their followers have been tremendous resources for creating peace. Secular organisations such as the United States Institute for Peace (USIP) have helped to bring religion into mainstream peacebuilding. With the move to include earth care in the peacebuilding agenda, the role of religion in environmental conservation and climate change has come under the spotlight. This chapter explores the potential of religion to influence the climate change agenda positively. It discusses how the various world religions' scriptures treat ecological issues and how the religious leaders, of each of these religions, have sought to interpret them, in light of the climate change debate. The focus of the chapter is to encourage African peace practitioners to seriously consider including the healing of nature and creation in their peacebuilding agenda and how they could use religion (religion has a stronghold in Africa) to encourage communities to play an active role in addressing the negative effects of climate change. The case of the African Initiated Churches (AICs) in Zimbabwe, is used to illustrate how one organisation effectively used religion in its 'earthkeeping' efforts.

Keywords Peacebuilding · Earthkeeping · Religion · Climate change · Healing · Ecology

---

Dr Dumisani Ngwenya a peacebuilding practitioner and academic who has been working in the field of conflict transformation, trauma healing at Grace To Heal Organisation, since 2003 in Bulawayo, Zimbabwe.

J. C. K. Kiyala and G. T. Harris (eds.), *Civil Society and Peacebuilding in Sub-Saharan Africa in the Anthropocene*, The Anthropocene: Politik—Economics—Society—Science 34,
https://doi.org/10.1007/978-3-030-95179-5_5

5.1 Introduction

From time immemorial, people have always understood themselves to be inextricably connected to the earth, and for Africans, in particular, it is only a few generations ago when this notion was fully embraced. However, due to industrialisation and globalisation, human beings have increasingly become disconnected from nature—with disastrous consequences. In Africa, the evidence of the adverse effects of climate change is becoming more pronounced with time. From the receding water levels in lakes, to the increased rate of desertification, unpredictable weather patterns, frequent and longer periods of drought, and severe storms and floods when it rains, it has become difficult to ignore the signs. Climate change has social, political and economic consequences which may contribute to violent conflict. Inter-communal and cross-border violent conflicts may result from competition over availability of renewable resources such as water. For example, conflicts between pastoralists and farmers in some Eastern and Western African countries are common and most likely to escalate as the effects of climate change increase (van Baalen/Mobjőrk 2017; Raleigh 2010; Furini 2019). Extreme droughts and flooding increase poverty levels and affect communities' well-being, resulting in unpeaceful situations (Dube/Phiri 2013). It is incumbent upon us, as peacebuilders, to play a role in making peace with the nature and creation – what Daneel (2011) refers to as 'earthkeeping'– that we, as people, have so badly abused.

As a person of faith and a peacebuilder, I see religion as having a role to play in combating the anthropogenic climate change the world is currently facing. Most world religions, from the Abrahamic faiths (Judaism, Christianity and Islam), and others such as Hinduism, Buddhism, Bahai faith and Confucianism, including African traditional religions (ATR), to varying degrees believe in the sacredness of the earth and emphasise the interrelatedness and interdependence of humankind and the rest of creation (Branton 2006). We also learn from the religions of the need to care for the earth for its own sake.

This short chapter seeks to provoke peacebuilders to consider their peacebuilding responsibilities as extending beyond the traditional peacebuilding mandates of healing human relations, to healing nature and creation, which has suffered violently at the hands of human beings. It is not meant to offer definitive answers but to spark a discussion about how we, as African peacebuilders, in particular, can start to think about our individual responsibilities to care for nature and creation as part of building peace in our continent and globally as well. This is not something that is at the grand and complex scale of environmental peacebuilding, but a call to examine whether we are good stewards of what has been entrusted to us, and how we as peacebuilders can contribute at a personal level to 'earthkeeping'.

5.2 World Religions' View of Human and Ecology Relationships

As noted above, religions of the world, in their scriptures, do speak, in one way or another, to the relationship between ecology and human beings. In this section I briefly survey some of them and their teachings related to their ecological views. I acknowledge that even within some of the religions discussed here, there are varying degrees of ecological beliefs and thoughts, but due to space constraints, this chapter will not dwell much on the nuances of these internal divergences. The purpose of this survey of the religions' cosmological and ecological perspectives is to show that every major religious belief offers opportunities that can help peacebuilders to think about their peacebuilding philosophy towards nature and creation.

5.2.1 Eastern Religions

5.2.1.1 Buddhist Approaches to Ecology

According to Buddhist teachings, the environment – that is inhabited, or the inhabitants – are both composed of basic elements: earth, wind, fire, water and vacuum, that is space. It is, therefore, "on the basis of these five basic elements that there is a very close inter-relatedness or interrelation between the habitat, that is the natural environment, and inhabitants, the sentient beings living within it" (Dalai Lama 1992). This emphasis on the interrelatedness of humans and nature seems to indicate that within Buddhism there is an inherent awareness of and responsibility to one's surroundings. According to Branton (2006, p. 213), it includes "an ecological vision and principle of interdependence that integrates all aspects of the ecosphere". The religion encourages its followers to extend love to all of nature and to avoid being cruel to other living beings. Earth is viewed as being sentient – that is, being alive and having feelings and awareness (Yŏ 2014, p. 484). According to Rakkhit (2014), a monk from Bangladesh, the *Vinaya* one of the Buddhist scriptures, requires monastics, as much as possible, to avoid harming nature. "Even when urinating in a remote area, he/she is advised to find a place that has no fresh grass or fresh leaves". He further suggests that one of the reasons for the establishment of the Buddhist rainy season retreats was to minimise destruction to the environment. This practice encouraged the monks to stay at the temple during the rainy season because during this time, in the rural areas, there was too much movement from the religious mendicants, which could result in agricultural damage.

It is out of this consciousness and realisation of the interdependence of human beings and nature that "Buddhist environmentalists demonstrate their concern for the total living environment, by extending lovingkindness and compassion beyond people and animals to include plants and the earth itself" (Branton 2016, p. 213).

5.2.1.2 Hinduism

Much like other Eastern religions, Hinduism also believes in the inextricable relationship between human beings and the environment. For most Hindus, all living things are sacred because they are part of God and the principle of *ahimsa*, nonviolence, means showing respect for all living things, humans, animals and plants. Nature cannot be destroyed without humans also being destroyed, because we need the natural world in order to survive (BBC Bitesize 2022). In the Hindu writings, the universe is described as an interconnected whole in which each part is interdependent with every other part. It is a living organism. The universe is understood by some to be God's body (Coward 1997). The mythology is deeply rooted in an ecological vision, where the human-nature relationship is at its core. The Hindu scriptures have been summarised by Kapila Vatsyanan (1992) thus,

> Man's (*sic*) life depends upon and is conditioned by all that surrounds him and sustains him, namely, inanimate, mineral and animate, aquatic, vegetative, animal and gaseous life. It is, therefore, Man's duty to constantly remind himself—in individual and collective life—of the environment and the ecology (cited in Coward 1997, p. 51).

Therefore, this vision permeates the biological, physical and spiritual dimensions of life. For Hindus, taking care of nature and creation is taking care of oneself and their community. One cannot survive without the other. Respecting the universe well and treating it well is an act of worship because God is the universe, and the universe is God.

5.2.1.3 Confucianism

According to Confucian thought, human beings exist in a harmonious relationship with nature. Even though people play a unique role in the environment, they must not become separated from nature, for, all the achievements of human culture have their foundation in nature (Christensen 2014). This view of a unified, interconnected, and interpenetrating universe emphasises an intricate awareness and deep reciprocity between humans and the natural world. Confucian practitioners appreciate nature as "intrinsically valuable and rely on history as a way of maintaining continuity and collective memory" (Branton 2006, p. 213). There is a harmonious existence and people are at the centre of this 'organic whole' with no separation between the natural world and humans. As Christensen puts it, "[if] humans do not fully realize that we are only one species among others, and that the whole ecological system exists in a mutually dependent relationship, then the result in the end will be obstruction of harmony between humans and the environment" (Christensen 2014, p. 283). He concludes by noting that,

> Human civilization should not be an enemy of nature, but should be seen as a part of the development of the world. Since people are considered to be a part of the natural environment as a whole, the action of not harming this 'whole' also implies an act of taking care of the people. As soon as the natural environment has been harmed, the natural resources available

> for future human generations will be reduced, and so their lives will be faced with dangers. This is contrary to the demand that we take care of the people's lives. Thus, one may say that taking care of the people must have protection of the natural environment as a prerequisite requirement (Christensen 2014, p. 283).

5.2.2 The Abrahamic Religions and the Environment

Judaism, Christianity and Islam are often referred to as the Abrahamic faiths because of their common belief in Abraham as the forefather of their respective religions. The perspectives of these religions on the relationship between human beings and the environment is different from the other faiths in that, although there is an interdependent relationship between nature and humankind, they are, nevertheless, different from each other. Be that as it may, the existence of humanity is intertwined with the existence of nature. In the same vein, God is above his creation, he is not nature and nature is not God.

5.2.2.1 Judaism

In terms of the Jewish traditional view of the environment, God is the originating power and unifying force of the universe. Creation is seen as a closed system which is complete, self-sustaining and dynamic. All actions in the universe are integrated and their impact can affect the balance and rhythm of life positively or negatively (Kopstein/Salinger 2001, p. 63). The Jewish Scriptures, the Torah, are written in the context of an agrarian society and many of the festivals the Jewish community is commanded to observe, are agricultural celebrations (Halpert 2012).

In the story of creation in Genesis 1 and 2, human beings are put in the garden to be 'stewards' of the earth. This idea carries with it a "notion of responsibility, that the human is responsible for the well-being of the earth". This concept is "founded on the conviction that the way humans till the earth relates to moral integrity" (Halpert 2012). Therefore, the misuse or lack of care for the environment is an indication of the loss of that moral compass. Furthermore, creation, as an act of God, is set apart (holy) and follows the law of God; and these laws, like the laws of physics, affect all without preference. Creation is intertwined in the web of life, and while how humans behave may affect the environment in ways that create changes, these changes are also felt within the larger scheme of things, resulting in repercussions that go beyond the immediate (Kopsten/Salinger 2001). According to the ancient commentary of the Hebrew Bible, the Qohelet Rabbah 7:28,

> When God created Adam, he showed him all of the trees of the garden of Eden and said to him: "See my works, how lovely they are, how fine they are. All I have created, I have created for you. Take care not to corrupt and destroy my universe, for if you destroy it, no one will come after you to put it right" (cited in Rakover 2003).

The responsibility to care for nature falls squarely on humans' shoulders. For they have been designated as its caretakers on behalf of God. It is within this context that the commandment to give the earth a rest every seven years (Leviticus 25–26; Deuteronomy 15) may be understood. It is necessary, according to the laws of the Sabbath, Sabbatical and Jubilee, for the earth to be allowed to rest from human interference. In the Rabbinic letter on climate crisis (2015), the seventh year of release (*Shmittah, Shabbat Shabbaton*), the Rabbis argue this point persuasively. The argument is best summed up by Rabbi Green (1998, pp. 102–103),

> We acknowledge the presence of God in the natural rhythms of past seasons. Our awareness of wind, rain, and dew, as daily miracles also serves to remind us that the purity of these gifts, so vital for our survival, must be maintained by human watchfulness. In thanking God for air, we assert our commitment to preserving them as sources of life and protecting them from life-destroying pollution.

The humans' responsibility to the environment, according to the *Talmudic* laws, is three-fold: *Ba'al taschit*, do not destroy—avoid plundering the earth, through reducing, reusing and recycling; *Shomrei Adamah*, guard the earth—from destructive human behaviour, and *Tikkun Olam*, repair the earth—repair harm done to earth and restore nature's balance (Kopstein/Salinger 2001, p. 71).

5.2.2.2 Christianity

There are varying and sometimes opposing schools of thought in the Christian world, on spirituality and ecology. This is especially true within the Evangelical branch of the Christian church. The evangelical church in America, is particularly divided on this issue with some supporting the ecological cause while others oppose it on the basis that the earth is there to serve human needs and human beings have dominion over it (Ronan 2017). Other branches of Christianity, such as the Roman Catholic and the Orthodox churches, have a more monolithic approach to eco-theology. This is because of the role their leaders have played in pushing for the inclusion of the environment as a spiritual issue. Pope Francis and Patriarch Bartholomew, in particular, have contributed significantly to the debate – even direction – through their letters to their followers.

This discussion will focus on those parts of the Christian church which support the proactive role in climate change mitigation efforts—what I might term 'ecologically friendly' Christianity. However, this approach does not diminish the views of the Christian groups that advance the notion that the earth is there to serve human needs and human beings have dominion over it. The importance of the earth is not in question, but rather the manner of its exploitation without regard to its preservation for generations.

The ecologically friendly church views environmental degradation as a moral failure on the part of human beings and strongly links ecological awareness to the idea of stewardship; seen as a sacred obligation to safeguard and preserve the earth (Ronan 2017). As Pope Francis (2015) has said in his *Laudato si'*, the damage to the environment is a reflection of the depravity of human beings' hearts. The Patriarch Bartholomew (1997 cited in Francis 2015, p. 8) put it more strongly and equated it to sin:

> For human beings… to destroy the biological diversity of God's creation; for human beings to degrade the integrity of the earth by causing changes in its climate, by stripping the earth of its natural forests or destroying its wetlands; for human beings to contaminate the earth's waters, its land, its air, and its life – these are sins… For, to commit a crime against the natural world is a sin against ourselves and a sin against God (see also Bennett 2008, p. 20).

As in Judaism, Christian theology on the environment starts with the creation story as stated in Genesis 1 and 2. God created the earth and gave it to humankind for their sustenance and he expected them to care for it in return. According to Genesis 1 v26-30, God entrusted human beings with the responsibility of caring for the earth which he had painstakingly created for them. The charge to fill, subdue and have dominion over the earth was not so that people could do as they pleased with it, but it was a charge to be stewards, to use it wisely on His behalf.

It is the responsibility of humans to look after the earth and everything in it, to take care of it, for its existence is not merely for humankind's use but it has intrinsic value in and of itself. "The ultimate purpose of other creatures is not to be found in us. Rather, all creatures are moving forward with us and through us towards a common point of arrival, which is God, in that transcendent fullness where the risen Christ embraces and illumines all things" (Francis 2015, p. 61).

First and foremost, nature reveals God's glory. "… his eternal power and divine nature, have been clearly perceived, ever since the creation of the world, in the things that have been made" (Romans 1:20; *cf* Psalm 8; Psalm 148: 3–5). In its pristine condition, nature shows the greatness of its creator. "Each creature possesses its own particular goodness and perfection… Each of the various creatures, willed in its own being, reflects in its own way a ray of God's infinite wisdom and goodness. Man must therefore respect the particular goodness of every creature, to avoid any disordered use of things" (Roman Catholic Catechism, p. 339). Therefore, the destruction of the environment, the extinction of a myriad of species, "means they can no longer give God the glory nor convey their message to us" (Francis 2015, p. 25). As nature reveals God's glory, so does He command that human nurture it to preserve His infinite glory.

Secondly, all of creation (including humans) has been created in such a way that it is all dependent on each other "to complete each other, in the service of each other" (Roman Catholic Church Catechism). "God has joined us so closely to the world around us that we can feel the desertification of the soil, almost as a physical ailment, and the extinction of a species as a painful disfigurement" (Francis 2015, p. 66). Thirdly, God actually cares about all of His creation – of His omnipotence and benevolence. God's command to give the earth rest every seventh year (Leviticus 25–26; Deuteronomy 15), is to avoid misuse and overuse, and so the earth can regenerate itself. Refusal to obey this command would result in forceful Sabbath rest for the land, through the children of Israel being taken into exile in foreign lands (Lev. 26: 33–35). In the New Testament, Jesus also highlighted God's care even for the "most insignificant creatures". In Luke 12:6 Jesus says, "are not five sparrows sold for two pennies? And not one of them is forgotten before God", and again in Matthew 6:26, "Look at the birds of the air: they neither sow nor reap nor gather into barns, and yet your heavenly Father feeds them…".

Therefore, human beings' relationships are not just among themselves and God and with their neighbours, but include a third dimension—between them and nature (Francis 2015). The Patriarch Bartholomew (2012) sees an even closer relationship. He urges people "to accept the world as a sacrament of communion, as a way of sharing with God and our neighbours on a global scale". Furthermore, it is his 'humble conviction that the divine and the human meet in the slightest details of seamless garment of God's creation, in the last speck of dust of our planet".

The World Council of Churches (WCC), and its individual member churches, have been advocating for action on climate change as far back as the 1980s (Toroitich/Kerber 2014; Hallman 1997). The theological underpinnings of the WCC's actions are very similar to those mentioned above. According to a 29-member consultation convened by the WCC to complete a study paper entitled "Accelerated Climate Change: Sign of Peril, Test of Faith", the evidence of global warming and distraction of the environment are not only a sign of peril but of test of the faith, and as such,

> [God as] Creator-Redeemer asks all who can see to read the sign of the peril that looms before the planet. Response to the peril becomes a matter of renewed faithfulness to "God [who] calls us to participate with God in God's own work of stopping the degradation of Earth's protective mantle and engaging in the other tasks of protecting and restoring the creation, human and nonhuman." From theological affirmation of God as Creator-Redeemer in the context of imminent peril, come urgent ethical imperatives, including among others: that we (1) "cherish the whole of creation"; (2) build communities in which all may "participate in obtaining and enjoying sufficient and sustainable sustenance from nature"; and (3) "protect this earthly habitat ... [for] our children's children's children, together with other life, on into the indefinite future" (Gibson 1994, p. 11.).

The WCC central committee statement of January 1994, released in Johannesburg, South Africa, affirmed that global warming and accelerated climate change were visible signs of the crisis being experienced by civilisation, and were rooted in the refusal to accept the boundaries of God's creation. It further noted that the social and environmental degradation caused by accelerated climate change was sin against God and a violation of creation. It called for the protection of the atmosphere as an essential and invaluable common good for current and future generations. Just like the other branches of Christianity, it proclaimed that radical change was only possible through repentance and changed life-styles. It then called on all Christians "to recognise the challenge to the life and witness of Christians that the crisis from accelerated climate change presents", and "to reinterpret Christian responsibility towards creation and to respond in faith and action to the peril in their own situation" (WCC 1994, pp. 247–8).

5.2.2.3 Islam

Much like the other two Abrahamic faiths, Islamic eco-theology also believes that earth and all of nature, has a higher purpose than simply to serve the needs of human beings. Creation exists to manifest the glory of God, as the 2015 Islamic declaration on global climate change says:

> God—Whom we know as Allah—has created the universe in all its diversity, richness and vitality: the stars, the sun and moon, the earth and all its communities of living beings. All these reflect and manifest the boundless glory and mercy of their Creator. All created beings by nature serve and glorify their Maker, all bow to their Lord's will. We human beings are created to serve the Lord of all beings, to work the greatest good we can for all the species, individuals, and generations of God's creatures (cited in Schaefer 2016, p. 11).

Furthermore, human beings and the environment are interdependent. This was affirmed by the organisers of the 'Islam and the Environment' conference held in Dubai in 2013, when they stated that,

> The environment lies at the core of the Islamic faith, and the underlying principal that forms the foundation of the Prophet Mohammed's […] holistic environmental vision is the belief in the interdependency between all natural elements, and the premise that if humans abuse or exhaust one element, the natural world as a whole will suffer (cited in Reuter 2015, p. 1224).

Therefore, the environment is viewed as being sacred, and according to Bilal (2017), it might even be equated to other forms of worship (*ibadah*) such as prayer and fasting. He further argues that, even if there was no threat of a resource crisis, Muslims are still expected to take good care of the earth and its resources, and to improve conditions of life on earth through due consideration towards "the environment with a sense of both duty as well as morality". Humans are to be the caretakers "of Earth, they are intended to savour the 'gifts' of Earth, and they are accountable to God for their actions" (Schaefer 2016, p. 12). As Torabi/Nori (2019, p. 346) have pointed out, "the environment offers humans profound and constant opportunities to be aware of God's presence"; therefore, any "maltreatment of the environmental factors would be tantamount to human neglect of giving due respect to God's clear signs". Human beings "as the guardians of Allah's creation, … have a responsibility to protect the environment" (Torabi/Nori 2019, p. 348).

5.2.3 *African Traditional Religions*

Although African Traditional Religions (ATRs) are not homogeneous, nevertheless, they have certain cosmological beliefs, rituals, and practices in common. According to Tarusarira Tarusarira (2017, p. 400), the African world usually,

> … exists in two spheres – the visible, tangible, and concrete world of humans, animals, vegetation, and other natural elements; and the invisible world of the spirits, ancestors, divinities, and the supreme deity. Yet it is one world, indivisible, with one sphere touching on the other. Its specific elements are basically the belief in the existence of God and/or

gods; the belief in spirits, both good and bad; and the belief in cultic prohibitions (taboos) and moral violations.

Indeed, animism is a common thread in all ATRs, as a result of this, certain animals, vegetation and natural phenomena such as hills, or rivers are deified (Mwale 2014–2015; Tarusarira 2017). In African cosmology, "the relationships between humans and nature, and spirit and nature" are not dual but are intertwined "into an interdependent system of existence" that is held together by spiritual interactions (Tarusarira 2017, p. 407). Therefore, African spirituality is inextricably linked to nature and, to a large extent, depends on it for religious rituals and people's everyday existence. Furthermore, trees are a source of fuel, building material, and used for medicinal purposes. Therefore, conservation of the environment is in the best interest of the spirituality, health, wealth, livelihoods and wellbeing of the community. There is a deep respect and reverence for the environment, and community members, from an early age, are taught to protect the natural environment. Trees should not be cut down willy-nilly and certain sacred ones should not be cut down at all or used as firewood, even if they were dry. Domestic animals and people were prohibited from exploiting or frequenting certain sacred places such as specified mountains and water pools.

As can be noted from the above discussion, most religions, at their core, have the inextricable and symbiotic interdependent relationship between human beings and nature. Nature is deemed to be important, and having a right to exist in its own right. Nature sustains human life, and humanity has a responsibility to care for, and use it responsibly for the sake of future generations. Furthermore, nature plays an important spiritual role in the cosmology of most major religions; either as reflecting the glory of the Creator—magnifying His greatness, as in the Abrahamic religions, or as embodying the Divine and being the dwelling place of the spirits, as in the Eastern and African traditional religions. Religion plays an important role in African lives, and bears influence over the vast majority of the population, and has often been used in peacebuilding efforts to address human conflicts. In the same manner, it could be drawn upon to assist the efforts of healing nature and creation.

5.2.4 Earthkeeping by Communities of Faith in Zimbabwe

A fascinating example of the use of religion to address climate change is the work of M. L. Daneel, professor Emeritus of Missiology at UNISA, through his organisation, the Zimbabwean Institute of Religious Research and Ecological Conservation (ZIRRCON), in the Masvingo Province of Zimbabwe, between 1988 and 2003 (Daneel 1998, 2011). He worked with both the traditionalists, under the banner of the Association of Zimbabwean Traditional Ecologists, and the African Initiated Churches (AIC), under the Association of African Earthkeeping Churches (AAEC). According to Daneel (2011, p. 130), the achievements of this project included 15000–18000 mother nurseries, some of which cultivated as many as 100 000 seedlings each year. Schools and women's groups also ran their own small nurseries, and during this period 12 to 15 million trees were planted throughout the province. A variety of tree species were planted which included:

- fruit trees in orchards for personal and commercial use;
- exotics such as eucalyptus for building operationsindigenous trees for firewood and the restoration of denuded land;
- leucaena for cattle fodder, firewood, and nitrate-fixing in arable lands; and
- indigenous hardwood, such as kiaat and pod mahogany, as a long-term investment for future generations (Daneel 2011, p. 130).

The earthkeepers became known for cultivating indigenous fruit tree seedlings, mountain acacia, thorn trees, etc, more than any other organisation in the country. Daneel avers that government officials, including President Robert Mugabe, used to attend and participate in their annual tree-planting ceremonies.

Conservation clubs were involved in the struggle against soil erosion by filling erosion gullies with stones and planting vetiver grass in the affected areas. Pupils from some 30 clubs in rural schools were taken on trips to learn about trees and birds, and from time to time, Parks and Wildlife officials would accompany the children to the national parks to teach them about big game and species that were no longer found in communal lands. On the other hand, spirit mediums and the elders assisted the chiefs by restoring the customary laws on the protection of trees and wildlife in the ancestral holy groves. Offenders were heavily fined and made to plant trees in denuded areas.

Another aspect of this project involved the ritualisation of the tree planting ceremonies, which took a traditional or Christian form depending on which group was leading the ceremony. For the AZTREC, the rituals resembled the rain-making ceremonies, which included ancestral beer and sacrificial addresses to the ancestors on the basis of the traditional cosmology. The seedlings were entrusted into the protective care of the guardian ancestors, after which, tree planting would take place. Although the Christian earthkeepers were present during these ceremonies, they would not participate in them and only joined the traditional earthkeeper for the actual tree planting activity (see Daneel 1998 for more details on the ceremonies).

The church earthkeepers held what Daneel calls the ‘tree-planting Eucharist’. This ceremony integrated the earthkeeping ministry with the traditional Christian sacrament of Holy Communion. As he points out, “this development was of pivotal importance, for it brought environmental stewardship right into the heartbeat of church life and biblically based spirituality” (Daneel 2011, p. 131). The AAEC’s ceremonies involved,

- Preparations of the woodlots, that is, the digging of the holes for the seedlings, fencing and naming the woodlots the ‘Lord’s Acre’, the equivalent of the traditional sacred groves.
- Dancing and singing around the seedlings to praise God, the great Earthkeeper.
- Sermons by the AIC bishops, speeches by representatives from the Forestry Commission, Parks and Wildlife and other government officials.

- Christian participants' public confession of their ecological sins of tree felling without replanting, poaching and promoting soil erosion.
- Confession followed by each member picking up the seedlings and going to the table where bread and wine were administered. This act symbolised the drawing of nature into the inner circle of communion with Christ the Redeemer, the head of the church and of all creation (Daneel 2011, p. 131).

After communion, the Christians would then be joined in the tree planting activity by the traditionalists, who, although present during the ceremony, would not participate in the above rituals. At the end of the tree planting, the Christian tree planters would then kneel before the prophetic healers for laying on of hands and prayer.

5.3 Towards an Ecologically-Sensitive Peacebuilding Paradigm

Generally, peacebuilding in Africa has focused on human relations, and rightly so, because of the many violent conflicts Africa continues to experience. Although this will continue to be an important aspect of our peacebuilding work, the threat of climate change to the environment and human existence calls for African peacebuilders to shift from the traditional concept of building peace to a holistic one that includes the environment. This work should not be left to conservationists alone.

Almost all religions seem to be in agreement that the degradation humans have caused to the environment over the years is violence against nature and creation. Although some religions view earth's primary function as the sustenance of human existence, there is also a common understanding that nature exists in its own right and for the glory of God or the Divine. There is an inextricable interdependence between humans and nature—a symbiotic relationship, as it were. The Creator's original plan, for a respectful and sustainable co-relationship between earth and human beings, has been shattered by the careless and irresponsible exploitation of nature by humans, and this threatens the very existence of both.

'Earthkeeping' is a spiritual act, and Africans tend to be spiritually conscious (whether, or not, they follow a particular religion). As such, it is incumbent upon African peacebuilders to seriously consider what role they ought to be playing in the protection and preservation of the ecological system. We need to think about developing a peacebuilding paradigm that encompasses the totality of our existential reality. That is to say, a peacebuilding approach that covers human relations, human spirituality and relations with nature and creation. For instance, from a Christian perspective (as in some of the religions discussed above), peacebuilding should happen on three dimensions: reconciling human beings with God (because of sin, we are in conflict with him), with one another (because of sin, interpersonal harmony and cohesion has been shattered) and with nature (the sustainable utilisation of natural resources, was overtaken by greed and recklessness).

The purpose of this reflection is to stimulate a practical discourse at personal and societal levels on: (a) how we as individual peacebuilders can adjust our behaviour and practices to reflect a life style that enables a healthy respect and care for the environment, (b) how we can make environmental conservation a peacebuilding agenda and mainstream it into our traditional programmes. How can we incorporate some of the principles espoused by the diverse religious persuasions to transform our own attitudes and practices and those of the communities with whom we work? Finally, how do we get governments to think long-term when it comes to the socio-economic policies they formulate?

5.4 Conclusion

As peacebuilders, we are concerned about conflict and violence, and we seek to foster positive peace in society. As amply demonstrated by the religious discourse above, misuse, overuse or abuse of the environment is violence towards an entity on which our very existence depends, and to which we are inextricably connected. We need to understand that we are fast approaching a point where a holistic peace will not be possible without a shift from an anthropocentric view of the environment. The field of peacebuilding is a fluid one that is ever changing in response to contemporary and arising conflictual issues. Climate change is a contemporary issue that threatens peace and harmony. African peacebuilders have to take up the challenge of 'earthkeeping,' and actively work for a more sustainable use, and better treatment of the environment. Pivotal to this is the awakening, within the African peoples of that spiritual connection that has existed between nature and them, and moving governments and business towards a more balanced and sustainable use of the continent's natural resources. In addition, as peacebuilders, we have our own individual contributions to make towards a healthy and less violated environment by how we live and treat the nature. We need to introspect on how we contribute towards the healing of nature and creation in our immediate environs first, before we can think about a more global contribution.

In short, a holistic peacebuilding approach should encompass the preservation of the environment as part of a toolkit for prevention of violent conflict. The environment is becoming an increasingly critical part of the peacebuilding ecosystem.

References

BBC Bitesize. (2022). What does Hinduism teach about the environment? https://www.bbc.co.uk/bitesize/guides/zbvrq6f/revision/3. Accessed 18 April 2022

Bennet, G.D. (2008). A comparison of green chemistry to the environmental ethics of the Abrahamic religions. *Perspectives on Science and Christian Faith*, 60(1), 16–25

Bilal, M.G. (2017). Islamic ecotheology: a religious call to protect ecosystem. http://www.khaleafa.com/khaleafacom/islamic-ecotheology-a-religious-call-to-protect-ecosystem. Accessed 4 Nov 2020

Branton, C.J. (2006). An introduction to religion and ecology. *Environmental Practice*, 8(4), 121–214

Christensen, J.E. (2014). Building an environmental ethics from Confucian concepts of Zhengming and Datong. *Asian Philosophy*, 24(3), 279–293. https://doi.org/10.1080/09552367.2014.960297. Accessed 3 December 2020

Coward, H. (1997). Hindu spirituality and the environment. *Ecotheology*, 3, 50–60

14th Dalai Lama of Tibet (1992). *A Buddhist Concept of Holiness*. Address on February 4, 1992, at New Delhi, India. https://www.dalailama.com/messages/environment/buddhist-concept-of-nature. Accessed 13 Oct 2020

Daneel, M.L. (1998). *African earthkeepers*, vol. 1: *interfaith mission in earth-care*. Pretoria: UNISA Press

Daneel, M.L. (2011). Christian mission and earth-care: an African case study. International *Bulletin of Missionary Research*, 35(3), 130–136

Dube, T., Phiri, K. (2013). Rural livelihoods under stress: the impact of climate change on livelihoods in South Western Zimbabwe. *American International Journal of Contemporary Research*, 3(5), 11–25

Furini, G. (2019). The influence of climate change of the escalating communal conflict between herdsmen and farmers: the case of the Fulani ethnic group in Nigeria. *JANUS.NET e-journal of International Relations*, 10(2), 33–52. https://doi.org/10.26619/1647-7251.10.2.3. Accessed 4 December 2020

Gibson, W.E. (1994). Energy, people and North-South equity: a WCC consultation on climate change. *Network News*, 14(1), 11

Hallman, D.G. (1997). Ecumenical responses to climate change: a summary of the history and dynamics of Ecumenical involvement in the issue of climate change. *The Ecumenical Review*, 49(2), 131–141

Halpert, J. (2012). Judaism and Climate change. *Yale Climate Connections*. https://yaleclimateconnections.org/2012/02/judaism-and-climate-change/. Accessed 15 Sept 2020

Islamic declaration on global climate Change (2015) International Islamic Climate Change Symposium, Istanbul, Turkey, 17–18 August. https://climatenetwork.org/2015/08/18/islamic-climate-declaration-calls-for-fossil-fuel-phase-out/. Accessed 26 Nov 2020

Kopstein, P., Salinger, J. (2001). The ecocentric challenge: climate change and Jewish tradition. *Ecotheology*, 6(1), 60–74

Mwale, N. (2014–2015). Traditional religion in the context of climate change: a Zambian perspective. *The Journal of Humanities*, 13, 1–24

Patriarch Bartholomew. (2021). Message of His All Holiness, Patriarch Bartholomew, for the day of prayer for the protection of the Environment (01/09/2016). *Laudato Si Movement 2021* Retrieved from https://www.laudatosi.org/ecumenical/day-of-creation/message-patriarch-bartholomew/ Francis (2015) *Laudato si': encyclical letter of the holy father Francis on care for our common home*

Rabbi Green, A. (1998). *Kol haneshamah: Shabbat eve*. Wyncote, PA: Reconstructionist Press

Rabbi Waskow, A. (2015). *Rabbinic letter on climate -Torah, Pope, and crisis inspires 425+ Rabbis to call for vigorous climate action*. The Shalom Centre. https://theshalomcenter.org/RabbinicLetterClimate. Accessed 14 December 2020

Rakkhit, P. (2014). Significance of Nature in Buddhism. *Buddhistdoor Global*. https://www.academia.edu/10330777/Significance_of_Nature_in_Buddhism. Accessed 6 December 2020

Rakover, N. (2003). Statement by World Jewish Congress. http://arcworld.org/faiths.asp?pageID=81. Accessed 6 November 2020

Raleigh, C. (2010). Political marginalisation, climate change, and conflict in African Sahel states. *International Studies Review*, 12, 69–86

Reuter, T.A. (2015). The green revolution on the world's religions: Indonesian examples in international comparison. *Religions*, 6, 1217–1231. https://doi.org/10.3390/rel6041217. Accessed 8 Sept 2020

Ronan, M. (2017). Religion and the environment: Twenty-first century American Evangelicalism and the Anthropocene. *Humanities*, 6 (92), 1–15

Tarusarira, J. (2017). African religion, climate change and knowledge systems. *The Ecumenical Review*, 69(3), 398–410

Torabi, M., Noori, S.M. (2019). Religious leaders and the environmental crisis: using knowledge and social influence to counteract climate change. *The Ecumenical Review*, 71(3), 344–355

Toroitich, I.K., & Kerber, G. (2014). *Diakonia*, sustainability, and climate change. *Ecumenical Review*, 66(3), 289–301

World Council of Churches (1994) Statement on global warming and climate change: a call to the Churches. *Ecumenical Review*, 46(2), 247–248)

Yŭ, D.S. (2014). Sentience of the earth: eco-Buddhist mendalising of dwelling place in Amdo, Tibet. *Journal for the Study of Religion, Nature and Culture*, 8(4), 483–501

Part II
Micro-level Case Studies

Chapter 6
Empowering Women to Build Peace: A Case Study of Grace to Heal Trust in Lupane District

Ntombizakhe Moyo-Nyoni and Jean Chrysostome K. Kiyala

Peace is not something you wish for; it's something you make, something you do, something you are, something you give away.
Robert Fulghum

Abstract Research on the role of women in peacebuilding and other community development projects suggests that they are listed among the absentees of peace processes and left out of development agendas. This perception is paradoxical to the role women play in society, and they are most affected by lack of peace and development (poverty, violence, social insecurity, etc.) This action research assesses the impact of empowering women in peacebuilding programs in the Mzola Central Ward in Lupane District, Matabeleland North Province, Zimbabwe by Grace to Heal Trust (GTH), a civil society organisation, in order to establish women's potential to engage in peacebuilding and development activities. Forty women were selected from various rural communities through purposive and convenience sampling, and they were given training in a series of peacebuilding programs via action research cycles. The overarching argument underscored by this chapter contends that equipping women with skills and enforcing gender inclusivity in resolving community problems and promoting a peace and development agenda has proven very productive and helped change the fallacious perceptions about women's agency and resiliency in peace matters. Furthermore, this inquiry reveals that women regained confidence and recognition in their active and interactive participation in responding to conflicts constructively; they also contributed to the development of their respective communities.

Dr. Ntombizake Moyo-Nyoni is a Lecturer in Peace Studies Department, Midlands State University, Gweru, Zimbabwe.

Jean Chrysostome K. Kiyala is a Senior Lecturer, International Centre of Violence, Durban University of Technology, South Africa; Associate Professor and visiting lecturer the University of Bandundu, Catholic University of Bukavu and Evangelical University in Africa based in Bukavu (DR Congo).

J. C. K. Kiyala and G. T. Harris (eds.), *Civil Society and Peacebuilding in Sub-Saharan Africa in the Anthropocene*, The Anthropocene: Politik—Economics—Society—Science 34,
https://doi.org/10.1007/978-3-030-95179-5_6

Keywords Civil society · Grace to Heal Trust · Mzola Central Ward · Peacebuilding · Women empowerment · Lupane · Zimbabwe

6.1 Introduction

The need for women's direct and active participation in peacebuilding is an established fact, as the promotion of gender equality and women's empowerment feature amongst the goals of the Beijing Platform for Action (The World Bank 2002, p. 2). In terms of development, The World Bank Policy Research Report, *Engendering Development—Through Gender Equality in Rights, Resources, and Voice* concluded: "Gender inequality retards economic growth and poverty reduction" (The World Bank 2002, p. 4). In a similar vein, citing The World Bank's (2002) report, Klot (2007, p. 2) asserts: "Gender equality brings to peace-building new degrees of democratic inclusiveness, faster and more durable economic growth and human and social capital recovery". Furthermore, this scholar argues:

> [...] peacebuilding may well offer the single greatest opportunity to redress gender inequities and injustices of the past while setting new precedents for the future. But these opportunities can be enhanced significantly – or constrained – by how the international community sets its priorities for recovery and uses its resources for peacebuilding (Klot 2007, p. 2).

In Africa, where injustices suffered by women—being side-lined in issues of development and dealing with violence that mostly affect them and children—it is extremely necessary and urgent to engage them in efforts to redress past and present unfair treatment to which they are subjected. Zimbabwe is no exception to this reality, as reports detail the nature and extent of atrocities experienced in intrastate brutalities such as the Gukurahundi violence suffered by women. Murambadoro (2015) gives accounts of the most atrocious brutalities in Zimbabwe, namely the Second Chimurenga violence, Gukurahundi egregious acts of inhumanity and the violence related to elections during the 2000s. Killander/Nyathi (2015) aver that "between 10 000 and 20 000 civilians were killed by state and state sponsored agents between 1982 and 1988 in Zimbabwe" (p. 463). In addition to these politically motivated executions, "torture, rape and other sexual offences, genital mutilations, assault, and arson" were common in Zimbabwe (Killander/Nyathi 2015, p. 463). Ever since, calls are often heard from human rights organisations and civil society organisations (CSOs) to bring the truth to light and hold perpetrators accountable. Similarly, the urgency of truth telling has become a prerequisite to national reconciliation (Murambadoro 2015), and the call for healing *the wounds of Gukurahundi* keeps surfacing in studies of conflict transformation and peacebuilding in Zimbabwe (Ngwenya 2017; Ngwenya/Harris 2015).

In this turbulent and troubling context, women have not been left behind in trying to address violence, seeking peace and even initiating processes to redress violence in the community. This has been the case in Mashonaland in the East of Zimbabwe (Chitando 2019), and women's resiliency has been observed through their participation in peacebuilding in rural Zimbabwe (Rukuni et al. 2016). In the overall climate of terror instituted in Zimbabwe and the repressive machinery of the ZANU-PF, women and girls have not been spared, and they are also victims of such violence (Zengenene/Susanti 2019).

These introductory paragraphs have laid the ground for igniting the debate and getting more peace effort trials underway, where women's potential and resiliency are tested and evaluated in order to establish that they can always be actors of society's destiny in terms of achieving peace and development. To examine and determine this, pre- and post-training evaluations were conducted with women, as well as with various community stakeholders, through focus group discussions, to assess the impact of interventions. The findings revealed that women were enriched with new skills, knowledge, and values which resulted in the transformation of attitudes and behaviour of women who became capable of identifying and finding solutions to the problems affecting their welfare. Women were motivated to act beyond the customary perceptions, as they took new peacebuilding initiatives aimed at reducing violence. One way to succeed in this agenda is by tackling gender equalities, starting from peacebuilding agendas, and expanding such efforts to development and economic growth. This problem cannot solely await the international community's intervention; building local capacities to support and involve women's action is equally important and could yield promising outcomes.

Initially, this article focuses on the peacebuilding intervention strategies adopted by a local civil society organisation (CSO) known as Grace to Heal Trust (GTH) to enhance women's peacebuilding activities at personal, family, and communal level. The GTH as a non-governmental faith-based CSO has the mandate of fostering a faith-based process of peace, reconciliation, justice and conflict transformation in Zimbabwean communities (Ngwenya 2017). GTH did not limit its work in Mzola Central Ward to women: interventions were held with forty community leaders, forty youth members and forty women. The aim of the interventions was to empower individuals and the community with peacebuilding skills to be used at personal, family, and communal level. However, this chapter focuses exclusively on programs conducted with women.

Secondly, the chapter focuses on evaluating the impact created by interventions. The evaluations revealed that the interventions had a transformative impact on the lives of the women, and the community was inspired by the new initiatives taken by women. The women were able to take new peacebuilding initiatives and, in one-way or the other motivated each other, their families and the community members to peaceful living. These findings were similar to those of a study conducted by Kache (2014) with Zimbabwean women, assessing the effectiveness of the work conducted by a civil society organisation on voter education, conducted with women prior to an election and which explored attitudes to and behaviour surrounding voting.

The chapter concludes by discussing the findings of the study and their implications. Against the backdrop of this introduction, we report the outcomes of action research which involved the training of women in Lupane District, Mzola Central Ward women in peacebuilding skills, seeing that conflict-handling skills were essential in reducing violence at personal, interpersonal, family, and communal levels.

Women empowerment is investigated from the perspective of civil society's engagement in peacebuilding and economic development. The authors share the views of Klot (2007) on numerous gaps that are discernible in women's absence and inability to leverage their interests in transition processes from military to civilian life after armed conflicts, in social protection, and where HIV/AIDs matters are concerned. Thus, women's participation in peacebuilding must increase (Klot 2007). To contribute to this agenda, this paper will focus on the case of women in Lupane District.

This chapter is subdivided into 8 sections. The introduction is followed by a section that provides the background and context of Lupane district as a research site. The next section reviews the literature and analyses the concept of civil society, women empowerment, and women and peacebuilding. This opens onto the section which presents the theoretical frameworks, based on two paradigms, that guide this research, namely: peacebuilding and grassroots mobilisation theory. The section that follows discusses research design, sampling, data, methods, and ethical standards observed in the course of this inquiry; and this is followed by the presentation of the results which consists of: (1) pre-training observations; (2) interventions (training sessions), (3) post-training evaluation (responses from women); and (4) post-training with different community stakeholders. The discussion of the results comes next, before the two last sections: notable recommendations and conclusions.

6.2 Background and Context

Lupane district, like most communities in Zimbabwe and Africa, is on a quest for positive peace. According to Galtung (1990) positive peace implies the absence of all forms of violence. Lupane was, and still is, ravaged by both physical and structural violence, manifested in different forms. Repressive violence tendencies have been experienced, ranging from the colonial era, the liberation war, the Matabeleland disturbances known as Gukurahundi (Ngwenya 2017; Stauffer 2009) to post-independence political violence experienced almost every election year (Coltart 2008). The district like most parts of the Matabeleland region, suffers from underdevelopment, marginalisation, illiteracy, poverty, male dominance, deprivation, exclusion, and limited livelihood alternatives among many other devastating experiences (Ndlovu-Gatsheni 2003). The situation in Lupane is like that raised by the Institute for Justice and Reconciliation (2008, p. 2) within the South African context, where it was observed that "[m]any post-conflict societies never succeed in resurrecting their communities. Instead, communities remain dysfunctional, violent and divided places and become the breeding ground for fresh cycles of violence".

A parochial political culture is upheld in Lupane; thus, citizens are narrow-minded and mostly concerned with local politics (Masunungure 2006). During the period of the current study, there were two dominant political parties, namely the Zimbabwe African National Union Patriotic Front (ZANU PF), the party that governs the nation and has been in charge since independence in 1980; and the main opposition party, the Movement for Democratic Change (MDC), which has been in competing for power and governance with ZANU PF (Coltart 2008). Owing to the fact that the region is within Matabeleland, most preferred not to disclose their allegiance to the Zimbabwe Africa Peoples Union (ZAPU) which was aligned to the people of Matabeleland before the formation of a unity accord after independence. Amidst the political tensions in which the ZANU PF and the MDC have been enmeshed, women have been left out of development projects and peacebuilding agendas. Lupane is in the Matabeleland North province of Zimbabwe, being considered the centre of the province. Lupane town is located 175 km from the city of Bulawayo, the second largest city of Zimbabwe and the major city of the Matabeleland region. Mzola Central Ward is one of the wards of rural Lupane, being located about 100 km from Lupane town.

Like in any other community exposed to both direct and indirect conflicts, the women of Mzola Central Ward suffer the most. In addition to the socioeconomic and political violence, women's suffering is widened by problems linked to structural male dominance and traditional practices that burden rural women (Southern African Development Community 2016b). The SADC Gender and Development Monitor keep track of the progress made by SADC Member States in the vision of achieving the targets and goals of this regional body's Protocol on Gender and Development (Southern African Development Community 2016a). Despite all efforts by the Zimbabwean government and the regional and global institutions aimed at eliminating all forms of inequality, injustice and violence against women these, as well as gender-based violence, remain prevalent. The United Nations Security Council Resolution (UNSCR) 1325 advocates for the protection of women, the prevention of violence against women and the participation of women in peacebuilding processes (Nduwimana 2006; Villellas 2010). Similarly, the African Union upholds Agenda 2063, focusing on achievement of gender equality, and the SADC has a protocol on gender and development focusing on women empowerment and gender equality. Zimbabwe has its own gender policies, such as the Zimbabwe National Gender Based Strategy 2012–2015, which is essential for sustainable development. Despite all these endeavours, Zimbabwe seems to be lagging, as noted in the SADC 2016 report, that Zimbabwe has a low number of women in local government, with 271 women out of 1,692, which is only 16%.

6.3 Civil Society and Women Empowerment

The concept of civil society is largely covered by philosophical, social, and political sciences, and this leaves it open to various connotations and interpretations. In most cases, it refers to activities that are undertaken independently from the state and private sector; its members commit themselves to serve various interests on a voluntary basis; they use their different organisations' capacity to defend what seems to be neglected by the state, and right what is regarded as unfair, in the areas of human rights, civilian protection, social welfare, health and sanitation, education, economic opportunities, labour rights, conflict resolution, peace processes and societal harmony. According to Paffenholz (2015, p. 108), "Civil society is generally understood as the arena of voluntary, collective actions of an institutional nature around shared interests, purposes, and values that are distinct from those of the state, family, and market". This description encapsulates the key definitional concepts and reflects the nature of CSOs' engagement. From the same perspective, according to The World Bank Social Development Department's Report issued in February 2007, CSOs are:

> The wide array of non-governmental and not-for-profit organizations that have a presence in public life, expressing the interests and values of their members or others, based on ethical, cultural, political, scientific, religious or philanthropic considerations" (The World Bank 2007, p. 3).

It is believed that civil society has an exceptional potential to contribute positively to peacebuilding and alleviation of conflicts; it can achieve this independently as an autonomous actor since peace processes are concerned and such programs are carried out by governments or even by the international community (The World Bank 2007, p. v). The World Bank warns that, despite the apparent success of its initiatives in peacebuilding, "Civil society is not a panacea" (The World Bank 2007, p. v). This resonates with the organisational dimension and capacity of civil society and the range of its activism and commitment that Paffenholz (2015) expounds by asserting:

> Civil society consists of a large and diverse set of voluntary organisations and comprises non-state actors and associations that are not purely driven by private or economic interests, are autonomously organized, typically show civic virtue, and interact in the public sphere (p. 108).

The scope of civil society's field of action is also extended to post-conflict challenges, locally and internationally, and The World Bank (2007) puts forward some indications to support civil society's engagement, especially through the UN Security Council's statement of September 2005: it highlights the comparative leverage of civil society on dialogue and providing community leadership. Civil society's field of operation is extensive and diverse in the context of militarised hostilities:

> During armed conflict, the provision of aid through civil society actors (mainly NGOs but sometimes also associations) increases tremendously as state structures are either destroyed or weakened. There is no doubt that this kind of service is extremely important to support war-affected populations (Paffenholz/Spurk 2006, p. 32).

In such a turbulent and volatile environment, local civil society organisations can be instrumental in facilitating dialogue or mediating between disputing parties that could be fighting at a local level (village or district). For instance, civil society representatives managed to obtain the release of citizens by armed groups and other fighting forces in Nepal; peace zone negotiation or "violence free days as churches negotiated during the war in El Salvador for a child vaccination campaign" (Kurtenbach/Paffenholz 1994); or talks between international or national aid agencies and the warring parties to ensure delivery of aid to their communities (in which country/ies) (Orjuela 2004).

The role civil society plays in peacebuilding is axiomatic (The World Bank 2007), and several studies demonstrate that this is the case both theoretically (Chandhoke 1995) and pragmatically (Belloni 2001), at local or country level (Barnes 2006; Cheema 2011), and also globally, in the international sphere (Pouligny 2005; The World Bank 2007).

Catalysts, such as interventions by CSOs, are essential in stimulating women into active participation in peacebuilding processes. Paffenholz (2015) identified seven functions of civil society in peace processes. These functions include protection, monitoring, advocacy, socialisation, social cohesion, facilitation, and service delivery. Taking a closer look at the grassroots women, one will agree that women are in one way or the other involved in community peacebuilding. Sadly, their works and commitment often go unnoticed, indicating a clear need for a change of mindset of both males and females (McKay/Rey 2001). However, others operate at the first level of conflict analysis and research—the "Macro level"—that is "dominated by the more academic approaches to the analysis of conflicts and actors" (Paffenholz 2004, p. 4). There are prominently visible women, such as Thania Paffenhol, Jennifer F Klot, and many others.

Notwithstanding the great strides that have been made by CSOs towards peacebuilding, "[c]ivil society and donors need to more strategically identify the objectives and demonstrate the relevance of the approaches to different phases of conflict/peacebuilding" (The World Bank 2007, p. v). Activities run by civil society require clear objectives, foreseeable effects, and strategies to deal with institutional constraints, without which CSOs' actions may end in failure (The World Bank 2007, p. v).

After exploring the concept of civil society that is pivotal in understanding the work of GTH, we turn now to discuss the theory of peacebuilding, because of its relevance to the activities of GTH and to determine how peace and healing can be attained through the work of women gathered in a CSO and being guided by the tenets of the search for peace and peace construction.

United Nations Security Council Resolution 1325 (2000) on Women, Peace and Security discusses extensively various options and areas of women's engagement in peace efforts; this includes, but is not limited to, interventions in peace processes in the context of political transition, in tackling gender-based violence, ensuring women are protected, etc. (Nduwimana 2006). The United Nations World Summit Outcome of September 2005, paragraph 116, underlines the agency of women in peacebuilding efforts as follows[1]:

> We stress the important role of women in the prevention and resolution of conflicts and in peacebuilding. We reaffirm our commitment to the full and effective implementation of Security Council resolution 1325 (2000) of 31 October 2000 on women and peace and security. We also underline the importance of integrating a gender perspective and of women having the opportunity for equal participation and full involvement in all efforts to maintain and promote peace and security, as well as the need to increase their role in decision-making at all levels (United Nations 2005a).

[1] United Nations World Summit Outcome, September 2005, par. 116.

Women's activism in the resolution of conflicts and building post-conflict peace is a legitimate responsibility they should not be denied. The successful participation of women in peace agendas can be enhanced by actions that promote gender equality and gender equity in responsibility sharing in matters that concern them the most, such as gender-based violence, the safety and security of women and children during and after armed conflict, human protection and creating equal social and economic opportunities for all. Gender equality is expected to be promoted also in the field of education (African Union 2004; Southern African Development Community 2016b). Gender equality and the empowerment of women, according to a Resolution adopted by the General Assembly of the United Nations on 16 September 2005, the World Summit Outcome, paragraph 58 entails eliminating "pervasive gender discrimination" in the following ways: getting rid of gender inequalities in primary and secondary education; protecting free and equal rights of property inheritance by women, and inheritance of property and housing; making accessible the right to reproductive health; promoting equal labour market access to women as well as sustainable employment and adequate labour protection; ensuring all forms of discrimination and violence against women and the girl child are eliminated; putting an end to impunity; and "promoting increased representation of women in Government decision-making bodies, including through ensuring their equal opportunity to participate fully in the political process" (United Nations 2005a).[2]

To support our earlier statement on empowerment of women, which is a more inclusive vision of United Nations Security Council Resolution 1325 (2000) on Women, Peace and Security (Nduwimana 2006), Article 143(d) of the Resolution adopted by the General Assembly on 16 September 2005, the World Summit Outcome, states that women empowerment should embrace the legal field, such as the application of the rule of law, as declared by the Heads of State and Government of UN member states: "[We] call upon States to continue their efforts to eradicate policies and practices that discriminate against women and to adopt laws and promote practices that protect the rights of women and promote gender equality" (United Nations 2005a).

It is evident that the International Community is concerned with the overall participation of women in peace tactics and strategies, and the need for their inclusion is not questioned. It is rather reinforced, as declared by the heads of states of SADC, in the Solemn Declaration on Gender Equality in Africa, African Union, July 2004, Article 2: [We hereby agree to] "ensure the full and effective participation and representation of women in peace processes including the prevention, resolution, management of conflicts and post-conflict reconstruction in Africa as stipulated in United Nations Resolution 1325 (2000)" (African Union 2004). These statements are being applied in countries such as Sierra Leone and Burundi, where, with the assistance of civil society organisations and in partnership with the UN, national peacebuilding agendas for women have been, established (Klot 2007).

[2] Paragraph 58(a-g) of the Resolution adopted by the General Assembly on 16 September 2005 [without reference to a Main Committee (A/60/L.1)]: 60/1. 2005 World Summit Outcome.

According to the SADC Gender and Development Monitor's 2016 Report, for tracking progress on the implementation of the SADC Protocol on Gender and Development, of a total of 1,187 peace process agreements signed between 1990 and 2017, only nineteen percent mention women, while only five percent refer to gender inequities. The domination by men indicates the serious need for mind-set transformation for both men and women in regard to involvement of women in peacebuilding processes and strategies for dealing with problems experienced by women (Southern African Development Community 2016b).

Unfortunately, such initiatives have been trailing the priorities of the Zimbabwean Government in terms of implementing the African Union's Solemn Declaration on Gender Equality (African Union 2004) as well as making the UN Security Council Resolution 1325 (2000) on Women, Peace and Security work. Taking the agenda of the Peace Building Commission (PBC) agenda to the Zimbabwean context that is being investigated serves the purpose of increasing the regional (SADC) capacity to respond to African women's desire, hope and ambition to be prominently active in peace efforts. Cases of women in South Africa and Burundi that we evoked earlier simultaneously warn about the perceptions of peacebuilding and the need for peace education among people in African communities where the need for peace, economic growth, social and political stability, security, and development are increasingly evident. But deepening a discussion on empowerment of women requires a theoretical and conceptual framework that channels the discussion towards specific and useful outcomes. Thus, empowerment of women in this study is framed within the concept of civil society, while employing peacebuilding theory.

6.4 Women and Peacebuilding

In the previous lines we showed the failures of the Zimbabwean Government to promote women's action in peacebuilding, and the political inequalities that women suffer because of unequal distributions of political responsibility, such as appointment of ministers in the Government. This situation where women are left out slows development. Furthermore, the exclusion of women from key roles in society violates their rights to equal access to social, economic, and political opportunities. When such injustices are experienced, civil society intervenes to redeem the affected social group that is made up of women and girls in this case.

The United Nations Security Council Resolution 1325 (2000) on Women, Peace and Security underscores the need for inclusive and gender-balanced efforts in peacebuilding, as stated in its Article 6, emphasising women's participation for durable peace: "Africa needs lasting peace in order to successfully enter the third millennium. For peace to be lasting it must, among other things, be inclusive. The time has come to understand that women are part of the solution. Together, peace is possible" (Nduwimana 2006, p. 6). The gender question appears in many UN documents and African Union declarations, and it is given a particular consideration, especially in peace (Klot 2007; Nduwimana 2006; United Nations 2005a). Many awareness campaigns need to be launched on the agenda of the Peacebuilding Commission (Klot 2007; Nduwimana 2006; United Nations 2005b).

Klot (2007, p. 1) observes:

> In making gender equality the PBC's only thematic mandate, a new doctrinal imperative was created for ensuring systematic attention and resources to advancing gender equality within transitional recovery, reintegration and reconstruction efforts.

The PBC was created as consolidated strategies by world leaders at the 2005 World Summit as they reiterated the significant role that women play in conflict prevention, resolution and building peace; these leaders demanded the entire and effective implementation of the Security Council Resolution 1325 on Women and Peace and Security (Klot 2007). That was the first time such an important initiative was undertaken, and they sought to create links across "development, peace and security and human rights" (Klot 2007).

The challenge many women face in peacebuilding range from their self-awareness and ability to act for peace to how to deal with the male dominating capacity in peace processes and the lack recognition of their endeavor in the more patriarchal African society (McKay/Rey 2001). These perspectives emerge from a study conducted by McKay/Rey (2001) involving 16 women who participated in a two-day workshop conducted in Cape Town to assess their perceptions of peacebuilding and their involvement in this field. Their views appear legitimate because they are often ignored in peace processes and this gives the impression that their male partners perform better, and their work is not adequately recognised like that of their male partners in peace work is. It is through education that these uneven opinions could change and bring more women and girls into the frontline of peace negotiations and get their inputs as partakers of these processes. Putting forward such self-awareness does not imply undermining or ignoring women's credentials and ongoing commitment to peace in their community.

Isike/Uzodike (2011) observer that women have been active and at the center of peacebuilding efforts in society in pre-colonial Africa, and their agency in peace endeavors has been associated with "their cultural and socio-political roles as well as their contributions to the overall well-being of these societies" (p. 32). This recognition has reflected communal perceptions that link women in peace action with their nature of peacemakers or certain stereotypes such as "woman are more pacific than men" (Isike/Uzodike 2011, p. 32). However, these authors observe that those feminine peacebuilding portrayals have dissipated in neo-colonial Africa, along the myths that sustained them and the sacred character of their being that was acknowledged in pre-colonial Africa (Isike/Uzodike 2011, p. 32). This erosion of past privileges enjoyed by women can be supported by the fact that they now experience social, economic and political marginalisation (Isike/Uzodike 2011, p. 32).

Their forceful and "voluntary" participation in hostilities, and their use as child soldiers and all the negative impact of their association with armed conflicts have further reinforced a fallacious image of their identity, and a global consciousness is needed to redeem or redress those wrong perceptions of women, associating them with violence. The contemporary scholarship on peacebuilding testifies to fact that the depiction of women as peacebuilders continues to prevail; this has been tested and proven correct (Khodary 2016; McKay/Rey 2001). Women continue to participate in peacebuilding and their agency in societal transformation is recognised (O'Driscoll 2017), and even advanced further and reinforced by the United Nations Security Council (Mutis et al. 2011; Nduwimana 2006). A solid characterisation and argument about women as peacebuilders is put forward by Isike/Uzodike (2011, p. 55):

> Women's peace agency is rooted in the values of their womanhood and an ethic of care that values relationships, inter-connectedness and empowerment from which springs forth empathy, co-operation, tolerance and love.

It follows that women and girls' visibility at local and international peace efforts is evident and susceptible of incentivising peace mobilisation and raising more awareness on the ignored and undiscussed peace issues in a systematic way; this would encourage many countries to have women in the driving seats—to initiate and mobilise communities to draw up national action plans that conform to the Peacebuilding Commission's Agenda (PCA) to support women's endeavor in developing durable peace projects and strategies. To attain this vision, both international civil society organisations and the UN should be involved, especially in weak or failed states. The importance of global efforts to spearhead the PCA is well articulated by Klot (2007, p. 7):

> Given the recent initiatives in a number of countries to develop national action plans to prevent SGBV, the PBC and the PBSO might assess the effectiveness of this approach and, in collaboration with UNFPA, UN Action, UNIFEM, UNDP, UNICEF and others, identify ways in which peacebuilding frameworks can more systematically address this concern.

Peacebuilding covers a vast domain of studies, and is also concerned with gender, uneven social relations, economic and political unfairness. Concerted efforts should be undertaken by PBC to consider diverse needs of women such as "water, sanitation and energy; and the PBC should advocate adequate transformations" (Klot 2007) in the areas where women are confronted with a precarious condition of life. This requires the commitment of institutions such as PBC, the Commission of the United Nations Development Fund for Women (UNIFEM), and the Peacebuilding Support Office (PBSO) (Klot 2007). Klot's analysis concluded with the following recommendations: boosting women's participation in peacebuilding, consolidating the capacity of institutions to investigate gender equality and peacebuilding, addressing the "gaps in knowledge and information on gender and peacebuilding", improving "coherence of UN Actions in support of gender equality and peacebuilding", and providing "adequate resources to mainstream gender equality into peacebuilding" (Klot 2007, p. 11).

6.5 Theoretical Framework

The plethora of violent experiences in the Mzola Central Ward clearly reveals that the work of Civil Society Organisations (CSOs) in peacebuilding is of paramount importance. Civil society bodies are voluntary organisations, comprising non-state actors and associations that are not purely driven by private or economic interests, but by the welfare of public sphere. CSOs play a fundamental role in reducing violence and facilitating peacebuilding. This view is supported by Paffenholz/Spurk (2006); they assert: "During armed conflict civil society can advocate on behalf of a peace agreement, against violence and human rights violations, for broad-based participation in the peace process as well as for specific issues" (p. 29). We considered peacebuilding theory and grassroot mobilisation model of activism to sustain the impact of women's action to build peace. These two paradigms are subsequently exmined in the next two sections.

6.5.1 *Peacebuilding*

The notion of peacebuilding emerged in the works of John Galtung, in its "Three Approaches to Peace" that are peacemaking, peacekeeping and peacebuilding (Galtung 1976), later developed by John Paul Lederach (1997) as a theory of constructive change resulting from conflict to avert the adverse effects of such conflict in interpersonal relations (Lederach 2015). The concept of peacebuilding has been embraced by peace researchers globally and is in very much in vogue among the theorists of conflict transformation and peace practitioners. It is described as "an overarching term to describe a long-term process covering all activities with the overall objective to prevent violent outbreaks of conflict or to sustainably transform armed conflicts into constructive peaceful ways of managing conflict" (Paffenholz/Spurk 2006, p. 15).

Peacebuilding is intended to prevent and manage armed conflict and foster durable peace in the aftermath of large-scale ordered violence; it should initiate conditions that are propitious "for economic reconstruction, development and democratisation, but should not be equated and thus confused with these concepts" (Paffenholz/Spurk 2006, p. 16). This observation simultaneously enlarges and restrains the secondary scope of the peacebuilding field from its initial association with "peacemaking and peacekeeping" (Galtung 1976). With Lederach, peacebuilding touches directly on human relations, the harm that conflict causes and the change that is also occasioned to reconcile society (Lederach 1997). Based on the Conflict Transformation Peacebuilding theory of Lederach/Paffenholz (2014) remarks:

> For Lederach, reconciliation comes from truth, justice, mercy and peace. Consequently, he stresses the need to rebuild destroyed relationships, focusing on reconciliation within society and strengthening its peacebuilding potential (p. 5).

Because transformation is described by Lederach (2015) as a way of looking and seeing specific issues at stake, it purveys a set of lenses through which social conflict can be understood; these lenses drive our attention to certain facets of conflict and assist us in bringing the complete meaning of the conflict into honed focus. To explain conflict transformation, he utilises the following procedural categories: "A transformational approach begins with two pro-active foundations: (1) a positive orientation toward conflict, and (2) a willingness to engage in the conflict in an effort to produce constructive change or growth". (Lederach 2015, p. n.p.). Thus reconciliation, justice, peace and mercy become the end of this process.

In the context of Zimbabwe, the search for peace should consider the transformation interrelations as the pathway to durable peace. How local initiatives engage in this process is what this inquiry will strive to determine. Lederach and the transformative peacebuilding school have driven efforts to take peacebuilding to local; however, according to Paffenholz (2014) this approach has been counterproductive, failing to yield desired peacebuilding results, because it has neglected to be "context specific" and "context sensitive, oriented towards the long term, inclusive or accountable to local constituencies" (Paffenholz 2014). It is therefore important, in grounding peacebuilding trials in a particular environment, peace actors should be cognizant of other contours of Lederach's theory such as the specificities of the context, and the interplay of elements that can lead to social change (Lederach 2015). That applied to the situation and context under which this inquiry was conducted.

It will take transformed thinking to view the involvement of women with optimistic perceptions of promoting desired change. Peacebuilding can be an opportunity to redress gender inequalities and injustices of the past while setting new perceptions of the future (Villellas 2010).

6.5.2 Grassroots Mobilisation Theory

This study was influenced by grassroots mobilisation paradigm—a theory assumes that leaders can cling to violating actions like war or adopt structural policies that harm and hinder the progress of community members. In such circumstances, there is a need to empower grassroots people with skills for opposing war and other violating systems. The moment the grassroots begins to take action discouraging the act of violence, leaders remain with no option other than to follow the stance taken by citizens. This theory influenced this study, as focus was given to empowering grassroots women with peacebuilding skills in order to reduce violence. The theory also emerges from the mainstream peacebuilding theory developed by Lederach (1997), and it is well-spelled out in his pyramid of conflict transformation. Level 3, identified as "Grassroots Leadership", is comprised of local leaders, leaders of indigenous NGOs, those in charge of community development, local health officials and leaders of refugee camps (Lederach 1997, p. 39). The range of activities and approaches adopted at this level of peacebuilding architecture encompasses organising local peace commissions, developing grassroots training programs, putting in place mechanisms to fight against prejudice, initiating psychosocial work, and acting upon experiences of postwar trauma (Lederach 1997, p. 39).

Grassroots mobilisation is called to respond to the needs created by conflict. Such an approach has been tested in Zimbabwe by Chivasa (2017) and yielded positive results, because it promotes community participation in different stages of action planning and implementing and opens the way to local ownership and financial self-reliance. Informal peace committees present additional advantages: for example, they provide social groups with equal opportunities to be represented and assist in meeting the needs and aspirations of the larger community (Chivasa 2017). The author contends that, on the one hand, informal peace committees are also confronted with several limitations, and the greatest one is that they lack official recognition by the government; and on the other hand, they also have strengths, like being self-initiated and representing the interests of the communities that offer space for action and they can be replicated (Chivasa 2017).

From a different vantage point, grassroots mobilisation can be linked to the idea propounded by Maxwell (2004), who advocates that communities who have been experiencing series of violence need "demilitarization of the mind"; it entails a process of transformation of people's orientation and understanding, through education with conflict management and conflict transformation skills (Maxwell 2004).

Lupane women were targeted as the starting point to work towards the transformation of Mzola Central Ward. CSOs in Zimbabwe and elsewhere have adopted this theory as they empower grassroots organisations with skills in preventing violence. According to Paffenholz/Spurk (2006), of seven tasks done by CSOS in peacebuilding referred to above, three of these seem very much in line with the grassroots mobilisation theory, namely the protection of civilians, socialisation of civilians and promotion of social cohesion. In a bid to influence the grassroots in Mzola Central Ward, GTH focused on training women with peacebuilding skills. The interventions included a needs assessment, a baseline survey, training workshops and evaluation. The needs assessment and the baseline survey also served as pre-workshop evaluations.

6.6 Design and Methods

6.6.1 Design

Paffenholz (2004) distinguishes three different approaches of conflict analysis, and these are put into separate categories: "between academic and practical, quantitative and qualitative, and participatory and non-participatory methodologies, as seen at all three levels of analysis (macro, meso and micro)" (p. 4). This inquiry employed the third level, namely, the "micro level" which Paffenholz (2004) describes as the state at which "conflictactorsmappings" are extremely relevant "on the local/project level of intervention, especially when implemented in a participatory manner" (p. 4). This current empirical study adopts a similar approach to implementation and intervention in a local community.

This inquiry is designed as action research focusing on a single unit of analysis—an intervention applied by Grace to Heal Trust (GTH)—namely, training women in peacebuilding, and assessing the impact of this empowerment action and women's collective capacity to enhance community peacebuilding. GTH is studied here as a local CSO (L-CSO). The interventions were clustered into three phases, namely: baseline survey/pre training evaluation, workshops/trainings and evaluations. The baseline survey revealed the situation of women and the community before the intervention, which was later compared with the post-training situation of the women and the community at large, revealing the impact of the training (Barbie/Mouton 2001). Furthermore, the baseline survey contributed to the interventions by shaping a series of planning, actions and reflections. After each intervention the organisation, through its personnel with comprehensive skills in peacebuilding programming and processes, reflected and planned interventions, also being influenced by the interactions, feedback, and suggestions from the women, as is the situation in action research.

The action research process involves five progressive activities, namely planning, acting, observing, reflecting, and evaluating the intervention or action undertaken to solve a problem. This reflects the way we approach the problem examined here. The impact of the program/action was assessed through a double evaluation conducted before and after the intervention by a few selected community stakeholders. Trained women in the program were involved at each stage of this cycle.

Action research (AR) design was used in this study. Action research is research based on the collaboration of researcher and the researched, anticipating resolving the problem experienced by the researched. Reason/Bradbury (2002) observe that action research is:

> A participatory, democratic process concerned with developing practical knowing in the pursuit of worthwhile human purpose grounded in a participatory worldview... [I]t seeks to bring together action and reflection, theory and practice, in participation with others, in the pursuit of practical solutions to issues of pressing concern to people, and more generally the flourishing of individual persons and their communities (p. 1).

From the definition proposed by Reason/Bradbury (2002), it is clear that action research is not about acquiring knowledge, but it is about balancing theory and practice. Thus, action research creates a platform where both the researcher and the participants learn from the process and learn from their capacities. Action research embraces several worldviews; it is reflexive and pragmatic (O'Brien 2001) and combines learning with action and research (Zuber-Skerritt 2001). This aspect seems to be fundamental in AR as described below:

> Some of the most important educational principles developed from this holistic philosophy are 'learning through discussion', 'learning by doing' and 'reflective practice', all of which also relate to the second of Revans' concepts of Action Learning: 'action' (Zuber-Skerritt 2001, p. 18).

Fig. 6.1 Action research cycle. *Source* Melinda Kolk[3]

Kolk (2021) suggests a full, detailed twofold process of inquiry and reflection:

- It starts by determining the location of the researcher, where he/she intends to be, and the means to get to the research site.
- Then follows the cycle of action that includes six stages, namely: identification of the problem and envisioned success, developing a plan of action, gathering data, analysing data, and drawing conclusions, adjusting the adopted theory and repeating the cycle, and finally reporting the results (Kolk 2021) (Fig. 6.1).

Action research involves a cycle of planning, acting, and reflecting. The GTH team were involved in several sessions revolving around cycles of planning, actions, and reflection throughout the interventions. The aspect of planning, actions and reflections was not only limited to the GTH team, but the same was done with the women, who became keen on problem identification, planning around the solutions of the problem, implementing the actions towards the solution and reflecting on problem, which then influenced the next level of planning, actions, and reflections. Additionally, the GTH team, some of whom were facilitators of these workshops, reflected after each intervention and planned interventions. From the onset, issues raised during the introductory meetings were debated, and continuously re-visited to develop relevant plans and actions to contribute towards the desired change, as is the norm in action research.

Action research was chosen because it creates a platform where practical problems can be solved while participants learn and act upon the issues at stake. It is a collaborative mode of investigation; it presupposes that those who participate are co-researchers and patterners, that each one's inputs are valued, and that it is transformative, as O'Brien (2001) explains:

[3] https://creativeeducator.tech4learning.com/v07/articles/Embracing_Action_Research.

> For action researchers, theory informs practice, practice refines theory, in a continuous transformation. In any setting, people's actions are based on implicitly held assumptions, theories and hypotheses, and with every observed result, theoretical knowledge is enhanced (p. 6)

This mode of inquiry is beneficial in peacebuilding research which demands that theory be met with practice to impact social or relational change, or conflict transformation (Kaye/Harris 2018). As focus is given to reflection and action, which are fundamental in peacebuilding work, we adopted it to conduct this study.

To this effect, women formulated four peer mediator groups, according to their villages. The peer mediators intervened in community conflicts at village level towards the end of their training and even beyond the period of the programs implemented by GTH. The hope of the GTH team was to see women developing a culture of resolving conflict non-violently and passing this approach on to children, who will pass it from generation to generation. After each peacebuilding initiative taken by peer mediators, feedback was shared during gatherings and the sharing became a resource of learning for the women.

6.6.2 Methods

Qualitative methods were applied both in collecting search conferences, the main source of action research data, and analysing and interpreting them, as suggested by O'Brien (2001). The results presented here were drawn from the interpretative paradigm in discourse analysis (Talja 1999) and content analytical methods (Hsieh/Shannon 2005). Talja (1999) remarks that discourse analysis provides for interpretation of data not depending largely on the amount of data collected (interviews) but the systematic interpretation of available information:

> The research results are not generalizable as descriptions of how things are, but as how a phenomenon can be seen or interpreted. The aim of discourse analysis is to produce interpretations that are intrinsically macro sociologic (p. 472).

We collected sufficient information to ascertain the validity and generalisability of our findings, based on the "subjective interpretation of the content of text data" (Hsieh/Shannon 2005, p. 1278). This approach requires a "systematic classification process of coding and identifying themesor patterns" (Hsieh/Shannon 2005, p. 1278). The process of coding was done simultaneously with thematisation of clustered data to categorise emerging themes, and contextualisation of participants' discourses and their context to allow an adequate interpretation. The same analytical patterns of content and discourse analyses prevailed at the pre-training observation and-post-training evaluation. Emphasis was put on peacebuilding interventions conducted with women and the methodology used, which was predominantly guided by the action research cycle, relying upon the case of women in Zimbabwe's Matabeleland North Province in Lupane district, Mzola Central Ward which provided reflections for CSOs and women empowerment.

The reports influenced the planning, interventions, reflection processes, reporting and scheduling of further interventions. The information for the pre-evaluation observation and post-evaluation was also handwritten, as it was observed that the women in other similar workshops were not comfortable with audio recordings. The usual tendency was that in cases where there were audio recorders, women would accept that the recorders be used, but they would not speak about personal and specific issues, but would only refer to general issues.

The evaluation of this action research study focused on the impact caused by the training on the attitudes, behavior and situation of the women and the community of Mzola Central Ward. Pre- and post-evaluation systems were used for evaluation. The gaps between the pre training and the post training were compared to ascertain the impact created by interventions. Seeing that there was a two-year period between the pre- and post-evaluations, there was a possibility that some factors came into play resulting in the transformation of attitudes, behavior, and contexts. However, observation and some issues shared by participants during the pre-trainings, training and the post-training evaluations can be attested to as signs of positive change.

The post-training workshops evaluation meetings with stakeholders were conducted with a total of forty-eight participants, who were divided into four focus groups which had between ten and eleven participants. The post-workshop evaluations sessions were held with three focus groups of eleven participants per group. The benefit that came with use of focus groups was that focus groups created a more natural environment, as participants interacted and discussed situations as they were. Additionally, the participants inspired each other as they raised practical issues that were linked to their welfare. The collected data was later classified thematically.

6.6.3 *Sampling*

Two non-probabilistic sampling methods were used, namely purposive sampling or subjective sampling, by which the selection of members of the target population depends on the researcher's judgement; and convenience sampling, though which data is collected by respondents. The uses of these two techniques were dictated by the existence of the GTH CSO that was the main source of data gathering. An information letter was sent to the Headman and the ward Councillors, who were both residing within Zola Central ward, and they gave the gatekeeper permission to conduct the inquiry within the ward.

An interesting dynamic was that the Headman, meant to be non-partisan, had a strong allegiance to ZANU PF by virtue of being part of national governance, while the councillor who was chosen through the ballot action belonged to the main opposition party (MDC) and was woman. These two held a dual conflicting relationship, that of belonging to two opposing political parties and that of having different sexes, as the community upholds a very high view of patriarchal dominance. In the Lupane context, being a woman was predominantly associated with inferiority.

The intended outcome of engaging these two leaders at a personal level was gaining a successful entry into the community. Acceptance by these leaders translated into being accepted by the entire community. This was attested to during the post evaluation, when one of the leaders who had not attended the baseline survey, for one reason or the other, opened up and said, "I appreciated a lot in being part of this meeting. At the beginning I did not want to come, today I came because I respected the person who invited me (the Headman), but now I realise the value of coming together as a community; there are many problems here, but based on what is said about women's work, I realise the solutions of these problems can be found". The Headman and the Councillor were tasked to invite participants after discussions with the GTH team on who should be invited. The sample can be regarded as purposive although it is worth noting that the selection was based on the perceptions of these two community leaders tasked by the GTH team. Concerning the selection of participants for the stakeholders meeting, the instruction was to invite people of influence from various sectors of the society, thus community leaders, church leaders, women, youth—including both females and males. A total of 40 participants were invited. For the women's workshops, a total of forty women were invited; these included five members of ZANU PF, five members of the MDC, five members of ZAPU, five young mothers, ten members from religious groups, five from the African Traditional Religions groups and five assuming community leadership roles, which included village health workers, childcare workers, and school committee members. Table 6.1 gives a summary of the women participants.

The main purpose of using these criteria for selecting participants was to create a balance among participants and support the concept that peacebuilding is about cooperation. However, the observation made was that it took a long time for the women to realise the intended purpose as they kept demonstrating contestants' political tendencies especially in the early stages of the intervention.

Furthermore, the engagement of both the Headman and the Councillor selecting participants was in line with conflict-sensitive approaches suggested by Fisher et al. (2000), who observe three elements to be considered by organisations working in

Table 6.1 Participants for women's workshops. *Source* Authors

Designation of women	Number of participants
Women in politics ZANU PF MDC ZAPU	5 5 5
Women representing churches	10
Younger women	5
Women in community leadership	5
Women representing African Traditional Religion	5
Total	**40**

conflicts which are understanding the context of operation, understanding interactions between interventions and the context, and acting upon the understanding of interactions to avoid negative impact and maximise positive impact.

6.6.4 Observed Ethical Standards

O'Brien (2001) notes: "Because action research is carried out in real-world circumstances and involves close and open communication among the people involved, the researchers must pay close attention to ethical considerations in the conduct of their work" (p. 11). To conform with ethical norms, all stakeholders, including local authorities in the research site, were consulted and permission was granted to conduct action research, all participants were treated with dignity and each one's viewpoints were acknowledged and given attention to. The inquirers abide by the guiding principles of confidentiality, liberty, beneficence, and anonymity where this was required and requested by respondents.

To maintain the ethical standard associated with AR approaches, the intentions of the trainings were communicated, and it was emphasised that participation was meant to be voluntary and that those not comfortable were encouraged to freely opt out. From the outset, notes were taken and later compiled into reports for the GTH files.

6.6.5 Data

The nature of this study called for use of empirical text data, which in this case, encompassed a socialisation approach (walking, talking and note-taking), participatory observation, focus group discussion and unscheduled interviews. These approaches rely extensively on qualitative methods and helped us to secure in-depth information concerning the attitudes, behavior and situations of the women and the community before, during and after the trainings. For example, during the first session, women did all they could to maintain their seating arrangements influenced by religious, political and friendship bases. The observation made was that towards the end of the program, women were coming to meetings walking and talking and exchanging stories with their neighbours belonging to different political or social groups.

6.7 Results

The intervention planned did transform the mind-sets of women who realised that peace is not something one wishes, but what a person has the potential to create, demonstrate and live up to in daily life. We shall present these finding in the subsequent sections.

6.7.1 Pre-workshop Observations

As indicated above, a pre-workshop evaluation was conducted during the needs assessment, all stakeholder meetings and the women's introductory meeting. Universally, the pre-evaluation sessions revealed that the culture of violence is widespread.

Human rights violations, segregation in political and religious affiliations, political violence, rape, murder, suicide, domestic violence and gender-based violence were reported to be prevalent. Causes of conflict were linked to intolerance of each other, alcohol abuse, substance abuse, male dominance, limited livelihood alternatives and limited skills in resolving conflicts.

Thus, conflicts were always negative, which resulted in poor management and produced destructive results. Due to such perceptions, peace was considered an impossible aspect. However, the discussions on the nonviolent conflict resolution strategies of negotiation and mediation revealed those women, and the community at large, were aware of these and they had used them in the past. These methodologies were used in various scenarios, such as in negotiations for marriage and in cases where someone's cows grazed on someone else's fields. However, despite being aware of these nonviolent ways of dealing with conflicts, women indicated that that violet method for resolving conflicts were usually employed, as they felt the methods produced quick results and were accepted by the majority. Examples given indicated that conflicts between parents and children, women and husbands, and among youth in community sporting activities were largely resolved through beating or fighting.

Relationships in the community were said to be broken, as indicated above. Community members were divided by past hurts, which were largely linked to political violence. The political climate played a major role in shaping relationships in the community; it is as if politics permeates the livelihoods of the majority of community members. The discussions we had with participants revealed that most people in Mzola Central Ward have exposed each other to different forms of violence in resolving political, economic and social conflicts. Killing through axing, stabbing, burning, beating and food poisoning were common forms used in eliminating opponents during conflicts. There was also a trend of the destruction of property of opponents, where homesteads were burnt, and setting death traps on the livestock of opponents, among other ills. The young and the old, males and females were dragged into the political divisions, such that identities were linked to political affiliation more than any other. For instance, during the early stages of interventions with women, women struggled to relate to one another as they were affiliated to different political parties.

As indicated above, people in Lupane were dependent on environmental agriculture for survival. Due to problems associated with climate change, community members remained with extremely limited livelihood alternatives, which then fuelled resource-based conflicts.

6.7.1.1 Stakeholder's Meeting

A community stakeholders meeting was held prior to the training of women. The session was held with forty participants composed of representatives of all stakeholders in the community, which included community leaders, church leaders, political party leaders, the youth, and the women. As already indicated above, the selection of participants was done through the Headman and the ward councillor. The purpose of the meeting was to introduce GTH and the peacebuilding work to the community. The platform served as a process where community needs were assessed. The meeting was also used to establish the pre-intervention context of Mzola Central Ward.

The initial stakeholders meeting, which also became part of pre-workshop evaluation, was conducted through four focus group discussions with ten participants per group. Subsequently, pre-training focus group discussions with women were held with four focus groups of ten participants per group. Seeing that participants were invited by community leaders, it was clarified during the introductory session that the political/community leaders had nothing to do with the meeting.

During the pre-evaluation meeting, four focus groups were conducted with 10 participants in each, where they were tasked to map the relationships of different entities in the community. The focus of the discussion was to enable community members to realise that there was an ideal community that they were envisioning, and it was spelled out that it was the goal of the GTH to assist the community to attain this ideal community envisioned through training in peacebuilding programs. This meeting became a baseline survey, which then resulted in the training of community leaders, women, and youth in various peacebuilding topics. This meeting mirrored the situation in the community, as tensions were demonstrated during the discussions.

The problems that were said to exist in the community included: issues of political divides, youth and parents conflicts, teen pregnancies, child marriages, abuse of alcohol, disunity among women, problems of underdevelopment, conflicts among community members, and male dominance, which was spoken about and observed during meetings as women could not adapt freely to the meeting setup. It was only the Councillor who could freely express her views. Youth also could not express their opinions freely; it was predominantly the men within the leadership structures who were in control of the discussions. Efforts were made to give everyone a chance, but it was clear that the culture of establishing a level ground for everyone was a missing link among community members.

6.7.1.2 Introductory Meeting with Women

The introductory meeting with women served dual purpose of introducing the program to the women and of conducting an assessment of the situation of the women before training. At the beginning, the women were given a chance to state their expectations and issues of concern. The expression of expectations and concerns was essential as it gave participants a chance to seek clarifications on the program. Objectives were shared, which gave participants a guideline of what was expected from them. Women were tasked with sharing their ideal community through a community mapping exercise. A safe space was created, as participants were divided into to four groups of ten based on their seating arrangements, to enhance sharing and participation. As the women came, it was observed that everyone wanted to sit close to those women they were comfortable with. The community mapping exercised was like that conducted during the community introductory meetings, where core relationships between community members were said to be broken. Women-to-women relationships, men-to-women relationships, children and parents' relationships were all said to be broken. Reasons to supplement the judgement on broken relationships were shared.

Women indicated that they envisioned a community with healthy relations. Their definition of a healthy community was similar to that suggested by the Institute for Justice and Reconciliation (2008, p. 31); it notes that "healthy communities are not communities that are problem free, but instead are communities that work together to ensure a better quality of life for all their members". A similar idea is propounded by Lederach (2003) who proposed that a healthy community is characterised by justice, human security, social justice, spirituality and trauma healing. The idea reflected by these two scholars was observed in participants towards the end of the interventions as they adopted nonviolent means to establish justice, worked together to prevent human insecurities and respected each other's levels of spirituality.

Discussions were held on augmenting a healthy community: women encouraged each other to be responsible for their personal, family and community well-being. Each woman was tasked with making a deliberate effort to lead a life contributing to a healthy society. An example given was that of the family set-up in a Ndebele society, where the man of the home is said to be the head of the family. The women completed the scenarios by saying women are necks supporting and controlling the head and feeding into the activities of the entire body, which resemble children. It was said that the women are the ones to work on their values to become positive, so that they will influence their behaviour and attitudes in a way that that their husband and children will learn from them so that they will go and share the good deeds with their peers. It was said that if the right morals were implemented, they will slowly contribute to the eradication of negative behaviour.

6.7.1.3 Workshops with Women

GTH had three main objectives for the women's program, which were to:

- Empower women with peacebuilding skills so that they will use them in their day-to-day living.
- Enable women to realise their potential, which was to boost their confidence as they were handling conflicts and taking peacebuilding initiatives.
- Instil a sense of ownership of community peacebuilding programs, so that women will be a community resource in peacebuilding.

The goals pursued by GTH programs constituted the underlying objectives set for conducting our inquiry with Lupane women.

In various situations within the globe, women are often the main victims of conflicts as they suffer physical, structural, and cultural violence. Women in Mzola Central Ward, like most women in Zimbabwe, have suffered from social, economic and political violence. Zengenene/Susanti (2019) assert that politically driven violence against women and girls in Zimbabwe, particularly in Harare, is prevalent. Women experience political brutalities not only in town, but in Zimbabwean rural settings as well, and this was found in the Shurugwi district where such violence takes the form of *militarised political violence* (Marongwe 2012). The women in Mzola Central Ward revealed that during the disturbances of the early 1980s, and in the electoral violence of 2002 and 2008, they suffered severely and there was no-one to rescue them. The women were physically beaten by soldiers, war veterans and youth militia in various time of political disturbance. In some cases, their husbands and children became victims, some became injured, and women nursed them. Women were raped, they had their properties (such as cattle and goats) seized, yet again they later bore the trauma associated with the enforced disappearances of their husbands and children. They cared for the wounded and they buried the dead and remained caring for the orphans. All these difficulties left women with inner wounds, which are re-opened by every act of violence. Discussions during training revealed that such experiences left women with limited or no hope of a peaceful society.

The main message during the trainings was to echo that a peaceful society was possible. The UNESCO Seville statement (1986) observed that war began in the minds of men; it is in the minds of men that defences of peace shall come. Training workshops were developed, which were aimed at facilitating the establishment of a healthy society characterised by peaceful living among community members. A healthy society indicates the presence of a culture of peace. A culture of peace is defined by UNESCO (1997) as a set of

> values, attitudes and behaviours that reflect and inspire social interaction and sharing based on the principles of freedom, justice and democracy, all human rights, tolerance and solidarity, that reject violence and endeavour to prevent conflicts by tackling their root causes to solve problems through dialogue and negotiation and that guarantee the full exercise of all rights and the means to participate fully in the development process of their society, … (p. n.p.).

Women were empowered in five main areas, which were: understanding conflicts, violence and peace; negotiation skills; mediation skills; and the role of women in peacebuilding. Additional topics were covered which were based on the requests made by women. These topics revolved around income-generating projects.

6.7.1.4 Understanding Conflict, Violence and Peace

The aim of the session was to enhance women's understanding of the major concepts underpinning peacebuilding, namely conflict, violence, and peace. Women were tasked with brainstorming these three terms, which were later discussed at length using various examples. Illustrations through storytelling, role plays, and drama were used to clarify the distinction between violence and conflict. The illustrations clearly corroborated the assertion made by the Institute for Justice and Reconciliation (2008) that acknowledges violence harms more people than those who initiate it: it rolls down the community. Emphasis was set on the idea that conflict is a demonstration for the desire for change, while violence can be a method used to force change; however, this kind of change will be destructive and temporary. Conflict analysis approaches were considered, setting emphasis on nonviolent methods of resolving conflict and violence prevention.

6.7.1.5 Nonviolent Conflict Intervention Strategies

Conflict intervention strategies of negotiation and mediation were expounded through various methodologies such as small group discussions, drama presentations and sharing of experiences. Women were encouraged to resort to nonviolent ways of resolving conflicts, which usually create room for engaging those with a stake in a conflict to contribute towards the solution of the problem.

6.7.1.6 The Role of Women in Peacebuilding

In all sessions women could identify the problems experienced in the community, considering the topic at hand. The topic on the role of women in peacebuilding, reminded the women of the problems expressed by women and the community throughout the trainings; the question for the women was what role women can play in the transformation of personal, family and community problems. Cases studies of women taking peacebuilding initiatives were shared and discussed. The session culminated in the formulation of peer mediator groups, where women were divided into four groups according to their village. Each group was given the task of selecting at least one problem in the village and working together to find a solution to the problem.

6.7.1.7 Peer Mediator Groups

The tasks given to women after the session on "the role of women in peacebuilding" was aimed at helping women realise their potential. The women were given a chance to share with the larger group their experiences of identifying and solving a problem in their villages. The platform became a learning point for the women, who realised that they have the capacity to use nonviolent ways in resolving problems. Women were empowered with the roles of peer mediator groups and encouraged to take peacebuilding initiatives. This was done to help women learn from their capacity. They learnt that they were a resource to each other, and the resource needed by the community in fostering peacebuilding and development.

The observation made was that building of relationships did not happen overnight, but it was a process that began from the very first moment of planning the interventions. Engagement and interaction created a workable platform where participants decided to make their coming together work, which resulted in the experiencing of the positive fruits of coming together.

6.7.1.8 Income-Generating Projects

The value of the training on income-generating projects was based on the premise that women indicated that most community and family conflicts are linked to economic instabilities. In a Zimbabwean context, Moyo (2009) observes that most women hold onto violent partners/marriages because they cannot sustain their livelihood without men. A strong relationship between economic gains and dependence on violent systems came out strongly during the training. It was said that much political violence involved youth and war veterans, who in one way or the other attain monetary gain as they are rewarded by their superiors for being involved in such activities. Helping women create livelihood alternatives could reduce violence, especially domestic violence, as most women depend on men who are frustrated because of their economic conditions (Moyo 2009). A similar idea is reflected in the works of Carney (1998: 43), who believes that livelihoods are sustainable when a community can cope with recovery from stress and shock; both enhance its self-capacities and assess both the present and the future, while not undermining the natural resources base.

Empowerment of women by means of income-generating projects implied an increase in women's capabilities of taking care of their social development as they ventured into different self-help projects. Participatory engagement dialogues, a dominant feature in action research, were instrumental in development of the women's skills in self-help projects. The approach was similar to the suggestions made by the Institute for Justice and Reconciliation (2008) which advocated that "dialogue is also recognised as a valuable tool with which break down barriers and to build trust. As a global practice, it has the potential to deepen levels of communication and understanding among each other" (p. 16). The observations made were that the topic of self helps projects stimulated most women to active participation as the topic addressed their perceived needs. The trust levels of women seemed to increase during the discussions, participants began to feel at ease with each other and began to relate to other maturely, as opposed to earlier in the session, where women seemed unkind and aggressive towards each other.

The other strength observed during training was that of having program facilitators who were female. This observation is like that made by Research and Advocate Unity (2014) who maintains that there is strength in having women facilitators to train women, as it enhances women's understandings and participation. The trainers created a platform, of which some participants said that if other women have the capacity of acquiring new knowledge and sharing it with others, then they have the capacity to do the same. On another note, having women facilitators became a source of hope to the community women, who confessed that they thought their girl children are growing up to become housewives and in some instances house maids, but they realised that they could do better if they took them to school. Women were motivated to take the education of their girl children seriously. Henceforth, women could speak openly about sexual violence and other women-related problems, being motivated by the fact that it was a women's session and other women could identify with them and give them an ear. Listening was one skill encouraged, seeing that active listening is essential in peacebuilding.

6.7.2 Post-workshop Evaluation: Responses from Women

Post-training evaluations with women were done through focus group discussions at the end of the training period. As indicated above, three focus groups of eleven participants were conducted with a total of thirty-three women. The idea was to have all women who participated in the workshops present, but some were absent as the meeting clashed with another community gatherings. What was learnt was that there is a need to conduct meetings on a day that is not booked for other community meetings. The discussions were guided by the two questions which were:

- What do you remember from the trainings conducted by GTH?
- What have you or other women done about what was learnt?

As indicated earlier, the responses from women were analysed and classified under the following themes.

6.7.2.1 Women Assuming Leadership Positions

Mzola Central Ward women said that GTH training helped them to realise that women were hard-working and having a passion for community peace and development. During the training, it was evident that women were left out in key processes of contributing to community peacebuilding and development. During the evaluations, women indicated that the situation after the training was better, as some women have seconded each other to leadership positions. The implication was that the training empowered women with a realisation of the desire to serve, and realised the need for voting for each other in community leadership positions. Another issue was that was that women have accepted the idea of gender equity regarding assuming community leadership: women now perceived women and men equal when considering community leadership.

However, the problem that emerged was that males are still lagging when it comes to understanding the concept of community leadership and women. It was raised that in most instances women were assigned "deputy" positions, and within such posts their voices were said to be ignored by their male counterparts. In the words of one participant, "we are now involved in the community leadership, but we are at times treated like place holders".[4] This implied that women were given positions, but they were not given a chance to function in those positions fully. In the words of another, "these men will always sit on your shoulders; they give you a position then they control your every single breath".[5] These indicated that power is given to women, but not fully given, which is a notion imbedded in patriarchism, which views only men as capable community leaders. One participant said, "The situation is like you are given a bus to drive, but at the same time given a baby to carry and feed while behind the steering wheel".[6]

6.7.2.2 Women Actively Participating in Community Meetings

The workshops enabled women to realise the available opportunities where they could bring their unique insights and skills in community peacebuilding. Women from Mzola Central Ward indicated that for years they thought attending community meetings was about physical presence and nothing more. In the words of one participant,

> it took us to be trained by GTH to realize that when one attends a meeting she is not going to listen and accept all, but to express her opinions on the subject discussed and contribute on the decisions made… long back leaders will come here and tell us everything and we will go home grumbling, but now it's a different story, we speak our hearts out on issues that affect us and our children.[7]

What was drawn out from this was that when women are given the space and encouraged to be involved in peace building processes, they have a capacity to contribute towards the desired constructive change.

Prior to the training, women seemed to be socialised to believe that males are the only ones with the capacity for solving problems. However, the problem noted during the focus group discussions was that the women trained still believed that their capacity ended at village level; thus, they do not have the capacity of going

[4] Comments made by participant LMCW_5 during focus group discussion (6/12/17).

[5] Comments made by respondent LMCW_2 during focus group discussion (6/12/17).

[6] Comment made by participant LMCW_1's comments during focus group discussion (6/12/17).

[7] Comments made by respondent LMCW_4 during focus group discussion (6/12/17).

beyond that. Putting it in the words of one participant: "our acts *as peer mediators* [italic added] are well known in our villages, anyone with a problem approaches us for help, but can't be found in the meeting of the Headmen contributing to the solutions, that side it's the Headmen's place and his team…"[8]

6.7.2.3 New Knowledge and Skills Attained

Women had their understandings on peacebuilding concepts and practices enhanced. During the training, women could not tell the difference between conflict and violence, and peace was understood as a farfetched phenomenon, which for some was understood as an eschatological concept, which could only be attained in heaven. The post-training discussions revealed that women had attained a new understanding of the major concepts which were fundamental in peace building. Women now spoke of conflict as a problem that needs those with a stake in it to resolve and bring to a desired end. Women had gone on to contribute towards the transformation of various conflicts at personal, family and community level. One point that kept coming up during the discussion was that cooperation is essential in handling conflict, as it involves the creation of a win-win situation and fulfils the needs of all parties involved. Women spoke of violence as a common approach still used by some to deal with conflicts, but they spoke of its negativity, as it does not bring a lasting solution to the problem as the loser will want revenge.

Women revealed that during preparation for elections there was always severe violence that occurs between the two main political parties, but during the evaluation women indicated that there was a decrease in political violence, as the women had been mediating in community and interpersonal conflicts which usually escalated to political conflicts during election periods. What was learnt from this was the strength of action research in preventing violence, as those with a stake in conflict were involved in the processes of transforming their conflicts. The implication of this was that inclusivity created a kind of ownership of the program, which had a high chance of creating a lasting impact, as it allows those with a stake in the conflict to take their time and work together to identify the underlying causes of the conflict and explore possible ways of resolving it.

6.7.2.4 Women and Men Working Together Non-violently against Conflicts

When men and women work together there are high chances of addressing positive as well as political issues without discrimination. The women in Mzola Central Ward avowed that their success was influenced by the active relationship they have with the male community leaders. In the words of one woman, "[o]our strength comes from our Village head who understands us; we update him of every step we are planning

[8] Observation made by participant LMCW_3 during focus group discussion (6/12/17).

to take, he listens to us and gives us all the support, and this has enabled villagers to value our input, to the extent that some of them have requested our mediation especially in their family problems".[9]

However, it was observed that the work done by women was hindered by socio-cultural barriers and patriarchal dominance. In some cases, women needed permission from their husbands to be involved in peacebuilding work. Yet again, some women were accused of being busy bodies, who liked to be involved in other people's affairs.

6.7.2.5 Initiatives Taken by Women After the Training

The discussions revealed that women have taken new peacebuilding initiatives since the time of their training. The mediation groups have taken a lead in identifying problems in the community and seeking solutions for bringing about the desired change. A notable example was the crisis at the school, where there were donations from an international non-governmental organisation for the construction of a classroom block, and the community members were reluctant to get the construction done, and so were passing the tasks to each other. The school development team, which was predominantly ZANU PF, could not push for the work to be done, as they said the resources were donated by an organisation linked to the opposition party seeking to overthrow the government. Having realised that the school was robbed of its right of enjoying the benefits of the school, women began to push to get the project of reequipping the school done. As they did so, the community members slowly participated in the project and with time the momentum picked up, and the work was accomplished.

Furthermore, the trained women went on to educate other women who were not part of the GTH trainings on insights learnt. The teaching was not only limited to women, but the trained women took responsibility for bringing together the female youth and trained them in nonviolent conflict handling skills, sewing skills and values of moral living. During the time of the evaluation, one group of peer mediators had started a project of sewing school uniforms and selling them to community members. The project was fruitful; as said by the headmaster of the school at a separate meeting, the project has resulted in benefiting several learners, as evidenced by the reduction of numbers without school uniforms. This clearly indicated that the trained women had a capacity of learning new skills and putting them into practice, even for economic gain.

Observations during the discussions revealed that women have gained confidence, which was largely reflected in the way they expressed themselves. Additionally, women appeared satisfied by what they have become and what they had accomplished by then. There was a huge change from how women viewed and related to each other and with the GTH team between the commencement of the interventions and the

[9] Remarks made by participant LMCW_2 during focus group discussion (6/12/17).

present. What can be learnt from the situation is that people can learn new ways of doing things and be satisfied by the outcome brought.

6.7.3 Evaluation with Different Community Stakeholders

6.7.3.1 Post-workshop Evaluation

The second session of evaluations was held with community members, who were also invited by the Headmen and the Councillor. A total of forty-eight participants were divided into four focus groups with ten to eleven participants each. A deliberate effort was made to invite representatives of all sectors of the community. The participants were ten youth (five males and five females), five young mothers, five women who were part of the training and five women who were not part of the training, three community men, ten representatives from the churches and ten representatives of community leaders.

The purpose of conducting this meeting was three-fold. Firstly, the evaluation was conducted to assess the impact created by the interventions held with women. Secondly, the evaluations created a platform where GTH received feedback from the community on the peacebuilding work done. Thirdly, the evaluations enabled the community to realise that women were trained in peacebuilding and had been taking new community peacebuilding initiatives. The unintended consequence was to create a platform where the community members learnt much about the work done by GTH, as CSOs at times were thought to be for bringing nothing else other than regime change agenda. Along these lines, sentiments expressed were, "we thought the organisation was here to formulate a new political party". Furthermore, the discussions revealed that participants gained new knowledge on peacebuilding issues during the evaluation, as it is the nature of focus group discussions that participants learn from each other. In the words of one participant: "I have never attended any of these meetings, I always hear from those who usually attend, but as I'm here today I have learnt through discussions that we should live at peace with each other, we should talk about our differences instead of fighting…"[10]

During the session, the community members were randomly categorised into four groups. Efforts were made to ensure that the distribution of participants was equal. Each group was given questions to discuss and bring feedback into the larger group. Each group had a representative of GTH who guided the discussion. Key questions included:

- What do you know about the work done by GTH with women in the community?
- What are some of the effects created by the interventions taken by GTH with the women in this community?

The feedback to the larger group revealed the following themes.

[10] Uttered by participant LMCW_7 during focus group discussion (03/02/18).

6.7.3.2 Women Adopting Noticeable Peacebuilding Achievements

It took the women to start peacebuilding initiatives for the men in Mzola Central Ward to realise the potential of the Mzola Central women. It seemed that the work done by women enabled men to change their attitudes and behaviour towards women, which resulted in women being, included in meaningful leadership positions. One outstanding task talked about was the initiative of establishing a new school block (discussed above). The argument that could be raised was the fact that the ward already had a female Councillor. In essence the Councillor was voted for based on what in most Zimbabwean communities voting is usually based—a specific political party, not a specific individual.

Women were said to be involved in various community committees, such as the school development committee, secondary school building committee and the gardening project committee. However, even though this was seen as progress by the community members, the women had a divergent view, as alluded to above. Women felt they were not given enough space within the leadership opportunities given to exercise their capabilities. One observation made was that when women are on their own, they excel very well in what they do, but when mixed with men they withdraw and allow men to dominate. However, the hope was that the act of letting women assume community leadership roles was a sign of the desired change and development. In the light of this change, it was realised that training can transform areas in which people are socialised. The community members were socialised to male leadership and dominance, but training of women enlightened them to realise the potential of women.

Based on the discussion of the impact of interventions with women, the community meeting recommended the need for training most community members so as to enhance peaceful living in the community. Some notable evidence was attested by one female participant, who was not part of the trained women, who commented:

> I have seen huge changes from those who were trained: my neighbour who attended workshops received fertiliser and some of us were not given *because we did not belong to the political party distributing the product* [italic added]; she went on to share her portion with me and the other neighbour. That action was not usual, but now I am realising it was because of these teachings…[11]

Following this assessment of post-training evaluation, we also investigated different insights of women who showed their entrepreneurship in peace processes that we present in the next section.

6.7.3.3 Women Taking New Peacebuilding Initiatives

The discussions revealed that women have been involved in mediating conflicts in the community and they had gained confidence in the work. Bronkhorst (2011) makes a similar observation: "the confidence for women needs to be built to encourage them

[11] Remarks made by Participant LMCW_6 during focus group discussion on (03/02/18)

to participate in conflict prevention and conflict resolution". The Headman revealed that over the years murder was a common practice, which was initially a taboo, but as time went on people got used to it as it became an order of the day. In the words of the Headman,

> Every Thursday, *sporting days,* [italics added] someone one could be killed or be beaten nearly to death during sporting activities…it was all because of soccer; members of different teams could target harming their opponents' way beforehand and they will settle the score after a soccer game…but these women intervened in such conflicts and these days there are less of such fights and we believed they will end.[12]

The community members could hold grudges, based on what happened in different community platforms, and settle them during the community sporting activities.

Other peer mediation tasks spoken about included marital conflicts and conflicts between neighbours—which were very common, as some neighbours could deliberately wrong their neighbour as a way of settling an unresolved conflict. In some cases, some community members were reported to deliberately let their cows onto the fields of their neighbours, as it was believed that cows fed from maize stock were really killed by disastrous hazards linked to drought. The women's discussions referred to these kinds of negative acts, which seemed to be slowly diminishing. They mediated issues of child abuse, where, in most cases orphans were abused by their caregivers by being assigned tasks which were more than they could bear. Issues of forced marriages were common, and women intervened and discouraged such practices. The common trend reported was that if parents saw their girl child walking with a male person (especially those considered to be well off economically), the parents would force out the girl to go and stay with the man and they would expect that to translate to marriage where the men would have to pay a fine for taking the girl to his home as well as the bride price.

There women were reported to have resolved parent/child conflicts, which were said to have been fuelled by failing to interpret the issue of the rights and responsibilities of children. In some instances, women had intervened in cases where parents were abusing their children, while in some cases there were children who were demonstrating rude behavioral traits such as abuse of alcohol and substance abuse. Very often, such behavior was countered with corporal punishment and other punitive ways of enforcing discipline. It was said that women's peer mediator groupswere engaged at various levels by community members to mediate conflicts, while in some cases they could see the need and intervene. An observation made was that peer mediators were valued as those who promote and encourage nonviolent ways of resolving differences. According to one woman, who was one of the peer mediators, peer mediator groups drew women together. She said, "We felt as a family again. It was difficult to at first to discuss freely with some villagers whom I had considered enemies, but with time we got used to it and we mended our friendship, and we do a lot together." These acts of women seem to tally with the findings of a report done by African Centre for the Constructive Resolution of Disputes (ACCORD) (2011) which states that "when women are soundly trained and effectively capacitated in the

[12] Concern raised by the Headman of the Ward, during his meeting with the research team (03/02/18).

discourses that they seek to influence, they are better equipped to address structural issues that inhibit full and effective participation in both local and high-level peace processes" (p. 44).

6.7.3.4 Women Demonstrating Acts Promoting Peace in the Community

Women were reported to have demonstrated acts that promote peace in the community. All groups reported that there was an elderly woman who had had her homestead burnt. Community women gathered clothes, food, blankets and other necessities from their homes and from the entire community for her. The women further went on to gather building material, to rebuild her house. It was indicated that similar misfortunes had befallen in the past and affected families were left to battle though recovery own their own. Not only so, but another group reported that some women have been seen encouraging other women to attend community meetings and other community gatherings. Two notable tasks alluded to, were the efforts made by women in the school development program of building a new classroom block and the expansion of the community garden. Regarding the community garden, one woman said: "This garden is a clear demonstration of the work done by women. This garden was once known as the ZANU PF garden, but now it is everyone's garden… I m realising that training women did not only influence minds, but it shapes attitudes, which resulted in acts promoting community cohesion.[13]

Among innovative skills uncovered from women were new ways of relating to each other that we present in the section that follows.

6.7.3.5 Women Improved the Way They Relate to Each Other

The discussion revealed that women improved relationships among themselves. In the words of one Village-head, "All these women had labels, known by everyone in the community. You could tell this one is a friend to this one…but now they are one, you can't tell who is who".[14] It was noted that disunity of women was previously observed during community gatherings, especially funerals. Women could relate to each other on political and religious affiliations such that if a member of a certain political party or church denomination led a song during a funeral procession, those from the different political party or church could not sing along. However, the situation has changed; women are now working in unity.

[13] Observation made by participant LMCG_1 during focus groups discussion on (03/02/18).

[14] View expressed by participant LMCV_1 during consultative meeting on 2(03/02/18).

6.7.3.6 Women Improved the Way They Relate with Girl Children

The evaluations revealed that the relationships between women and the female youth improved. During the training, the conflict between parents and the youth were continually discussed. Key issues were focussing on problems associated with moral living, school dropout rates and early pregnancy. It turned out most women were no longer discussing issues of moral living with their children, as they claimed it was the job of the aunties and grandmothers. In some instances, teen girls were considered major enemies of the women, as girls were referred to as "husband snatchers".

As alluded to above, women conducted a session where they empowered youth with issues of moral living. This discussion confirmed that women conducted a training session with the female youth, where they discussed issues of moral living and conflict resolution and taught them sewing skills. The effort of training girls was noted as a victory by the Ward Councillor, who said the meeting had had many challenges, but it had finally materialised. At one point the meeting was announced through ZANU PF platforms, which resulted in ZANU PF members capturing the meeting, which later discouraged the non-ZANU PF youth from attending such meetings. The councillor indicated that they had to try to conduct the meeting in a neutral place, which promoted the inclusion of everyone. Sadly, it appeared that there was no empowerment held with boy children, who for sometime have become victims of structural systems, as their pleas were left unaddressed as attending was directed at elevating the girl child.

6.8 Discussion of the Findings

6.8.1 Pre-training Observations

The findings revealed that the training of women by GTH resulted in the transformation of values, attitudes, and actions of women which improved relationships and the way of life of the Mzola Central community. These findings are similar to those reported by Moyo (2017) and Harris (2013) who validates that training of individuals in peacebuilding is bound to bring about the desired transformative action. The implication of this is that human beings have a capacity for learning new ways of doing things; thus, despite the widespread use of violence in the management of differences, there is a possibility of educating people in nonviolent ways of dealing with differences and these can be learnt and be put to practice.

Greed, corruption and politicisation of aid were commonly reported to influence distribution of food aid and farming inputs from the government and other non-governmental entities, not only in Lupane, but in most parts of Zimbabwe (Desportes/Moyo-Nyoni 2020).[15] Moreover, the division of communities was clearly

[15] Forthcoming paper, Journal for Southern African studies (2021).

seen during social gatherings such as funerals, community meetings, weddings, and church services. What was sad was that prior to the interventions by GTH, the women had accepted all these situations as normal.

The aspect of community members lacking in conflict resolution skills linked well with observations made by Bronkhorst (2011), who advocates that that "lack of legitimate and functioning conflict resolution tools and mechanisms contributes to human insecurity, by affecting the ability of communication to deal with their own vulnerabilities" (p. 6). Yet again, the pre-workshop evaluation with women indicated that women could not make a clear distinction between violence and conflict as they were of the opinion that both conflict and violence were negative and bad.

The pre- and post-evaluations held with two different groups aimed at measuring the differences which were created by training women in peacebuilding. It is almost certain that without this kind of training held by GTH with grassroots women of Mzola Central ward, these changes were not going to occur. Of course, one remains unsure of what the situation was going to be like in the Mzola Central Ward without the peacebuilding work done with women. The pre-workshop evaluation clearly revealed the possible scenario of how the situation was without the training of women. The other question to ask might be: was the situation going to worsen or to transform for the better without GTH's interventions? It is possible the situation was going to worsen, as it is the case that violence repeats it if left untransformed. In the *Programme of Action on a Culture of Peace* (UN Document A/53/370), eight characteristics of a culture of violence are identified: belief in power that is based on force, having enemies, authoritarian governance, secrecy and propaganda, armaments, exploitation of people, exploitation of nature and male domination. The exposure of women in this study to training helped them voice out what they perceived as a vital need in their lives—being empowered to initiate and cooperate in community peacebuilding and development projects.

What emerged from focus group discussions indicates that women can receive training in peacebuilding issues, and they are also able to implement what they learn. This opinion is also expressed in the Iraq context by O'Driscoll (2017), who advocates that lack of women's participation in peacebuilding, was linked to lack of capacity. From the comment made by participant LCMW_1, the idea raised by Mzola Central women was similar to that of Khodary (2016) who observes that state building can provide women with the opportunity to mobilise, be heard and become more politically active due to opening of the democratic space. However, this might not be the case, as the space can also end up being dominated by men. The implication of this is that there is a need to educate men also on the aspect of equality and fairness in the distribution of community roles, so that women will be stimulated to maximise their potential. The idea of inclusion of those with a stake in a conflict in working for peace is fully reflected in the work of Hove/Harris (2019) who observe that "the development of infrastructures for peace (peer mediator groups) at all levels offers the greatest potential to build peace and that strategic planning is a necessary tool to help these infrastructures to be effective" (p. 2).

The participatory nature of this study was encouraged by participants. For instance, participant LMCW_3 made a comment about the availability of GTH and their participation in searching for solutions to their problems, whereas it was not possible to join the Ward Leader to discuss issues and offer paths of remedy.The implication of this is that there is a need to widen the scope of women, when it comes to understanding their operations, not as those limited to boundaries, but to teach them that they can apply the same principles used at village level even with large groups. In the words of O'Driscoll (2017), women are hugely underrepresented in peace processes as core actors and when they do participate their role is limited and the quality of their participation of extremely important. This then speaks to the need for a reorientation of community members, community leaders, national leaders, and the global systems on the involvement of women in peacebuilding, so that they will consider women as a valuable resource at various levels.

Furthermore, participants pointed out that open dialogue and partnership with men who are committed to peacebuilding and a development agenda would consolidate the progress they envision—to transform their respective communities. This view tallied with the findings made by Klot (2007), who noted that the involvement of women in peace building and creating an environment where gender inequalities are broken will accelerate the success of peace efforts and economic growth. Citing the *Integrating Gender into the World Bank's Work: A Strategy for Action (January 2002)*, she remarks: "Gender equality brings to peace-building new degrees of democratic inclusiveness, faster and more durable economic growth and human and social capital recovery" (p. 2).

The appetite to rebuild the community grew in the discussion. The case in point is that which relates to the school that was robbed and vandalised, and women initiated a project to restore this school and re-equip it. The whole community rallied behind these women to give their supportive action. The action echoes the sentiment expressed by Sister Usha Jevan, of the Durban Brahma Kumaris group, in Moyo (2016) who stated, "People should be encouraged to concentrate on doing right, which will slowly eliminate the desire to do wrong, just as it is in the life of a growing onion, which naturally peels off the outer old layers when new layers develop inside" (p. 95).

6.8.2 Post-training Evaluation

Following a preliminary examination of women's commitment to peacebuilding and developmental projects, it was necessary to gather their views regarding the change that undergoing GTH intervention has made in their lives. Such a process demanded that they were assembled in different focus groups to assess the impact of the action. The necessity of reporting is to establish the effects or the effectiveness of the intervention with a view to adjusting the strategy for the subsequent or alternative or improved action cycle. In this regard, Fisher et al. (2000) assert: "The purpose of giving feedback is to maintain and enhance relationships by telling someone the effects of their behaviour on you" (p. 37). Looking at the responses shared by both groups during the post-evaluation session, it is clear without any doubt that the peacebuilding work conducted with women yielded positive results.

Both groups had tangible actions or activities which were alluded to as signs of transformation of attitudes, behavior, and situations. There were very clearly practices that were highlighted as signs of learning. For example, women being able to challenge the status quo and express their opinions during a stakeholder's evaluation meeting, which was different from the stakeholders introductory meeting. This is a clear sign indicating that the self-esteem of women was stimulated by their training, which influenced their views of self, other women, and the entire community. Another interesting point was on the reference made to how women have learnt to express themselves over issues that concern them and their children. The very fact that the views of the two groups seem to tally, suggest that there is no reason to doubt the impact created by program.

The opinions of the community group provide a strong means of validating the views of the women. The community groups, like the women's group, kept raising points validating the transformation of attitudes and actions of the women. Yet again, views from the community group expressed the need for engagement of the entire community in future similar training, and this on its own reveals the impact of the intervention.

Based on the findings of this study, two issues of reliability need to be considered. Some participants for both pre- and post-evaluation within the group of stakeholders were different; however, there was convincing evidence that the interventions resulted in change of attitudes, behavior and the situation of the women. Secondly, there is a possibility that the participants knew that positive feedback was expected by the evaluators and therefore aimed at pleasing them. However, the responses given by the two groups during the post-training evaluation seemed linked and genuine, as participants openly spoke about issues in the presence of neighbours and villagers, showing that what was said were authentic and practical. The responses indicated that women learnt new ideas and were keen to adopt them for life. The evaluations were done at the end of the duration of GTH's work with the Mzola Central community, and it is not certain what the situation will be like afterwards. However, what is certain is that at the time of the evaluation, the culture of violence seemed to be diminishing as the culture of peace was slowly introduced.

6.9 Recommendations and Conclusions

The interventions held by GTH with Mzola Central Ward women were aimed at building safe, integrated communities, which can work together to transform conflicts for social justice and maintain a community fabric that aims for sustainable development. Peacebuilding requires transformation of minds, hearts behaviours, and human relations. The women in Mzola Central demonstrated their capacity to learn new concepts and skills and the ability to implement them through taking new initiatives. The involvement of women in peace building is a fulfilment of the United

Nations Security Council Resolution 1325, which advocates for the prevention of violence, protection of women and participation for women in peace processes (Nduwimana 2006). It has been observed that such interventions go undocumented. Thus, providing a systematic record of a locally-based trial of the UN Resolution 1325 in the context of Zimbabwe opens the learning platform for women, communities, CSOs and other social movements to appreciate women's agency, resiliency, and determination to arise from defeating stereotypes and fallacious perceptions that ignore their potential, and resolve to better their communities.

It is worth noting that to foster durable peace and sustainable development programs, there is also a need to audit the obstacles women have met and what have been global past failures, and to draw historical lessons with a view to embracing new values and principles and using ongoing research to empower women and move the peace and economic agenda forward (Mutisi et al. 2011; The World Bank 2007). It goes without saying that addressing issues around gender inequality (Klot 2007), and recognising women's natural inclinations to peace, communal care, and their aspirations for societal well-being (Isike/Uzodike 2011) are essential factors that could optimise their productivity and these can yield tremendous peace and development goods, as the Lupane women have demonstrated.

It is evidenced that GTH contributed greatly to enhancing the capacity of women in Mzola Central Ward and their involvement in peacebuilding. Women developed an interest in personal, family and community peace. However, a great deal more should be done to ensure that women have a space to maximise their potential. It is also clear that interventions done by CSOs in communities are of fundamental importance and effective in community peacebuilding. In the light of this, the study makes following recommendations:

- Grassroots men should be trained in peace building as they are a key instigator of violence in communities, and they are a major hindrance in the desire of peaceful living expressed by women.
- Violence in the community is widespread: therefore the community might benefit from extensive violence prevention workshops, starting with children, so that they will not grow up with a culture of violence.
- Various trainings should be conducted with women, seeing that they have the capacity to learn new ideas and later implement new initiatives based on what was learnt. Furthermore, women should be given more support as they participate in peacebuilding processes.

The limitations of this inquiry were testing the GTH in Mzola Ward only and using the action research cycle only in one socio-economic setting. However, it still carries weight as asserted earlier: the nature of text data gathered for this research and the analysis and interpretation of facts were acceptable scientific procedures that ensured the reliability and validity of the outcomes reported here. The body of evidence tabled here and the subsequent engagement with existing scholarship in this field corroborate the established premise that women's empowerment is a key driver of societal economic growth and communal durable peace. Women need financial support, space and opportunities to display their full potential.

This chapter concludes by recommending that civil organisations like GTH should extensively conduct training with various stakeholders in communities, including grassroots men, for the transformation of the mind-set of individuals and the community, thereby contributing towards a culture of peace and the reduction of violence as a way of dealing with differences.

Appendix 6.1: Schedule of Activities and Participants

Period	Activity	Participants involved	Number of participants	Purpose of intervention	Member responsible
October 2014	Initial contact	Key community leaders	2	Expression of interest in conducting peacebuilding work in the community	GTH director GTH associate (a peace builder whose origin is in Mzola)
February 2015	Needs assessment	Various community members	13 homesteads were visited	Assessment of the levels, extents and effects of violence	GTH team
March 2015	Mobilization of participants for stakeholders' workshop	Headman and the councillor	2	Discussing the recruitment in preparation for the stakeholders' workshop	GTH Director GTH youth officer and GTH women's officer
March 2015	Stakeholders' meeting		40	Baseline survey Introduction of the organisation to the community Spelling out the objective of the workshop	GTH Director GTH Women's program officer GTH Youth Officer GTH programs officer
April 2015	Mobilization for women's workshops	Headman and the councillor	2	Discussion on the recruitment criteria for the women's workshops	GTH women's programmes officer GTH associate

(continued)

(continued)

Period	Activity	Participants involved	Number of participants	Purpose of intervention	Member responsible
May 2015	Women's introductory workshop	Women from various villages of Mzola central	40	Introducing GTH and its programs to the women Establishing a baseline survey Pre workshop evaluation	GTH women's programmes officer GTH associate
May 2015 – Nov 2017	Workshops with women	Community women	Numbers ranged between 30 and 44	Empowering women with skills of peacebuilding	GTH women's programme officer GTH associate
Dec 2017	Post-workshop evaluation with women	Women who attended workshops	33	Assessing the impact effected by interventions made	GTH women's officer GTH youth officer GTH programs officer GTH programs officer GTH member on internship
Feb 2018	Post-training evaluation with community stakeholders	Various community stakeholders	48	Evaluation of the impact created by the GTH trainings	GTH programmes manager GTH women's officer GTH youth officer Two Associates of GTH

References

ACCORD (2011). Resolution 1325 in 2020: Looking Forward, Looking Back. https://reliefweb.int/sites/reliefweb.int/files/resources/1325_in_2020.pdf. Accessed 14 May 2021.

African Union (2004). Solemn Declaration on Gender Equality in Africa. https://www.mrfcj.org/pdf/Solemn_Declaration_on_Gender_Equality_in_Africa.pdf Accessed 10 April 2022.

Barbie, E. & Mouton, J. (2001). *The Practice of Social Research.* Cape Town: Oxford University Press.

Barnes, C. (2006). Agents for change: Civil society roles in preventing war & building peace. The Hague: European Centre for Conflict Prevention.

Belloni, R. (2001). Civil society and peacebuilding in Bosnia and Herzegovina. *Journal of Peace Research, 38*(2), 163–180.

Bronkhorst, S. (2011). *Climate change and conflict: lessons for conflict resolution from the southern Sahel of Sudan.* Umhlanga Rocks: ACCORD.

Carney, D. (1998) Implementing the sustainable rural livelihoods approach. Paper presented to the DfID Natural Resource Advisers' Conference. London: Department for International Development.

Chandhoke, N. (1995). *State and civil society: Explorations in political theory.* Thousand Oaks California: SAGE Publications.

Cheema, G.S. (2011). Engaging Civil Society to Promote Democratic Local Governance: Emerging Trends and Policy Implications in Asia. https://icld.se/app/uploads/files/forskningspublikationer/icld-wp7-printerfriendly.pdf. Accessed 14 May 2021.

Chitando, A. (2019). From victims to the vaunted: Young women and peace building in Mashonaland East, Zimbabwe. *African Security Review, 28*(2), 110–123.

Chivasa, N. (2017). Efficacy of informal peace committees to peacebuilding: Evidence from Seke district, Zimbabwe. *African Evaluation Journal, 5*(2), 1–11.

Coltart, D. (2008). A Decade of Suffering in Zimbabwe: Economic Collapse and Political Repression under Robert Mugabe. Centre for Global Liberty and Prosperity Development Policy Analysis No. 5., Washington, D.C.: The CATO Institute.

Desportes, I., & Moyo-Nyoni, N. (2021). Depoliticizing disaster response in a politically-saturated context: the case of the 2016-2019 drought in Zimbabwe. *Journal of Southern African Studies,* (forthcoming 2021).

Fisher, S., Matovic, V., Ludin, J., et al. (2000). *Working with Conflict 2: Skills and Strategies for Action.* London: Zed books.

Galtung, J. (1976). Three approaches to peace: Peacekeeping, peacemaking and peacebuilding. https://www.galtung-institut.de/wp-content/uploads/2016/06/galtung_1976_three_approaches_to_peace.pdf. Accessed 14 May 2021.

Galtung, J. (1990). Cultural violence. *Journal of Peace Research, 27*(3), 291–305.

Hove, M., & Harris, G. (2019). *Infrastructures for Peace in Sub-Saharan Africa.* Cham: Springer.

Hsieh, H-F., & Shannon, S.E. (2005). Three approaches to qualitative content analysis. *Qualitative Health Research, 15*(9), 1277–1288.

Institute for Justice and Reconciliation (2008) *Community healing participation training material.* Cross Print, Cape Town.

Isike, C., & Uzodike UO (2011) Towards an indigenous model of conflict resolution: Reinventing women's roles as traditional peacebuilders in neo-colonial Africa. *African Journal on Conflict Resolution, 11*(2), 32–58.

Kache, C. (2014). Does empowering women to register and vote make a difference? A preliminary report on women's experiences with 2013 elections. Harare: RAU.

Kaye, S., & Harris, G. (2018). Participatory Action Research For Peacebuilding. *Peace Review,* 30(1), 62–68.

Khodary, Y.M. (2016). Women and peace-building in Iraq. *Peace Review,* 28(4), 499–507.

Killander, M., & Nyathi, M. (2015). Accountability for the Gukurahundi atrocities in Zimbabwe thirty years on: prospects and challenges. *Comparative and International Law Journal of Southern Africa*, 48(3), 463–487.

Klot, J.F. (2007). Women and peacebuilding https://www.un.org/peacebuilding/sites/www.un.org.peacebuilding/files/documents/wgll_backgroundpaper_29_01_08.pdf Accessed 14 May 2021.

Kolk, M. (2021). *Embrace Action Research: Improve classroom practice with action research ... and tell the story.* https://creativeeducator.tech4learning.com/v07/articles/Embracing_Action_Research. Accessed 14 May 2021.

Kurtenbach, S., & Paffenholz, T. (1994). Kirchen können in Kriegen vermitteln. El Salvador und Mosambik zeigten die Chancen kirchlicher Schlichtung. *Der Űberblick, 3*(1994), 116–119.

Lederach, J.P. (1997). *Building Peace: Sustainable Reconciliation in Divided Societies.* Washingston, D.C.: United States Institute of Peace Studies.

Lederach, J.P. (2003). *The Little Book of Conflict Transformation.* New York: Good Books

Lederach, J. (2015*). Little book of conflict transformation: clear articulation of the guiding principles by a pioneer in the field.* New York: Good Books.

Marongwe, N. (2012). Rural women as the invisible victims of militarised political violence: the case of Shurugwi district, Zimbabwe, 2000-2008. D Phil. University of the Western Cape, South Africa.

Masunungure, E. (2006). Why Zimbabweans won't rebel: Part 1. http://archive.kubatana.net/html/archive/opin/060922em.asp?sector=LAB%26year=2006%26range_start=31 Accessed 10 April 2022.

Maxwell, A.M. (2004). The role of education in demilitarising sub-Saharan Africa. *Journal of Peace Education, 2*, 73–89.

McKay, S., & Rey, Cdl. (2001). Women's meanings of peacebuilding in post-apartheid South Africa. Peace and Conflict: Journal of Peace Psychology *7*(3): 227–242.

Moyo, N. (2009). The impact of structural violence and the sprees of HIV/AIDS among women and girls in Bulawayo. MCOMM thesis, University of KwaZulu-Natal. https://researchspace.ukzn.ac.za/. Accessed 26 Feb 2021.

Moyo, N. (2016). Peacebuilding among ex-prisoners and their families: Enhancing the impact of the Second Chance Rehabilitation Centre, Zimbabwe. PhD Thesis, Durban University of Technology.

Moyo, N. (2017). Using restorative justice approaches with ex-prisoners and their families: an action research project in Zimbabwe. In: Kaye S, Harris GT (eds.), *Building peace via Action Research: African case studies* (pp. 207–222). Addis-Ababa: UN University for Peace.

Murambadoro, R. (2015). 'We cannot reconcile until the past has been acknowledged': Perspectives on Gukurahundi from Matabeleland, Zimbabwe. *African Journal on Conflict Resolution, 15*(1), 33–57.

Mutisi, M., Ogunsanya, K., Ettang, D. (2011) Resolution 1325 in 2020: Looking Forward, Looking Back. https://media.africaportal.org/documents/1325_in_2020.pdf. Accessed 14 May 2021.

Ndlovu-Gatsheni, S.J. (2003). The post-colonial state and Matabeleland: Regional perceptions of civil-military relations, 1980–2002. In R. Williams, G. Cawthra, and D. Abrahams (Eds.) *Ourselves to Know: Civil-Military Relations and Defence Transformation in Southern Africa* (pp. 17–38). Pretoria: Institute for Security Studies.

Nduwimana, F. (2006). United Nations Security Council Resolution 1325 (2000) on Women, Peace and Security Understanding the Implications, Fulfilling the Obligations. New York: UN Office of the Special Adviser on Gender Issues and Advancement of Women (OSAGI).

Ngwenya, D. (2018). *Healing the Wounds of Gukurahundi in Zimbabwe: A Participatory Action Research Project*. Cham: Springer.

Ngwenya, D., & Harris, G. (2015). The consequences of not healing: Evidence from the Gukurahundi violence in Zimbabwe. *African Journal on Conflict Resolution 15*(2), 35–55

O'Brien, R. (2001). Um exame da abordagem metodológica da pesquisa ação [An Overview of the Methodological Approach of Action Research]. In R. Richardson (Ed.) *Teoria e Prática da Pesquisa Ação [Theory and Practice of Action Research].* Universidade Federal da Paraíba, João Pessoa, Brazil (English version).

O'Driscoll, D. (2017). Women's participation in peacebuilding and reconciliation in Iraq. https://reliefweb.int/sites/reliefweb.int/files/resources/239-Womens-Participation-in-Peacebuilding-Iraq.pdf. Accessed 14 May 2021.
Orjuela, C. (2004). *Civil society in civil war. Peace work and identity politics in Sri Lanka.* Gothenburg: Department of Peace and Develoopment Research.
Paffenholz, T. (2004). *Designing transformation and intervention processes. Transforming Ethnopolitical Conflict.* Cham: Springer.
Paffenholz, T. (2009). *Civil society and peacebuilding.* Geneva: The Graduate Institute of International and Development Studies.
Paffenholz, T. (2014). International peacebuilding goes local: analysing Lederach's conflict transformation theory and its ambivalent encounter with 20 years of practice. *Peacebuilding, 2*(1), 11–27
Paffenholz, T. (2015). Civil society and peacebuilding. *Development Dialogue* 63, 108–118.
Paffenholz, T., & Spurk, C. (2006). Civil society, civic engagement, and peacebuilding. *Social Development Papers: Conflict Prevention and Reconstruction,* 36, 10–55.
Paffenholz, T. & Spurk, C. (2010). A comprehensive analytical framework. In T. Paffenholz (Ed.) Civil Society and Peacebuilding: A Critical Assessment (pp.65–78). Boulder: Lynne Rienner.
Pouligny, B. (2005). Civil society and post-conflict peacebuilding: Ambiguities of international programmes aimed at building 'new'societies. *Security Dialogue, 36*(4), 495–510.
Reason, P., & Bradbury, H. (2002). Action Research: an opportunity for revitalizing research purpose and practices. *Qualitative Social Work* 2(2), 155–175.
Rukuni, T., Shanyisa, W., Madhuku, J., & Musingafi, M. (2016). Enhancing women participation in peacebuilding and decision making process in Zimbabwean rural communities. *Information and Knowledge Management, 6*(9), 54–60.
Schirch, L. (2004). *The little book of strategic peacebuilding.* New York: Good Books.
Southern African Development Community (2016a). SADC Gender and Development Monitor 2016. https://www.sadc.int/issues/gender/sadc-gender-and-development-monitor-2016/. Accessed 14 May 2021.
Southern African Development Community (2016b). SADC Gender and Development: Tracking progress on implementation of the SADC protocol on gender and development monitor 2016. SADC, Harare: SARDC.
Stauffer, C.S. (2009). Acting out the myths: the power of narrative discourse in shaping the Zimbabwe Conflict of Matabeleland, 1980–1987. University of KwaZulu-Natal, Unpublished.
Talja, S. (1999). Analyzing qualitative interview data: the discourse analytic method. *Library and Information Science Research 21*(4), 459–477.
The World Bank (2002). Integrating Gender into the World Bank's Work: A Strategy for Action. https://genderinsite.net/sites/default/files/worldbankstrategypaper.pdf. Accessed 14 May 2021.
The World Bank (2007). *Civil Society and Peacebuilding: Potential, Limitations and Critical Factors.* Washington, D.C.: The International Bank for Reconstruction and Development / The World Bank.
UNESCO (1986). The Seville Statement On Violence, elaborated by 20 signataires from 13 countries at the 7th CICA, Seville, May 1986, disseminated by decision of the General Conference of Unesco at its twenty—fifth session Paris, 16 November- 1989. https://www.researchgate.net/publication/232418248_The_Seville_Statement_on_Violence/link/594b916caca2723195e130fd/download. Accessed 3 Oct 2021.
UNESCO (1997). UN Documents Gathering a body of global agreements, Document A/RES/52/13, Resolution adopted by the General Assembly [without reference to a Main Committee (A/52/L.4/Rev.1 and Add.1)] 52/13. Culture of peace 20 November 1997, Fifty-second session, Agenda item 156. http://www.un-documents.net/a52r13.htm. Accessed 3 Oct 2021.
United Nations (2005a). Resolution adopted by the General Assembly on 16 September 2005 [without reference to a Main Committee (A/60/L.1)]: 60/1. 2005 World Summit Outcome. https://www.un.org/en/development/desa/population/migration/generalassembly/docs/globalcompact/A_RES_60_1.pdf. Accessed 14 May 2021.

United Nations (2005b). "Resolution A/Res/60/1 adopted by the General Assembly on 16 September 2005 [without reference to a Main Committee (A/60/L.1)]: 60/1. 2005 World Summit Outcome https://www.un.org/en/development/desa/population/migration/generalassembly/docs/globalcompact/A_RES_60_1.pdf. Accessed 14 May 2021.

Villellas, M. (2010). Background paper The role of women in peacebuilding: Proposals for the implementation of 1325 resolution. https://www.researchgate.net/publication/283425254. Accessed 14 May 2021.

Zambara, W. (2015). Community Healing: A Training Manual For Zimbabwe. Cape Town: Institute for Justice and Reconciliation / Peacebuilding Network of Zimbabwe.

Zengenene, M., & Susanti, E. (2019). Violence against women and girls in Harare, Zimbabwe. *Journal of International Women's Studies 20*(9), 83–93.

Zuber-Skerritt, O. (2001) Action Learning and Action Research: Paradigm, Praxis and Programs. In S. Sankara, B. Dick, Passfield (Eds.), *Effective Change Management through Action Research and Action Learning: Concepts, Perspectives, Processes and Applications* (pp. 1–20). Lismore: Southern Cross University Press.

Chapter 7
Peacebuilding Through Indigenous Knowledge Systems: Lessons from Civil Society Organisations in Zimbabwe

Ashton Murwira

Abstract Since the attainment of independence in 1980, Zimbabwe has remained troubled with the question of how to effectively build peace. This question has been tackled through different peacebuilding approaches and actors yielding mixed results. The liberal peacebuilding model has been dominant from 1980 to the present. Unfortunately, the nation remains stuck in negative peace. Against this background, a crop of civil society organisations (CSOs) emerged in peacebuilding using the Indigenous Knowledge Systems (IKS) which is housed under sustainable peacebuilding theory. The CSOs have made use of local cultures, language, idioms and practices mainly in the rural areas with the aim of transforming conflict hotspots into zones of peace. However, there is scarce conventional literature on the lessons, experiences and results of CSOs that have used IKS in peacebuilding in Zimbabwe. The lessons drawn in this chapter, help peacebuilders on the extent to which IKS are relevant or can be effective in building durable peace. The chapter unbundles CSOs peacebuilding experiences using documentary review and in-depth interviews with peacebuilding practitioners, local communities, and academics.

Keywords Peacebuilding · Sustainable peace · Civil society · Indigenous knowledge

7.1 Introduction

The quest for peace remains a daunting task for both academics and peacebuilding practitioners. Societies are at one point at peace or at war and again in search of solutions to resolve and transform conflicts. Against this background, civil society orgnanisations (CSOs) have occupied the middle space (between communities and the government) with an effort to building peace in societies. The chapter begins with a brief background of the nature of conflicts that confront Zimbabwe, exposing the weaknesses of past peacebuilding efforts that have resulted in conflict resurgence or the continuation of fragile and negative peace. The section highlights the categories of CSOs that are found in the peacebuilding space. The next section of the chapter

Dr Ashton is a lecturer in the Department of Political and Administrative Studies with the University of Zimbabwe.

J. C. K. Kiyala and G. T. Harris (eds.), *Civil Society and Peacebuilding in Sub-Saharan Africa in the Anthropocene*, The Anthropocene: Politik—Economics—Society—Science 34, https://doi.org/10.1007/978-3-030-95179-5_7

examines peacebuilding theory and locates the inception of Indigenous Knowledge System (IKS). The chapter then turns to discuss lessons drawn from CSOs building peace through IKS. Lastly, a conclusion is drawn. Pseudonyms which include Civil Society Official A (CSOA), Civil Society Official B (CSOB), Civil Society Official CCS C (CSOC), PeacebuilderA, Community LeaderA are used to present the findings from interview respondents.

7.2 Unpacking Zimbabwe's Broken Past

Zimbabwe has experienced many conflicts since the colonial era. The major and recurring conflicts are structural in nature. During colonial rule from 1890-1980, the successive white minority governments retained power using inhumane methods such as violence, oppressive and discriminatory laws against the black majority. The ills of the colonial system led to African resistance that morphed into an armed struggle that resulted in the death of many people and left the society divided. The liberation struggle, a series of negotiations and international pressure led to the independence of the country in 1980 (Laakso 2003, p. 2; Munhande/Nciizah 2013). Unfortunately, Zimbabwe inherited a system of authoritarian rule from the white colonial government (Masunungure 2011). While CSOs were present during the Mugabe era, they failed to make an impact.

The peacebuilding space has often been dominated by the political elite, with CSOs playing a secondary role. At independence, there were calls made by the political elite in Zimbabwe African National Union (ZANU) led by Robert Mugabe, to reunite and reconcile with 'former enemies'. The 'enemies' were the Rhodesian Front (RF) of the Smith government and later Zimbabwe African People's Union (ZAPU) led by Nkomo. In the reconciliation call, the political elite was concerned with state- building as opposed to peacebuilding, hence, the reconciliation was aimed at gaining control over the government by the black majority. Sachikonye (2012) argues that the reconciliation between whites and blacks was based on interests rather than attitudes. As a result, positive peace was not achieved since it requires a change in attitudes, beliefs and perceptions about the other person or group. The high politics (power politics) during this era made it impossible for CSOs' voices to be heard by the state and the governed.

Aside from the black and white tensions, the black-to-black conflict also emerged in the first decade of independence. The country witnessed major human rights atrocities that were committed by the state in Midlands and Matabeleland areas during Operation *Gukurahundi*. The affected areas lie in the central and southern parts of Zimbabwe, where the main inhabitants are the Ndebele-speaking people who are the second-largest ethnic group in the country. The violations left huge scars which are yet to be addressed up to the time of writing. In fact, there are still debates on how to effectively heal the wounds (Mashingaidze 2005, p. 85; Sachikonye 2012, p. 19; Cameroon 2017, p. 1). The calls for reconciliation were elite-centric. With the polarisation, CSOs are also divided on how the 1983–6 conflict can be resolved and transformed into durable peace. However, the post-Mugabe era has seen opening up on the *Gukurahundi* conflict and results are yet to be seen on how CSOs will effectively tackle it. The black-to-black reconciliation between the political elite in the ZANU and the ZAPU ended the violence and resulted in the Unity Accord signed in 1987. The terms of reconciliation in the Unity Accord resulted in the ZAPU joining the ZANU in the government (Mashingaidze 2005, p. 86). However, the process was more of political pragmatism in search of stability and power consolidation than promoting genuine reconciliation. The indigenous peacebuilding approaches were conveniently ignored.

Apart from the first decade experiences, the country has been torn and polarised by a series of election-related violence and conflicts over land and natural resources. The failed reconciliation efforts, epitomised by recurring conflicts, indicate the need for a holistic peacebuilding strategy in Zimbabwe. Paffenholz (2010) states that peacebuilding is a process aimed at achieving peace in which structures and institutions are developed based on justice, equity and cooperation; and the aim of peacebuilding is to prevent and manage armed conflict and sustain peace after large organised violence has ended.

7.3 Civil Society Organisations and Peacebuilding in Zimbabwe

Gyimah-Boadi (1996) defines civil society as the realm between the household, family and the state. According to Sachikonye (1995, p. 7) civil society relates to "an aggregate of institutions whose members are engaged primarily in a complex of non-state activities, economic and cultural production, voluntary associations and household life and who in this way preserve and transform their identity by exercising all sorts of pressures or controls upon state institutions". According to Spurk (2010, p. 6), civil society "consists of a huge variety of mainly voluntary organisations and associations that maintain different objectives, interests and ideologies." From these definitions, civil society (loosely or tightly organised) perform an assortment of activities that are intended to fill the gaps left or created by the state, family and the business sector. The CSOs are therefore formed to give an expression and direction to the social, political, spiritual and cultural needs of its members, including peacebuilding (Barnes 2005, p.7).

The CSOs are not homogenous entities both in composition and outlook (Zigomo 2012; Sachikonye 2012) and can be categorised according to type. While several categories exist, Sachikonye (2012, p. 133) groups the CSOs into the developmental, humanitarian and governance related clusters. Masunungure (2014, p. 9) puts CSOs in Zimbabwe into three main generations which include: first generation (comprised of humanitarian based CSOs); development-oriented CSOs; and governance-oriented CSOs. The CSOs found in the development generation are those that were born in the first decade of independence in Zimbabwe. These include the

Zimbabwe Congress of Trade Unions (ZCTU) and the Association of Women's Clubs (AWCs) (Masunungure 2014; Sachikonye 2012). The third generation of governance-based CSOs emerged in the second decade when the country began to face economic problems (Masunungure 2014, p. 9). Amongst these are student-led groups such as the Zimbabwe National Students Union (ZINASU) (ibid). This chapter adopts the typology used by Masunungure in understanding the evolution of CSOs in Zimbabwe. From the generations of CSOs, I focused on the first generation which includes CSOs that are found in the humanitarian cluster.

The chapter is guided by the peacebuilding theory. Paffenholz (2010, p. 45) states that peacebuilding is a process aimed at achieving peace whereby structures and institutions based on justice, equity and cooperation are developed. Peacebuilding addresses the underlying causes of a conflict and prevents their transformation into violence (Galtung 1975, p. 297). Two branches emerge from this theory: liberal and sustainable peacebuilding. Liberal peacebuilding is premised on democratic rebuilding of states after an armed conflict (Paffenholz 2010, p. 46). Liberal peacebuilding is also traced from the ideas of Immanuel Kant in his 1795 'Perpetual Peace' where peacebuilding could occur between states embracing democratic values (Kant, 1795). It placed emphasis on bringing shattered states into conformity with the international system's prevailing standards of domestic governance (Paris 2002, p. 638). The key issues that emerge from liberal peace models are democratisation, economic liberalisation, human rights and rule of law (Donais 2012, p. 23). The CSOs that fall within the category of democratisation and governance has assumed these functions in their advocacy roles in bringing change in the Zimbabwean context. The chapter builds upon the weaknesses of the liberal model that lack input from 'below'. These approaches reflect the universal solutions to conflicts without considering the context-specific realities.

Apart from the liberal peacebuilding approach, this chapter locates the efforts of the CSOs in peacebuilding within the sustainable peacebuilding framework. Sustainable peacebuilding has a wider focus of peacebuilding while liberal has a short-medium term focus akin to state-building (Paffenholz 2010, p. 49). Proponents of sustainable peacebuilding include Lederach, who argues that it involves the creation of structures, processes and training of people within a generation (Lederach 1997). Paffenholz (2010, p. 47) observes that Lederach's definition is premised on promoting sustainable reconciliation within societies. He further argues that sustainability is a proactive process that can result in a spiral of peace and development as opposed to a spiral of violence and destruction. Sustainable peacebuilding involves transformation of the attitudes, behaviour and differences of groups into an engine of peace (Peinado 2003). In addition, sustainable peacebuilding implies a complete transformation of the state and society with a focus on socio-economic rebuilding aimed at addressing the underlying causes of a conflict (ibid).

Another dimension of a sustainable peacebuilding model is that it focuses on democratising the political institutions in order to bring a new spirit into the political system that is legitimate, effective and capable of responding to social frictions by allowing peaceful resolutions (Paffenholz 2010). It can be noted that sustainable peacebuilding is a multi-pronged issue that encompasses physical infrastructure, re-launching of the economy, addressing national reconciliation, establishing new institutional and political reforms for resolving conflicts peacefully (Peinado 2003). There are the CSOs that can pay attention to building and reforming norms and institutions that bridge electoral violence, structures based on insecurity and fear to structures based on security and trust that can sustain peace and democracy (Saki/Katema 2011). Such roles can be played effectively by the local CSOs who have greater proximity to the local people, thereby forging lasting peace (Harpviken/Kjellaman 2004).

7.4 Indigenous Knowledge Systems as a Tool for Peacebuilding

In the sustainable peacebuilding framework, I noted that some of the CSOs have built peace using IKS. The IKS is an alternative to liberal peacebuilding (Close 2016, 2017). It embraces longer-term bottom-up initiatives that seek to build inclusive social capital necessary for durable peacebuilding models (Brown 2016). In the Zimbabwean context, in *Shona* culture, the IKS is rooted in the concept of *Hunhu* or *Hunhuism*. The concept is moulded on an African vision that strives for an upright and virtuous individual (Sibanda 2014, p. 26). To be virtuous in the *Shona* context, one must respect the cultural values, morals and norms. They are uncodified laws that are embedded in natural law. The spirit of *Hunhu* is understood best by the local people with the use of their own language. For example, a *Shona* idiom which says *Mwana ndewa amai kana ari mudumbu asi kana abuda ndewe munhu wese* loosely translated means: “A baby belongs to the mother whilst it is still in the womb but once it is born it belongs to everyone”. This shows that everyone has a duty to look after and take care of the others in the community. Sibanda (2014, p. 26) argues that *Hunhu* is being human. It implies rationality and recognises that one cannot exist without the other or the society as whole. Mapara (2009, p. 142) defines it as what indigenous people know and do or what they have known and done for generations. The IKS involves long historical practices that are applied to a defined local community; it is firmly rooted in customary law or a set of rules and prohibitions followed by indigenous people (Adeosun 2015). From these definitions, Africans and Zimbabweans have their own local ways of relating socially, politically and economically. These local ways have tied groups together, for example, in community projects and in some cases resolving communal problems. The core values in IKS include promoting a culture of respect, love, unity, cooperation and tolerance. The absence of this has led to conflict resurgence in Zimbabwean societies. The CSOs have played a role in educating and socialising or reawakening *hunhuism* which is an appeal to moral values of the people. The incorporation and extent of success by CSOs is discussed in this chapter.

7.4.1 Indigenous Knowledge Systems and Peacebuilding

The practice of building peace through IKS is not new to humankind. Each society has its own unique way of resolving conflicts and establishing peace. While this is as old as history, little is known in conventional academic literature about the practice of IKS by the CSOs in peacebuilding in Zimbabwe. The IKS have been preferred in peacebuilding in that they create a sense of ownership and breeds legitimacy of the initiatives and programmes. In this respect, CSOs have tapped into local culture, beliefs, proverbs and practices to advocate, socialise and strengthen local knowledge that promotes peace and unity within communities. This section details the experiences by CSOs in building peace through IKS.

7.4.2 Envision Zimbabwe Women Trust Experience in Peacebuilding

CSOs have made use of IKS in peacebuilding extensively in rural areas and somewhat in the urban settings. Envision Zimbabwe Women Trust is one such organisation that has embraced IKS in building a culture of peace in the rural parts of the country. The organisation embraced the concept of inclusivity and working with the local people to reconcile broken relationship after the 2008 election violence. The CSO ran a peacebuilding programme called 'Building bridges' where it engaged traditional leaders who were once perpetrators of violence in the area (www.peacedirect.org). Under the programme, the organisation held workshops with the local community together with the traditional leaders who after the programme transformed to agents for peace and active mediators in the community (www.peacedirect.org). Local people who participated in the workshops noted that there were remarkable changes in the community as people could now trust each other and even eat together during community engagements (www.peacedirect.org).

The key issue that made the programme a success was the involvement of the traditional leaders to be the champions of peace in the community. Local people trust more their local leaders and the trust becomes stronger if they become agents of positive change (interview with a peacebuilderA).[1] In the rural areas, chiefs, traditional leaders and village elders have a local command that is tied to the customary beliefs.[2] It is believed that when one breaks the instructions from the community elders, a bad omen will occur to the offender or community. The CSO further succeeded in sowing peace by making the local people co-authors of the programme, it was a method of 'doing with rather than doing for'. Maphosa/Keasley (2014)view it as peace from within. In this regard, embracing inclusivity and co-authoring peace with the local community resonated with the principles of IKS that led to positive change in the

[1] Interview with PeacebuilderA, in Harare, 5 November 2020.

[2] Ibid.

community. The CSO was sensitive to local needs, managed to work with the local resources and actively engaged the people. Engagements I had with the local people in the community revealed that the community was now peaceful to the extent that since 2008 there have been very few if not none cases of election-related violence. The peaceful electoral environment in the 2013 and 2018 elections appears to show that election-related violence has subsided in the rural community (interview with a local community leaderA).[3] Accordingly, the use of IKS strengthened the trust and unity in the community, which are essential in creating a base for durable peace.

7.4.3 Peacebuilding Experiences of National Youth Organisation

Another CSO that has managed to build peace using IKS principles is the National Youth Association of Youth Organisation (NAYO). The organisation is youth-oriented and aims to bring peace and active engagement of the youths in matters of public life. Through stakeholder and area mapping, the organisation organised festivals and later created of peace clubs in Harare (urban area). While they came up with idea of festivals and local peace clubs, they managed to have a smooth cooperation from the youths by playing a facilitatory role. Funk (2012, p. 392) argues that when engaging the bottom-up voices, CSOs must be "facilitators or midwives rather than headmasters in building peace". The aim of the CSO was to enhance local skills in the youth to deal with conflict scenarios emanating from their communities. Solutions to conflicts were context specific and relevant. Building peace through this approach ensures that CSOs do not impose or bring generalised solutions to a problem. In the past such approaches resulted in fragile peace as they lack local legitimacy and ownership. Consultations with the local thus bolstered the CSO's approach in peacebuilding.

Apart from festivals and the creation of local peace clubs, the organisation implemented a Joint Work Group among the urban youths. The aim was to depolarise the community which was divided along party line. The Joint Work Group facilitated interactions and education among the youths involved. The interactions resulted in the youth breaking the polarisation and working together. The success was also attributed to the Government of National Unity (GNU) that had been created by the main political gladiators at the national level (www.peacedirect.org). Unfortunately, the end of this union saw the youths reverting to their political camps and engaging in intra-party factional conflicts (ibid). The experiences of building peace through IKS with the urban youths were strongly affected by the prevailing economic environment. The GNU managed to bring some economic stability whose gains cascaded to most citizens. Under this scenario the motivations to participate or being recruited to engage in political violence will be very minimum, the reverse of this is high in an unstable economic environment. In the later, youth are easily recruited as they

[3] Telephone interview with Community leaderA, 7 November 2020.

will be in desperate need of money or food. This scenario could only be averted if the youths had managed to be socialised effectively to develop a culture of peace or zones of peace as in the case of community work done in rural areas.

Experience of the CSOs working in rural areas demonstrates that community projects help to unite people and generate bonds even in the face of economic challenges. This reality manifested in an interview I had with one CSO project manager.[4] Their CSO works in rural parts of the country and embrace IKS in building peace. The CSO managed to build peace using local knowledge of the elderly people who told them about enhancing the *Zunde ramambo* concept (a practice where local communities work in a field to produce food that will cater for the whole community in times of hunger). Since food is a need to all, the CSO mobilised people with the assistance of local leaders (Chief and Headman) to till and plant maize for the benefit of the community. They managed to preach messages of peace and forgiveness which saw victims, perpetrators and instigators working together in the field. The effect of this was seen when outside political players wanted to sponsor violence in this community. The project brought unity, trust and a sense of collectiveness. The projects owned and run by the local people created a zone of peace and sustainability since the project is still running in the absence of the CSO. The practice by the CSO confirms Funk's (2012, p. 392) argument that external players are supposed to localise peace through networking and partnering with local actors, tap into indigenous peace resources. Indeed, networking with the local elders and traditional leaders helped to bring an idea of building durable peace through IKS under the *Zunde ramambo*.

7.5 Why Indigenous Knowledge Systems Work

Local language is one of the key components of IKS embraced in peacebuilding. Peacebuilding programmes must be in the local language. Language makes a programme more appealing and leaves permanent marks in people's lives and communities. This approach has also been embraced by the CSOs in their peacebuilding exercises. In an interview, a CSO member stated that:

> They make use of local idioms in peacebuilding campaigns. This demonstrates an element of sensitivity to local context that CSOs should aim for when working with the local communities in African societies. For example, people are familiar with idioms like "a person is a person because of other people". This brings an element of tolerance and valuing another person within the community.[5]

The interviewee gave a similar local *Shona* idiom that they use in building peace, that is *kugara hunzwana* meaning "people can live together well if they understand each other"; this encourages cooperation within communities. For example, for a community to be called a community the people within it are supposed to be understanding and be tolerant to each other. This comes as an individual realises that

[4] Interview with a member of civil society (CSOA), Harare, 7 November 2020.

[5] Interview with CSOB, in Harare, 8 November 2020.

people are not able to live and survive in isolation. Tapping into local language is a realisation that every community has its own ways of starting and resolving conflicts through their own language. A peacebuilding practitioner[6] concurred in an interview that, local language leads to easy understanding and appreciation of a peacebuilding campaign. Past peace interventions have shown that local peace languages leave lasting memories in people's lives in rural areas. Ubuntu is defined as a sense of responsibility to care for others who are in need (Maphosa/Keasley 2004). Funk (2012) postulates that: peace has a cultural dimension; every cultural community has its own vernacular language for conflict resolution and the use of it gives peace substance and legitimacy. It follows that, the CSOs peacebuilding approaches have to have the presence of the local language for the outcome to be appealing and durable. Lack of appeal and legitimacy explains conflict relapse.

Apart from spoken local languages, the CSOs have made use of art and images in local communities to sow peace messages in communities. An interviewee[7] highlighted that in their training workshops they would make use of pictures and posters with images about which they would ask people to think and link with their day-to-day experiences. The posters would show the power of soft strength in the face of any situation. For example, one poster had a hard rock which was slowly being eroded by a stream of water. The water resembled soft strength and the hard rock the difficulties one could face in life. They argued that though soft strength might be slow in bringing change, it had a long-term positive effect that would eventually lead to the hurdles being washed away. Each picture or poster had a local theme promoting love, respect, tolerance and collective responsibility which are linked to IKS in peacebuilding.

While peacebuilding through IKS has its own merits towards cultivating sustainable peace, the model also has its flaws. An academic argued in an interview that, the CSOs need to be aware of some of the negative practices found in IKS which in a way contradict peacebuilding. This has been witnessed in some proverbs taught and come to be accepted by societies as normal yet they suffocate other people's rights. A peace activist[8] gave an example of a *Shona* saying that, *kurwa rudo* loosely translated to 'fighting is love.' This local norm has resulted in societies in rural areas accepting or tolerating domestic violence which has mostly affected women and children. Although the local norm is part of the IKS, the CSOs need to find ways of deconstructing such norms that infringe the rights of women and children.

Key lessons that can be drawn from the use of IKS by CSOs in peacebuilding are summarised below:

- Simple, cheap and easy to follow, especially on the part of the local people who will be aware and familiar with their own ways of addressing an issue. It breeds context specific and relevant solutions to a problem. For example, in certain conflicts there are specific people who can resolve them such as the community elders. Equally

[6] Interview with PeacebuilderB, in Harare 8 November 2020.

[7] Interview with CSOC, in Harare, 9 November 2020.

[8] Interview with PeacebuilderC, in Harare, 9 November 2020.

the community elders have the potential of abusing the local command they have to instigate or cause violence within their communities.
- It can strengthen or bind relationships between the people and the CSOs working in an environment. Once the bond is created, so is trust. In this way, local people will cooperate or have a "buy-in" on a programme brought in by the CSOs.
- Local turn or bottom-up initiatives work as an antidote to donor-dependent programmes or activities which in the long run have the effect of creating a dependency syndrome. Donor dependent programmes can be hijacked by professional CSOs who are driven by money.
- Working with the local promotes ownership of a programme. The local people can then be able to drive it alone with their local means and knowledge. This is called self-sustaining peace (Maiese 2004).
- In addition, working with the local turn has an effect of strengthening public opinion on an issue. Once people have a common understanding of peace issues, it is easier to create peace constituencies or zones of peace that can counter any violent forces that may arise within a community.

7.6 Ways Forward

The CSOs need to begin by carrying out a thorough analysis of the needs and fears within an affected community. This minimises the chances of faulty assessment, plan and intervention. Sustainable change can be reached through collaboration among CSOs, local people and experts. Osamba (2001) notes that, in the context of emotional oneness in the community, emotional scars and wounded relationships are healed. Further, CSOs can build coalitions or networks to help address financial challenges. Members can pull resources together for sustainable interventions instead of competing for donor funding.

For the CSOs to engage successfully with the local people, they need to assume key roles that can enable a democratic way of working together with the people. The CSOs need to create an open and inclusive platform to incorporate the views of the people. They can do this by, fostering an atmosphere of trust, mutual respect, and shared aspiration in which everyone may fully and openly contribute to the achievement of common goals (TTM Associates 2017). At least the CSOs need to assume the role of facilitators or catalysts (Maiese 2004) in their peacebuilding programme. They should not assume that the local people have no useful input but rather should demonstrate their willingness to learn and work together on what they already know. Through facilitation, they become bridges to connect people with other stakeholders, using their local ways towards addressing a problem. Where they fail to facilitate a programme, they can also act as advisors who only make recommendations in consultation with local people rather than imposing pre-tailored ideas. By performing these roles, the local people and other stakeholders feel involved and that they own a programme that a CSO might have. It encourages collaboration. Collaborative processes incorporate diverse ideas from other stakeholders in the peacebuilding process, bearing in mind that it must be grounded in the local input. In this way, a demand-side peacebuilding programme can be realised as opposed to a supply-side initiative.

Maiese (2004) posits that culture is a seedbed for the development of a peacebuilding model. Burton (1990, in Osamba 2001) argues that culture is a satisfier which makes it a critical part in conflict resolution. When working with the local inputs, the CSOs may also want to understand and learn the local culture, tradition or practice within a context. This should then be incorporated in their peacebuilding campaigns. For example, local idioms, language, folktales, symbol and sayings that have elements of peacebuilding. Again, chances are high that there will be buy-in from the local people. In some cases, where the CSOs have been denied access to a rural community can be explained in the sense that the CSOs program is seen by some chiefs as a threat to their beliefs, culture, and practices (Maiese 2004). Designing a peacebuilding programme rooted in local culture is like following pathways in ethnic wisdom (Augsburger 1992, in Osamba 2001). Donais (2012, p. 58) argues that the "State is about order, civil society is a producer, repository and distributor of social norms and practices that generate community". It is the richness of local wisdom that makes a programme more appealing and effective in bringing change within communities.

Lastly, the CSO should not only involve local people during planning and implementation of an intervention. Rather, the democratic and all-inclusive process should be carried over to evaluation of a project. When evaluation is done with the people and there is feedback, they will know how to improve their peacebuilding exercise. If a CSO decides to vacate an area, the people can sustain the intervention. In some cases, a CSO can go further to publish their findings for the benefit of other CSOs or interested stakeholders who may want to carry out future similar exercises in the area through IKS. This creates continuity and sustainability of a programme. The emphasis on local turn or input is that CSOs cannot build peace alone but they build with others comprised of the local people, researchers and other experts in peacebuilding.

7.7 Conclusion

Building peace through IKS by CSOs is still alive in Zimbabwe. The chapter has noted that success stories have been recorded in the rural parts of the country where the CSOs have embraced IKS in building peace. Lasting memories and structures have been established as local people were the co-authors of the peace, thus enhancing legitimacy of a programme. Partial successes have been recorded in urban settings where the people have been exposed to modern peacebuilding approaches and in some cases the stability of an economic environment shapes the outcome of peace interventions by the CSOs. However, where IKS have been appropriately used, durable peace has been realised. The IKS which are grounded in bottom-up approaches are critical in promoting, resolving and transforming conflict by CSOs when they assume a midwifery role. The IKS promote local ownership and generates legitimacy among the people in peacebuilding. Ownership and legitimacy are key in realising durable or sustainable peace.

References

Adeosun, H.O. (2015). Indigenous knowledge systems and dispute resolutions: The Yoruba example. In M. Mawere, and A. Nyamekye (Eds.), *Between rhetoric and reality: The state and use of indigenous knowledge in post-colonial Africa* (pp. 241–260). Cape Town: Langa RPCIG.

Barnes, C. (2005). Weaving the web: civil society roles in working with conflict and building peace. In P. van Tongeren, M. Brenk, M. Hellema, and J. Verhoeven (Eds.), *People building peace 2: success stories of civil society* (pp. 7–24). London: Lynne Rienner Publishers.

Brown, G.K. (2016). Civil society building in post-conflict countries. In A. Langer, and G.K. Brown (Eds.), *Building sustainable peace: Timing and sequencing of post-conflict reconstruction and peacebuilding* (pp. 110–123). Oxford: Oxford University Press.

Cameron, H. (2017). The Matabeleland massacres Britain's willful blindness: The International History Review http://www.tandfonline.com Accessed 26 Dec 2020.

Donais, T. (2012). *Peacebuilding and local ownership: post-conflict consensus-building*. New York: Routledge.

Funk, N.C. (2012). Building on what is already there: valuing the local in international peacebuilding. http://www.jstor.org Accessed 12 Nov 2020.

Galtung, J. (1975). Three approaches to peace: peacekeeping, peace-making and peacebuilding. In J. Galtung (Ed.), *Peace, war and defence-Essays in peace research* (pp. 282–304). Copenhagen: Christian Ejlers.

Gyimah-Boadi, E. (1996). Civil society in Africa. *Journal of democracy*, 7(2), 118–132.

Harpviken, K.B., and Kjellman, K.E. (2004). Beyond blueprints: civil society and peacebuilding. https://steps-for-peace.org/wp-content/uploads/beyond-blueprints-prio.pdf Accessed 24 Nov 2020.

Laakso, L. (2003). Opposition politics in independent Zimbabwe. *African Studies Quarterly (online)*, 7(2). http://www.africa.ufl.edu Accessed 21 Nov 2020.

Lederach, J.P. (1997). *Building peace: sustainable reconciliation in divided societies*. Washington, D.C.: United States Institute of Peace Press.

Maiese, M. (2004). Elicitive training. www.beyondintractability.org. Accessed 10 Nov 2020.

Mapara, J. (2009). Indigenous knowledge systems in Zimbabwe: Juxtaposing post-colonial theory. *Journal of Pan African Studies*, *3*(1), 140–153.

Maphosa SB. and Keasley A (2014) Building peace from within: an introduction. http://www.researchgate.net Accessed 11 Nov 2020.

Mashingaidze, T.M. (2005). The 1987 Zimbabwe National Unity Accord and its aftermath: A case of peace without reconciliation? http://www.codesria.org. Accessed 22 Nov 2020.

Masunungure, E. (2011). Zimbabwe at crossroads: challenges for civil society. http://www.osisa.org. Accessed 20 Jan 2021.

Masunungure, E.V. (2014). The changing role of civil society in Zimbabwe's democratic processes: 2014 and beyond. https://library.fes.de. Accessed 14 Jan 2021.

Munhande, C., & Nciizah, E. (2013). Perpetuating colonial legacies: Reflections on post-colonial African State's development trajectories; Observation from Zimbabwe. Gweru: Midlands State University.

Osamba, J. (2001). Peacebuilding and transformation from below: indigenous approaches to conflict resolution among the pastoral societies in borderlands of Eastern Africa. www.accord.org.za. Accessed 13 Nov 2020.

Paffenholz, T. (2010). Civil society and peacebuilding: a critical assessment. In: Paffenholz T (ed) 2010. Civil society and peacebuilding: a critical assessment (pp 43–64). London: Lynne Rienner Publishers.

Paris, R. (2002). International peacebuilding and the mission civilisatrice. *Review of International Studies*, 28(4), 637–656.

Peinado, M.M. (2003). Role of NGOs and civil society in peace and reconciliation processes. http://www.ceipaz.org. Accessed 20 Jan 2021.

Sachikonye, L. (1995). *Democracy, civil society and State*. Harare: SAPES.

Sachikonye, L. (2012). *Zimbabwe's lost decade, politic and development*. Harare: Weaver Press.

Saki, O., & Katema, W. (2011). *Voices from civil society*. In: Murithi et al (ed) Zimbabwe in transition: a view from within. Auckland Park: Fanele.

Sibanda, P. (2014). The dimensions of 'Hunhu/Ubuntu' Humanism in the African sense the Zimbabwean conception. Journal of Engineering. http://www.iosrjen.org. Accessed 29 Nov 2020.

Spurk, C. (2010). Understanding civil society. In T. Paffenholz (Ed.) *Civil society and peacebuilding: a critical assessment* (pp. 3-29). London: Lynne Rienner Publishers.

TTM Associates (2017). Political Intelligence: from skills to behaviors. https://www.ttmassociates.com/wp-content/uploads/PQ-1.pdf. Accessed 10 April 2022.

Zigomo, K. (2012). 'A Community-Based Approach to Sustainable Development: The Role of Civil Society in Rebuilding Zimbabwe', 2 April, Solidarity Peace Trust. http://www.solidaritypeacetrust.org Accessed 21 Jan 2021.

Chapter 8
Civil Society and Peacebuilding in the Kivu Provinces of the Democratic Republic of Congo

Georges Bomino Bosakaibo

Abstract A robust and organised civil society is vital in preventing violence. Congolese civil society organisations (CSOs) have supported citizens' interests since the time of the Second Republic from 1965 to 1997. The civil war eruption in the 1990s in the Democratic Republic of Congo (DRC) motivated the CSOs to engage in political dialogue for sustainable peace. Despite this dynamism, the CSOs weaknesses limited the impact of their action against violence in the eastern provinces. This chapter explores how effective or not has civil society been in ending armed conflict and building sustaining peace in Kivu provinces via the use of seven civil society's peacebuilding functions. The findings are drawn from desk research, mainly exploring data from books, journals, reports, and official documents; and the assessment reported here is conducted within civil society's functions that the literature in this field provides. This chapter contends that CSOs still exert a weak influence in building peace due to various factors that hinder the fulfilment of peacebuilding functions aiming for a sustainable peace.

Keywords Civil society · Peacebuilding · Conflict · Violence · Kivu Provinces · Democratic Republic of Congo

8.1 Introduction

Between 1998 and 2003, the Democratic Republic of the Congo (DRC) experienced the so-called world's most violent conflict since World War II between 1998 and 2003 (Malan/Porto 2004). Many national armies from Africa were involved in the war as well as Congolese and non-Congolese armed groups. It caused the death of 3.9 million people while displacing over 8 million more both within and outside the country (Coghlan et al. 2006, p. 49). The war resulted in the division of the DRC into several autonomous parts controlled by armed coalitions focusing on natural resources to sustain their war efforts and make a return on their 'war investment' (International Crisis Group 2000, p. 66). Civil society emerged not only to denounce exactions and crimes caused by the war, but also to call for a negotiation to resolve

---

Dr. Georges Bomino Bosakaibo is an assistant professor in the Department of Policy Studies, Faculty of Policy Studies, Nanzan University, Nagoya, Japan.

J. C. K. Kiyala and G. T. Harris (eds.), *Civil Society and Peacebuilding in Sub-Saharan Africa in the Anthropocene*, The Anthropocene: Politik—Economics—Society—Science 34,
https://doi.org/10.1007/978-3-030-95179-5_8

the conflict as protagonists in the war explored ways to defeat their opponents on the field using force.

During the Inter-Congolese Dialogue (ICD) held in South Africa from 2002 to 2003, civil society participated in the peace negotiations. Subsequently, civil society had representatives for all transitional institutions. The strategical conception to avoid a potential polarisation of the transition in case it only consisted of former warring groups led to direct inclusion of civil society in the transitional institutions. Despite the persistent and recurrent instability in the two Kivu provinces, the transition process (2003–2006) was successful as it led to the first multiparty elections in the DRC in 41 years. The role of civil society can be investigated while several Congolese and non-Congolese stakeholders contributed toward the success of the transitional process and the resolution of the conflict, even though currently, nowadays the recurrent instability and killings of civilians continue in the eastern part of DRC, particularly in the Kivu provinces.

Despite a certain dynamism of the restoration of a lasting peace, the civil society has shown many weaknesses that limit the impact of its actions in contributing to finding solutions against the violence in the persistent conflict. Indeed, this situation in DRC, especially in its eastern provinces suffering from the recurrent violence, is addressed by the following research question: How effective has civil society been in ending armed conflict and sustaining peace in conducting the seven peacebuilding functions of protection, monitoring, advocacy, socialisation, social cohesion, facilitation, and service delivery in Kivu provinces? This research relies on the secondary qualitative data from books, journals, reports, official documents. The frame of analysis, as presented in Thania Paffenholz's (2010) critical work on civil society and peacebuilding, is considered a lens through which the data are explored. This chapter argue that the civil society, despite its slight development, continues to exert a weak influence in building peace due to various factors that hinder the fulfilment of peacebuilding functions and that need to be identified and addressed in order to create mechanisms that can impact its action towards a sustainable peace.

This study explores the role played by civil society in peacebuilding with emphasis being placed two Kivu provinces. The scope of the study extends from the beginning of the war in 1996 to the current situation in the eastern parts of DRC in 2021.

This chapter is articulated as follows: after this introduction, Sect. 8.2 gives a conceptual framework; Sect. 8.3 historical background to the civil society organisations (CSOs), particularly in the context of Eastern Congo; Sect. 8.4 offers the findings identified as the weaknesses of the civil society; Sect. 8.5 analyses the findings with regards to the weaknesses of the civil society and their impact on the peacebuilding process; and the final section, Sect. 8.6 the conclusion for the chapter. The subsequent section clarifies the concept of civil society in the peacebuilding process.

8.2 Civil Society and Peacebuilding

8.2.1 Conceptual Framework

There is no general consensus between observers on the importance of civil society to peacebuilding. Civil society and civic culture have the tendency to frustrate change and progress towards a just and equitable society according to supporters of the Marxist ideology. International community faced the complexities of peacebuilding efforts that accompanied the proliferation of armed conflicts in the 1990s. However, according to Ekiyor (2008), civil society actors have increasingly become vital forces in discourses, initiatives and programs that foster peace and security across the world.

Sub-Saharan Africa has experienced different civil wars and intrastate wars in the recent decades, including countries such as DRC. These wars have resulted into an acrimonious number of deaths, displacements of communities, and the proliferation of small arms and light weapons (SALW), thus rendering growth and development stagnant. The civil society has been at the forefront of dismantling authoritarian regimes to replace them with democratic governance, initiating and promoting reconciliation processes, carrying out localised peacebuilding initiatives, and advocating for the adherence to peace agreements and the tactical building of capacities through peace education (Mbayo 2012).

The vital involvement of Civil society actors as primary providers of basic social services in war affected areas cannot be overemphasised, where the public institutions and state apparatus had ceased to function or had been considerably weakened due to the conflict. In these situations of anarchy, CSOs have often played the role of mitigating conflict and building peace (Mbayo 2012, p. 41). However, Ekiyor (2008) underscores the fact that civil society is not a homogenous group and, therefore, not all members of civil society can be regarded or defined as peacebuilders.

In recent years, there has been an increase of interest in discovering how strengthening civil society can contribute to conflict resolution and peacebuilding rather than the activities of civil society that have led to its recognition and many more of its envisaged and potential roles. In the move towards the capacity building process of CSOs in conflict societies, the international community has provided significant support, both financially and physically to achieve this goal. It is interesting to note however, that despite the massive rise in peacebuilding initiatives that is aimed at strengthening civil society, these initiatives have not been backed by a systematic research and documentation agenda. This incongruity has led to minimal exposition of the role of civil society in peacebuilding and its potentials in reducing conflict, ending violence, ending armed conflict, and the building of peace (Paffenholz 2010) in Africa, especially for the eastern provinces' CSOs in the DRC.

As a contested concept, civil society has various meanings shifting according to rhetorical needs (Purdue 2007, p. 1). The concept initially gained momentum in the late eighteenth century with the emergence of capitalism and was originally understood as the ability of individuals to deal with strangers without using force, particularly in urban centres according to various authors (Purdue 2007; Edwards 2009; Pishchikova/Izzi 2011).

Although the practical involvement and scholarly discussion of the role of civil society in peacebuilding emerged in the 1990s, the "context of the global war on terror focus shifted again from peacebuilding to military intervention, from local civil society to the role of the state and its institutions in reforming governance and promoting development" (van Leeuwen 2009, p. 50). In 2010, the World Bank and Paffenholz worked for the re-emergence of civil society (2010) in academic literature partly, through the work of the Global Partnership on the Prevention of Armed Conflict.

In the context of peacebuilding, promoting the development of independent CSOs was part of the typical formula for liberal peacebuilding along with the promotion of civil and political rights and marketisation (Paris 2004, p. 19). The discussion of civil society participation in peacebuilding within the liberal tradition tends to focus on the inclusion of CSOs in peace negotiations, as well as their advocacy function in relation to the state. Models and specific types of civil society engagement are presented in peacebuilding which also assume their existence within a liberal state (Paffenholz 2010, 2014; Richmond/Pogoda 2016).

The functional approach suggested by Paffenholz/Spurk (2010) remains one of the most prevalent definitions of civil society. It identifies seven key functions of civil society in peacebuilding contexts including: protection, monitoring, advocacy and public communication, in-group socialisation, social cohesion, intermediation and facilitation and service delivery (Paffenholz 2010, pp. 65–75). This approach represents a liberal definition of civil society. These seven key functions, discussed in further detail in the following subsection, are considered the frame of analysis for this chapter.

8.2.2 Functions

The seven key functions of civil society in peacebuilding contexts are explained as follows: *Protection, Monitoring, Advocacy, Socialisation, Intergroup social cohesion, facilitation and mediation, and Service delivery.*

- *Protection* refers to the provision of security needs by civil society actors, either acting alone or in cooperation with other agencies. The main provider of this function is the state; however, in cases of acute state fragility and conflict, the relationship between state and society can break down. During and after conflict, protection becomes a precondition for other civil society functions, as civil society actors are substantially hindered from taking up peacebuilding roles when threatened by violence.

  Civil society protection is often associated with specialised protection non-governmental organisations (NGOs), such as Peace Brigades International, that support local actors, either directly or indirectly. During a civil war, a number of

local human rights organisations can monitor human rights violations and send all information to the National Human Rights Commission, the media and Amnesty International (AI). These bodies have used the data to successfully lobby, at the international level, for the establishment of a United Nations (UN) monitoring mission. Local civil societies can negotiate 'zones of peace', within which arms are not allowed, and have occasionally taken over responsibility for human security initiatives such as de-mining, disarmament, and demobilisation when official programs have been found wanting.
- *Monitoring* concerns international and local civil society groups that monitor relevant issues such as the human rights situation, or the implementation of agreements, and provide recommendations and information to decision-makers or human rights and advocacy groups. Such monitoring can work to hold governments and armed groups accountable for abuses or substandard performance and can also serve as an early warning system. Monitoring activities is most effective when designed to harmonise with protection and advocacy initiatives.
- *Advocacy* refers to agenda-setting and the application of pressure by CSOs. Civil society actors can push for the commencement of negotiations and the implementation of negotiated agreements, or work against the recurrence of warfare. Equally important are global international advocacy campaigns that lobby. Advocacy can be divided into public and nonpublic forms. Public advocacy can involve petitions, demonstrations, press releases, social media, or public relations campaigns. Non-public advocacy is generally backchanneled and operates through informal dialogues and relationships. The impact of advocacy initiatives is increased when organisations have campaigning knowledge, base their advocacy on results of monitoring initiatives, and know how to use the media to support their cause.
- *Socialisation* refers to in-group bonding that supports democratic behaviour and promotes tolerant and peaceful values within society. This is realised through the active participation of citizens in various associations, networks, or movements. Socialisation takes place only within groups, not between former adversary groups.

Every national or local association that practises peaceful coexistence contributes to this function. There are two main types of interaction or socialisation: socialisation for peace and in-group identity-building. Socialisation for peace involves activities that promote a culture of peace whether in society at large or within a single group. In-group identity-building is an important way for marginalised groups to develop a sense of political identity that allows them to operate peacefully in the political space available.

The key institutions in society that influence how people learn democratic and conflict-response behavior are families, schools, religious groups, secular and cultural associations, and the workplace. In most countries in conflict, these socialisation spaces tend to reinforce existing divides. The overwhelming focus of socialisation initiatives has been on conducting short-term projects with NGOs, which, due to their limited reach and access, have no real power to socialise people.

- *Intergroup social cohesion* focuses on social capital between groups as it is invariably degraded or destroyed during wars between those groups. Therefore, it is crucial to build 'bridging ties' across adversarial groups as well as (peaceful) 'bonding ties' within specific groups. The objective of social cohesion is to help these groups learn to live together in peaceful coexistence.

  Social cohesion is an area where CSOs face challenges in making an impact. As explained in the discussion on socialisation above, divided societies have many strong institutions for interactions, including families, schools, and religious organizations. When these institutions are polarised and hostile, few social cohesion initiatives can be highly effective.

  Social cohesion initiatives may generate more impact when they aim at bringing people together to work for a common cause (for example, joint water management) rather than focusing only on reconciliation. Long-term systematic initiatives have been more effective than short-term scattered ones, especially when they have focused on a wide range of societal cleavages and also bridged the gap between difficult groups.
- With regard to *facilitation and mediation,* civil society can function as a facilitator to help bring parties together in a peace or transition process. Facilitation can take place both at the local and at the national level. At the national level, prominent civil society leaders, international NGOs and research institutions are occasionally engaged in mediation or facilitation. This facilitation can also be issue-oriented, as when civil society groups facilitate violence-free days to secure access for service delivery (vaccinations, food programmes, etc.). Hence, facilitation can operate in support of both protection and service delivery.
- As far as *service delivery* is concerned, during armed conflict, state structures are either destroyed or weakened, and the population may be starved of essential services. Civil society actors (mainly NGOs, but sometimes associations as well) can and do step forward to provide aid and social services. There is no doubt that this function is extremely important to help the war-affected population and to support reconstruction of the state and society at large. However, service delivery can only have an impact on peace processes only if agencies create entry points for other functions such as protection and social cohesion, especially when large-scale violence ends. Then, the next section presents civil society historical background in the DRC context.

8.3 Historical Background of Civil Society in DRC

Given the emphasis on context and local knowledge in the post-liberal peacebuilding literature, many case studies have been published considering locally driven peace initiatives, including in the DRC. This is likely related to Severine Autesserre's argument in her book entitled 'The Trouble with the Congo' where she claimed that the dominant international peacebuilding culture shaped the intervention in the

Congo in a way that precluded action on localised violence such as individual land conflicts, ultimately dooming externally-led efforts. She argues as follows:

> The eastern part of the Congo remained so violent during the transition not only because of regional and national tensions, but also because of the presence of distinctively local problems. These included conflicts over land, mineral resources, traditional power, local taxes and relative social status of specific groups and individuals (Autesserre 2010, p. 176).

In addition, Vlassenroot/Raeymaekers (2004) edited a book arguing that efforts to end the war in the Eastern DRC should be grounded in the causes of conflict that exist below the international and regional level. They highlight that:

> The Congolese conflict can only be fully understood with reference to the ways in which conflict, together with a legacy of colonial and state policy that preceded and informed it, has created a situation in which the rational pursuit of individual livelihood ends up reproducing the collectively irrational phenomenon of war (Vlassenroot/Raeymaekers 2004, p. 290).

A significant amount of the literature on the DRC concerns on the history of and the regional and national causes of the conflict (Prunier 2009; Stearns 2011, 2014). Although this literature does not advance the participation of local actors in peacebuilding and peace formation processes, it provides a historical context for both local peacebuilding efforts and it helps to situate local efforts in relation to the state.

For instance, in 2013 the Congolese state still enjoyed a high degree of legitimacy despite being powerless in an administrative and in a coercive sense (Raeymaekers 2013, p. 600). Thus, Raeymaekers concludes that "political legitimacy remains rooted in a tight association of political protection, military coercion and economic capital between violent elite networks" as opposed to popular embeddedness or alternative justice (2013, p. 614). In addition, the structural, political, and historical causes of state-building failures in the DRC result in the powerlessness of local, national, and international stakeholders to influence Congo's unmanageable political landscape (Trefon 2011, p. 122).

Some of the more recent literature on the DRC addresses the themes like local approaches to peacebuilding and civil society. Stearns (2014) questions the validity of addressing local causes of conflict such as land and ethnic disputes rather than national causes of conflict including political cleavages and military elites, which he indicates are broadly perceived by experts on the conflict as the main sources of conflict.

Further, a small amount of locally published literature written by Congolese authors exists regarding the local causes of conflict and community level conflict resolution and peace agreements in the Eastern DRC. This literature addresses on the local causes of conflict as well as traditional conflict resolution mechanisms which exist at the low level of administration called "les chefferies". Nonetheless, it is important to remember that such works are also embedded in local power and conflict dynamics.

The socio-political organisation of rural communities is composed of chefferies as well as secteurs where 70% of the population live in these spaces and their leadership are responsible for governance and development. The core difference between the two is that secteurs are governed in accordance with the laws of the state whereas chefferies are governed through customs and by traditional leaders (Maneno 2014, pp. 177–178).

More scholarly work also needs to be undertaken to understand the nature and design of CSOs in the DRC as well as highlighting their contributions to peacebuilding. There is minimal literature discussing how the Congolese civil society functions within the context of a fragile state. There is also no existing literature which outlines the dynamics within these organisations in addition to whose interests they represent and how. Further, there is limited information on how these organisations engage with national and international partners, and there is not any literature on how they interact with grassroots structures with whom they have formal and informal relationships.

The point of departure for this chapter is as follows: while the international community's role is known, in peacebuilding in the DRC, only little is known about how peacebuilding priorities are negotiated between local and international actors. Thus, in DRC, one can speak about civil society with its challenges in the course of its history.

Under the second Republic in Mobutu regime between 1965 and 1990 some forms of organised civil society existed, but the full emergence of CSOs was the result of the democratisation process decreed by President Mobutu on 24 April 1990. This emergence of civil society occurred according to a policy similar to the one that freed the political opposition parties allowing them to function again since their ban in 1965.

The convening of the National Sovereign Conference which took place between 7 August 1991 and 6 December 1992 brought together a certain coalition between opposition political parties and CSOs. This National Sovereign Conference considered to be a national political dialogue brought together 2 842 delegates, representing all strata of society, with the aim of identifying the causes behind the country's post-colonial failure and making recommendations on a new path for the country's development (Nzongola-Ntalaja 2002, p. 190).

The first instance of the politicisation of civil society in the DRC was the abovementioned conference, particularly after reaching the agreement for the conference to be responsible for the establishment of political institutions needed to manage the country during the transition (Nzongola-Ntalaja 2002, p. 190). Civil society soon emerged as an influential force between the irreconcilable government and political opposition. Unsurprisingly, the presiding elected officer of the conference was one of its members, Archbishop Monsengwo (from the Roman Catholic Church). At this time, the civil society movement split into two factions. The first faction was composed of organisations that joined the main opposition coalition known as the Sacred Union of the Radical Opposition (USOR) which was subsequently renamed to the Sacred Union of the Radical Opposition and Allies from Civil Society (USORAS). The second faction comprised of organisations that held a conciliatory view vis-à-vis the regime (Koko 2016, p. 118). Civil society remained aligned to political parties, considering itself as both a (political) power broker and contender until 1997 when the Mobutu regime was overthrown.

State-civil society relations had a new dynamic with the rise of Laurent-Désiré Kabila to power in May 1997. Throughout the war period, Laurent-Désiré Kabila and his movement called the Alliance des forces democratiques pour la liberation or Alliance of Democratic Forces for the Liberation of Congo (AFDL) showed the mistrust and hostility toward CSOs accusing them to be accommodative of Mobutu regime. For their part, the CSOs made their criticisms toward the AFDL for the group's role in human rights violations committed against civilian populations, especially Rwandan Hutu refugees (Koko 2016). The CSOs were also concerned with the prospect of losing the space they had gained during the transition should the AFDL have emerged victorious on the battle front (De Villiers/Omasombo, 1998, p. 57). They openly advocated for a negotiated settlement of the war as well as the AFDL's absolute dissolution.

The AFDL embarked on a process of reducing civil society's space of operation when it took over state power in May 1997. Thus, it set thus the stage for future confrontations between the two entities. Indeed, leaders of NGOs and churches presented a different type of challenge for the regime according to the International Crisis Group (ICG 1999, p. 13). In principle, "their policy options were similar to those of the non-violent opposition parties. They urged for rapid moves toward elections and democratisation, and they affirmed the legitimacy of the National Sovereign Conference … which was the one time in the Congo's constitutional development, when they participated formally" (Koko 2016, p. 118).

Two major factions of the civil society held competing views regarding the AFDL regime during the National Sovereign Conference and the subsequent period: the Congolese Civil Society (CCS) the origins of which may be traced back to the group that formally drew closer to the Mobutu regime – appeared more conciliatory toward the regime and opted for 'constructive engagement' and the civil society of the DRC the origins of which can be traced to the bloc that associated itself with the political opposition during the first transition – adopted a 'confrontational approach' (Koko 2016, p. 119). This latter group was more effective with dynamic networks such as the National Council of Development Non-Governmental Organisations (NCD-NGOs) with its vibrant provincial structures known as Regional Councils of Development Non-Governmental Organisations (RCD-NGOs).

Notably, over 250 delegates representing all the country's provinces attended CNONGD's successfully organised national conference on reconstruction, held in Kinshasa in June 1997. Among other issues, the conference 'declared its concern for the protection of fundamental liberties' and addressed "the absence of political dialogue, indiscipline in the army, the absence of a clear-cut division between the state and the AFDL, and the absence of a constitutional framework' (ICG 1999, p. 13). A month later, the Roman Catholic Church, through its Bishops' Permanent Committee, echoed NCD-NGOs complaints in expressing concern over the new authorities' decision to reject the constitutional project adopted by the people at the National Sovereign Conference and 'noted that certain acts were being committed by members of the new regime which did not respect the dignity of the human person' (ICG 1999, pp. 15–16), acts that negated the rule of law. In response, the regime attempted – albeit unsuccessfully – a number of strategies to gain control over civil society including directing through government channels all NGOs' financial and other aids coming from abroad, arresting activists, centralising the NGOs' registration process through the ministry of justice and dissolving NGOs that failed to comply fully with the new legislation (Koko 2016, p. 119). However, the Kabila regime adopted a more conciliatory approach in its engagement with civil society after the eruption of the Second Congo War in August 1998. The following section presents the findings.

8.4 Weaknesses of Civil Society in the Kivu Provinces in DRC Citizen Protection Strategies

Communities tend to understand their own problems and needs, and they confront conflict together. Mahony's respondents mentioned many collective community protection practices in the Kivu provinces: going to fields in groups to minimise risk; using whistles to alert each other to danger; reorganising agriculture in order to cultivate crops at a safer proximity to their village. Some mentioned examples of the non-Rwandophone population protecting Rwandophones. Answer to the question "Who helps communities mobilise these protection strategies?" included church leaders, community social groups, cooperatives, and essentially any existing structures where people would meet to talk about problems (Mahony 2013).

There exist also larger structures extending beyond single communities. Inter-ethnic traditional structures such as Baraza la Wazee try to bring together moderate leaders of multiple ethnic groups to try to confront interethnic conflict and promote unity. Although some of Mahony's respondents expressed concern about the politicisation of such inter-ethnic structures, and although there is some risk of bias or exclusion, Baraza La Wazee spokesmen emphasised that high-level examples of ethnic collaboration can influence people's thinking (Mahony 2013, p. 20). Multi-ethnic leadership delegations were able to negotiate with authorities or armed groups and express people's needs.

Churches constitute another vital mobilising structure among the population. The Catholic Church network, as the largest, reaches to nearly every remote corner of the country. Although some of Mahony's respondents expressed concern that the churches were factionalised and competitive, or that their leaders had lost some credibility due to their prominent role during the election period, their legitimacy is nonetheless broadly recognised. Church networks are important sources of information and analysis. The Catholic Church has a "Justice and Peace" network with a commitment to confronting human rights issues (Mahony 2013, p. 20). A Catholic Church delegation led by the bishop travelled to Rutshuru, for instance, to speak with the M-23 rebel group about their concerns for the population. There are hundreds of small organisations, many more than have existed in the past, especially due to increases in the availability funding for their work (Gouzou 2012). Although single charismatic leaders dominated many small and weak organisations, other organisations have significant networks of activists, technical skills, and ideas for political change in the Congo. Unfortunately, the prevalent dynamic of security threats also constrained their activism and advocacy efforts. The killing or threatening of so many human rights activists occurred to such a large extent in the past several years that United Nations organisation for stabilisation mission in the Congo (MONUSCO) human rights division established a special Protection Unit for Human Rights Defenders, Journalists and Witnesses (Mahony 2013).

The mobilisation of power bases for action shows that the weaknesses of CSOs tend to attract more programs, and over time, more programs, and many international actors have over time given up on partnering or working closely with the Congolese civil society. The acknowledgement of change in the DRC requires mass-based support and functional civil society structures to mobilise it. International organisations themselves cannot mobilise a Congolese popular base for change, but they can think strategically, identify opportunities for such mobilisation and strengthen and protect the organisations and individuals who might be promising. They support the convening of Congolese discussions and even participate in the debate. The humanitarian community helped this process by engaging more with civil society forces outside the professional NGO "assistance" community (Mahony 2013, p. 39).

Another important area of mobilisation recommended by Mahony's several respondents was to encourage greater involvement of women in the work for political change. Experiences in other countries, such as Liberia, has shown how women can use symbolic action and mass mobilisation to effectively break through traditional attitudes about conflict and violence as well as pressure armed actors to alter their behavior (Mahony 2013, p. 40). To date, there does not appear to be any large-scale movement of women in the DRC working on conflict reduction, but any broader mobilisation strategies must ensure that they are inclusive of women at the leadership level to take advantage of these possibilities (Mahony 2013, p. 40).

8.4.1 Human Rights Monitoring Problems

Human rights activists and humanitarian actors as well as other civil society leaders in North Kivu suffered harassment after having denounced the arms distribution and subsequent abuses against civilians in North Kivu, specifically attributing responsibility for these acts to the governor and his associates. Several also asserted that Rwandan army troops had supported soldiers linked to the *Rassemblement congolais pour la démocratie* (RCD) [Congolese Rally for Democracy] based in Goma town (RCD-Goma) rebel movement. The consequence of these statements involved several activists who had signed them receiving anonymous telephoned death threats and warnings by unidentified armed men. As a result of such threats, four leading human rights activists fled Goma the capital city of North Kivu province.

When a Human Rights Watch researcher raised these cases with provincial security officials in early February 2005, they admitted that security agents might have threatened the activists but denied having ordered them to do so. They promised to investigate and to prevent future threats. However, later, another activist and his wife were beaten and robbed in Goma. Another instance of harassment took place in Goma on 23 April 2005 against a human rights defender by unidentified armed men who threatened his family and beat one of his relatives.

The work of humanitarian agencies was disrupted, and soldiers threatened staff and looted the property of humanitarian organisations in Masisi, Rutshuru and Lubero, disrupting the delivery of assistance to civilians. Soldiers linked to RCD-Goma looted two health centers and caused damage at a third, and 10 health centers had to restrict their operations for more than a month. Other many cases of looting occurred during the combat at Nyabyondo.

Local actors and authorities are rather different from one another but they share certain characteristics that are different from those of other peace initiatives in eastern DRC, International Alert (2012, p. 45) presents the difference in the following ways: They both provided for significant involvement by local actors in defining the content and implementation of the solutions required to resolve their problems in the long term. These actors were chosen on the basis of their legitimacy, influence, and expertise in the area concerned. They are thus directly contributing to establishing "social contracts" and government accountability in DRC. In doing so, they help to create a more inclusive style of governance that is transparent and democratic. The main problem in using for this approach is encouraging the actors and authorities involved to embrace the dialogue and follow up on the commitments reached.

École de la Paix organisation led the initiative in Bunia in sharing certain characteristics. It involved rehabilitating and addressing the trauma of former child soldiers caused by opposing armed groups through creative activities such as singing, drawing and drama. The initiative, which is entirely based on the children's first-hand experiences, adopts a playful and creative approach to transforming their trauma into a positive force for social change. This is an aspect that is often lacking in more traditional initiatives designed to help victims (International Alert 2012, p. 45).

The initiatives led by civil society actors led initiatives offer particularly relevant and interesting alternatives to the more traditional large-scale programs. Nevertheless, they remain very much on the fringes of the response seeking to bring an end to conflict in Eastern DRC. The principles that form the basis of these initiatives should inspire the entire peacebuilding sector, whether it be government, international or local actors (International Alert 2012, p. 46)

8.4.2 Advocacy Challenges for Peace

Few civil society initiatives have been found as an alternative and grounded approach to peacebuilding. Such an approach represents a promising counterpoint to the technical interventions that took place under the stabilisation and reconstruction program for areas emerging from armed conflicts or the National Stabilization and Reconstruction Program (STAREC) and The International Security and Stabilization Support Strategy (ISSSS) frameworks (International Alert 2012, p. 44). The first initiative is advocacy work led by the *Forum des Amis de la Terre* (FAT) [the Friends of the Earth Forum], the *Syndicat de Défense des Intérêts Paysans* (SYDIP) [the Farmers' Defence Union] and the *Fédération des Organizations de Production Agricole du Congo* (FOPAC) [the Federation of Agricultural Production Organizations of Congo], as part of the negotiations that led to the new agricultural code. The second is the participatory and dialogue-based action-research project put in place by partners of the *Institut Vie et Paix* (IVP) [the Life and Peace Institute], the *Action pour la Paix et la Concorde* (APC) [the Action for Peace and Concord], the *Réseau d'innovation organisationelle* (RIO) [the Organizational Innovation Network] and the *Action pour le Développement et la Paix Endogènes* (ADEPAE) [the Action for Endogenous Development and Peace]. Lastly, the *École de la Paix* [the School of Peace] in Bunia developed a program with a significant interest in addressing trauma and rehabilitating conflict victims using artistic activities, according to International Alert (2012, pp. 44–45).

The advocacy work in relation to the new agricultural code was carried out by a coalition of peasant unions and CSOs. Various consultation documents in North Kivu contain their demands and recommendations which were integrated into the agricultural code. APC, RIO and ADEPAE organisations led the second initiative which involved participatory action-research that led to the implementation of a bottom-up dialogue process, bringing together local actors and the authorities (International Alert 2012, p. 45). Its implication is that the authorities and other actors take responsibility for monitoring the way in which the solution is put in place.

A common challenge for CSOs in conflict settings has involved engaging with decision makers. In DRC, for instance, CSOs struggle to engage in constructive dialogue with political authorities and to secure their participation in peace initiatives, reforms, and efforts to increase accountability, according to Ellis (2017). In other case studies, CSOs have felt they would be more effective if international actors supported them to develop locally appropriate advocacy strategies and to build cross-cutting local, national, regional, and international advocacy networks and platforms (Stephen 2017, p. 36).

The subject of civil society and women in peacebuilding is another challenge as the DRC is emblematic of how violence and conflict produce different gender impacts. Congolese women and girls suffered disproportionately during and after the country's 1998–2003 war (Adekeye/Paterson 2010, p. 28). The country's peacemaking and post-conflict reconstruction processes provided to women the opportunities to engender societal structures, especially with regard to political participation, transitional justice, and security sector reform. Some Congolese women formed a caucus to participate in the ICD, in which they played a paramount role in lobbying for an agreement between the feuding parties. It is important to note that women's networks and non-NGOs also actively engaged in campaigns advocating the recognition of women's rights during and after the DRC's constitution making processes (Adekeye/Paterson 2010, p. 28).

The advocacy resulted in the 2006 constitutional provision of a 50% quota for women's participation in government and called on the government to respond effectively to sexual violence. Women's organisations also helped to mobilised and train women to participate in political processes, boosting the women's voting at the 2006/2007 polls. It is necessary to place gender at the heart of the Congolese peacebuilding processes as a pre-condition for succeeding in effecting a broader transformation of structures, policies, norms, and relationships, which are still shaped by ideas of patriarchy (Adekeye/Paterson 2010, p. 28). However, to this day, this 50% quota has never been achieved.

Rather than being a fundamental right that is necessary for the development and implementation of effective peacebuilding strategies, the perception of gender equality in the DRC is often being that it is solely a "women's issue" and a matter of political correctness. There is a general belief that women are entirely responsible for incorporating gender issues into peacebuilding programs. The widely held perception that women and civil society can best deliver the gender aspects of projects has further undermined collective efforts at achieving gender equality in peacebuilding processes. However, some progress has been made in engaging men to a greater extent in the struggle for gender equality, and to engendering peacebuilding and social justice concerns in the Congo. Nevertheless, activists from the sector have contributed significantly to governance, security sector reform, Disarmament, Demobilisation and Reintegration (DDR); peacebuilding, and state building while civil society's peacebuilding efforts may have been relatively weak so far. Civil society in the DRC has also played a critical role in mobilising social capital to hold the government accountable and make it more responsive to public needs. Furthermore, civil society groups have delivered services to prevent and resolve conflicts, and have promoted democracy, human rights, and the rule of law (Adekeye/Paterson 2010, pp. 7–8).

The burden of sexual and gender-based violence persists in the DRC despite the presence of the UN mission (MONUSCO) presence, and the Congolese government's endorsement of international conventions such as UN Security Council Resolutions 1325 (2000), 1820 (2008), and 1888 (2009) on women, peace, and security. More than 8,000 Congolese women were raped in 2009 during fighting between warring factions, according to the United Nations Population Fund (UNFPA) and to date, the rape is still ongoing. Sexual violence is particularly widespread in the war-torn eastern region of the country, considered as the "rape capital of the world" (UNFPA 2010, p. 28). Meanwhile, the lack of a fair and effective national justice system has made it difficult to address the rights of rape survivors and victims of other sexual violence (Bosmans 2007).

Since the DRC's independence in 1960, religious organisations and other NGOs have played a pivotal role in the country's peace processes, including in the run-up to, and after, the historic democratic election of 2006/2007 (Adekeye/Paterson 2010, p. 29). Such groups continue to be integral to peacebuilding processes in delivering services. The international donor community has often channeled aid through civil society actors in a political climate in which the central state is weak due to internal and national conflict as well as poor governance practices. However, such actors often lack capacity and cannot be a substitute for building an effective state, which must remain a key priority (Adekeye/Paterson 2010, p. 29).

8.4.3 Socialisation Hindrances

Furthermore, International Alert underscored the basic weakness of civil society which is due to the difficulty it has in positioning itself in a coherent, clear and constructive manner in relation to three groups: its 'base', i.e. local populations, political authorities, and international bodies. These relations have prevented CSOs from playing their role for change. This problem derives from the civil society, which is highly politicised, organised within ethnical ties, and the manipulated by donors imposing their priorities instead of basing programs on strict analysis of the specific context (International Alert 2012, p. 44).

The base is not neutral due to its close proximity to an ethnic territory where associations for peace and development are created for the defense of its particular interests in Minembwe, Bunyakiri, and Fizi. This attribute enhances the competition and power struggles that are identified in parts of civil society, as each association tends to work primarily for its own area or community (International Alert 2012, p. 44). In addition, civil society depends financially on international organisations that limit its role in the program execution of the international agencies (International Alert 2012, p. 44). Such dependency hampers Congolese organisations in the development of innovative strategies for peace.

Trust and the deficit of trust within civil society in conflict settings is another issue because the levels of trust are very low and people rely on personal connection and social groups to survive due to the religious, and ethnic identity mobilisation by conflict actors. Civil society often reflects and replicates these patterns. In DRC, conflict dynamics are based on geography as different refugees settled around different local communities for which they had an affinity. Subsequently, many CSOs emerged from within local communities to serve their needs therefore limiting the reach of service to reach other conflict-affected areas.

Thus, to some degree, CSOs working at the local level in DRC reflect the way in which conflict has shaped society (Ellis 2017; Stephen 2017). This limited opportunities for dialogues between communities lead to a trust deficit that hampers their effectiveness within CSOs. They often have 'limited opportunities to raise funds or build joint platforms and alliances to engage in advocacy, which can curb their ability to contribute to broader goals such as ending violence and supporting peaceful political change and sustainable development' (Stephen 2017, p. 26). This trust deficit in turn tends to reinforce the fault lines along which the conflict has developed.

8.4.4 Social Cohesion Issues

Highly politicised Civil society is an important factor in Eastern DRC. It is difficult to distinguish the boundaries between civil and political society due to political competition and formal mechanisms for political representation being weak. CSOs are targeted by political actors for their own interests while civil society to acts to foster its own agendas (Stephen 2017, p. 49). When the Rwandan refugees arrived in Congo, the civil society mobilised against Rwandan backed armed groups. The mobilisation involved nationalist and anti-Rwandan discourse with effect of solidifying divisions and tensions between communities. This politicisation placed the peacebuilding in jeopardy (Stephen 2017, p. 49). Bilak (2009) affirms the following:

> The civil society politicization fact indicates the problematic relationship between its members and those in power. Civil society is a locus of political competition that is synonymous with opportunity and power as well as difficult to distinguish from the political sphere (p. 157).

International Alert (2012, p. 44) describes it as an "incestuous" relationship with those in power and which often acts as an advantage to a political career. The power struggles have affected the civil society and prevented it to implement real strategies for advocacy or constructive dialogues with the authorities, and they have undermined its neutrality and legitimacy.

Finally, politicisation prevents the preservation of expertise and key skill sets within civil society, as the most competent personnel are almost systematically absorbed by international organisations. This hierarchical relationship of dependency is a factor in civil society remaining in a position of structural weakness. This situation raises a real problem in terms of the long-term impact of the international community's presence (International Alert 2012, p. 44).

8.4.5 Dialogue Facilitation Deficiency

Civil society represents a significant dimension of the peacebuilding sector. For example, it has played a central role in political discussions and peace accords from 1991 to 1992, the Sun City Accord in 2002, and the Goma Conference in 2008. To date, it continues to engage in peace initiatives across all levels of society to the grassroots level. Many CSOs civil society organisations are active in the peace sector in North and South Kivu alone, and are often local organisations working with support from international NGOs. Activities include increasing awareness, peace education, skills development (training), advocacy and mediation, reconciliation, and arbitration (International Alert 2012, p. 43).

However, the failure of the peace processes in DRC derived from the misunderstanding of the nature of civil society in the Kivu provinces and exclusion of grassroots' representatives according to Santoso (2010, p. 35). Research concerning the role of civil society in DRC has suggested that international organisations involved in the Congolese peace process had tendency to assume that civil society in the Kivu provinces viewed its Western counterpart where civil society represents the needs of the people to the state and keeps the state accountable to the people. Civil society in the Kivu provinces, however, developed rather differently, with today's distinctive social and bureaucratic structures having been shaped by the colonial administration of earlier years (Santoso 2010, p. 35).

According to Santoso (2010, p. 35), civil society with its networks of national, regional, and provincial entities is also considered to include patron-client networks entities throughout the country. The influx of international aid and humanitarian intervention throughout the conflict as a second step of conflict resolution shows how international actors use local actors and associations without considering the needs of the local population (2010, p. 35). In addition, the inclusion of civil society in any peace process seems to distinguish whether or not civil society representatives are truly connected to the population or if they are simply the puppets of political elites or international organisations. If this is the case, then the civil society is not able to play its pivotal role in the peacebuilding process.

There was a failure to include grassroots civil society groups as the peace process catered primarily to the élite in the country and had little effect on realities on the ground. Thus, it is important to affirm, in lines with Santoso's statement that 'peace processes need to occur both from the top down and from the bottom up' (Santoso 2010, p. 35). However, there is evidence in the Congolese case proving that various forms of the peace process that tried to create mechanisms for inclusion of civil society in the peace process agreements with militia leaders and senior political leaders before 2008, the Goma Agreement in 2008. Hundreds of civil society leaders attended a conference to present their concerns for consideration.

The grassroots civil society members were marginalised and had little access to the peace process even though they were the most important stakeholders in the process. According to Santoso (2010, p. 35) representatives of groups present at the peace talks, as well as some who had been excluded, were from a range of groups. Some belonged to the official 'Civil Society' bureaucratic network while others were not part of this 'official' network but were leaders of NGOs either run by international organisations based in Europe or local to the area.

Santoso's research on the structure of civil society and understanding the degree of participation reached the following consensus that, in the Congolese context, the peace process has had little success not only because it has faced immense challenges due to the regional dynamics of the conflict but also because it has primarily existed in a top-down form and failed to integrate bottom-up processes. It has failed not only to include the right members of civil society but also to ensure that the shape of the peace process matches the society to which it hopes to bring peace (Santoso 2010, p. 35). Notably, the bottom-up processes have been successful in some areas like 'Somaliland where indigenous attempts at bottom-up peacebuilding succeeded in generating a relatively peaceful and stable situation' (Autesserre 2010, p. 248).

For Santoso (2010), civil society in the context of the Kivu provinces has developed separately from its European and North American counterparts as a hybrid of deeply entrenched patrimonial associations, transnationally controlled NGOs, and small local associations. It is important to take these realities into account when designing mechanisms for civil society inclusion in a peace process.

8.4.6 Frustration in Service Delivery

Impact of donor funding patterns on CSOs is illustrated by the research in DRC that highlighted the dependency on foreign donor funds. In addition, the desperation for financial support has led CSOs to adapt their focus to reflect donor priorities. They also reported frustration with international partners who did not support organisational development through fundraising activities to help CSOs achieve greater autonomy and sustainability (Ellis 2017; Stephen 2017). Furthermore, smaller CSOs have been marginalised by a reduction of transaction costs. Notably, donors and international NGOs continuously seek to decrease the transaction costs of aid. They often give fewer, larger grants to a small number of CSOs (often based in the capital city) that have shown their capacity to manage resources and meet complex reporting requirements. However, this condition impedes them supporting smaller grassroots CSOs that lack the capabilities to absorb and manage large grants, even though they may be in the optimal conditions to work with marginalised or remote communities (Stephen 2017, p. 40).

The professionalisation of CSOs has the potential for weakening their ties with communities and is another weakness. International actors seek partnerships with CSOs in conflict settings to push CSOs and social movements towards professionalisation, to make them able to meet complex donor reporting requirements. Over time, this can result in community-focused and/or politically active organisations becoming transforming into donor-facing technocratic service delivery organisations, and in the process, weakening their ties and engagement with local communities, a connection which is vital for their effectiveness in violent conflict settings (Pouligny 2005).

In DRC, local CSOs have worked with international actors to provide access to specific localities to deliver aid, but their role was mainly limited to implementing predefined initiatives by the later. Indeed, local CSOs report having little opportunity to help shape projects based on their understanding of local needs, which in turn affects their relationships with local communities and means they are unable to represent their needs (Stephen 2017, p. 40). The following section involves the discussion through which the effectiveness of civil society can be determined in the context of Kivu provinces.

8.5 Civil Society' Role and Effectiveness in Peacebuilding

In this discussion, the seven central functions, as suggested by Paffenholz/Spurk (2010: 65–75), are used as a lens to examine and to understand the civil society's role and its effectiveness in peacebuilding in the east of DRC particularly in the Kivu provinces. It is worthy to start with the first key function.

8.5.1 Protection of Citizens

Considering protection, the civil society has to provide security to people or in cooperation with other agencies. Notably, DRC government, as main provider of basic human rights, has been unable to do so due to the conflict and limitations in the ability to reach the war area being uncontrolled by the government. In this case, the civil society is the appropriate actor to reach the area and take care of the population in need to secure its basic human rights. However, 'the CSOs also need protection as they face an increasingly difficult working environment in order to play their role in peacebuilding where violence is prevailing' (Paffenholz 2010, p. 360).

During or after conflict, communities try to bring solution to their problems and needs. In the eastern provinces of Kivu, the existing structures like Church leaders, community social groups, and cooperatives help mobilise for the protection strategies using various collective community protection practices but these practices are usually limited to the protection of a certain ethnical community. This is a factor that hinders the structures from functioning properly.

In addition, the collaboration between the international actors and CSOs is also needed for mass mobilisation and encouraging the greater involvement of women to enhance protection in peacebuilding process. The women's role is not only wished in the Eastern Congo but also, they can learn from women involved in conflict and violence resolution in Liberia. Currently, even though women, unlike their male counterparts are involved but not yet fully integrated in the peacebuilding process in the Kivu provinces, they have much to offer with regard to constructing peace in the eastern provinces. The CSOs including women are supposed to become involved in humanitarian security to promote the international lobby but how this effectively carried out in the Kivu provinces is difficult to ascertain as it remains a sensitive issue among actors involved in the conflict. For the protection to be effective, its implementation needs to be monitored.

8.5.2 Monitoring of Relevant Issues

In this function, international and local civil society groups have the duty to monitor relevant issues which mainly involve human rights. However, in the context of eastern provinces, particularly in Kivu, sometimes, the international and local human right activists who are supposed to monitor human rights themselves become the target and victims. These civil society leaders and human rights activists denounced abuses against civilians, and subsequently suffered harassment by some government authorities who were very likely accountable for these human rights abuses. In my opinion, in order to play a monitoring role in a prevalent conflict dynamic, it is necessary for the local civil society actors to have security and be protected, but this is not the case in Kivu provinces. This linkage can allow the function of monitoring to proceed smoothly despite the delicate situation of human rights sensible issues.

Several reports have addressed the Rwandan army's involvement in supporting RCD-Goma rebel group in DRC. However, many activists who signed the publication of these statements received anonymous telephoned death threats, warnings to their families by unidentified armed men, to stop them from denouncing abuses. Consequently, those leading the protection of human rights could not stay in the area due to this threat on their lives, and they had fled out Goma to preserve their lives. This situation is a similar case for all Somalian organisations playing a monitoring role who have been vulnerable to pressure, threats, attacks and even assassination (Paffenholz 2010, p. 336). This situation illustrates the fact that the lack of protection for the activists or civil society actors can hinder the monitoring function if some important precautions are not considered.

The activists monitoring human rights, as reported in Kivu provinces, also encounter risks when they hold governments and armed groups accountable for abuses or substandard performance, and they can also serve as an early warning system. The denunciation of security agents involved in the threats made against activists resulted in the neglect of investigations by the provincial security officials who had promised to investigate the cases to prevent future threats. Surprisingly, different activists and their respective families were beaten and robbed without investigation and punishment of their perpetrators. These cases demonstrate how the government officials, as decision makers placed under a greater pressure, are themselves involved in human rights abuses (Paffenholz 2010, p. 337), but who are supposedly working hand in hand with human rights' activists. The denunciation bring fear to the official governments who condone the threat and try to silence the human rights activists.

This situation becomes complicated because the local official governments who are supposed to investigate the denounced abuse and bring the perpetrators to justice are themselves involvement in the abuse. If, after the government actors have promised to carry out investigations, the human rights watch activists or local civil society activists receive threats, then the complicity of the decision makers becomes evident; they are corrupted or silenced by obscure forces. The provincial officials may see human right activists as antagonists of their plans when their own interests, or those of individuals who support them to keep the status quo, are undermined by the denunciation of human rights abuse in conflicts areas for obscure interests.

The connection between the local civil society in Kivu provinces and international agencies for the victims and their own protection is crucial for a better motoring function in the peacebuilding process. There is also the question that comes to the mind of any observer regarding the human rights abuse in the Eastern Congo concerning the reasons human right abuses are committed vis-a-vis the provincial officials and UN peacekeepers, who are known and continue to live free with impunity. This shows the complicity of various stakeholders who prevent the actions of CSOs actions for their own interests. The monitoring function has a link to the advocacy function.

8.5.3 Advocacy for Peace Promotion

The advocacy function calls for CSOs to set the agenda and press for certain important issues. Civil society actors can push for the commencement of negotiations and the implementation of negotiated agreements, or work against the recurrence of warfare. In the Kivu provinces, the alternative and grounded approach to peacebuilding took place in the same way the advocacy work related to the new agriculture code. The push of the negotiation and the initiation of the participatory action research led to accomplishing the implementation of a bottom-up dialogue process contrary to the up down model that excludes some of the grassroots actors.

These initiatives brought local actors and authorities together to make them conscious of their responsibility for monitoring peacebuilding. Putting pressure on and convincing key actors to engage in conflict resolution and mobilising popular support for peace are necessary step to end conflicts nonviolently (Paffenholz 2010, p. 310). However, bringing decision-makers together is a challenge in conflict settings due to civil society struggling to engage in constructive dialogue with political authorities to participate in and secure initiatives and efforts to increase accountability (Ellis 2017). In this struggle, the civil society needs the support of international actors to develop adequate advocacy strategies to build advocacy networks and platforms. Subsequently, the collaboration is necessary to foster strategies.

While local and international actors need to work together, the consideration of women is another challenge as they are the mostly victims of the violence and conflict in the Kivu provinces. They have been becoming involved in the peace building process to use their skills to engender societal structures for various reforms in political participation, transitional justice, and security sector reform. In the case of DRC particularly the Kivu provinces, women's participation in campaigns advocating the recognition of women's rights and the ICD resulted in securing the quota of 50% for women's participation in the government at all levels and achieving the agreement between the rival parties for peace.

As a fundamental right the gender equality is not a women's issue but a carrier for incorporating and delivering gender aspects in the peacebuilding process. Just as men are involved in peacebuilding processes, women deserve to become involved and strengthened in peacebuilding efforts in Kivu provinces because:

> The promotion of women's peacebuilding programs will not be successful while the decision-making systems and mechanisms are still dominated by men and if the women themselves support the existing norms and practices and are not aware of their rights' (Angom 2018, p. 187).

Nevertheless, in Kivu provinces, women's exclusion is revealed in their 'quasi-absence from peace negotiations which have taken place to end try and end the conflict, despite a strong mobilisation of women's organisations to insist on participation in peace negotiations' according to Jane Freedman (2015, p. 114). The case of sexual violence calls not only on men but also on women to become involved in peacebuilding by meeting with victims. This approach can help women victims to alleviate the stigma and trauma of the negative impact of rape. In doing so, the socialisation can take place for the reintegration identity in-groups.

8.5.4 Socialisation In-Group Identity

Considering the base in the Kivu provinces as in-group bonding that promotes democratic behavior and peaceful values within society, sometimes the base is not neutral as it is often linked to an ethnic territory because the formed associations or the community aim to defend only their own ethnic interests, implying non-security for those external to the ethnic groups. When the base is connected to an ethnic territory it loses the value of in-group bonding where socialisation can take place. Then it is impossible to go beyond this ethnical boundary to reach out to other people who need socialisation to bridge out the ties within the groups. This is that which the Kivu provinces are experiencing as a hindrance to this function.

This experience implies that the base structure does not learn to live together in peaceful coexistence with other ethnic groups as it is biased with regard to ethnic protection in the Kivu provinces such as the protection of the Tutsi ethnic group. This way of operation is contrary to Coleman's (1988) argument stating that the institutionalised social interactions can become available to individuals (Coleman 1988 cited in Howell/Pearce 2001, p. 28). Indeed, in this case, tolerant and peaceful values cannot be promoted in an ethnic territory that defend only its own interest.

Learning to live together in peaceful coexistence requires building trust to improve the trust deficit within the civil society and between the other people. In the conflict area, such as in the eastern parts of Congo, level of trust between communities are very low where people need social interactions, connections, alliance to survive. Civil society in the DRC reflets the way communities are connected and are dispersed geographically. As most of civil society came from within local communities to serve local needs there is a limit to exposure to other affected communities. Without this exposure to other affected communities in Kivu provinces, the trust and deficit of trust cannot be restored and each member is locked to their community to the detriment of socialisation.

However, CSOs are called to go beyond ethnical protection to consider everyone who needs protection and socialisation within and outside of the ethnic territory. This leads to the consideration of large structures that are necessary to reach out to those in need through their leaders. The obvious and vital mobilising structure in DRC is the catholic church. In Eastern Congo, the catholic church plays a central role in peace building despite the complaints of people about churches in general. The complaints imply that they are biased and compete with others with the consequence that many leaders lose their credibility due to their involvement in politics or taking sides in the electoral period. Their neutrality in the electoral period is necessary as it is one of the democratic consolidating pillars and values. This bias and competition in electoral period do not support the gaining of people's trust church leaders. The reason resides in the fact that people who benefit from their help have trust on them and want them to remain neutral in order to reach out to everyone beyond ethnic group boundaries in the peacebuilding process.

One mentioned salient case is that of the catholic church with its Justice and Peace network that confronts human rights issues with regard to protection. These existing structures like the churches and other significant networks are also under prevalent dynamics of security threats that limit their action. Some of them have been assassinated, for example, the case of Bishop Munzihirwa who was slain in Bukavu (South Kivu) in 1996 during the conflict. Therefore, the protection is necessary to foster their activities for people's security in order to practice peaceful coexistence in socialising in families, schools, religious groups, secular and cultural associations, and the workplace, as is required by the socialisation function. The mobilisation of power bases for action is very important because the fragile situation of conflict can always downplay the function of socialisation. It is then difficult to evaluate the socialisation function in the cases of the Eastern Congo provinces due to certains issues such as division, competition, and the politicisation of CSOs which are supposed to be neutral. The socialisation function always operates alongside social cohesion in the peacebuilding process.

8.5.5 Intergroup Social Cohesion

The degradation or destruction of social-capital groups call for the social cohesion to function in order to bridge the ties to reunite adversarial groups divided by the war (Paffenholz 2010, p. 366). Social capital generates an inevitable public good, and it is an unwilled outcome of institutionalised social interactions which then becomes available to individuals, according to James Coleman (1988), and Putnam argue that social capital becomes a property of the groups or even nations (Putman 1993 cited in Howell/Pearce 2001, p. 28). Once the conflict destroys the social capital, it becomes necessary to reconstruct it for public good.

In Kivu provinces, the social capital is partly hindered where civil society operates in a highly politicised area as boundaries with political spaces are not clear. Political actors advance their interests and civil society moves between the two spaces. This can be illustrated in the way civil society in Kivu has mobilised against the presence of Rwandan-backed armed groups with nationalist and anti-Rwandan discourse. This politicisation has affected the civil society and is a threat to peace that needs to be eradicated by both international and local actors in order to contribute to peacebuilding and social cohesion for the public good. This positioning of civil society is a basic weakness because it is not coherent nor carried out in a constructive manner in relation to groups.

The relationship between the three main groups the local population, political authorities, and international bodies has not allowed CSOs to play a role for change. The rationale behind this comes from the characteristics of civil society, which is highly politicised with regards to power struggles, and organised according to ethnic groups and donors' tendencies to control civil society to reflect their priorities.

Many civil societies, within themselves live in contradiction to social cohesion as they do not promote dialogue. Consequently, there is no trust between them. This lack of trust within or between them hinder their effectiveness. With the deficit of trust, it is difficult for them to engage in fund raising to contribute to broader goals (Stephen 2017, p. 26). The opportunities for dialogue for tackling issues related to the common interest throughout the divided areas enable the reversal of the trust deficit in the affected areas in Kivu. Therefore, the social cohesion function is very important as it brings people together for a common cause.

8.5.6 Facilitation and Mediation

This function is also linked to the other functions, particularly social cohesion, to bring together opposing parties or communities during the wars. To bring people together, civil society has to work as a facilitator at all levels. This facilitating function is a central task for UN as a world organisation and involves its charter mandate for the peaceful settlement of disputes (Boulding 1994, p. 229). In the DRC, not only UN but also CSOs play a role in its context. Playing this role in the context of conflict is a big challenge. It is important to note that the representation by civil society actors as a key component to making a process such as peace negotiations more legitimate is problematic due to the conceptual difficulty regarding civil society and the power differentials embedded in such a representative act (Zanker 2018, p. 35). Despite this challenge, the facilitation function has been partly effective in DRC as the CSOs have played a pivotal role in different political discussions such as the peace accords from 1991 to 1992, the Sun City Accord in 2002, and the Goma Conference in 2008, and many other engagements in peace initiatives across all levels of society to the grassroots level.

Many CSOs are active in the peace sector in North and South Kivu alone, with most of them working with the support of international NGOs. The facilitation of civil service concerns also other issue-oriented violence-free days. Those activities include increasing awareness, peace education, skills development (training), advocacy and mediation, reconciliation, and arbitration (International Alert 2012, p. 43). It is important to note that facilitation is linked to and supports the protection and service delivery functions.

In addition to the above actions, a problem of perception exists where the nature of civil society is sometimes misunderstood, and the grassroots representatives are excluded, which has led to the failure of the peace process failure in DRC. The relationship between international organisations and civil society involved in DRC have certain perceptions of each other. International organisations think about the role of civil society in Kivu provinces as representing the needs and, accountability of the people to the state. However, civil society has developed differently with social and bureaucratic structures (Santoso 2010, p. 35). The perception is notably different from reality. This can sometimes prevent them from working toward a common end.

Often, International actors run local NGOs and associations which do not satisfy the needs of the local population. However, there are also grassroots local NGOs which do address local needs and are primarily funded by their own members from income generated by second or third jobs (Santoso 2010, p. 35). It is necessary to address the needs of local population are necessary and international actors sometimes ignore these needs and marginalise local NGOs that address these needs.

Notably, it is worthy to call on each actor involved to consider the role of the other in order to facilitate the process of service delivery reaching the local population. Each actor, according to their perception, works differently: the international actor follows the top-down approach, while their counterpart the local actor prefers the bottom-up method. In order to develop a useful process for the facilitation function, both approaches the top-down and bottom-up approaches can be used to reach out to the needy during or after the unrest.

8.5.7 Service Delivery Domination

During armed conflict, the population may be starved of essential services because the state structures are either destroyed or weakened. At this point the Civil society actors (mainly NGOs, but sometimes associations as well) can and do step forward to provide aid and social services. Related to protection and facilitation, the service delivery function is necessary. Nevertheless, in the case of the eastern parts in Kivu provinces, the CSOs show dependency on the foreign donor funds to foster their activities. This dependency forces the CSOs to adapt their focus to reflect donor priorities. This might not be the correct attitude because there is no freedom of thinking and using the service delivery for the appropriate or targeted needy population. Financial support becomes an obstacle for greater autonomy and sustainability of CSOs in the Kivu provinces. The CSOs have to work to find their own financial supplies in order to curb this dependency which can hinder their service delivery activities due to lack of finance.

The issue of the financial dependency is also connected to the marginalisation of the smaller CSOs (to reduce the costs of transactions) by the international NGOs. International actors have chosen a small number of CSOs based in the cities and they have the capacity to manage their resources. This manner of working puts aside the true population that can benefit from the aid due to the criteria based on the donor's priorities. Meanwhile, grassroots actors who are unable to manage large funds cannot obtain grants even though they can reach the marginalised population in the remote communities (Stephen 2017, p. 40). Unless, international donors consider these issues that hamper the service delivery in the grassroots level, mainly in Kivu where the conflict is still ongoing, the service delivery will not reach the well targeted persons in need.

Another important element is the professionalisation of CSOs by the international donor. It can be observed that that which happened in Kivu regions shows the difficulty to bring the local and international actor together for a common service delivery function. The international donors tend to dominate the function due to their possession of finances and impose their views and methods on the local CSOs, as the latter lack funds. In addition, the international actors push the CSOs towards professionalisation to meet donor reporting requirements. This places the peacebuilding process in jeopardy as the objective to assist the needy depends on meeting those requirements.

If the international actor has to dominate and decide on all the procedures with the consequence of the grassroots people not receiving aid, the domination can weaken the CSOs' ties with the communities and the CSOs cannot have an impact in peace building process as they depend mostly on the international donors. This service delivery function needs facilitation such that it avoids the professionalisation of the CSOs and places more importance on reaching the population and providing for their needs. Many CSOs have a limited role in delivering aid as they lack opportunities to help shape projects based on their understanding of local needs, which in turn affects their relationships with local communities and means they are unable to represent their needs (Stephen 2017: 40). If there is a will to help the marginalised population during the conflict, the assistance must be sincere and reach the targeted population in need. Having understood how the seven key functions work in peacebuilding particularly in Kivu provinces, a conclusion is drawn from the study in its entirety.

8.6 Conclusion

This study objective was to examine how effective has civil society been in ending armed conflict and sustaining peace in conducting the seven peacebuilding functions of protection, monitoring, advocacy, socialisation, social cohesion, facilitation, and service delivery in Kivu provinces? Several important elements have been identified. Concerning the peacebuilding functions, protection is characterised by the mobilisation of protection strategies by various leaders in the existing structures and mobilisation of power bases including women. These strategies have helped to minimise the risk of violence but they do not ensure total peace. The monitoring is marked by the harassment of human rights activists and their families; local and international actors; and authorities' differences in their involvement in their problems, solving them sometimes without consensus.

In the advocacy function, various CSOs initiatives considered to be alternative and grounded approaches to peacebuilding have been effective. In addition, CSOs struggle to engage in constructive dialogue with political authorities, and women's engagement in peacebuilding is a challenge as their participation has not brought about an efficient response to the persistent sexual and gender-based violence that still occurs today. The issues of CSOs base related to the local population, political authorities, and international bodies as well as the trust and trust deficit within and external to civil society, were revealed in the function of socialisation. These factors constituted the hindrance to these peacebuilding functions.

In addition, concerning the social cohesion, the high degree of politicisation in civil society, not only indicated by the problematic relationship between its members and those in power but also by localised conflicts dynamics, is an obstacle to peace building functions. It is worth to note that CSOs as facilitators have seen their involvement in various peace discussions and accords being unsuccessful, where the exclusion of grassroots representatives has been detrimental and has of their inclusion that led to peace processes failure in the Kivu provinces. This is illustrated by the CSOs being considered as the branches of patron-client networks and the influx of international aid and humanitarian intervention with local NGOs being dominated by the international actors imposing their priorities instead of considering those of the people in the fields. This has led to a conflict between up-down and bottom-up approaches in the conflict resolution carried out by CSOs in the Kivu provinces. Furthermore, some factors have been revealed in the service delivery function regarding the CSOs' dependency on foreign donors' funds, where their desperation for financial support has led to the reliance on the international donors and submission to their priorities. Furthermore, a CSO professionalisation by international actors has led to the weakening of their ties with communities due to the imposition of applying international donors' priorities in the course of the peacebuilding process.

All the above-mentioned identified factors confirm this chapter' argument that the civil society, despite its slight development, is still weak in building peace due to various factors that hinder the fulfilment of the key peacebuilding functions. These identified factors have contributed to the hindrance of the seven-key peacebuilding functions, varying substantially from function to function in weakening the CSOs' various actions in peacebuilding processes despite a few positive initiatives being detected in each function and their impact in the Kivu provinces.

These identified factors constitute the challenges to the functions of peacebuilding process for the conflict resolutions carried out by the CSOs. Therefore, it is imperative and worthy to address these challenges first by making mechanisms that can impact positively on the CSOs' actions to stop and eradicate the recurrent conflict and move toward a sustainable peace in the Kivu provinces in DRC. Unless, these identified peacebuilding factors are addressed or eradicated as a pre-condition, sustainable peace cannot be attained with the consequence being a resurgence of conflict for the population of the Kivu provinces in the eastern parts of DRC.

References

Adekeye, A., & Paterson, M. (Eds.) (2010). *Post-Conflict Reconstruction in the Democratic Republic of Congo (DRC).* Cape Town: Centre for Conflict Resolution

Africa's World War: Congo, the Rwandan Genocide, and Making of a Continental Catastrophe. Oxford: Oxford University Press

Angom, S. (2018). *Women in Peacebuilding in Northern Uganda.* Cham: Springer Nature Switzerland AG. https://doi.org/10.1007/978-3-319-75883-1

Autesserre, S. (2010). The Trouble with the Congo: Local Violence and the Failure of Peacebuilding. New York: Cambridge University Press

Bilak, A. (2009). La Société Civile Face à l'État: Vers une Transformation Positive des Conflits. In T. Trefon (Ed.), *Réformes au Congo (RDC). Attentes et Désillusions* (pp. 1–290). Tervuren: L'Harmattan

Bosmans, M. (2007). Challenges in aid to rape victims: the case of the Democratic Republic of the Congo. *Essex Human Rights Review, 4*(1), 1–12

Boulding, E. (1994). *Building Peace in the Middle East*. Boulder & London: Lynne Rienner Publishers

Coghlan, B., Brennan, R., Ngoy, P. et al. (2006). Mortality in the Democratic Republic of Congo: A nationwide survey. *Lancet* 367– (9504), 44 51

Coleman, J. (1988). Social Capital in the Creation of Human Capital. *American Journal of Sociology* 94, (supplement), S95–S120. https://www.crcresearch.org/files-crcresearch/File/coleman_88.pdf. Accessed 8 Oct 2021

Ellis, C. (2017). DRC Case Study Report. In M. Stephen, Partnerships in Conflict. Oxfam International. https://policy-practice.oxfam.org/resources/partnerships-in-conflict-how-violent-conflict-impacts-local-civil-society-and-h-620359/. Accessed 21 Nov 2020

Ekiyor, T. (2008). The Role of Civil Society in Conflict Prevention: West Africa Experiences. https://gsdrc.org/document-library/the-role-of-civil-society-in-conflict-prevention-west-african-experiences/ Accessed 15 Feb 2021

Freedman, J. (2015). Gender, Violence and Politics in the Democratic Republic of Congo. London: Routledge

Gouzou, J. (2012). Study on the Role of Civil Society in Governance Processes in the Democratic Republic of Congo (DRC), CARE International, ISaC and PSO

Howell, J., & Pearce, J. (2001). *Civil Society and Development: A Critical Exploration*. Boulder: Lynne Rienner Publishers

International Alert (2012). Ending the deadlock: Towards a new vision of peace in eastern DRC. https://www.international-alert.org/sites/default/files/publications/201209EndingDeadlockDRC-EN.pdf Accessed 5 Oct 2021

International Crisis Group (1999) How Kabila lost his way: The performance of Laurent-Désiré Kabila's government. ICG DRC Report Number 3https://www.files.ethz.ch/isn/28673/008_performance_kabila.pdf. Accessed 5 Oct 2021

International Crisis Group (2000). Scramble for the Congo: Anatomy of an ugly war. ICG Africa Report No 26, Nairobi/ Brussels. https://www.files.ethz.ch/isn/28694/026_scramble_for_the_congo.pdf. Accessed 24 Aug 2021

Koko, S. (2016). The role of civil society in conflict resolution in the Democratic Republic of the Congo, 1998-2006: An appraisal. https://www.ajol.info/index.php/ajcr/article/view/144736. Accessed 21 Nov 2020

Mahony, L. (2013). Non-military Strategies for Civilian Protection in the DRC, Fieldview Solutions. eliefweb.int/sites/reliefweb.int/files/resources/Non-military-protection-in-the-DRC.pdf. Accessed 20 Nov 2020

Malan, M., & Porto, J. (2004). *Challenges of Peace Implementation: The UN mission in the Democratic Republic of the Congo*. Pretoria: Institute for Security Studies

Maneno, B.K. (2014). L'Enjeu de la Décentralisation en RDC: Organisation et Gestion des entités de base "secteur" et "chefferie. Paris: L'Harmattan

Mbayo, A.S. (2012). Civil Society and Peacebuilding in Africa: An Assessment of the Contributions by Local Women's Civil Society Organisations to Peacebuilding in Sierra Leone. PhD Thesis, Hiroshima University

Nzongola-Ntalaja, G. (2002). *The Congo from Leopold to Kabila: A people's history*. London/ New York: Zed Books

Paffenholz, T. (2010). What Civil Society Can Contribute to Peacebuilding. In T. Paffenholz (Ed.), *Civil Society and Peacebuilding: A Critical Assessment* (pp. 381–404). Boulder: Lynne Rienner Publishers

Paffenholz, T., & Spurk, C. (2010). A Comprehensive Analytical Framework. In T. Paffenholz (Ed.), *Civil Society and Peacebuilding: A Critical Assessment* (pp. 65–78). Boulder: Lynne Rienner Publishers

Paffenholz, T. (2014). Participation of Civil Society in Peace Negotiations. *Negotiations Journal,* 30(1), 69–91

Paris, R. (2004). *At War's End: Building Peace After Civil Conflict. Cambridge.* New York: University Press

Pishchikova, K., & Izzi, V. (2011). Engaging Civil Society in Peacebuilding. In R. Marchetti & N. Tocci (Eds.), *Conflict Society and Peacebuilding: Comparative Perspectives* (1st ed., pp. 47–71). New York: Routledge India. https://doi.org/10.4324/9780367818050

Purdue, D. (2007). Introduction: Dimensions of civil society. In D. Purdue (Ed.), *Civil Societies and Social Movements: Potentials and Problems* (pp. 1–18). London: Routledge

Pouligny, B. (2005). Civil society and post-conflict peacebuilding: Ambiguities of international programmes aimed at building 'new'societies. *Security dialogue, 36*(4), 495–510. https://doi.org/10.1177/0967010605060448

Raeymaekers, T. (2013). Post-war Conflict and the Market for Protection: The Challenges to Congo's Hybrid Peace. *International Peacekeeping* 20(55), 600–617

Richmond, O. P., & Pogodda, S. (2016). Introduction: The contradictions of peace, international architecture, the state, and local agency. In *Post-liberal peace transitions: Between peace formation and state formation* (pp. 1–17): Edinburgh: Edinburgh University Press

Santoso, A. (2010). Civil Society and Peace Processes in Kivu. FMR, 36, 35 36. https://www.fmreview.org/sites/fmr/files/FMRdownloads/en/DRCongo.pdf Accessed 30 Jan 2021

Stearns, J.K. (2011). *Dancing in the Glory of Monsters: The Collapse of the Congo and the Great War of Africa.* New York: Public Affairs

Stephen, M. (2017). Partnerships in Conflict, Oxfam International, October 2017. https://policy-practice.oxfam.org/resources/partnerships-in-conflict-how-violent-conflict-impacts-local-civil-society-and-h-620359/. Accessed 21 Nov 2020

Trefon, T. (2011). *Congo Masquerade: The Political Culture of Aid Inefficiency and Reform Failure.* London: Zed Books

United Nations News Centre (2010). Tackling Sexual Violence Must Include Prevention, Ending Impunity – UN Official, 27 April 2010 https://news.un.org/en/story/2010/04/336662 . Accessed 15 Feb 2021

Van Leeuwen, M. (2009). *Partners in Peace: Discourses and Practices of Civil-Society Peacebuilding.* Surrey: Ashgate

Vlassenroot, K., & Raeymaekers, T. (2004). Introduction. In K. Vlassenroot, & T. Raeymaekers (Eds.), *Conflict and Social Transformation in Eastern Congo* (pp. 13–39). University of Ghent Conflict Research Group

Zanker, F. (2018). *Legitimacy in Peacebuilding.* London and New York: Routledge

Chapter 9
Zimbabwe's National Peace and Reconciliation Commission and Civil Society: Partners in Peacebuilding?

Lawrence Mhandara

Abstract The chapter examines whether civil society has forged partnerships, formal and informal, with the NPRC to build peace in Zimbabwe. Using Spurk's functional model, the findings of the study evinces that civil society mostly carries out parallel activities suggesting a weak partnership with the NPRC. This weakness arises from two factors framing civil society activity: confrontation; and the loss of empirical independence by aligning with opposition politics. Nonetheless, when performing the functions of protection and monitoring, independently or with the symbolic involvement of the NPRC, civil society has tended to be effective compared to engaging in facilitation, socialisation and community building.

Keywords Civil Society · National Peace and Reconciliation Commission · Peacebuilding · Partners · Zimbabwe

9.1 Introduction

Africa has been characterised by a significantly robust civil society enabled by an expanding political space, improved education and a discourse promoting democratic activism. Correspondingly, the growth of civil society organisations has been hinged on questions of democracy, human rights and the particular need for peace and reconciliation against the background of episodes of social and political schisms. This research comes against a background of democratic enthusiasm that has advanced the idea that organisations outside state institutions have a role to play in peacebuilding (Orjuela 2003; Pouligny 2005; Barnes 2009; Paffenholz 2009a, b; Spurk 2009). The emergence of non-traditional, diverse and complicated conflicts has been most notable in the post-Cold War era. This has also witnessed the inclusion of non-traditional players in peacebuilding. Thus the involvement of CSOs in peace processes has been most visible since the demise of the Cold War. To this end, Daley (2006, p. 304) notes that "Making peace is now pursued by a veritable industry of international, regional, state and non-state actors." The incessant calls for broad-based peace processes prevalent in the post-Cold war era have strong foundations

Dr. Lawrence Mhandara is a senior lecturer in the Department of Political and Administrative Studies at the University of Zimbabwe.

J. C. K. Kiyala and G. T. Harris (eds.), *Civil Society and Peacebuilding in Sub-Saharan Africa in the Anthropocene*, The Anthropocene: Politik—Economics—Society—Science 34,
https://doi.org/10.1007/978-3-030-95179-5_9

in the record of weak performance by states in peacebuilding. The consequence of this has been the agitation for the active involvement of the CSOs not to replace but to complement the seemingly failing traditional institutions of peacebuilding. This is premised on the presupposition that CSOs are not only the reliable merchants of democracy but also of peace.

In Zimbabwe, civil society has demonstrated its value in pushing for democracy and respect for human rights. Nonetheless, the vibrancy and innovativeness exuded has gone largely unnoticed in the domain of peacebuilding, mainly because the roles of CSOs in peacebuilding have attracted minimal scholarly interest among researchers compared to their role in democracy. To that extent, Paffenholz (2009b, p. 5) observation is handy. He argues that "Although there has been a massive rise in peacebuilding initiatives aimed at strengthening civil society, these initiatives have not been accompanied by a systematic research agenda. As a result, we have known little about the role of civil society in peacebuilding." The argument becomes more accurate with respect to a functional partnership between the civil society and NPRC which has almost escaped scholarly attention with a few exceptions (for example Zambara 2018). To address this knowledge gap, this research analyses whether there is any partnership between the civil society and the NPRC and if such a partnership exists, to what extent is it encouraging complementarity than competition in peacebuilding? Several attempts to promote peace, healing and reconciliation have not been successful resulting in the establishment of the NPRC in the 2013 constitution to undertake the constitutional responsibility of transforming the country from a culture of violence to a culture of peace. With the operationalisation of the Commission in 2018, and the subsequent adoption of a five-year strategic plan (2018–2023), it is appropriate to ascertain the extent to which civil society is supporting the NPRC mandate. The study uses the framework developed by Spurk (2009) based on a functionalist approach to civil society peacebuilding, utilising the seven potential functions of civil society in peacebuilding. Their functions are then assessed on the basis of their execution with or without the involvement of the NPRC.

The first section provides the conceptual foundation of the chapter in which the concepts of civil society and peacebuilding are clarified. An attempt to explain the vital link of civil society to peacebuilding is also made in the section. The next section lays the historical involvement of civil society in peacebuilding in Zimbabwe in response to episodes of direct and structural violence. The subsequent section employs Spurk's functional model to examine the specific roles of civil society in peacebuilding. Simultaneously, the linkages and partnerships with the NPRC are also ascertained. The final section provides the concluding remarks.

9.2 Conceptualising Civil Society and Peacebuilding

The reinvigorated interest in democratic processes and practice is believed to have propelled the concept of civil society into prominence. The concept is, however, hardly new as it can be traced as far back as the political thoughts of both liberal and Marxist scholars such as Locke, de Tocqueville and Hegel (Bratton 1994; Hyden et al. 2003). Despite this traceable long history of existence, the term civil society has attracted varied interpretations. Yet there is consensus among scholars that certain elements can be used as the basis for distinguishing civic organisations from state institutions (Moyo 1993; Bratton 1994; Sachikonye 1995; Hyden et al. 2003; Alexander 2006; Daley 2006). Taylor (1990 cited in Moyo 1993, p. 2) proffers a comprehensive description of the concept which he presents in three senses: First, a minimal sense in which there are free associations that can be empirically shown to be free from the control of the state. Second, a strong sense in which civil society (CS) has the ability to organise and coordinate their activities without the control of the government. Third, the strongest sense in which civil society can be identified as a conglomeration of associations that has the political and organisational capacity to both coordinate and influence public policy.

Basing on these conceptual constructs, a civil society organisation can therefore be identified as an association of individuals within a given society or state who seek to articulate, advance and influence public policy without the control of the state. Civil Society is thus the realm and range of voluntary organisations and associations which occupy the space between the family and the government, which ideally exists independent of the state (Harbeson et al. 1994), to articulate and pursue shared interests, purpose and values (Spurk 2009, p. 7). The CSOs are distinct from political parties in that they have no immediate aim to gain political office or to exercise political power. Inherent in this characterisation is the conjecture that the state and CSOs exist in separate environments and therefore serve different political purposes. According to Moyo (1993, p. 2) "assuming that all state apparatus are distinguishable from the wider societies in which they are to be bound, scholars have tended to view civil society as that part of society which is outside the state sphere. There is thus a presumed basic duality between the state and civil society as existing in separate social reality." Although the duality of the CSOs and the state is widely acknowledged and recognised, the supposed role of the two is complementary rather than contradictory as the characterisation seems to imply. To this end, "the state and civil society should be treated as intertwining parts of the same social reality...the dichotomy between the state and civil society is based on a false dualism which negates the fact that civil society means the same thing as a political community" (Moyo 1993, p. 2). This is important to emphasise amid an intense scholarly campaign for the expanded role of civil society to cover the non-traditional ground that includes peacebuilding. Civil society organisations straddle the ethnic, religious, professional, labour, gender, human rights, political, and student groups excluding political parties, businesses and the media (Molutsi 2000).

The activity that is known as peacebuilding is not new. Its evolution has been explained in existing works (e.g. Galtung 1976 cited in Spurk 2009; Ncube 2014; Ryan 2015). However, the concept entered the mainstream discourse of peace in 1992 when it appeared in the UN document *An Agenda for Peace*. In 1994, the application of the concept was extended to development. The UN *An Agenda for Development* published in 1994 underscored that peacebuilding offers a chance to establish new institutions – social, political and judicial, that can give impetus to development.

Since then, the notion of peacebuilding has increased in importance, appearing and dominating both literature and policy on development, security, peace and conflict. Peacebuilding is essentially the process of achieving peace. Peacebuilding is defined by the Alliance for Peacebuilding (2012) as:

> a set of long-term endeavours undertaken continuously through multiple stages of conflict… and involving collaboration at several levels of society… peacebuilding emphasises transformative social change that is accomplished both at the process-oriented level, and through tools such as negotiation, mediation, and reconciliation, and on the structural level, through the development of resilient institutions and social processes that allow conflict to be resolved through political, rather than violent means (p. 12).

The import of the definition is its emphasis on collaboration in the kaleidoscopic activities of peacebuilding. The collaboration acknowledges the involvement of both state and non-state actors like civil society to achieve the goals of social transformation through peacebuilding. Beyond, this collaboration, the goal is to build positive peace by aiming not only to proofing the society against direct violence but also structural violence hence a focus on both the processes and structures or institutions. The definition also emphasises that building peace should be a long-term activity involving political, structural and social interventions that must guard against conflict recurrence. Within this continuum of activities, we can locate the role of civil society.

Two dominant approaches characterise the debate on peacebuilding (Paffenholz 2009, p. 45). The first is liberal peacebuilding. It narrowly applies to the democratic rebuilding of states emerging from armed conflict. Its focus is eliminating direct violence or achieving negative peace. Liberal peacebuilding has also extended to democratic transitions in states under authoritarian rule. It is an approach based on the understanding that peacebuilding is based on values: the higher the level of democratisation, the higher the level of peace. The second is sustainable peacebuilding, attributed to the works of John Lederach, based on the establishment of structures, processes and training of people within a generational-long time. Its goal is positive peace, and goes beyond the elimination of direct violence but establishing frameworks and institutions that undercut the non-violent forms of social injustice. In these processes and actions, civil society facilitates the participation of citizens beyond the voting process, and it is viewed as a basic pillar of peacebuilding (Spurk 2010, p. 3). Hence there is a general agreement among scholars on the centrality of civil society in enhancing and consolidating democracy (Diamond et al. 1998; Nyang'oro 1999), and peacebuilding (Pfalzenhoff 2000; Ncube 2014 etc). Theoretically, the crucial role of civil society in transitions from destructive conflicts to peace has impelled analysts to consider civil society as the vital link in the sustainability of peacebuilding. With peacebuilding theory and practice, the involvement of civil society seems to be undisputed (Paffenholz 2009, p. 43): "…the practice of peacebuilding shows that civil society has a role to play and is an accepted player in peacebuilding initiatives."

This recognition of civil society in peacebuilding goes back to the 1990s (Paffenholz/Spurk 2006). This civil society-peacebuilding discourse is largely informed by the theorisation of the comparative advantages of civil society as agents of change in general, and the functions that civil society ought to play in peacebuilding as gleaned from democratic theory, development discourse, peacebuilding theory and various in-the-field experiences (Ncube 2014). In essence, there are convincing arguments that recognise the centrality of civil society in peacebuilding (Orjuela 2003; Pouligny 2005; Barnes 2009). The civil society's role in peacebuilding is translated into practical peacebuilding outcomes through the performance of specific functions. Thus, Spurk (2009) proposes a functionalist approach and argues that the functions of civil society in the democratisation field should inform the civil society peacebuilding agenda. Spurk's model is a synthesis of the Merkel and Lauth, and Edward civil society role models combined to the functions contributed by development cooperation practice. The model has seven functions that expose the in-depth appreciation of civil society's role in political, social and development process. The seven functions are summarised hereunder:

- The protection of citizens' lives, freedoms and property from the actions of the government or its agents.
- Monitoring the activities and functions of the government and its institutions to ensure accountability. The subject of monitoring may be human rights, public spending, corruption and any other matter of public concern.
- Advocacy and communication involving the articulation of interests on behalf of the society and being able to create channels of communication to promote the interest of various groups onto the public agenda create awareness and encourage public debate.
- Socialisation through the formation and practice of democratic attitudes among citizens. They encourage the transmission of habits of tolerance, mutual trust and compromise in the context of the democratic process.
- Building communities through providing an arena where voluntary participation is possible. This may strengthen bonds among citizens of diverse social backgrounds and encourages cohesion while minimising social cleavages.
- Intermediation and facilitation in which the CSOs attempt to balance the power of the government by engaging it at various levels. Beyond that, it negotiates and communicates with the state on behalf of various interest groups.
- Service delivery in situations where the government falls short on providing necessities such as shelter, food etc.

The functional model is useful to develop in-depth scrutiny and appreciation of civil society in peacebuilding. In the case of this contribution, we examine how these functions are performed to complement the mandate of the NPRC.

The findings presented in the subsequent sections are based on qualitative research. 12 interviews were conducted in Zimbabwe from 13 October to 10 November 2020. Of these, six respondents were drawn from civil society organisations, three from the NPRC and two from academia. Field research was preceded by a review of literature and reports on civil society and peacebuilding in both Zimbabwe and beyond.

9.3 Historical Involvement of Civil Society in Peacebuilding in Zimbabwe

Direct and structural violence in Zimbabwe have deep roots in colonial politics. To manage the racial conflict between the blacks and whites, violence became a culture among the institutions created by the colonial government. This culture largely explains the nationalists' response to colonial rule through organise an armed struggle that enjoyed institutional backing from the Organisation of African Unity (OAU). The establishment of the Organisation of African Unity (OAU) in 1963 entrenched the principle of self-determination and majority rule in the Organisation's Charter. The pledge was to eradicate all forms of colonialism from Africa with success coming through negotiations in most countries except for most of the former Portuguese colonies and self-governing, white-ruled colonies like Southern Rhodesia and South Africa. The solution to the problem came in the change of strategy from engagement to armed resistance. This was executed through the active support from the OAU's Coordinating Committee for the Liberation of Africa which became the incubator for violent resistance to colonialism. The Committee became the bona fide conduit for arms and training for armed resistance against colonial authority. One prominent characteristic of the liberation movement was not just its confrontation of the white minority government through violence. Nonetheless, this response was rational given that colonial state-organised violence was comprehensive and brutal with the sole objective of containing nationalism by targeting the nationalist leaders, guerrillas and their collaborators and the civilians for their moral and material support to the liberation movements. Detentions, arbitrary arrests, severe floggings, curfews, killings, abductions, disappearances and torture (Sachikonye 2011), and killed including thousands murdered by the Rhodesian forces at Nyadzonia and Chimoio. These were decisive instruments in the colonial state's 'toolbox' of repression to intimidate, demoralise, humiliate and traumatise the Africans. The experiences of this period of deliberate violence and trauma have not been adequately addressed but what is certain is that it left a permanent scar on society. It is also not untrue that during the same period, civil society did not have a weak role in promoting respect for the rights of the people except a few church organisations affiliated with the Catholics.

At independence, the new government inherited the institutions that abused Africans with minimal structural changes. The reasons for continuity were largely to do with political stability and nation-building goals. The legal frameworks that sustained the colonial conquest were largely unaltered to be compatible with the new political order. The peacebuilding efforts were mainly around issues of inter-racial reconciliation which was premised on the 'forgive and forget' policy. The efforts were top-down but enjoyed support from the uncritical civil society (Ncube 2014). However, civil society concerns began to emerge after the first major armed conflicts after independence. These were recorded as early as 1980 in the assembly points dotted around the country that hosted guerrilla fighters from ZIPRA and ZANLA. Fierce fighting between the two forces first occurred at Entumbane in 1980 and spread to Gweru and Harare by 1981. The clashes were a preface to Operation *Gukurahundi* that started around 1982 and stopped with the signing of the Unity Accord in 1987. Both the government forces and the dissidents were involved in dehumanising acts. Unlike in the epoch of colonial rule, civil society footprints in peacebuilding were beginning to be visible during this conflict. The CCJP gathered information from victims and survivors of the clashes contained in the report *Breaking the Silence: Building True Peace.* The report has become the major reference point on the issue in the absence of official public documents. The organisation would return to prominence in the post-2000 period when the civil society ganged with the opposition to confront the government accusing it of human rights violations.

In the late 1980s, civil society groups such as student unions were at the forefront of steering the debate about political reforms with emphasis on issues that concern good governance, human rights in response to Mugabe's approbation of a legislated one-party state. The renewed activism among civil society, buoyed by events happening elsewhere in the world, forced the abandonment of the one-party state agenda. Indeed, the civil society contribution motivated the emergency of the Zimbabwe Unity Movement (ZUM) led by Edgar Tekere.

The second decade of independence is famed for introducing anxious moments among the people, more especially in economic terms. Mugabe's capitulation to capitalism demands to liberalise the economy forced his government to adopt the neo-liberal Economic Structural Adjustment Programme (ESAP) which had far-reaching socio-economic and political implications. Economic liberalisation was effected through a phased structural adjustment programme. The first phase was implemented from 1990 to 1995. The programme was aimed at opening the economy through monetary policy and trade liberalisation, withdrawal of subsidies among other austerity measures. The austerity measures created the basis for popular alienation and political discontent to which the government has been accused of a harsh response among the consumer associations and the hard-hit low-income groups (Sachikonye 2011). In the 1990s, CS engaged with obtaining economic challenges as they affected peace. In 1996, the Ecumenical Support Services (ESS) initiated a debate on the effects of the Economic Structural Adjustment Program (ESAP). It organised a multi-national three-day workshop in Harare which led to the writing of the *Zimbabwe Kairos Document: A call to prophetic action.* The document opened the discussion about a better future and a peaceful transformation of Zimbabwe into a prosperous and democratic country. In 1997, the Zimbabwe Council of Churches initiated the National Constitutional Assembly. Thus "The relationship between the church and state has been to a large degree collaborative although the church occasionally helps and capacitates civil society to confront the state on fundamental issues of basic human needs and human rights" (Maguwu 2006, 13). Later in the decade, civil society peacebuilding work revolved around democratisation and the constitutional reform agenda, through organisations such as the National Constitutional Assembly (NCA) (Ncube 2014).

The formation of the Movement for Democratic Change out of the NCA in 1999 presented a cogent threat to Mugabe's power, especially in the aftermath of his party's defeat in the February 2000 constitutional referendum. Thereafter, radical economic and political policy tenor ensued, marking the genesis of the political polarisation between the ruling party and the opposition that has subsisted to the time of writing this chapter. The polarisation has centred on the questions of elections, legitimacy and rights abuses. To that extent, the post-2000 peacebuilding agenda revolved around mitigating the power struggles between the ruling party and the opposition. Ncube (2014) observes that this period has seen the expansion of the civil society organisations involved in peacebuilding (examples include the Bulawayo Agenda, Habbakuk Trust, the Centre for Conflict Management and Transformation (CCMT), the Zimbabwe Peace and Security Programme, the Zimbabwe Human Rights Non-Governmental Organisation (NGO) Forum, and the Zimbabwe Peace Project). The civil society organisations have held various transitional justice and peace activities that include the 2006 *The Zimbabwe We Want'* organised by the ecumenical churches – the Zimbabwe Catholic Bishops' Conference (ZCBC), the Evangelical Fellowship of Zimbabwe (EFZ) and the Zimbabwe Council of Churches (ZCC). The *'Save Zimbabwe Campaign'* held in 2007 was organised by the Zimbabwe Christian Alliance with civil society and the opposition. The initiatives continued throughout the period of the inclusive government up to the time when the NPRC was operationalised in 2018.

9.4 Civil Society Peacebuilding Roles in the Post-Mugabe Period

The most important constitutional development as regards peacebuilding after the departure of Mugabe from office has been the operationalisation of the NPRC in February 2018. The NPRC has a responsibility to foster peace and reconciliation in the country. The NPRC was established under the 2013 Constitution which leaves the Commission's legal status beyond rebuke. The 2013 Constitution enjoyed trilateral support from the three political parties that constituted the inclusive government that ruled between 2009 and 2013. Section 251 (1) of the constitution states: 'For a period of ten years after the effective date, there is a commission to be known as the National Peace and Reconciliation Commission...'. In terms of Section 252 of the Constitution, the NPRC has at least ten responsibilities all focusing on peacebuilding. The Commission has clear and written terms of reference and the Commissioners are guided on the mandate of peacebuilding and national reconciliation. Among the guiding principles of the five-year strategic plan is the NPRC's will to foster strategic partnerships with key stakeholders including civil society organisations and ensure coordination of partnerships in accordance with the NPRC Act. The NPRC has embarked on a number of activities since its creation operationalisation. In 2018, it commenced the outreach programmes. In 2019, it began consultations on Operation *Gukurahundi* as well as establishing provincial peace committees. Meanwhile, civil society has continued with its peace promotion functions.

If civil society failed to operate freely and unfettered because of government action during Mugabe's rule since he began to pursue the radical redistributive policies, the actions of Mnangagwa's government after November 2017 protected and guaranteed the expansion of the civil society space. Since Mnangagwa's inauguration speech on 24 November 2017, there have been legal reforms to amend laws that infringed on freedoms such as the Public Order and Security Act. Mugabe had adopted a hostile tenor toward the CSOs most of which were openly aligning or campaigning for the main opposition party, MDC. The adversarial relationship that developed between the civil society and government was based on the invisible line between the opposition and the former which the latter regarded as one and the same. The tensions were exacerbated by the perception that the CSO agenda is at the behest of the donor countries (Masunungure 2014). Mugabe's government was subsequently sensitive and intolerant to any criticism by CSOs leading to very difficult relations that survived on confrontation. But the situation in the second Republic changed in a positive trajectory with more freedoms and space for the CSOs (Mawarire/Kode 2017). How have the CSOs contributed to peacebuilding in the period? We examine the most prominent actions in the major sectors of the civil society guided by the functionalist model highlighted in the earlier section. These are protection, monitoring, facilitation, socialisation and community building.

9.4.1 Protection and Monitoring

Civil society organisations have advocated and mobilised people to participate in the process of democratic reform as the foundation for peace in the country. Civil society coordinated various policy dialogues and research projects to unpack the democratic and security sector reforms and underscored the importance of genuine national healing. Some organisations campaigned and lobbied for the fuller alignment of outstanding laws to the 2013 constitution. They also continued with the monitoring and documentation of human rights violations and abuses by the government only and not all political players. CSOs have produced reports on human rights issues, including the contentious issue of abductions. Groups such as the Human Rights NGO Forum, the Zimbabwe Peace Project that cogently appeared on the scene during the height of the redistributive policies and the subsequent radicalisation of public policy in the 2000s have maintained their advocacy on the respect for human rights. After the post-election violence that was triggered by the opposition protests against the Zimbabwe Election Commission (ZEC) alleged delay in releasing the results of the presidential vote, civil society advocacy was mostly on the response of the security forces to the violent protests where they addressed the civil and political rights of the protesters. Although a commission of inquiry was established to investigate the circumstances leading to the 1 August 2018 violence, civil society continued to expose violations.

The Zimbabwe Human Rights (NGO) Forum, a coalition of 22 organisations, produced reports accusing the security forces of harassing victims and their families, and demanded the safety and protection of the targets (Chibhamu 2019). Equally, the CSOs have emphasised the same after the January 2019 violent fuel protests. Their activities are directly in contact with the people, raising awareness and encouraging them to participate in safeguarding human rights. The Zimbabwe Human Rights (NGO) Forum has also expanded the knowledge of citizens on the link between democratisation and peace through its briefings. The only challenge could be the limited reach to the audience because of the restrictions on physical activity imposed by COVID-19 (Interview). However, their communications to the state institutions like the Zimbabwe Human Rights Commission (ZHRC) and the NPRC are not direct but through communiqués and reports that are uploaded on their websites. The ZPP produces monthly reports on violations that are downloadable from its website but these are not formally communicated to the NPRC. Similarly, the Zimbabwe Democracy Institute (ZDI) has produced reports on what they have described as the militarisation of the state. In parallel, the Research and Knowledge Management Department of the NPRC conducts research on the same issues, namely nature, scope, extent and causes of the conflicts and the intervening strategies. The Department executes its functions by carrying out evidence-based research through data collection, storage, analysing, documentation and dissemination, and archiving. The department has been working with state universities and individual research experts but not CSOs.

The Zimbabwe Election Support Network has performed the protection and monitoring function through early warning. It has been observing and monitoring electoral-related violations before, during and after elections the 2018 elections and the by-elections held since then. Civil society has also campaigned for the respect of human rights during the lockdown measures to contain the spread of COVID-19 in force since the end of March 2020 (Rivers/Ndlovu 2020). The ZPP has been monitoring human rights violations which it uses to produce early warning information in the form of monthly reports. Conflict prevention requires careful monitoring of indicators of rising tensions and taking measures to ease them (Haider 2014 cited in Rowderder 2015). Violence can be prevented if the right information is delivered to the right stakeholders, at the right time, in the right format, enabling the stakeholders to take the right actions. The ZPP does not formally engage the NPRC throughout the performance of the early warning function. The activities have been carried out outside the Thematic Committee on Prevention and Non-Recurrence, under the department of conflict prevention, management, resolution and transformation, chaired by former Commissioner Patience Chiradza. The Committee has membership from political parties represented in the Parliament of Zimbabwe and civil society organisations. The weak civil society collaboration with the NPRC accounts for the failure of the CSOs systems to be transformed into a preventive response. Specific response plans have not been developed as part of the early warning efforts.

Further, civil society organisations have frequently met to evaluate the National Peace and Reconciliation Commission's (NPRC) work, the NPRC Act, prospects for transitional justice in the country and their role in the process. The NTJWG is made up of 46 Zimbabwean non-governmental organisations, including ZimRights, representing various transitional justice stakeholders, as an interface platform between transitional justice stakeholders and official transitional justice processes. The NTJWG demanded the abandonment of the operational narrative of the new dispensation represented by the 'let bygones be bygones' pronunciation, arguing that impunity discourages efforts to build a just society. The NTJWG has challenged the government's call for the involvement of traditional leaders in the NPRC process. As the NPRC was scheduled to start consultations in Gwanda and Bindura on February 9, 2018, an urgent court application by the Human Rights NGO Forum on Februrary 6, 2017 interdicted the Commission from carrying out any work before the appointment of a substantive chairperson.

Confrontation has been the mode of executing the protection and monitoring function by most of the CSOs. The confrontation tone among CS has been accompanied by appeals for nonviolent and inclusive approaches. However, the impact of militancy has been to harden positions, engender intolerance and increase the risk of direct violence. Some CSOs have supported anti-government discourses of human rights. For instance, in early August 2020, in a pastoral letter signed by ZCBC president Archbishop Charles Ndlovu, Archbishop Alex Thomas (ZCBC deputy president), and bishops Paul Horan (ZCBC secretary and treasurer), Michael Bhasera (Masvingo), Albert Serrano (Hwange), Rudolf Nyandoro (Gokwe) and Raymond Mupandasekwa (Chinhoyi) the the Catholic Bishops Conference attacked the government for perpetrating abuses against opposition supporters through abductions and creating a crisis of similar to the *Gukurahundi* operation. This was met by an equally combative response from the government. The Catholic Bishops were likened to MDC activists hiding behind religious titles by the Minister of Information and Publicity, Monica Mutsvangwa. The Minister strongly rebuked the bishops: "Its (the letter) evil message reeks with all the vices that have perennially hobbled the progress of Africa. It trumpets petty tribal feuds and narrow regionalist agendas. That he (Archbishop Ndlovu) hopes to sow seeds of internecine strife as a prelude to civil war and national disintegration."

Civil society has exploited the media as a conduit for public discourse around issues of human rights and injustices. In an environment dominated by relative freedom and minimal fear, journalists have published stories on subjects that were previously considered as sensitive to the government. However, the media has remained divided along partisan lines, with private papers published by the Associated Newspapers Zimbabwe (*The Daily News* and *The Daily News on* Sunday) and the Alpha Media Holdings (*News Day, Independent and Standard*) clearly advancing the peace agenda from the perspective of the CS aligned to the opposition while the state-owned Zimpapers such as *The Herald, The Chronicle* and *The Sunday Mail* counteracts. The divisions in the media have been starkly open in the case of the post-election mechanisms such as the Political Actors Dialogue (POLAD) and the value of the NPRC itself. The private media has dismissed both initiatives as lacking efficacy, preferring institutions that can impose retributive justice on the ruling party politicians. The politicised civil society has invoked credibility questions. Their affiliation to political parties is inimical to their independent action and diminishes their influence on the peacebuilding work of the NPRC. CSOs have exploited the independent media to disseminate information and public campaigns as well as encourage public debate on peace issues. Likewise, the NPRC utilises the state-controlled media to release information on peacebuilding activities.

9.4.2 Facilitation, Socialisation and Community Building

Politics by its nature is competitive and elections heighten the stakes. The acquisition of power is the sole reason why political parties exist. It is this process of acquisition of power that generates conflict and tensions that strain relations between the ruling party and the opposition groups. The resultant adversarial relationship imperils peace and peacebuilding. The electoral system in Zimbabwe is heavily coloured by a culture which approaches elections as a zero-sum competition. Civil society has attempted to facilitate dialogue between the opposition and the ruling party with a view to building consensus on the rules of the political game as well as on matters of national interest. Civil society however excluded from the multi-party liaison committee, an important platform to resolve inter-party disputes during the campaign period. This kind of strategy has however failed to help build confidence between the ruling party and the opposition during and after elections or between the government and the opposition or reduce needless confrontations.

The religious community in Zimbabwe remains one of the oldest and influential civil society groups. The religious leaders have spoken out against abuses and injustices in society. The church has done this through pastoral letters and sermons at churches and on social media. The priests and Bishops have invoked their moral authority in articulating their concerns on human rights and peacebuilding. The church has also continued to offer itself for conflict resolution and mediation as in the case of the offer made by the Zimbabwe Council of Churches to broker talks between the incumbent and Chamisa.

The socialisation and the rebuilding of communities' role by civil society aimed particularly to address grievances that emanated from the post-2018 election violence and alleged human rights abuses that occurred during the January 2019 protests and the arrests of organisers of the 31 July 2020 protests. Their activities also extended to building a culture of peace, to alter or shift conflict attitudes, to transform structural causes and consequences of conflict, to build social cohesion, and to mediate and facilitate state–society relations that have been strained since 2000. Civic education has been used by civil society to socialise the citizens on the culture of peace. A great deal of information and education of the public has taken place with an emphasis on the benefits of peace to both the nation and the individual. Civil society has been campaigning for a new culture associated with democratic values and practices. The intention is to encourage the emergence of political socialisation and the development of civic culture. This education has mostly undertaken through periodic workshops and campaigns in mainstream media, and targeted programs for demographic groups such as the youths and women. However, civic education has tended to reinforce the rural/urban imbalance in that CSO activities are largely concentrated in urban areas.

The desired impact of civic education on national ethos may not be realised if the majority of the people are excluded.

The CS also engaged in the capacity-building to build upon the momentum created by the community-mobilising activities by training targeted community leaders – such as traditional chiefs and the youth leadership in communities and in schools – in conflict analysis, mediation and resolution skills. The CSOs also carrying out peace education and dialogue projects aimed at information sharing, in particular with regard to human rights bills and statutes in the constitution and in various United Nations conventions. Through capacity-building programmes, civil society has come up with training programmes and structures to promote peacebuilding work at the community level; namely constituency peace monitors, peace committees and 'citizen journalists. This information is then collated and documented into reports for publicity and advocacy campaigns. The CS formulates peacebuilding mechanisms that serve as policy inputs into national peacebuilding processes, thus ensuring that peacebuilding activities at the grassroots level are linked to national peacebuilding approaches. Organisations such as the Heal Zimbabwe have also produced manuals on community-based dialogue as a means of preventing conflicts. CS through public and advocacy campaigns put out press releases, recommends and lobbies for appropriate responses at a national level by the NPRC.

With respect to community building, civil society has selected influential individuals such as politicians and community leaders for training as champions of peacebuilding. This has resulted in the establishment of peace committees made up of traditional leaders, church leaders and supporters of different political parties in different areas. Heal Zimbabwe Trust has established peace committees as well as promoted traditional methods of conflict resolution such as *Nhimbe* in provinces such as Masvingo. Meanwhile, on 4 July 2019, the NPRC completed the setting up of Provincial Peace Committees. Each established PPC is composed of between 25-30 members drawn from a wide range of stakeholders including government departments, civil society, traditional leaders and church organisations. One of the responsibilities is to promote peace within the province, create and or facilitate dialogue between groups and communities and exchange ideas on issues that may threaten peace and stability within the community (Marimbe 2019). Although civil society is represented in the committees, they continue to establish their own committees suggesting a lack of confidence in the NPRC structures. The motive behind the committees is to encourage tolerance, peaceful coexistence within local communities. As Chivasa (2017) notes, implementation and day-to-day operations of peace committees guarantee their sustainability even without external funding. He further avers that "one of the comparative advantages of informal peace committees is that all social groups in the community have equal chances of being represented, thus helping to meet the needs and aspirations of the community at large."

Although civil society is essentially seen as civil or good and inevitably contributing to peacebuilding in a positive way (Paffenholz 2009, p. 43), the performance of the roles reviewed above has not always promoted peace. Interviewees from both the NPRC and the civil society acknowledged the role of the CSOs in peacebuilding but questioned their effectiveness in sustainable peacebuilding. The notion that all civil society organisations are inherently progressive in opposing the injustices perpetuated by governments and in advancing the protection and the monitoring function is inaccurate in the context of Zimbabwe as most of them have been negating or glossing over abuses committed by the opposition political actors but never miss the commissions of the ruling party. The contributions, activities and programmes of civil society are not entirely driven by the values of genuine local representation but reactions to external donor pressures. Except for a few programs, the lack of grassroots representation in civil society activity fails to empower citizens to act on their own other than through their benevolence. Most of the activities (except for a few outreach programmes) have been undertaken in workshops and conference rooms far removed from both the NPRC and the beneficiaries of their programmes. Similar challenges have been reported in developing countries where research has shown that CSOs tend to crowd out local efforts and actors, as donor-driven initiatives have limited the capacity to create domestic social capital and ownership for the peace process (Paffenholz 2009).

The context of Zimbabwe strongly influences the activities of civil society and constrains its overall effectiveness. Paffenholz (2009, p. 22) avers that "Civil society tends to be a mirror of society. Thus, it is not astonishing that civil society organisations are just as divided as society along power, hierarchy, ethnic or gender lines, and can show moderate, as well as radical, images and behaviours". The influence of donors aligned to the opposition political parties is a key contextual factor undermining a strong civil society partnership with the NPRC. The opposition and donors are influencing the peace activities using their political influence and material support to push for the civil society to confront the government institutions on behalf of the opposition. While civil society is part of the political community it should not be part of the partisan community (Masunungure 2014). Intra-party dynamics within the main opposition have been ignored in favour of the inter-party or government (Interview). Factionalism in the MDC Alliance and the associated violence has gone unnoticed. One wonders whether civil society peacebuilding is all about the government and ZANU-PF only.

Yet "Disregarding other cleavages and tensions in societies… has proven to be dangerous and may lead to future outbreaks of violence" (Paffenholz 2009, p. 7). The uncivil side of the CS was noted as some have encouraged undemocratic and intolerant attitudes as well as inciting people to revert to primary groupings such as kinship and tribal structures (Interview). On that basis, the observation that CSOs are diverse, including the bad and ugly (Nyong'oro 1999) is applicable to Zimbabwe. The implication is that officials with links to political actors are increasingly instrumentalising their organisations on the basis of ethnicity, regionalism and political affiliation, preaching hatred against others. The role of civil society organisations in sponsoring and advocating for protests the government has been particularly a cause for concern to peacebuilding. Civil society has become sources of affluent divisions in society. There are CSOs in peacebuilding that are alien to the values they purport to advance. Indeed, "Civil society is a powerful tool in peacebuilding, but some can also have a negative effect on peace" (Last 2008, 4). Clubs and associations outside have become vehicles for socialising young men and preparing them for violent action as in the example of the Tajamuka/Sijikile which has incessantly gained fame for inciting rebellion against the government. Instead of encouraging peace, it has endangered it. This suggests that the CSOs' role in peacebuilding diminishes when it fails to respect the peacebuilding norms. In focusing on the role of CSOs in peacebuilding, it is critical that one examines organisations that best express the need to establish, practice and preserve peacebuilding values. The key to identifying CSOs as agents of peacebuilding is to identify those with agendas pushing for views and actions that encourage sustainable peace not those presenting themselves as anti-ZANU-PF movements.

9.5 Conclusion

Using Spurk's functional model, our research stresses that civil society has the potential to play an effective role in peacebuilding. The involvement of the CSOs in peacebuilding in Zimbabwe has not been decisive. Civil society mostly carries out parallel activities suggesting a weak partnership with the NPRC. This weakness arises from the political context in which most of the CSOs have lost their empirical independence by aligning themselves with opposition politics. Although executed independent, or with symbolic involvement of the NPRC, functions of protection and monitoring have been effective when executed. Conversely, efforts aimed at facilitation, socialisation and community building have not been as effective as polarisation, violence, intolerance and hate speech persists. This is largely because of the confrontation paradigms preferred by civil society which spoils both the peacebuilding goals and opportunities for partnership with the NPRC.

References

Alliance for Peacebuilding (2012). Peacebuilding 2.0: mapping the boundaries of an expanding Field. https://static1.squarespace.com/static/5db70e83fc0a966cf4cc42ea/t/5e960cd128d1592862e72d17/1586891988458/AfP-Mapping-Report_online_FINAL.pdf Accessed 30 Sept 2021.

Alexander, K. (2006). *The future of democratic politics in Zimbabwe*. Cape Town: Institute of Justice and Reconciliation (IJR).

Barnes, C. (2009). Civil society and peacebuilding: Mapping functions in working for peace. *The international spectator, 44*(1), 131–147.

Bratton, M. (1994). Civil society and political transition in Africa. IDR reports 11, no.6. https://africanphilanthropy.issuelab.org/resources/19673/19673.pdf. Accessed 5 Oct 2020.

Catholic Commission for Justice and Peace (CCJP) and Legal Resources Foundation (LRF) (1997). Breaking the silence: building true peace. Harare: CCJP and LRF.

Chibhamu, A. (2019). Relatives of August 1 army shooting victims, witnesses claim victimisation by state agents. *New Zimbabwean*, October 6. https://www.newzimbabwe.com/relatives-of-august-1-army-shooting-victims-witnesses-claim-victimisation-by-state-agents/. Accessed 5 Oct 2020.

Chivasa, N. (2017). Efficacy of informal peace committees to peacebuilding: Evidence from Seke district, Zimbabwe. *African Evaluation Journal* 5(2), 1–11.
Daley, P. (2006). Challenges to peace: conflict resolution in the Great Lakes region of Africa. *Third World Quarterly* 27(2), 303–319.
Diamond, J.L., & Martin, M.L. (1998). *Democracy in developing countries*. Boulder: Lynne Rienner.
Harbeson, J., Rothchild, D., Chazan, N. (Eds.) (1994). *Civil society and the state in Africa*. Boulder: Lynne Rienner.
Hyden, G., Court, J., Mease, K. (2003). Civil society and governance in 16 developing countries. https://odi.org/en/publications/civil-society-and-governance-in-16-developing-countries/. Accessed 16 March 2022.
Last, D.M. (2008). Civil Society and Peacebuilding. https://www.researchgate.net/publication/262375509. Accessed 5 Oct 2020.
Lederach, J.P. (2005). *The moral imagination: the art and soul of building peace*. Oxford: Oxford University Press.
Maguwu, F. (2006). The role of civil society in building peace in Zimbabwe: focus on the church. http://archive.kubatana.net/docs/cact/farai_maguwu_role_of_civil_society_zim_061001.pdf. Accessed 6 Oct 2020.
Marimbe, D. (2019). A review of provincial peace committees: The NPRC's step towards sustainable peacebuilding in Zimbabwe. *Focus on Peace Building* 1, 1–6. https://kubatana.net/wp-content/uploads/2019/08/Focus-on-Peacebuilding-August-7-2019.pdf Accessed 30 Oct 2021.
Masunungure, E. (2014). The changing role of civil society in democratic processes in Zimbabwe: 2014 and beyond. https://library.fes.de/pdf-files/bueros/simbabwe/13718.pdf. Accessed 5 Oct 2020.
Mawarire, T., & Kode, D. (2017). What future for civil society in Zimbabwe? https://www.opendemocracy.net/en/democraciaabierta/what-future-for-civil-society-in-zimbabwe. Accessed 5 Oct 2020.
Molutsi, P. (2000). Searching for a role: civil society in Southern Africa. Paper presented at the International Peace Academy. Gaborone: African Renaissance Institute.
Moyo, J. (1993). Civil society in Zimbabwe. *Zambezi*, XX (i), 1–13.
Ncube, C. (2014). Civil society and peacebuilding during Zimbabwe's government of national unity, 2009–2013. *African Security Review* 23(3), 283–294.
Nyang'oro, J. (1999). *Civil Society and Democratic Development in Africa*. Harare: Mwengo Press.
Orjuela, C. (2003). Building peace in Sri Lanka: A role for civil society? *Journal of Peace Research, 40*(2), 195–212.
Paffenholz, T. (2009a). CS and Peacebuilding. In T. Paffenholz (Ed.) *Civil society and peacebuilding: a critical assessment* (pp. 43–63). Boulder: Lynne Rienner.
Paffenholz T. (2009b) *Summary of results for a comparative research project: civil society and peacebuilding*. Geneva: Graduate Institute of International and Development Studies.
Paffenholz, T., Spurk, C. (2006). Civil society, civic engagement, and peacebuilding. Washington, D.C.: World Bank.
Pouligny, B. (2005). Civil society and post-conflict peacebuilding: ambiguities of international programmes aimed at building 'new' societies. *Security Dialogue* 36(495), 510–496.
Rivers, M., & Ndlovu, P. (2020) COVID raises the stakes for Zimbabwe's civil society movement. https://www.usip.org/publications/2020/06/covid-raises-stakes-zimbabwes-civil-society-movement. Accessed 6 Oct 2020.
Ryan, S. (2015) The evolution of peacebuilding. In R. Mac Ginty (Ed.) *Routledge handbook of peacebuilding* (pp. 25–32). London: Routledge.
Rohwerder, B. (2015) Conflict early warning and early response. https://gsdrc.org/wp-content/uploads/2015/02/1195_-Conflict-early-warning-and-early-response.pdf. Accessed 6 Oct 2020
Sachikonye, L. (1995). *Democracy, Civil society and the state*. Harare: Sapes Books.
Spurk, C. (2009). Understanding civil society. In: Paffenholz T (ed) *Civil society and peacebuilding: a critical assessment* (pp. 3–28). Boulder: Lynne Rienner.

Chapter 10
Civil Society and Peacebuilding in the Democratic Republic of Congo: Opportunities, Challenges and Recommendations

Jean Chrysostome K. Kiyala

Abstract This chapter investigates the extent to which civil society can be more productive in the context of the Democratic Republic of Congo (DRC); it consists of a cross-sectional empirical case study that largely relies upon findings that originate from secondary data (reviewed literature) and primary data (focus group discussions and surveys) that were conducted at the Catholic University of Goma-La Sapientia in June 2019, in North Kivu Province (DRC). The sample was comprised of graduate and postgraduate students, academics, members of civil society organisations and security services. *Training of civil society's actors and leaders was* singled out as the most important factor that could enhance civil society's effectiveness: it represents the highest mean of 1.3429. Finally, a four-layer strategy emerged as the way to strengthen civil society's capacity to be more responsive and high-yielding: this starts with training civil society's leaders and actors, integrating peacebuilding programs at all levels of educational courses, developing and increasing partnership between local and international civil society actors, and prioritising a contextualised civil society's modus operandi in the DRC.

Keywords Civil society · Elections · Democratic Republic of Congo · Democratization · Peacebuilding · Political transition

10.1 Introduction

Since gaining independence from Belgium on 30 June 1960, the Democratic Republic of Congo (DRC) has been a theatre of violence during which, under three successive authoritarian regimes, quelling rebellions and fighting civil wars consumed most of the energies and resources that could have served to combat poverty and misery and to build functional and modernised infrastructure and thus prompt the country's development and the pursuit of the common good. Efforts by governmental and non-governmental organisations have demonstrated sustained interest in preventing

---

Jean Chrysostome K. Kiyala is a Senior Lecturer, International Centre of Violence, Durban University of Technology, South Africa; Associate Professor and visiting lecturer the University of Bandundu, Catholic University of Bukavu and Evangelical University in Africa based in Bukavu (DR Congo).

J. C. K. Kiyala and G. T. Harris (eds.), *Civil Society and Peacebuilding in Sub-Saharan Africa in the Anthropocene*, The Anthropocene: Politik—Economics—Society—Science 34,
https://doi.org/10.1007/978-3-030-95179-5_10

conflict and managing violent conflicts, but unfortunately their initiatives have been fruitless, other than being, in some instances, "… a mere pile of peacebuilding stones rather than a sustainable peace" (Reychler/Paffenholz 2001p. xiii). This assertion is justified by recurring and ongoing insecurity and violent conflict even in areas controlled by UN troops (Copeland 2012). Violence relating the inadequacy of electoral processes and the inability of the DRC's Government to respect the legal terms of transitional agreements (such as setting up a truth and reconciliation commission to deal with past and recurrent domestic human rights violations, and infringement of international human rights and humanitarian law) (Kiyala 2018). Electoral violence and unifnished transitional justice are similtaneously key contrintors to chronic violence and consequences of structural violence in the DRC (Kiyala 2018).

Against the backdrop of this introduction, in the volatile political environment of the DRC, what role can civil society play? How does civil society respond to the needs of peacebuilding in the context of problematic democratisation and political transition? To what extent do actions undertaken by civil society organisations (CSOs) signal progress and drive optimism towards attaining successful democratic processes and durable post-political transition peace? What is the salient impact of local citizens' movements and CSOs on deterring the drivers to conflict in the DRC? What are the contextual factors that hinder the effectiveness of such interventions, including cooperation among international non-governmental organisations and national non-governmental organisations, local dynamics, and the geopolitics of the region?

The overall objective of this study is to investigate the role that civil society plays in peacebuilding, specifically in the context of democratisation and political transition, with the purpose of gathering key components of the interventions that are needed to increase the effectiveness and usefulness of civil society in the DRC. An in-depth exploration of the multidimensional crisis of the DRC was essential to lay the foundation for a better understanding of CSOs' relevance in this context, prior to overran overview of the role civil society plays in peacebuilding, specifically in the context of democratisation and political transition. This study explores the opportunities and the challenges and proposes ways of redress to make civil society more engaged and effective in its vision to enact social, political, and economic transformation for the attainment of the common good and lasting peace. Peacebuilding and democracy theory are adopted as the paradigmatic framework to guide this work, which is designed as an exploratory case study and used a pluralistic methodological approach. It is structured as follows: the introduction is followed by a socio-political background to the DRC, an overview of civil society's evolution in this country, the theoretical framework that accompanies this work, research design and methods, results, discussion and finally, conclusions and recommendations.

10.2 Overviewing the Concept of Civil Society

Paffenholz (2015) defines civil society as "the arena of voluntary, collective actions of an institutional nature around shared interests, purposes, and values that are distinct from those of the state, family, and market" (p. 108). It consists of "a large and diverse set of voluntary organisations and comprises non-state actors and associations that are not purely driven by private or economic interests, are autonomously organised, typically show civic virtue, and interact in the public sphere" (Paffenholz 2015, p. 108).

The contemporary meaning of civil society is traceable to the Age of Enlightenment, between the 18th and the 19th centuries, when civil society was identified with the state (Carothers/Barndt 1999). A further distinction is made by Keane (2009) who writes: "Contrasted with government, civil society meant a realm of social life – market exchanges, charitable groups, clubs and voluntary associations, independent churches and publishing houses – institutionally separated from territorial state institutions" (Keane 2009, p. 1). After World War II, the concept of civil society gained impetus in the face of a world plagued by totalitarian regimes, and it became associated with a domain that opposed tyranny. This movement was spearheaded by the Marxist theorist Antonio Gramsci (Carothers/Barndt 1999). This conceptual evolution approximates the contemporary thinking behind civil society—as the world undergoes a revolution against authoritarianism, dictatorship, and excessive *dirigisme*, especially in Africa and Asia, "[i]t was in the midst of disenchantment with the overreach of the state – in Africa as well as Asia – that the concept of civil society took hold of the imaginations of both the left and the right. It promised an exit from bureaucratic inefficiency and political indifference" Chandhoke (2007, p. 612). From the same perspective, Keane (2009, p. 2) notes: "For the first time in their history, the political languages of democracy and civil society have become conjoined". Furthermore, arguing on the same note, Keane (2009) alludes to Thomas Paine's pamphlet *Common Sense* (1776), to tie together civil society discourse and democracy, and theorising on the assertion that despotic power can be contested by establishing the earthworks of civil society (Keane 2009, p. 3).

Paffenholz (2015, p. 108) defines civil society as "the arena of voluntary, collective actions of an institutional nature around shared interests, purposes, and values that are distinct from those of the state, family, and market". It consists of "a large and diverse set of voluntary organisations and comprises non-state actors and associations that are not purely driven by private or economic interests, are autonomously organised, typically show civic virtue, and interact in the public sphere" Civil society refers to the "sector of voluntary action within institutional forms that are distinct from those of the state, family and market" (Paffenholz/Spurk 2006, pp. 2–3). It can help develop sustainable peace strategies.

10.3 Civil Society in the Democratic Republic of Congo

The emergence of CSOs in the Republic Democratic du Congo is traced back to the era of the Independent State of Congo (ISC) whose governance fell directly under Leopold II, King of the Belgians (1885–1908), and which was guided by the following normative and legal framework:

- The Decree of 28 December 1888 which allowed private associations to operate.
- The Ordinance of 18 March 1912 which acknowledged the existence, validity and legality of the Chambers of Commerce and Industry of Congo.
- The Decree-Law of 1 March 1914 which granted civil personality to Civil Society organisations.
- The Decree of 19 July 1926 which authorises the creation of non-profit public utility establishments (Congo Forum 2009).

These four legal documents paved the way for the future development and functioning of CSOs, which contrasts with a certain view that situates the origins of CSOs in the DRC at the time of the Sovereign National Conference (CNS) of 1990–1992 (Congo Forum 2009) (Fig. 10.1).

The real leap of Congolese civil society occurred when President Mobutu ended the mono-party regime and introduced multipartyism on 24 April 1990. In this context, civil society became more visible in the political arena and exerted a significant role when Archbishop Laurent Monsengwo Pasinya, a Catholic Prelate, took

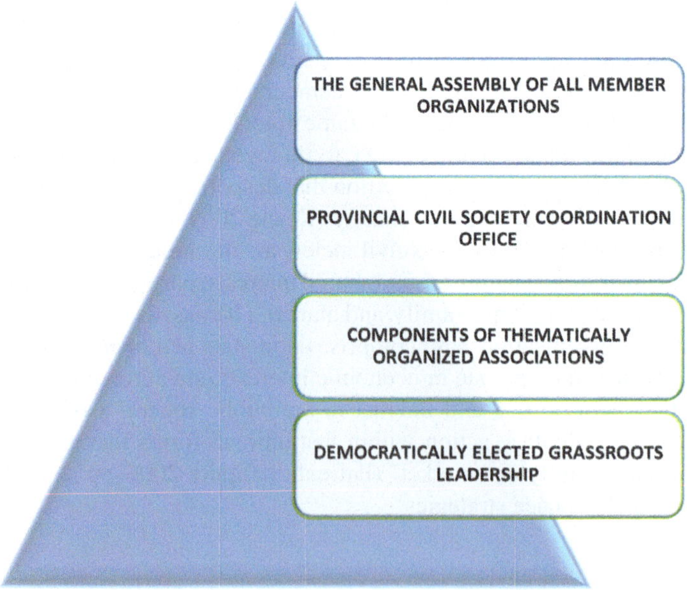

Fig. 10.1 Structure of Civil Society in the DRC *Source.* Designed from Congo Forum 2009's report

the reins of the Sovereign National Conference and became the Speaker of the High Council of the Republic / Transitional Parliament of the Republic of Zaïre, a function he held between 1991 and May 1997 after Laurent Kabila overthrew Mobutu and renamed the Country the Democratic Republic of Congo.

It is important to mention that civil society's actions have been met with brutal responses by former dictatorial rulers: many activists have been assassinated and there has been no accountability for their murder. The brutalities experienced by civic movements and other civil society platforms have aimed to muzzle outspoken militants, silence the voice of the people, perpetuate human rights violations, and continue with undemocratic voting systems. Activists have been victimised (Chic 2018), and brutalised by the police to contain their protests (Clowes 2016) but they continue to participate in political transformation through nonviolent resistance (Perera et al. 2018), and despite the killing of their fellows, such as Rossy Mukendi and others (Yahaya/Bello 2020), CSOs remain committed to advancing their agenda of advocacy, protection, and overall political and institutional transformation.

10.3.1 Civil Society and Peacebuilding

The investigation of civil society's role in peacebuilding is framed within two main theories, namely peacebuilding and democracy. In the discussion it will become clear that there are linkages between civil society, democracy theory and peacebuilding. Understanding this interrelation is key to developing a more realistic and efficient strategy that could prompt civil society's civic commitment to embracing peacebuilding principles and democratic ideals in its ongoing transformative action.

Paffenholz/Spurk (2006, p. 15) define peacebuilding as "an overarching term to describe a long-term process covering all activities with the overall objective to prevent violent outbreaks of conflict or to sustainably transform armed conflicts into constructive peaceful ways of managing conflict". This definition is not exhaustive of what peacebuilding's methodology, approaches, resources, actors/agents, beneficiaries, objectives, and outcomes are. However, it gives a conceptual understanding that can be completed by including its aims and expected outcomes. The aim of peacebuilding is to prevent and manage armed conflict and consolidate peace in the aftermath of large-scale systematic violence (Paffenholz/Spurk 2006). In a similar vein, Oberschall (2013) asserts that enhancing cooperation among adversaries is needed to build lasting peace, and can occur through power sharing; reconstruction of the economy; pursuing justice for victims and accountability for perpetrators; reaching settlement measures for refugees and populations forced into displacement; and establishing law, order and security, including reinsertion of belligerents into civil life and fighting criminal activities. For Last (2010), peacebuilding refers to the process of doing away with obstacles to durable peace by minimising the opportunities for violence, whether physical or structural. All these definitions are not exhaustive, but rather necessary to be considered in a study that aims at building peace.

Several scholars have worked on establishing how civil society is essentially linked to peacebuilding (Belloni 2001; Lederach 1997; Paffenholz 2015). Due to the proliferation of conflicts in the 1990s and the complex nature of the interventions required from the international community and donors, civil society's response is more and more sought (Paffenholz/Spurk 2006). The two authors define the role of civil society in peacebuilding in relation to civil society's action, namely its response to the areas of *protection, monitoring peace processes and accountability*, its commitment to *advocacy and public communication*, in the sphere of *socialisation and a culture of peace, conflict-sensitive social cohesion,* in its response to *intermediation and facilitation* of peace processes, and in the context of *service delivery* (Paffenholz/Spurk 2006).

Civil society's share of engagement in peacebuilding is embedded in the three types of peacebuilding actors illustrated in Lederach's pyramid and the three approaches adopted by these actors at each of the three levels (Fig. 10.2) (Lederach 1997, p. 39). At the top of the pyramid, civil society is represented by religious leaders; at the intermediate level, the leadership class, according to Lederach, is comprised significantly of civil society, namely ethnic and religious leaders, academics or intellectuals and humanitarian heads or those of NGOs; and finally, at the bottom layer civil society is involved through leaders of indigenous NGOs, leaders who champion community development, and those who lead refugee camps (Lederach 1997, p. 39) (Fig. 10.3).

In this model, drawing from the experience of Ethiopia, El Salvador, and Cambodia, Lederach argues that, in this top-down framework of peace-search and -building, or "trickle-down" approach, "[...] a process of 'national' transition is initiated involving the political leadership in creating a framework that will lead to democratic elections" (Lederach 1997, p. 45). Bottom layer leadership at grassroots level can exert an important influence towards peace and conflict resolution by

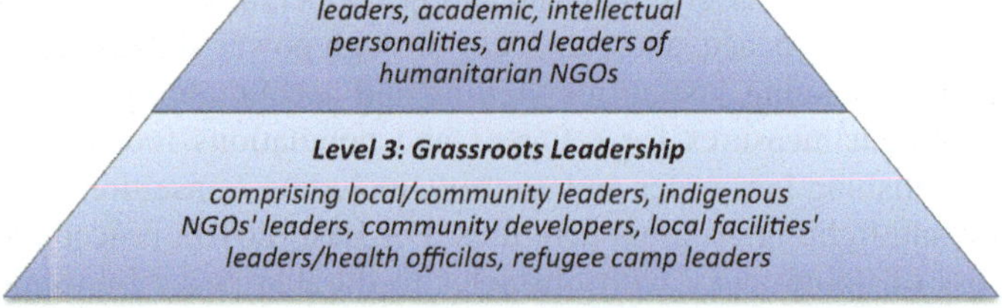

Fig. 10.2 Three types/levels of peacebuilding's actors. *Source* Author's adaption Lederach's (1997, p. 39) peacebuilding pyramid

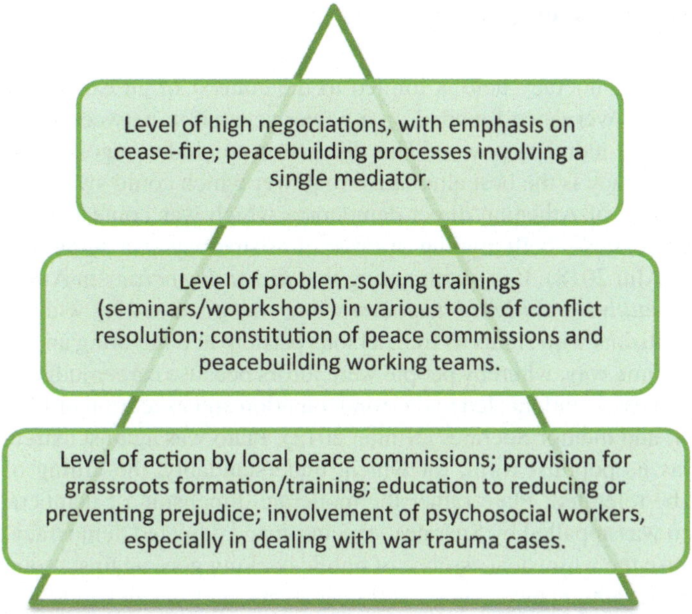

Fig. 10.3 Three levels of peacebuilding. *Source* Author's adaption of Lederach's (1997, p. 39) peacebuilding pyramid

providing training to communities about "reducing prejudice and enhancing community decision making" (Lederach 1997, p. 54). Local leaders are instrumental in making reconciliation possible starting from the grassroots level and attending to the immediate needs of vulnerable populations in the aftermath of violent conflict and civil war, even in transitional periods.

To sum up, it will be helpful to use the results of a scientific study conducted by Nilsson (2012) on the durability of peace efforts that involve civil society. She found that, while the inclusion of civil society actors did not affect the durability of peace nor reveal negative effects, the involvement of civil society actors combined with the participation of political parties meaningfully leveraged the longevity of peace. It was further established that the prospects of lasting peace in nondemocratic societies demanded that civil society actors be implicated in peace agreements (Nilsson 2012). This last finding and the overall outcome of Nilsson's study sustain the necessity of getting civil society involved in peace processes to ensure that durable settlements are obtained, especially in the context of young democracies or in countries known for their legacy of non-democracy or generating and proclaiming illegitimate electoral results. In this regard, the DRC is no exception.

10.3.2 Civil Society and Democracy

Theorising on democracy here is limited to the context of an existing representative system of governance founded on a power separation between the legislative, executive and the judiciary powers. This discussion precludes ongoing debates about whether democracy is the best alternative to power which could sway our argument back to the time of Athenian direct democracy which was contested by Plato and his mentor Socrates—both arguing in favor of aristocracy as a better alternative to democracy (Aliu 2018). It is worth noting that *direct democracy* in Athens differed from *representative and liberal democracy* that was governed by wider nations. In Athens, all citizens with rights as free persons participated in voting and justice took place in the same way, whereby people were jurors because career judges and prosecutors did not exist, and this led to the condemnation and execution of Plato's friend, interlocutor, and mentor Socrates (Ninian 2012). Plato was against extreme democracy such as its populist form, for which, understandably, the killing of Socrates would not be tolerated; Plato rather the moderate dimension of democracy (Rowe 2001). Plato was appalled by Socrates' execution and depicted democracy as *hysterical*—because the whimsical essence of public backing goes against "sound reason" and generates "fatal inconsistency" as time goes by; and saw it as rule by *ignorance* because leaders tend to ignore the "inconvenient truths", in politics [in *The Republic* Book VI] (Ninian 2012). By disapproving of the vices mentioned earlier, Plato was apprehensive that the Athenian society would engender immense conflict: a political type that would yield hunger for an autocratic leader or a tyrant (Rowe 2001). Our society has suffered from the same dangerous realities that Plato feared most, and this has sustained the existence of and civic engagement by civil society.

Plato's views are relevant to this inquiry to examine why CSOs exist and stand outside the state. This is naturally a response to the attitudes adopted by politicians, from which Plato dissented, namely the lack of "unencumbered love of wisdom" which should oppose falsehoods, physical pleasures, material pleasures, meanness, and cowardice that corrupt leaders [in Book VII of *The Republic*] (Ninian 2012). All are desires and tendencies that threaten to corrupt leadership. What is pertinent about Plato's political philosophy is the contemporaneous relevance of his views on immigration as embraced by contemporary leaders, and the danger of leaving voting processes in the hands of inexperienced and untrained voters. This serves as a caution in terms of the ethical foundation of democratic ideals and political life: morality and political life are inseparable because "unencumbered love" of truth/wisdom is the light of a politician—who for Plato should be a philosopher (Rowe 2001).

Civil society is essentially connected to democracy because it represents a sphere where diverse groups freely engage with one another in the pursuit of projects that integrate their interests (Chandhoke 2007). The author argues that the absence of civil society corresponds to the non-existence of the democracy and freedom that are essential for democratic engagement. The linkage between her views and Plato's conception of democracy as an exercise of skillful and knowledgeable citizens and civil society is reflected in the following assertion by Chandhoke (2007, p. 613):

"By asserting civil society, people demand that regimes recognise the competence of the political public to chart out a discourse on the content and the limits of what is politically desirable and democratically permissible". In the context of political transition, Bratton (1994) notes that political actors grapple with putting in place regulations that allow redistribution of collective resources beyond the immediate power struggle. Civil society plays a significant role in the critical moment during the transition that intervenes between the fall of the incumbent regime and the takeover by the newly elected actors, following the period during which political space experiences independence and freedom that prompts the formation of political parties, as Bratton asserts (1994, p. 10): "Instead of providing a refuge of last resort for dissident politicians, actors in civil society are freed to take on truly 'civic' functions for which they have a more natural aptitude". In a similar vein, Carothers/Barndt (1999, p. 21) point out: "An active, diverse civil society often does play a valuable role in helping advance democracy. It can discipline the state, ensure that citizens' interests are taken seriously, and foster greater civic and political participation".

Since the strength of liberal and constitutional democracy lies in power separation between the three branches of state, namely the executive, the legislative and the judiciary powers (Krause 2000). It is important to note that the differentiation of various functions is merged with "a system of checks and balances" (Krause 2000, p. 235). Civil society ensures that there is no abuse of power in this system of governance, and its engagement is not limited to one of these institutions but is seen all the three.

10.4 Design and Methods

This inquiry was designed as an exploratory case study using various sources and consequences of conflict in the DRC, while focusing substantially on the DRC's protracted conflict to show why stability created by a successful democratisation and political transition can enhance durable peace. The role of civil society in the pursuit of peace is clearly the focal point of this investigation. Designing this inquiry both as case study and exploratory study allows us to carry out in-depth research which is susceptible to generate valid and reliable knowledge of the impact of civil society on peacebuilding in the context of democratisation and political transition in the DRC.

While the reliability and generalisability of a single case study is contested, a triangulation of data sources and methods employed in this inquiry ensures its outcomes are valid and generalisable. Triangulation in this study refers to the use of "two or three different methods to explore the same subject" (Davies 2007, pp. 35–35). This will be done by combining methods (qualitative and quantitative) of data collection, analyses and interpretation which are based on constructivist realism as an ontology through which a positivist approach is merged with a constructivist worldview to produce solid, valid and reliable knowledge (Cupchik 2001). Quantitive analysis was employed to draw statistics about the characteristics of the sample, whereas qualitative data comprise literature-reviewed secondary data and primary data originating from focus group discussions provided in-depth information for this study.

Seventy respondents out of eighty-three participants were involved in answering survey questionnaires, which were converted to variables measured at ordinal scales between *I strongly agree,* and *I strongly disagree*, and these variables were coded, thematised and analysed via the IBM Statistical Package for the Social Sciences (SPSS) Version 27. Firstly, Cronbach's Alpha Test [α] (or coefficient alpha) was performed to evaluate the reliability of the set of scale. Secondly, descriptive analyses were executed to determine frequency distribution of the sample's features and various institutions involved in this inquiry.

This study uses non-probability sampling techniques, namely convenience and purposive sampling. Convenience sampling is based on taking what is available where you can get it, while purposive sampling refers to a strategy where the researchers target a population which is believed to be representative of that being studied (Davies 2007). The available data was provided during a workshop conducted at the Catholic University of Goma-La Sapientia (CUG-LaS) from 11th to 14th June 2019. Participants largely were taken from academic and research institutions (graduate and postgraduate students, lecturers, professors, and researchers) as components of civil society, as well as local CSOs. Two main sources of data taken here are group discussions and survey questionnaires. Several responses from 70 participants were coded and converted to variables measured as nominal and ordinal scales via the SPSS tool. These 70 participants were key informants because of their experience on the way civil society operates in the DRC. Permission was obtained from the gatekeeper to conduct this study at the CUG-LaS and respondents participated freely. Of 83 attendees, 70 responded favorably to the survey after consenting to the ethical considerations explained, based on the principles guiding research in social sciences according to Bless et al. (2013), particularly non-maleficence, beneficence, autonomy, fidelity, respect for participants' rights and dignity, informed consent and voluntary participation, anonymity and discontinuance. Confidentiality and appropriate referral were not significant because the nature of the open forum and group discussions which generated responses to survey questionnaires did not require this. Ethics in analysis and reporting research outcomes, and reporting back to participants (Bless et al. 2013) were also observed.

Figure 10.4 represents the main features of the sample as far as occupations of participants is concerned .

Participants in the study came from three countries: two from Burundi (2.9%), two from South Africa (2.9%), and 66 from the DRC (94.3%). The highest representation came from graduate and postgraduate students pursuing various academic studies at the CUG-LaS, among whom there were thirty-two from the Faculty of Philosophy, nine from the Faculty of Theology, six from the Faculty of Social Sciences and three from the Faculty of Psychology. In addition, there were eighteen academics, including the participating universities' rectors, professors and lecturers, one police officer and one member of a local NGO. Table 10.1 captures the frequency distributions

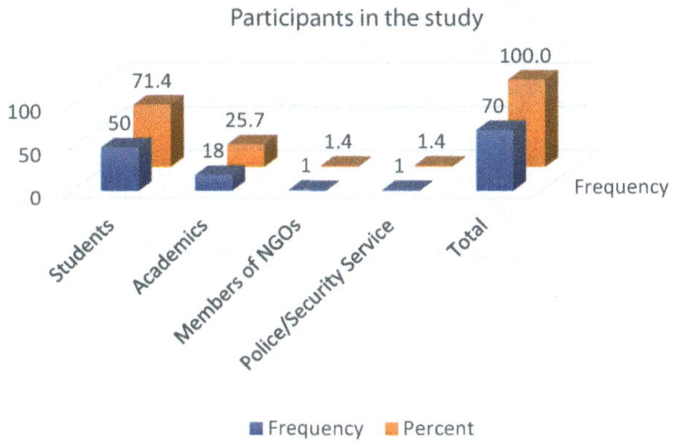

Fig. 10.4 Features of the sample. *Source* Author

Table 10.1 Organisations to which participants belonged

		Frequency	Percent
Valid	Evangelical University of Africa	3	4.3
	Catholic University of Goma _ La Sapientia	56	80.0
	Catholic University of Bukavu	2	2.9
	Sub-Saharan University of Africa	2	2.9
	Great Lakes School of Theology and Leadership	1	1.4
	Durban University of Technology/International Centre of Nonviolence (ICON)	2	2.9
	Rema Burundi	1	1.4
	Other	3	4.3
	Total	70	100.0

and percentages of individual categories of participants in the inquiry based on six groupings.

10.5 Challenges, Opportunities and Recommendations

Civil society is confronted by a myriad of challenges, mainly those related promoting civil liberties, ensuring the respect of the rule of law, improving governance system, depoliticising civil society participation in democratisation and service delivery, protecting citizens' movements and human rights activists (see Appendix 10.1).

Participants identified several opportunities through which civil society can be more relevant, especially in monitoring peace processes, implementing peace agreements to end armed conflicts, redressing social inequalities, promoting gender equality and equal access to education opportunities, and helping demilitarised areas to achieve durable peace (see Appendix 10.2).

From the list included as Appendix 10.3, 10 recommendations were made by all the groups involved in group discussions:

- Train CSOs' actors and executives
- Call on international experts and NGOs for exchange programs
- Review the education system by including the concept of conflict resolution in the national curriculum at all levels
- Draw on African social realities to create and submit our own conflict resolution theories
- Create a group (an autonomous and independent think tank) to investigate further all drivers to conflicts and assist local organisations to tackle identified root causes of conflicts
- Integrate conflict management courses into the curriculum of primary and secondary schools
- Introduce peacebuilding into the higher learning syllabus and encourage scientific research in conflict resolution and peacebuilding
- Organise conferences on conflict resolution at various levels of communities (rural, urban and in higher learning environment)
- Organise listening sessions where people can share their frustrations and worries relating to conflicts
- Create radio broadcasts on conflict resolution

10.6 Discussion

The density of materials supplied by participants in this inquiry translates the passion which animates all respondents for whom peace is a vital need. The discussion of findings will focus on empirical data resulting from the recommendations made by participants, and this is subdivided into three components: (1) Training and education of civil society actors and leaders (2) Introducing peacebuilding programs at all levels of educational courses (3) Sharing of expertise with international civil society organisations/NGOs (4) Glocalization and contextualisation of training models and interventions strategies (Fig. 10.5).

Firstly, training of civil society members, actors and leaders emerged as the most valuable of all the factors that could enhance the effectiveness of civil society. Several recommendations share the same trends, which encompass integration of peacebuilding programs at all levels of educational programs. Scholars such as Lederach (1997) encourage training of leaders at the grassroots level because of the responsibility they carry in enhancing reconciliation and delivering immediate and urgent services to the people most affected by conflict. Similarly, Plato observed that political actors need expertise and ethical standards so that they can exercise meaningfully their democratic rights and lead an ethical political life (Ninian 2012).

Secondly, the lifespan of the DRC's tumultuous political life has caused an erosion of the traditional values of conviviality and socialisation and impacted social capital, as many communities confronted each other, military or violently, thus destroying the

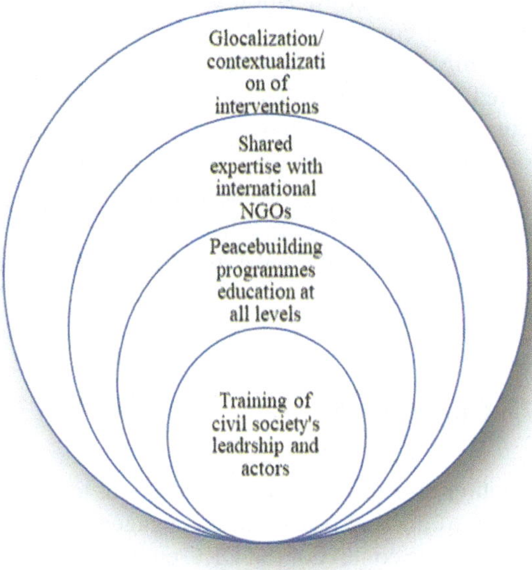

Fig. 10.5 Layers of Civil Society's effectiveness enterprise. *Source* The Author

social fabric, which needs to be repaired. Education for peace requires both infrastructure and the resolve from Congolese of all generations to embark on educational programs that inform on democratic ideas and peacebuilding—to break the vicious circle of dragged-out conflict and prolonged violence.

Thirdly, the need for international cooperation between external actors and local CSOs is much favored. Such a partnership and the assistance of international experts are pivotal in helping local CSOs to deal with the challenges involved in conducting fair and transparent elections, preventing electoral violence, and building post-electoral sustainable peace, before, during and after transitional phases. External CSOs could train locals in democracy, reconciliation, and peacebuilding, while learning and possibly taking advantage of locally-oriented peacebuilding mechanisms, such as the Baraza—a traditional restorative justice model of conflict resolution and indigenous jurisprudence in the eastern DRC (Kiyala 2016).

Lastly, glocalisation and contextualisation of interventions was evoked as an underpinning of an efficient response of civil society. The need for ongoing education and the expertise of international civil actors is justified by the desire to expand the understanding and knowledge of civil society's engagement worldwide. The expression *glocalisation* incorporates two concepts, namely *globalisation* and *localisation.* It refers to bringing the concept of civil society, democracy and peacebuilding from the global context and adapting its functionality, aims, agenda and modus operandi in the complex realities of the DRC. This reasoning for taking a glocalisation approach could be justified by the saying "there is no one-size-fits-all", in approaching violence related to democratisation and transitional processes. The diversity of contexts and responses to conflict transformation and peacebuilding, as elucidated by Lederach (1997), supports the importance of adaptability of strategies and interventions to fit the context. Political transition and the search for peace are distinct engagements that take into consideration local/domestic realities and which are not uniform, by requiring the adoption of local-based wisdom, and getting indigenous people embarked on this project. The fact is that most conflicts in the DRC are entrenched up-country, in rural and remote areas, and not in big cities. In this context the bottom-layer leaders are key players in any peacebuilding, providing training on democratic ideals and transitional processes. This view is supported Dah: "An essential element of democracy, it is argued, is that people have access to information and differing opinions so that they can make informed decisions" (Dah 1989, cited in McLaverty 2002, p. 308).

10.7 Conclusion

This inquiry aimed to investigate the extent to which civil society could effectively contribute to durable peace via its engagement in democratisation and transitional processes in the context of the DRC emerging from the legacy of authoritarian rules and a history of rebellions, civil wars and protracted armed conflicts. It was found that civil society emerged at the time when the DRC was the Independent State of Congo, then a Belgian colony. However, its real development and visibility were discernible in the 1990s following President Mobutu's speech which ended the mono-party system and opened the DRC's political space to multipartyism. Several civil society platforms, including civic movements, adopted a non-violent approach to pursuing the fight for the advent of a DRC which is truly democratic and where the rule of law frames and defines the system of governance and ensures that peace and security prevail. Unfortunately, many civil society activists lost their lives to defend democratic ideals under brutal and despotic rulers. The resilience of civic movements has been rewarded by a new era marked by the hope that CSOs could operate in a free and secure environment and continue to play their advocacy role, remain committed to the protection and human rights and civil liberties, and foster the democratisation agenda and respect of legal provisions that sustain peace and justice mechanisms in transitional societies—a legal framework that protect the rights of victims and upholds the rule of law.

It emerges from both literary data reviewed and empirical data that the DRC's civil society is confronted with a myriad of challenges that have to be overcome, but there are many opportunities too where CSOs can operate to consolidate peace and democratic institutions. Amidst an extensive record of strategies suggested to make civil society more effective, there came up, simultaneously, 10 factors identified by the findings which carry substantive potential to affect the leading role of CSOs in taking the democratisation agenda further and ensuring the DRC's citizens participate in electoral processes that are fair, peaceful, and transparent, in which they participate freely, and the voting results of which reflect the truth of the electorate's will. This, so far, has been a serious problem in the country following the last three nationwide elections (2006, 2011, and 2018). The new political dawn characterised by freedom of the media and the independence of CSOs could be a turning point towards the vision of a DRC as a mature democracy which is unequivocally democratic and where everybody abides by the rule of law. Reaching such heights demands a synergistic implementation of the four layers of an efficient civil society, and this should be supported by a determined collaboration between CSOs and the DRC Government.

Appendix 10.1: Challenges

• Promoting freedom of expression	• Disdain towards civil society
• To facilitate the search for consensus in the process of national decision-making	• The politicization of civil society • Violence carried out by unknown attackers on activists
• Police brutalities	• Killing of activists without justice
• Facilitating conditions for free, democratic, and transparent elections • Vulnerability of civil rights activists • Complicity of the population in securing civil society • Autonomy of civil society to facilitate peace dialogues	• Difficulties in fighting the hypocrisy of the political elites • Lack of protective measures to secure CSOs' actors and leaders • lack of resources in the delivery of services • Lack of cooperation by rival political parties • Nepotism and sentimentalism
• Improving the functioning of the judiciary	• The State's rejection of certain civil society demands
• Improving governance system • Lack of power to protect citizens	• Socialization and building social capital
• Creating conditions for freedom of expression	• Tackling the clash over customary and land laws
• Creating the conditions for the peaceful resolution of electoral disputes	• Lack of a spirit of consultation among civil society organizations for actions of common interest
• Facilitating civic engagement without confrontation and disruption of public order • Presence of insecurity • Incapacity to deal with increasing ongoing insecurity • Poor job management	• Lack of healthy relations among mutual associations • Neglect by the authorities in the face of brutal treatment of CSO actors • Lack of an appropriate framework for arbitration of political disputes • Distrust in the ethnic, cultural, and religious groups of the CSO groups
• To be the voice of the people in making decisions that have a national reach	• Easing of intergroup, intercommunity, inter-ethnic social cohesion
• Civil society's dependence on politicians who may exert their influence, which impedes the neutrality of civil society	• Tribalism and gender discrimination • Non-existent legal protective framework to ensure safety of human rights and civil rights activists

(continued)

(continued)

• Promoting freedom of expression	• Disdain towards civil society
• Absence of the principle of meritocracy among civil society actors • Impotence to enforce the principles of political alternation • Camouflaged agenda of some members of civil society	• Lack of an adequate framework for exchange between communities • Lack of an appropriate framework for mediation • Facilitation and mediation between rival political groups • Land disputes
• Repealing laws that favor only small groups	• Acceptance of others as viable players in civil society
• Ethical failure of civil society representatives	• Lack of conviviality among communities
• Politicization of civil society • Lack of training equipment on individual freedoms	• Non-applicability of pre-established laws • Lack of intergroup social cohesion

Appendix 10.2: Civil Society's Opportunities

• Monitoring peace and easing disputes in the context of increasing political conflicts and military confrontations, and promoting interaction among citizens' movements
• Mobilizing the resources available for development and ending recurring armed conflicts
• Opportunity for research on conflicts related to ethnicity and land control
• Fighting the mercantile spirit of economic operators that exploit children and fuel mineral resources conflicts
• Monitoring and implementation of peace agreements in inter-community and identity conflicts
• Redress of social inequalities and tackling failures of service delivery and ongoing protests
• Fighting unequal access to education, promotion of human rights and gender equality
• Safeguarding democracy and peace in the DRC
• Fighting against mismanagement or misuse of natural resources in organizations
• Militating for the independence of civil society, its neutrality, and cooperation among NGOs and citizens' movements

Appendix 10.3: Suggested Ways to Overcome Identified Challenges

Collective action	Individual/personal commitment to conflict resolution and peace management
• Change of mentality and political will • Respect for the principles of democracy • Dependence on legal institutions • Training and informing citizens in important areas • Restriction of civil society for consideration in the face of citizens • Train the population on the missions and functions of civil society • Raising awareness and motivating people to participate in an integral way for a new image of civil society • Highlight the importance of each group in the process of building worthy values • Instill a spirit of cooperation and exchange between communities, • The reinvention of civil society acceptance • The appropriation of this structure (civil society) by all segments and categories of population • There has to be awareness; thus, the awareness by each group of civil society in the context of peacebuilding. • It is necessary to proceed by the unification of civil society (collaboration, coordination, and order in the actions of civil society) • Good organization and specific goals • Respect for legal texts	• Multiply training and seminars on conflict resolution • Awareness and consultation of all social strata for inclusive participation; reinvention of the image of civil society • Everyone's commitment to peacebuilding and conflict resolution • There is a need to intensify awareness and awareness campaigns on the promotion of peace • Introduce courses on peace education in the national education curriculum giving them a special importance (i.e., giving a high weight) • Strong awareness and awareness of all • Providing fair justice • Raising awareness among the members here • Awareness of the members here ?? • Creating a sense of listening • Everyone's commitment to peacebuilding and conflict resolution • Good governance and respect for human rights • Financial independence of civil society • Setting up a communication cell within civil society • Security, understanding between political groups • Educating members of civil society through seminars, training, and the like

References

Aliu, Y. (2018). Original Paper Comparison of Plato's Political Philosophy with Aristotle's Political Philosophy. *Urban Studies and Public Administration 1*(1), 31-45. https://doi.org/10.22158/uspa.v1n1p35

Belloni, R. (2001). Civil society and peacebuilding in Bosnia and Herzegovina. *Journal of peace Research, 38*(2), 163–180.

Bless, C., Higson-Smith, C., & Sithole, S. L. (2013). *Fundamentals of social research methods: an African perspective* (5th ed.). Cape Town: Juta & Company.

Bratton, M. (1994). Civil society and political transition in Africa. *IDR Reports, 11*(6), 1–21.

Carothers, T., & Barndt, W. (1999). Civil society. *Foreign policy, Winter, 1999–2000*(117), 18–29. https://www.jstor.org/stable/1149558 Accessed 28 Oct 2021.
Chandhoke, N. (2007). Civil society. *Development in Practice, 17*(4–5), 607–614. https://doi.org/10.1080/09614520701469658
Chick, K. (2018). In Democratic Republic of Congo, activists face uphill struggle: Despite risk of arrest or death, many groups have sprung up to press for political and social change. *International Women's Media Foundation, December 17, 2018 10:00 PM*. Retrieved from https://www.thenationalnews.com/world/africa/in-democratic-republic-of-congo-activists-face-uphill-struggle-1.803860 Accessed 3 Feb 2021.
Clowes, W. (2016). Police Thwart Youth Protest Against DRC's Kabila. https://www.voanews.com/a/police-thwart-youth-protest-against-kabila/3613179.html Accessed 21 July 2021.
Congo Forum. (2009). La société civile en R.D.Congo. *Congo Forum, 03 march 2009.* https://www.congoforum.be/fr/2009/03/la-societe-civile-en-r-d-congo/ Accessed 3 March 2021.
Copeland, T. F. (2012). *Civilian protection in the Eastern DRC: Evaluation of the MONUSCO Peacekeeping Operation.* (Masters). Naval Postgraduate School Monterey CA California. https://apps.dtic.mil/dtic/tr/fulltext/u2/a560583.pdf Accessed 2 Jan 2020.
Cupchik, G. (2001). Constructivist realism: an ontology that encompasses positivist and constructivist approaches to the social sciences [33 paragraphs]. *Forum Qualitative Sozialforschung/Forum: Qualitative Social Research, 2(1), Revised 7/2008, 2*(1).
Davies, M. B. (2007). *Doing a successful research project: using qualitative or quantitative methods.* New York: Pagrave MacMillan.
Keane, J. (2009). Civil society, definitions and approaches. *International encyclopedia of civil society*, 461–464.
Kiyala, J. C. (2016). Utilising a traditional approach to restorative justice in the reintegration of former child soldiers in the North Kivu Province, Democratic Republic of Congo. *Africa Insight, 46*(3), 33–50.
Kiyala, J. C. K. (2018). *Child Soldiers and Restorative Justice: Participatory Action Research in the Eastern Democratic Republic of Congo.* Cham. Springer International Publishing. https://doi.org/10.1007/978-3-319-90071-1
Krause, S. (2000). The spirit of separate powers in Montesquieu. *The Review of Politics, 62*(2), 231–265.
Last, D. M. (2010). Civil Society and Peacebuilding. In: Routledge Encyclopedia of Peace. http://dx.doi.org/https://doi.org/10.2139/ssrn.2459768.
Lederach, J. P. (1997). *Building Peace: Sustainable Reconciliation in Divided Societies.* Washingston D.C.: United States Institute of Peace Studies.
McLaverty, P. (2002). Civil society and democracy. *Contemporary Politics, 8*(4), 303–318. https://doi.org/10.1080/13569770216068
Nilsson, D. (2012). Anchoring the peace: Civil society actors in peace accords and durable peace. *International Interactions, 38*(2), 243–266. https://doi.org/10.1080/17448689.2018.1484004
Ninian, A. (2012). Why Plato Hated Democracy: The Republic's clues about modern leaders and their popularity. https://bigthink.com/scotty-hendricks/why-socrates-hated-democracy-and-what-we-can-do-about-it. 27 Dec 2020.
Oberschall, A. (2013). History and peacebuilding. In S. Campbell & J. H. Peterson (Eds.), *Routledge handbook of peacebuilding* (pp. 184–195). Howick Place, London: Routledge.
Paffenholz, T. (2015). Civil society and peacebuilding. *Development Dialogue, 63*, 108–118.
Paffenholz, T., & Spurk, C. (2006). Civil society, civic engagement, and peacebuilding. *Social Development Papers: Conflict Prevention and Reconstruction, 36*(2006), 10–55.
Perera, S., Kambale, V. A., & Bussy, J. M. (2018). Youth Participation and Non-Violent Resistance in the Democratic Republic of the Congo: The Case of LUCHA. DLP Research Paper 55, Birmingham: Developmental Leadership Program / Dar-es-Salaam. https://media.africaportal.org/documents/Research-Paper-54.pdf. 30 Sept 2021
Reychler, l., & Paffenholz, T. (Eds.). (2001). *Peace-building: a field guide.* Boulder: Lynne Rienner Publishers.

Rowe, C. J. (2001). Killing Socrates: Plato's later thoughts on democracy. *The Journal of Hellenic Studies,* 121(2001), 63–76.

Yahaya, J. U., & Bello, M. M. (2020). Democratic Republic of Congo and Crisis of Leadership (2016–2018). *Journal of Economics, Finance and Management Studies, 04 April 2020*(3), 50–60.

Chapter 11
Promoting Democratisation in DR Congo: The Case Study of Act for Transparent and Peaceful Elections

Jean Chrysostome K. Kiyala and Gerard Mpang'de Bisambu

Abstract This chapter examines the role played by a civil society organisation (CSO), Act for Transparent and Peaceful Elections (AETA). This is a platform that supports the democratisation process and peaceful political transition in the Democratic Republic of Congo (DRC). The outcomes drawn from analysis of documentary sources (reports, meetings, conferences, interactions between AETA, the DRC Government and other CSOs) reveal that AETA's commitment to the Quadruple-level Democratization Monitoring Process has been instrumental in supporting general elections and the peaceful transfer of power. This framework has taken place at four levels: (1) reflexively, for the benefit of its members and other stakeholders; (2) for decision-makers and other key players; (3) with grassroots dynamics; and (4) for communities. These different levels of action reflect AETA's approaches to the democratization of the DRC, despite the hostile political environment, the prevailing insecurity and the perilous conditions under which CSO actors operate in this country. Thus, the institution of AETA, alongside the free activism of citizens' movements, remains a key factor for the advancement of democracy.

Keywords AETA · Civil society · Elections · Democratic Republic of Congo · Transition · Peacebuilding · Elections

11.1 Introduction

Since gaining national and international sovereignty from Belgium in 1960, the Democratic Republic of Congo (DRC) has faced a myriad of multifaceted crises, and the frenzied development of the most appalling poverty of its people mainly due

Dr. Jean Chrysostome K. Kiyala is a Senior Lecturer, International Centre of Violence, Durban University of Technology, South Africa; Associate Professor and visiting lecturer the University of Bandundu, Catholic University of Bukavu and Evangelical University in Africa based in Bukavu (DR Congo).

Mr Gerard Bisambu is the Secretary General of the Non-Governmental Organization (NGO) Agir pour les Elections Transparentes et Appaisées [Act for Transparent and Peaceful Elections] (AETA).

J. C. K. Kiyala and G. T. Harris (eds.), *Civil Society and Peacebuilding in Sub-Saharan Africa in the Anthropocene*, The Anthropocene: Politik—Economics—Society—Science 34,
https://doi.org/10.1007/978-3-030-95179-5_11

to the resignation and / or almost total absence and failure of the State. This context involves state paralysis and crippled institutions where citizens' survival becomes commensurate with Charles Darwin's theory of evolution known as natural selection that attributes the survival of organisms, species, and their offspring to their adaptation to the ecological context in which they find themselves. The declining socio-political and economic environment of the DRC, characterised by the degenerating provision of education, security, health care, employment, and livelihoods, epitomises the reality of a collapsed state where the Government has resigned from its obligations towards its citizens. This situation has made civil society more relevant and ever committed in the pursuit of the democratisation process and soothed political transition which are deemed indispensable in raising the sick elephant.

Election-related violence, intimidation, creating an atmosphere of terror, tracking and brutalising have become part and parcel of electoral scheme, as has been the case in Kenya, Nigeria, Zimbabwe and Ghana (Lauterbach 2015), and in Burundi, where such violence caused hundreds of deaths and refugees and the displacement of thousands of Burundians (Colombo et al. 2019). While election violence is endemic on the African continent, the recent polls in the USA in November 2020 reveal a transnational dimension of brutalities engendered by democratic and transitional processes worldwide. This reality makes the search for peace during voting procedures and power transfer a thorny question that begs for strategies to build peace and maintain the legitimacy of this political exercise.

In the backdrop of this dark picture of the DRC's peacebuilding challenges, it becomes evident that for true democratisation of the country, good governance requires an inclusive and holistic approach to which civil society organisations relentlessly strive to contribute, as will be outlined in this document. This chapter is written under the following assumptions:

- Civil society organisations have impacted peacebuilding in the context of societies in transitions, and during electoral processes (Paffenholz/Spurk 2006).
- It has been established that making civil society organizations part of peace processes augments the durability of peace settlements, and bringing together civil society actors and political parties has the potential to allow peace to prevail (Nilsson 2012).

However, it is necessary to look at civil society broadly—to include both locally based initiatives and international or external interventions. Such a consideration is important because in the context of structural changes that consist of building peace and advancing democratisation agenda, fostering social cohesion after large-scale conflicts, and building peace, external actors' understanding of local social and political stakes is often limited. A study by Belloni (2001) underscores such complexities in the context of Bosnia and Herzegovina where international interventions to build civil society for the purpose of fostering peace were unsuccessful. That investigation was useful to help us focus on the efforts of a civil society platform in the DRC, known as *Agir pour les Elections Transparentes et Appaisées* [Act for Transparent and Peaceful Elections] (AETA).

While Civil Society Organisations (CSOs) are globally involved in advocacy and lobbying for the advancement of democratisation and political transitions in countries exiting large-scale conflicts or from the legacy of protracted human rights violations, the impact of local CSOs on political transformation in the DRC is understudied. This chapter explores the role played by CSOs in the DRC, particularly AETA, in leading the transformative revolution aimed to create a politically healthy environment for the growth and safeguard of democratic processes, democratisation of all institutions, facilitation and nonviolent transfer of power in transitional contexts, and ensuring that a power balance between the three branches of the Government (Executive, Legislative and Judiciary) is observed and shielded by the Constitution.

The question that this chapter intends to answer is: how and to what extent have CSOs or have they not contributed to democracy development and political transition in the DRC, and how has the AETA platform been or not been relevant in this process? This investigation is conducted under the assumption that CSOs in the DRC have been active in the political transformation of the DRC and AETA's action epitomises the global patterns of citizens' movements in the pursuit of democratic principles and their embodiment in the DRC's general context.

The aim of this essay is to thoroughly present, through the achievements of AETA and other CSOs, the role, challenges, obstacles, opportunities, and perspectives of civil society in relation to its involvement in the implementation of the political transition and democratisation of the DRC. This reflection on the commitment of civil society will cover the period from 2003 to 2019. This will be done through literary and documentary research, as explained earlier, and will basically be founded in the review of the literature of civil society in theory and practice and the use of documents that embody the overall activities and achievements of AETA. To attain this, this chapter is dissected into the following main sections: (1) the introduction, which outlines the methodological procedure, the objective and rationale of this chapter; (2) the background of civil society's involvement in the democratic process and political transition in the DRC, and an overview of the tumultuous political context of the DRC including protracted conflicts and their devastating consequences, impunity and the need / implication for civil society to seek redress; (3) a discussion of the theoretical framework—the concept of civil society in theory and practice and its engagement in democratisation processes and political transitions; (4) an exploration of civil society in the DRC; (5) a presentation of the efforts put forth by AETA in the democratisation process and peaceful elections; and (6) conclusions drawn for the core argument.

11.2 Theoretical Framework

The referential context of civil society's action in this investigation covers the areas of democratisation and political transition to which we apply the peacebuilding theory. This methodological pluralism is deemed appropriate to anticipate and respond to the complexities associated with efforts to bring about political transformation and ease political transition in the aftermath of many decades of authoritarian rule and protracted armed conflicts.

11.2.1 Peacebuilding Theory

Peacebuilding is a process of exploring, identifying, and putting in place mechanisms that could help prevent violence or the recurrence of hostilities, in the context of volatility, negative peace or post-large-scale violence. This approach is the most comprehensive and may, to some extent, incorporate aspects of both peace-making and peacekeeping because it is a preventive conflict management paradigm. The peculiarity of peacebuilding is that it can freely tap into local resources and capacities to be used in achieving its goals. It traces the causes of conflict that are to be dealt with and uprooted, and stabilises the socio-political and economic environment. This entails that locally oriented alternative dispute resolution mechanisms, indigenous jurisprudence and cultural platforms attributed to CSOs are capable of building and maintaining community peace. There is a close link between peacebuilding and civil society which is its tool. It is important to define that which peacebuilding is and implies and the different aspects in civil society that are concerned with the process of building peace.

In the context of this paper, civil society is understood as "the sum of voices not controlled by the government, speaking for individual and collective interests" (Last 2010, p. 1), while peacebuilding consists of "the process of removing obstacles to lasting peace by reducing the opportunities for both physical and structural violence" (Last 2010, p. 1). Paffenholz/Spurk (2006, p. 15) define peacebuilding as "an over-arching term to describe a long-term process covering all activities with the overall objective to prevent violent outbreaks of conflict or to sustainably transform armed conflicts into constructive peaceful ways of managing conflict". This definition is not exhaustive with regards to what the peacebuilding methodology, approaches, resources, actors/agents, beneficiaries, objectives, and outcomes are. However, it gives a conceptual understanding that can be completed by including its aims and expected outcome. According to Paffenholz/Spurk (2006, p. 16), "[p]eacebuilding aims at preventing and managing armed conflict and sustaining peace after large-scale organised violence has ended", and "[p]eacebuilding should create conducive conditions for economic reconstruction, development and democratisation, but should not be equated and thus confused with these concepts" (Paffenholz/Spurk 2006, p. 16).

Peacebuilding can be both a pre- and post-conflict initiative the primary aims of which may consist in redirecting conflicts towards a social good and building peace by breaking conflict progression. In this struggle, the role of educators and activists to "conscientise" the masses is critically important, because as people's conscience is awakened regarding the privations of their legitimate needs and rights, and start to utter that, the conflict moves to another level which may become confrontational (Curle 1971). The role of civil society in peacebuilding is thus clear—to prevent escalation of that which is an open face-off. This aspect of peacebuilding implies that CSOs, citizens' movements and all activities should be educated towards peace to lead their actions and campaigns to the desired outcomes.

To demonstrate how the grassroots or the masses are involved in building peace, Lederach (1997, p. 94) notes that "An important task in the development of a framework for sustaining reconciliation is to build a peace constituency within the setting". His pyramid shows three levels of leadership and peacebuilding action (Level 1, referring to the "Top-Level Leadership"; Level 2, representing the "Middle-Ranged Leadership"; and Level 3, relating to the "Grassroots Leadership") (Lederach 1997). The grassroots leadership is placed at the bottom left-hand; it is made of various peacebuilding actors, and includes various social groups such as local leaders, leaders of indigenous NGOs, community developers, local health officials, refugee camps leaders, etc. This leadership utilizes various approaches to transform conflicts and enhance peace. Their strategies revolve around multiple tasks such as building locally based peace commissions, developing grassroots training, reducing prejudice, being involved in psychosocial work, and assisting with relief programmes to community members affected by post-war trauma (Lederach 1997).

Analysing and interpreting the pyramid from a 'bottom-up' perspective of peacebuilding via reconciliation breaks the common perception of peace as emanating simply from above (Level 1) or it is negotiated at the middle-range degree (Level 2). Lederach adopts a more communitarian, inclusive and participatory strategy to validate the agency of each category in the attainment of durable peace. He argues: "What stands out in all of these examples is the effort to provide an opportunity for grassroots leaders and others to work at the community or village level on issues of peace and conflict resolutions" (Lederach 1997, p. 55). The strength of CSOs is discernible in the pyramid where, at each level, civil society is represented: At Level 1, religious leaders are involved; at Level 2, leaders who enjoy respect in various sectors such as traditional, ethnic, and religious leaders, and academic and humanitarian institution leaders, are notably visible; and finally, at Level 3, most of the readers emanate from civil society. In the continuum designed on the pyramid, Lederach (1997) indicates how affected people in times of conflict could be any of the three leadership categories. It follows that peacebuilding becomes a tool for achieving reconciliation and healing affected members of a community, thus restoring relationships in various leadership positions becomes an imperative.

It is worth noting that the role of civil society in peacebuilding is evidenced and discussed by an extensive scholarship which examines the theories and practice of peacebuilding and the responsibility that CSOs take in the pursuit of peace (Lederach 1997; Paffenholz 2015; Paffenholz/Spurk 2006). CSOs' engagement is globally perceived in post-war nation-building and in transitional societies (Belloni 2001, 2008; Jeong 2005). Peacebuilding serves as an instrument of prevention and transformation of conflict. This condition is made possible by disarming, demobilising and reintegrating belligerents to civilian life, remaking government institutions, enhancing the rule of law, transforming unjust structures, providing emotional and psychological healing, addressing issues pertaining to race and ethnic inequalities, and enhancing economic and civil society institutions that can be necessary tools for the sustainability of peace.

11.2.2 Political Transition

Transition from one regime to another, or from one practice or ideology to another, can be a tasking and a complex process which requires justice (Andreevska 2013); it may involve social and economic justice (Arbour 2007); emerging transitional justice processes may include local approaches and reconciliation (Baines 2007); and transitional justice may also incorporate prosecutorial conventional justice (United Nations 2004, 2009). It has become evident that civil society is ostensibly present in political transition by enforcing democratic principles as the tool for peaceful transition (Belloni 2008; Bratton 1994).

In the context of the DRC, political transition is referred to as the intermediate period that begins with breaking from a dictatorial power to gradual steps taken towards conducting the polls, and that is followed by the peaceful transfer of power to the winner. Political transition in the DRC has carried the burden of many decades of economic, political, and social upheavals; security problems; armed and unarmed insurrections; protracted armed conflicts; as well as the ups and downs of diplomatic relations caused by the calamitous state affairs management since the early 1990s. This period has been characterised by lawlessness and global chaos despite the apparent peace which has been totally absent. This reality is a typical characteristic of societies in transitional, according to Tarif/Vircoulon (2016), who describe it as a moment of ultra-dependency of the country on the international community and the infusion of the state by external policies and influence that weakens and reduces the sovereignty of local authorities. It is a period during which the community applies artificial survival techniques to save the failing state that has a ruined economy and a drained state budget; it is a time of rivalry between international players (a matter of power and influence for the international actors involved); and it is marked by moments of political short-termism (it does not allow long-term choices and lasting reforms) (Tarif/Vircoulon 2016).

The political and social depiction of transition carried out here by Tarif /Vircoulon (2016) suggests that political transitions are not the time of national unity, but rather a period of security restoration, of state institutions' reforms and reformation, and of re-establishing the rule of law; it is a time for reconciliation between communities and of development resumption. Unfortunately, in the DRC, corruption and impunity have been rampant through and through, and there has been no transitional justice nor reconciliation and reparation for past crimes—a requirement for effective transition and nation-building after large-scale violence. Impunity is endemic in the DRC (Hall/LaRocco 2012; Yonekawa 2012), and it is extensively entrenched in all state institutions in such a way that it has sustained the perpetuation of international human rights and humanitarian rights law violations.

Since the fall of the dictatorial regime of President Mobutu Sese Seko's Second Republic on 17 May 1997, the DRC has undergone four political transitions. The first, which was initiated by the junta in 1997, lasted five years until January 2001 and came to a sudden halt with the assassination of the late President Laurent Désiré Kabila. The second occurred under the aegis of President Joseph Kabila Kabange and covered the period from 2001 to 2003, while the third was inaugurated by the Political Agreement for the Consensual Management of Transition in the DRC signed on 19 April 2002 in Sun City and was established from 2003 until the inauguration in 2006 of Mr. Joseph Kabila Kabange as the democratically elected president after the 2006 polls. A fourth transition was *de facto* following the postponement of the 2016 elections. The latter covered two years until the inauguration, on 24 January 2019, of President Félix-Antoine Tshisekedi democratically elected on 30 December 2018.

All the four political transitions' circumstances and contexts corroborate, either partially or totally, the complexities of transition discussed earlier. The DRC civil society has always played a central and sometimes front-line role through its engagement for the harmonious and peaceful management of electoral processes and outcomes, and it did the same in the above-mentioned transitions. The transition that brought about the end of the mono-party political rule of President Joseph D. Mobutu was worked out internally and externally, and civil society organisations were tremendously involved to weaken the brutal Mobutu's rule and cause its moral destabilisation that caused its implosion and mass protests initiated and sponsored by the Catholic Church and other mainline churches, especially during the *Marche de l'Espoir* [March of Hope] on 16 February 1992. Leroux (2018) reports that on 16 February 1992, tens of thousands of people took to the streets in the twenty-four communes of Kinshasa. The marchers, who had responded to the call of a group of Kinshasa priests, were marching for real democratic progress. Despite the ban on the march by the authorities in Kinshasa and the absence of official support from the hierarchy of the Catholic Church, the movement was overwhelmingly followed. Faced with this march, the repression was extremely harsh, and immediate. Security forces shot at the crowd, and many victims went unaccounted for from the massacre of 14 February 1992.[1] According to the official report, thirteen people were killed, an underestimated figure for the *Voix de Sans Voix* (VSV) [the voice of the voiceless] for Human Rights, which mentions the number of 35 victims (Leroux 2018).

In a similar move, in 2018, the Congolese Catholic Church called for new actions from its faithful, during a mass in tribute to the victims of the repression of the January 21 March 2018. Human Rights Watch (2018) reports that peaceful marches were called for by the Catholic Church lay leaders in Congo to entreat Congo's political leaders to respect the terms of the peace agreement mediated by the Catholic Church, signed in December 2016, which demanded that presidential elections be held by the end of 2017 as a disposition to ease political tensions in this second biggest country in Africa.

These commitments have largely been ignored, however, as President Joseph Kabila has held on to power through repression and violence. For more than a month, protesters had been urging President Kabila to publicly declare that he would not run for a third term (Mulegwa 2018). They played a crucial role in terms of raising awareness about the foreseeable violation of the constitutional mandate of the Head

[1] This statement is from Kiyala Chrysostome (a co-author of this chapter) who took part in the 16 February 1992 March of hope; he is a eyewitness of the events that happened as he describes them.

of State and mobilising the Congolese people to protest against any attempt by President Joseph Kabila to remain in power. The Catholic Church and mainline churches in the DRC have shaken people's consciousness and awakened them to throw away fear and resist the political regime of terror instituted by Joseph Kabila and his predecessors. The Roman Catholic Church's activism has been twofold: (1) the hierarchy is outspoken and privileges dialogue and lobbies for external and internal support to broker peace accords, and (2) the laity, through its CLC, adopt non-violent and peaceful marches as a means of attaining peace and easing democratisation progress. The lobbying and the church conduct with the international community is traced back to the era of the Sovereign National Conference during which Archbishop Laurent Monsengwo Pasinya, a Catholic Prelate, chaired this assembly and later became the President of the High Council of the Republic and Transition Parliament (HCR-PT) between 1992 and 1997.

The DRC's civil society, through the mediation of Archbishop Monsengwo, secured a peace deal to rescue the transition period already under severe pressure and threatened to be disrupted by President Mobutu and his cronies. Regarding the peace agreement, Babunga (2018) notes that this new agreement between the government and the opposition was the result of negotiations that started in September 1993. Despite the fact that the Union for Democracy and Social Progress (UDPS) of Etienne Tshisekedi wa Mulumba, belonging to the Sacred Union of Radical Opposition (USORA) Platform, refused to sign this agreement with the Presidential Movement, Étienne Tshisekedi himself calling the deal a *constitutional coup* by President Mobutu (Babunga 2018). Thus, on 19 February 1994, the UDPS called for a *ville morte* [Ghost/dead city] day in Kinshasa. Its members took to the streets to protest against that deal despite the challenge with which the DRC's transition was confronted. However, the HCR-PT was installed on 23 February 1994, with Bishop Laurent Monsengwo as president (who was already president of the High Council of the Republic). The HCR-PT was made up of 735 members called "Councilors of the Republic" (the equivalent of national deputies today); it would be dissolved in May 1997, upon the arrival of Laurent-Désiré Kabila (Babunga 2018).

In 2003, the DRC's civil society was one of the key players in various national consultations that led to Sun City's main agreement for the re-foundation of the nation's unity. Some of its actors became members of the consensual management of institutions, the government of national unity, the national assembly and mainly the two citizens' institutions—the Independent Electoral Commission (IEC) and the Audio-Visual Commission.

11.2.3 Democracy

Cvetek/Daiber (2009) define democracy by using the 1863's definition given by former US President Abraham Lincoln in the following essential terms: "The governance of the people, by the people and for the people", a principle dating back to the foundation of the Roman Republic (Hölkeskamp 2000). This definition has always retained its richness and substance over the years across world democracies. On the other hand, it is distorted and mutilated today by various political pragmatists of modern democracies. Bentahar (2014) points out that the modern concept of democracy, as the power of the people for the people and by the people, is clearly not limited to free and fair elections, allowing a majority to be defined, nor to the undivided rule of that majority. Moreover, every poll does not necessarily lead to democracy. Each of the prepositions included in this definition of democracy is full of meaning and demands which can be described as follows:

- Power *of the people* refers to the capacity, competence, and recognition of the people as the crucible of power and its holder. Power belongs to them. The *people* are understood here as every human being forming and sharing with other humans a community with the connection of belonging to a delimited geographical space to which it is spiritually and materially bound.
- Power *by the people* reflects the sovereignty of the people in the detention and free exercise of power, directly or indirectly, in the management of the *res publica*. The people have not only the right but also the duty to participate conscientiously in the management of the country to avoid being accomplices of any abuse in the management of state affairs. *Res publica* involves all social, economic, political, cultural, and environmental matters that are supposed to benefit all the people of a specific society as a matter of the natural rights of the people. In this way, citizens bear and share political, economic, social, and cultural responsibility of the management of their respective countries' issues; they can exercise power directly or indirectly. The direct exercise of power is done through the positive contributory intervention of the people in the functioning of the management of the state. The indirect exercise is done through the representatives to whom the people transfer power through voting. As a power order, the people also reserve the right to control and sanction the way in which power is exercised. The legitimacy of power requires that representatives are handed it by the people for whom they are to vote.
- Power *for the people* implies that power is exercised for the benefit of the people and exclusively for their well-being.

Referring to Aristotle et al. (2009) observe that democracy is a republican principle that all power must be exercised in the service of the general interest—power comes from below and is exercised on behalf of all. It is basically a modern democracy, where social positions are open, where power is self-control, and where governance is respected. This reflects Aristotle's humanist political conception that needs to be emphasised all the more since modern democracies are rendered tasteless and touch on twilight. Zabondo (2019) highlights the way in which African countries comment on democracy—they are happy with a cosmetic [aesthetic] democracy. Countries are formally disguised as democratic institutions, but the powers that govern them function exactly as dictatorial regimes (monarchies or oligarchies) (Zabondo 2019). Often, the behavior of African political actors does not resemble those of the democrats; there is a problem of appropriation of the notion of democracy by political actors, civil society and the population as a whole (Zabondo 2019).

The democratic drift stems from the fact that its ideological and pragmatic forms of modern invention are incompatible, and if not, do little to respond to the cries of misery that overwhelm the populations and the quasi-universal aspirations of the latter. Beyond its material and social responses, democracy is understandable, therefore, as a quest for values in which every man believes (Zabondo 2019). Democracy rests on the values of freedom, justice (fairness), the dignity of every man, social justice (distributive justice), and the participation ("free") of all (directly or indirectly) in the management of common life. Democracy is therefore the obligation of the values that contribute to the realisation of well-being. It thus becomes the project of an "ethnicisation" of the governance of the public aspect by the state (Zabondo 2019). It is in this context that Mr. James O.C. Jonah, former Under-Secretary-General for Political Affairs at the United Nations, affirmed that if there was the opportunity to choose, every human being would choose to live in a democracy (Zabondo 2019). To respond to the increasing and urgent demand for a better or true democratisation of countries all over the world, especially in Africa, civil society's actors and leaders should aspire to the emancipation from traditional views of leadership in the time of empires and kingdoms, and help leaders overcome dominion over their nations by embracing democratic principles of power sharing and separation of power. This is the way to move away from dominations that hinder the realisation of the well-being of the people and the best development of each individual; democracy then takes on the fundamental meaning of *emancipation* (Zabondo 2019).

Faced with the drift in democracy's values with which the political world is now ablaze, not only in Africa but also in the USA where recent attacks on the electoral process have been levelled by President Donald J. Trump alleging that their presidential elections were rigged or that there had been massive fraud, civil society is called to rise and be more vocal to safeguard faith in democratic tenets and stand against populism. The current state of 'national populism' variously materialises according to adaptations to the respective socio-political environment. Notably, according to Zacharie (2019), populism is a strategy that aims to oppose the interests of the 'people' to those of the 'elites', to whom it opposes new leaders supposed to meet popular demands. The populist strategy is based on the socio-cultural feeling of identity withdrawal. It is a backlash that would upset some traditional foundations of democracy. It deeply embodies a democratic feeling the effectiveness of which is only possible by the 'bottom up' approach.

Hence the need for a strong civil society—which is autonomous and independent of the power of the state, family and business influence, economics, ethno-sociological and political influence, and ties with the political establishment—is more than an urgent matter in the DRC, and it should be an effective counter-power and fight for freedom space in its action and thus impose itself in the restoration of democracy. In this regard, Bentahar (2014) brings up the challenge that we must ask ourselves in more directly political terms whether the ideas and political forces that appeal to the people are still democratic: is-it in the name of the left, the people, the working class, and democracy itself that democracy has been destroyed in the Eastern countries? To respond to this test, we shall now examine the platform known as AETA.

11.3 Methods

This chapter is designed as a case study and employs documentary and literary methodology that are simply qualitative. To respond to the questions that are posed here to guide this inquiry, besides general reading pertaining to the DRC's political context and the development of civil society, this chapter will focus specifically on the documents produced and used by AETA which is examined as the main CSO in this writing. Both literary and documentary sources are analysed and interpreted qualitatively using the constructivist paradigm which is interpretivist, and which holds contrasting assumptions to the positivist and empiricist tradition (Allen 1994). Although neutrality and objectivity are deemed minimal in this constructivist worldview, we strive to utilise phenomenological reduction or *Epoche*—by bracketing experiences and providing a description of a phenomenon as it appears to our senses, and related to the truth that we highlight, in the relation between the phenomenon and the self, as explained by Moustakas (1994). This approach is necessary to minimise bias in the interpretation of the documents and the facts that they contain, and that which our reading and interpretation tries to uncover and on which it reports.

For greater rigor and efficiency, it analyses civil society's action which is documented. The methodological approach adopted here is based on literary and documentary research. We reply upon a good range of scholarship to apply the methods appropriately, noting Ahmed (2010) in particular, and these involve the four components of documentary research (*data collection, data reduction, data display, and data drawing and verifying conclusions*) inspired by Miles/Huberman (1994). We proceeded by gathering all-inclusive data from CSOs and AETA in the DRC, then we conducted a *data reduction* by bringing data to a manageable aggregate of relevant information; this was followed by *data display* by gathering pertinent data that we subjected to analysis; and finally, we drew conclusions after interpretation of the collected findings.

As literary research, this chapter draws substantively from both written and electronic materials which are considered as important data sources from this type of inquiry. The main tools of literary research, and the guiding principles to utilising written and online resources, were observed (Towhee 2009). The authors of *The Handbook to Literary Research* (Correa/Owens 2010) note: "The ultimatum aim of most literary research is to produce some critical, theoretical, or historical writing. In most cases such work takes the form of a thesis or dissertation" (Correa/Owens 2010 p. 3). As documentary research, this inquiry seeks to find solutions to problems or answers to questions (Ahmed 2010). Several researchers argue that this mode of investigation is context-based and needs to be placed within a theoretic frame of reference to render its content intelligible, and it is observed that documentary research is good and sometimes has a higher cost effectiveness than social surveys and in-depth interviews, and even participant observation (Ahmed 2010, p. 2).

The conclusions drawn here derive extensively from collected data and the archives of AETA. Access to the AETA library was duly obtained after the gatekeeper letter was received and answered by the relevant authorities of this institution.

11.4 Civil Society Areas of Action

This section discusses the historical and contextual development of civil society in sociological and philosophical traditions. It then underlines how civil society is organised, its components and areas of action, which are all relevant to this investigation.

Civil society is made up of organised and structured cultural groups, sports or religious associations, mutual aid and solidarity groups, clubs, unions, societal initiatives or citizens' movements, action groups, groups of scientists or academics, voluntary organisations, Non-Governmental Organisations (NGOs), etc. All are non-political and not-profit platforms.

Civil society organisations work in various sectors such as the social, cultural, environmental, economic, and political sectors. Specifically, these may include women's and children's rights, the rights of people with reduced mobility, tenant rights, consumer rights, health, education, environment and climate, financial and economic governance, development, mining resources, democracy and elections, tourism, land, etc. These themes represent the different vital rights inherent in the human person. Cvetek/Daiber (2009, p. 13) categorise them into global themes such as environmental and climate issues, the global capitalist economy, terrorism, migration or global health crises that are shaking humanity and in narrow themes such as violence against women, discrimination of people with reduced mobility, the integration of ethnic minorities, social protection of marginalised children and mining children, etc.

The form of the constitution of civil society organisations, their mode, the space and nature of their actions, the objectives that they pursue, the sources of funding for their actions and the nature of their relations with the state and the private economies make them intrinsically special. According to Cvetek/Daiber (2009, p. 14), civil society organisations share the common characteristics of autonomy, self-regulation or self-discipline, volunteerism, and solidarity, as well as various fundamental democratic values (freedom of opinion, freedom of expression, etc.), considered as the basis of organisation.

11.4.1 Dynamics of Civil Society Development in DR Congo

It is important to give a brief background of the Congolese civil society at this stage since this chapter explores the involvement of one of its organisations that works to promote fair democratic processes and peaceful political transition. The evolution of civil society laid out in the previous section, particularly its development as embraced by social and political sciences, reflects how the DRC has adopted civil society civic and political engagement in the longstanding struggle to democratisation and nation-building.

The first emergence of civil society is traced back to between 1885 and 1908, though actors involved in it had close ties with the Belgian colonial power who assured its patronage and provided funding for its activities (Congo Forum 2009). A further report in the wake of the DRC's independence notes that political leaders acknowledged the works of civil non-profit organisations (Congo Forum 2009). Thus, on September 18, 1965, a decree-law was promulgated which regulated the organisation and functioning of member associations of civil society. Many Congolese took an interest in it and created many lucrative aimless associations which were rooted at the grassroots level and took care of development works, the promotion and defense of human rights, trade unions, employers' associations, young people, women, religious denominations, etc. These different components of civil society can no longer currently be excluded from national and local questions in the management of the DRC. The role of civil society is always that of a counterweight to power (Congo Forum 2009).

The same report notes that CSOs of the first era (1885–1908) belonged solely to Europeans who worked for the colonising power which provided these organisations with the finances needed to carry out their activities, without any involvement of indigenous Congolese, either in the management or as members or administrators. But later, the awakening of the black Congolese elite known as "The Evolved/The Elite" triggered a global consciousness, and they became influential. According to Congo Forum (2009), a Decree of 15 April 1958 allowed mutualist associations of a purely tribal nature to function, but they received subsidies from the colonial government. The leaders of these associations were democratically elected by the native members, and this was done under the supervision of Catholic missionaries. Most of the political and social leaders of the Congo during its accession to independence had emerged from mutual societies and emerging political parties.

The DRC civil society is conceived as pyramid, where grassroots leadership features at the bottom whereas the national structure is placed on the top.

Civil Society therefore cannot be an instrument of propaganda at the service of any Government or business. It was a decree-law of 1 March 1914, which granted civil personality to civil society organisations, while another decree of 1926 authorised the creation of non-profit public utility establishments. At that time, the management of these civil society organisations fell to Europeans who worked for the colonising power (Kouassi 2012).

According to a report by Congo Forum (2009), the main components are the following: group of development associations; group of women's associations; group of associations and organisations of youth; association group for the promotion and defense of human rights; group of trade unions and corporations; orders of doctors, pharmacists, engineers, economists and financiers; psycho-pedagogues; group of philanthropic associations and humanitarian organisations; group of learned associations, universities, and research; group of employers' organisations; group of cultural, sports and hobby organisations; and group of recognised religious confessions (Catholic, Protestant, Kimbanguist and Islamic community). In addition, civil society's mission in the DRC entails the preservation, promotion and dissemination of the democratic tradition, without excluding the coordination of all CSOs' activities and advocacy in various sectors of life. It exists for the purpose of conquering power, though its leadership and committed actors strive to ensure that they understand issues relating to social, economic, political, and security aspects of the country. Thus, CSOs' leaders keep touch with the grassroots and give heed to what local ordinary people convey to them for transformative action in the social, political and economic life of the country. Civil society works in partnership with the DRC Government and engages the state to pursue the common good. It follows that civil society bridges grassroots (ordinary citizens) and top leadership (people's representatives) for coordinated action to improve the living conditions of the DRC's citizens (Congo Forum 2009).

The expansion of civil society and local ownership became manifest in the era of the Sovereign National Conference popularly known under the acronym CNS which refers to its French version—the *Conférence Nationale Souveraine* (1990–1992). In 2001, President Joseph Kabila signed to law the framework within which civil society is understood and allowed to operate in the DRC. This alludes to Law No. 004/ 2001 of 20 July 2001 which lays down the provisions for civil society's statute and sphere of activities, and it should operate as non-profit associations (Hamuli Kabarhuza et al. 2003). Article 1 of this law states: "The non-profit association is one which does not engage in industrial or commercial operations, except on an ancillary basis, and which does not seek to provide its members with material gain. The non-profit association is apolitical" (Democratic Republic of Congo 2001, n.p.).

The emotional speech delivered by President Mobutu as he ended his leadership of the *Movement Populaire de la Revolution* [Popular Revolutionary Movement] which he created in 1967 and which had become the sole party of the Republic of Congo (renamed Republic of Zaire in 1971), opened the political space to many voices in civil society organisations. After the civil wars which cost the lives of more than 5 million people between 1996 and 2002 (Heaton 2010; Mayeda 2012), civil society was at the negotiation table which halted over 7 years of murderous war. This occurred at the Inter-Congolese Dialogue at Sun City in South Africa, which was signed on 16 December 2002 in Pretoria and became active in January 2003.

Article 2 of the same law provides the definitions of NGOs that apply to civil society, as it follows: "The non-profit association is by its nature and purpose: 1. An association of a cultural, social or educational or economic nature; 2. A non-governmental organisation NGO, in acronym; 3. A denominational association" (Democratic Republic of Congo 2001, n.p.). This legal framework defines the scope of non-profit associations' commitment and permits them to implement public policy. Kabarhuza et al. (2003, p. 30) observe that the DRC's civil society is working according to the above-mentioned law despite the rigorous legal provisions of that law.

As far as the components of civil society are concerned, Mushi/Kamavu (1999) list the following: traditional churches and NGOs working in the fields of development, human rights, democracy, economic governance, budget monitoring, workers' and employers' unions, corporations and professional associations, orders (doctors, architects, pharmacists, nurses, etc.), peasant and cooperative associations, cultural associations, scholarly associations, women's associations, humanitarian and philanthropic associations, etc. With the start of the democratic process in 1990, these various associations demonstrated civic responsibility by mobilising to support the democratisation efforts and the people's participation in the sovereign national conference (Hamuli Kabarhuza et al. 2003).

11.4.2 Dynamic Development

A diachronic analysis of civil society institutional growth since the 1990s in the DRC shows that it went through six dynamic phases which correspond with six models that followed a paradigmatic logic of responding to the contextual challenges posed by the Congolese society during various periods. These models are made up of the following:

- The first model dates back to as early as 1990, and corresponds to the challenge of promoting and protecting human rights against the violations that prevailed in a Congolese state which was deeply weakened by the totalitarian power of Mr. Joseph Désiré Mobutu. It is the first paradigm, and during this time, the Human Rights League (LDH) was created by Mr. Bwana Kabwe and that was the first civil society organisation to fight the many horrific recurring violations of citizens' rights. The struggle, led by local human rights organisations, involved actions taken to promote and defend civil liberties, and individual and collective rights. It was already a form of School of Democracy and the aspiration for the rule of law. The LDH was joined and reinforced by the VSV initiated by the late Floribert Chebeya[2] and many other organisations in this sector.
- The second model, which is intimately linked to the former, is naturally related to the struggle against dictatorship through the establishment of multi-party electoral democracy. Civil society organisations such as the League of Voters, the network of civil education organisations for all Zaïre 2000, and the Centre for Study, Documentation and Civic Animation (CEDAC) have played a leading role in this issue. Later, the Civic Education Production Unit (UPEC), the Network of Human Rights and Civic Education Organisations of Christian Inspiration (RODHECIC), the Jeremiah Group, etc., were born. The organisation called Acting for Transparent and Soothed Elections (AETA), which we will discuss in the form of a case study, is part of the development of this paradigm.

[2] Floribert Chebeya was a human rights activist in the Democratic Republic of Congo (DRC). At the end of the 1980s, he created the NGO 'La Voix de Sans Voix' (VSV). The association defends political prisoners, and touches on sensitive subjects, such as the abolition of the Military Order Court, instituted by Laurent-Désiré Kabila. He was assassinated on 2 June 2010. When he died, several organizations and United Nations officials called for an investigation, such as Amnesty International or Ban Ki-Moon. Five police officers were sentenced in 2011 at the end of it, considered sloppy by the civil parties (https://www.jeuneafrique.com/personnalites/floribert-chebeya/).

- The third model is formalised around the great challenge of the pacification and stabilisation of a DRC torn apart, broken and partitioned by what began as a war of aggression to turn into diverse rebellions caused by the ongoing activism of armed groups and self-defense groups. This problem has mobilised the entire civil society around the many initiatives to combat it. The National Campaign for Sustainable Peace in the DRC, a platform for organisations committed to democratisation, development, and peace, was a pilot programme that brought together almost all civil society organisations in 1999.
- The fourth model was developed concurrently alongside the first three, in the context of excessive poor governance, endemic civil wars and rebellions, the worsening social and economic crisis, and political and social unrest. This situation demanded an urgent and inevitable response. This model is described as the paradigm of social development which coincides with the emergence of organisations such as the Association for Integrated Development of Kivu (ADI-KIVU) that operate in Kivu, and those that used to be grouped in the National Council of Development NGOs (NCD-NGOS) at each level of the eleven provinces of the DRC, before these provinces were subdivided into 26, in the Regional Council of Development and NGOs (RCD-NGOs).
- The fifth model has emerged recently with the democratisation of the country's institutions and is threefold. These include economic and financial governance, environment and climate change advocacy, and transparency in government that is characterised by free access to reliable public information. The fifth model is supported by organisations such as the Citizens' Action Group for the Monitoring of the Transition/Citizen Observatory of Public Action (GAT/OCAP), organisations that are committed to good management of natural resources and bringing awareness about global warming, and others.
- The sixth model is complementary to the five previously mentioned. It reflects the commitment of the Congolese civil society to engage in all areas of the country's governance and sectors of life.

At present, membership of associations and organisations has increased exponentially, and they have taken various forms to the extent that enumerating CSOs has become more and more difficult. A survey of civil society organisations was carried out in 1996 by the National Council of Development NGO which provided statistics showing that the number of NGOs in the DRC was approximately one thousand, and they were spread throughout the country. More than two decades later, the figures provided in those statistics most likely tripled in the light of the freedom space that several associations enjoyed expressing themselves in the DRC over the last few years, and the role that civil society is increasingly engaged in in various spheres of human rights, advocacy for fair and transparent polls, and the social, political, and economic life of the Congolese people. Among these organisations are citizens' movements.

11.4.3 Citizens' Movements

In addition to the development of civil society, the path taken by the DRC towards putting in place democratic institutions since 2005 has been navigated by many movements and campaigns among which some have come out and established themselves as pressure groups and 'watch dogs' of democratic and electoral processes and of good management of public affairs. Although the literature on these associations is very limited in the DRC, it is necessary to name a few that have been mostly non-violent and energised by the youthfulness of their actors. They are also regarded as precursors of the new waves of pro-democracy actions, catalysts of transformations, and actors of human rights protection. These movements include *La Lutte pour le Changement* [the Fight for Change] (La LUCHA), *Comité laïc de coordination* [Lay coordinating committee] and Filimbi ('whistle' in the Kiswahili language).

La LUCHA was founded in May 2012 in Goma in North Kivu, a region ravaged by long civil wars, and claims exclusively for non-violent resistance inspired by the battles of Gandhi, Martin Luther King and Nelson Mandela (Ritimo Network 2018). It was in the wake of the Tunisian spring to support its claims or oppose state violence; it resorted to massive sit-ins and made a commitment to refrain from responding to violence with violence; and it uses lobbying in the event of arrests of demonstrators (Ritimo Network 2018). It also relayed the need for political alternance or change, in particular vis-à-vis President Kabila who wanted to run for a third term in an unconstitutional manner, while his regime was marked by numerous human rights violations, corruption, etc. (Ritimo Network 2018).

In this context, La LUCHA joins with other movements such as the *bye-bye Kabila* campaign (Ritimo Network 2018). *Bye-Bye Kabila* refers to the slogan that was chanted by the youths commanded by La LUCHA, and this was done in protest against President Joseph Kabila who intended to remain in power beyond his constitutional mandate by seeking a third term of its presidential mandate. Filimbi was officially born on 15 March 2015 in Kinshasa in the presence of representatives of *Y'en a marre*[3] [Fed up], *Balai Citoyen*,[4] Lucha and other personalities and journalists to raise awareness about the endless and devastating suffering of a large part of the Congolese population. La LUCHA and Filimbi operate as a non-violent activist group (Mbombo 2018; Perera et al. 2018).

According to Ritimo Network (2018), Filimbi refers to the alerts that the inhabitants of various neighborhoods and villages give themselves in the event of a threat and the signals that they send out calling for mobilisation, but Filimbi also has an additional meaning—it touches on the referee, the person who whistles against the failures of those in power to work for people's welfare (Ritimo Network 2018). Filimbi and La LUCHA maintain ties with their Senegalese and Burkinabe counterparts and with those from other countries, with whom they share visions and

[3] 'Y'en a Marre' ('Fed Up') is a group of Senegalese reporters and journalists, created in January 2011, to protest the ineffective government and register youth to vote.

[4] The Citizen's Broom or the Civic Broom, also called simply Balai Citoyen, is a political grassroots movement in Burkina Faso, which was part of the opposition against President Blaise Compaoré.

principles. As pressure groups, La LUCHA and Filimbi were instrumental in getting Joseph Kabila to step down when he wanted to stand for a third term. These citizens' movements opposed his second term beyond 2016 on 19 December 2016 (Clowes 2016).

The Comité Laic de coordination [Lay Coordination Committee] (CLC) is a lay association that militates for fair and transparent democratic processes and the peaceful transfer of power, originally in the context of Joseph Kabila's attempts to prolong his presidential term of office beyond 2016. The CLC organised peaceful marches to demand full respect of the DRC Constitution pertaining to the head of state's mandate and term of office. It intervened to support and accompany the transition peace agreement brokered by the Congo Catholic Bishops' Conference (CENCO) on 31 December 2016—a peace deal also known as the New Year's Eve or the Saint Sylvester Accord, signed between the government and the opposition to have landscape democratic elections in the DRC in 2018 and not beyond. The CLC is first a group from the Center for Information and Missionary Animation (CIAM), which responded to the call of the Church, by launching three conference-debates in the Congolese capital Kinshasa. The first was held on 27 October 2017 at Saint-Joseph parish on the theme "December 2017, issues and challenges (Malu-Malu 2018).

Hope was reinvigorated by the martyrdom of the latest activists and soldiers such as Rossy Tshimanga , Floribert Chebeya[5] and others. In a report to *Jeune Afrique Magazine* by the President of the League of Electors in the DRC, and the Deputy Secretary General of the International Federation for Human Rights, Nsapu (2018) writes that, on 2 June 2010, two days after being reported missing, the lifeless body of Floribert Chebeya was found in his car, suffocated. His colleague and friend, Fidèle Bazana, has still not been found. In March 2018, the country flares up again: in Kasai, the uprising is crushed in blood. The east of the country is once again the prey of militias and armed groups, fought by a national army known for its deadly violence. Added to this is the crackdown on peaceful protests Joseph Kabila during which hundreds of people were killed (Nsapu 2018).

Civil society activism has increasingly gained energy and attracted young people who display eagerness to pursue societal, political, and economic transformations via citizens' movements. We see unleashed energy for political and social change struggles by civil society. Through various strategies of pressure and resistance, citizen's movements have been true counter-powers who have succeeded in flexing, without fear, the rigidity and stubbornness of political power in the DRC.

[5] Rossy Mukendi Tshimanga is a Catholic and pro-democracy activist from the Democratic Republic of Congo who was allegedly gunned down by a police officer on 25 February 2018 during a peaceful protest in Lemba/Kinshasa.

11.4.4 National Framework for The Dialogue of the Civil Society of Congo

At the 2009's symposium of civil society organisations and the DRC's first national civil society forum held from 17 to 22 July 2013 in Kinshasa, the Congolese civil society decided to create the National Framework for consultations Civil Society (NFCCS).[6] During the proceedings of the above-mentioned forum, MBILIZI Jean-Baptiste (a member of the Mutual Christian Aid Association Mbilizi) and Eva Bondjele (member of the Association of People Living with Handicap) note that the NFCCS is above all concerned with covering the lack of the synergy of actions and of specialisation at the head of the Congolese civil society organisations, and this could be redressed by setting up a unifying framework that accounts for civil society's officials and organisations at the provincial, national, and international levels (National Technical Secretariat of Civil Society 2013). The synergy of civil society organisations that are part of the consultation framework is streamlined and systematised through specific thematic synergies that should interact to impact national policies. The acts of the first forum underline that thematic working groups form the basis for involving civil society in the formulation of public policies, in commitment to develop its expertise and professionalism, with the view of making well-documented counter-proposals on various issues of the DRC's governance to the Technical and Financial Partners for Sustainable Development (National Technical Secretariat of Civil Society 2013).

The national framework lists twenty-one thematic working groups that relate to the following sectors: army, police, gender and gender-based violence, justice and human rights, local governance and decentralisation, economic governance, public service, voting procedures, parliament, new information technology, media and culture, statistics, infrastructure and transport, water and energy, mines, agriculture and rural development, education and scientific research, health, HIV AIDS, environment, sanitation, water and forest, protection and employment, community dynamics, business climate, public and private partnership as well as workers' unions. Each thematic working group is led by a team leader named from among the member organisations of the corresponding specific domain. It should be noted that each thematic group has a correspondent at the government level to facilitate the coordination of civil society's involvement in the policy-making process and the implementation and the monitoring of national programs within that department.

[6] National Civil Society Consultation Framework is the national umbrella structure for CSOs in the Democratic Republic of Congo.

11.4.5 Apprehensions about the National Civil Society Consultation Framework

It is unfortunate that the idea of monolithic national coordination and structuring of civil society does not fit in with the DRC's civil society organisations. Some large organisations object to the notion of a unifying coordination of civil society such as the consultation network, and therefore remain very critical of it.

In a reflection published on this subject, Guy Mudumbula Makola, the Executive Secretary of the Provincial Council of Development NGOs of the city-province of Kinshasa, thinks that the DRC civil society has been of unparalleled commitment through tireless citizens' efforts to push towards political pluralism—without civil society being structured as a pyramid (Mudumbula 2013). This statement refers to the period dating back to 30 June 1960 when the DRC gained its independence and the subsequent struggle that opened the way to a liberal democracy based on political pluralism, and the path to subvert President Mobutu's dictatorship to peace processes amidst civil wars and political instability (Mudumbula 2013). According to Mudumbula, the victories achieved by previous struggles for independence, peace, and democratisation of the DRC were the result of a dynamic and credible civil society that was nurtured and driven rather by the promotion of shared common values than the monolithic civil structure with a national president (Mudumbula 2013). It is all the thematic groups that worked in perfect synergy, with trust in each other, without racing for power or competition for leadership at the level of the NFCCS.

A point that deserves being mentioned here concerns the negative role played by the nostalgic political leaders of the one-party era in Mobutu's authoritarian rule who, becoming aware of their crumbling hegemony in favor of the power of the people, began to infiltrate civil society organisations with agents to attain macabre and cynical strategy to weaken the voice and action of civil society organisations. Mudumbula (2013) believes that this scheme simply intends to muzzle civil society organisations. Considering this critical view, the pyramidal approach we alluded to earlier could sustain the vulnerability of civil society's actors and keep them malleable and instrumentalised—a risky condition for a free and effective civil society.

11.4.6 The Paradigm of Moral Authority

The monolithic conception of civil society has gone beyond the representation of civil society by a single President in whom all civil society organisations would recognise themselves. At the end of the first national forum, as mentioned earlier, the DRC civil society adopted the model of "*moral authority*". Indeed, the moral authority of the DRC civil society is a religious figure to whom one should always refer for any question. This personality provides guidelines, recommendations, and gives injunctions and calls for action to be undertaken by the grassroots. His/her word is a dogma and cannot be the object of contradiction. This figure enjoys the full confidence of civil society organisations and the candid recognition by social actors of political power and social influence. The "moral authority" serves as an institution that stands beyond the political manipulations. The character of the "moral authority" is, to paraphrase, an "epistemic authority" which means here an ethical politician with a solipsist knowledge of *res publica*. The concept is epistemic authority is borrowed from (Zagzebski 2015, p. 109) who explains it as follows: "an epistemic authority is someone who does what I would do if I were more conscientious or better than I am at satisfying the aim of conscientiousness – getting the truth". This notion of moral authority has expanded by imposing itself in political parties and groups in the DRC such that their leaders are given the embodiment of political, social, and moral power while that is not the case.

Former President Joseph Kabila Kabange and the current President Félix-Antoine Tshisekedi Tshilombo are given the designation of moral authority by their political platforms, the Common Front for Congo (FCC) and the Cape for Change (CACH), respectively. The attribute of moral authority is a sign of power and political power over other political actors moving under them which does not have any substantive relation with morality or ethics which is neither the characteristic nor the aspiration of those rewarded with it in the DRC.

11.4.7 The National Order of Electoral Experts (NOEE)

The National Order of Electoral Experts (NOEE) stands as a relevant illustration of efforts to make civil society more professionalised. The three electoral cycles of the Democratic Republic of Congo, in 2006, 2011, and 2018, have been the basis for CSOs' resurgence and their commitment to democratic processes by ensuring that the *Commission Electorale Nationale Independence* (CENI) [the National Independent Electoral Commission] remains independent, and the polls are conducted in respect of the DRC's Constitution. Thus, building the capacity of electoral commissioners requires providing them with skills and good ethical standards. This remains both an aspiration and a challenge of CSOs because skills to run fair elections and proclaim the true winners are not enough, but the CENI's agents must hold high moral principles for voting results to be credible and acceptable. For this reason, the *École de Formation Électorale en Afrique Centrale* (EFEAC) [The Electoral Training School in Central Africa] has been established in Kinshasa, the DRC's capital city, and it is one of the most effective solutions for professionalising electoral procedures. It has put electoral experts in the job market by training those who hold master's degrees in electoral cycle management who are able to work as professionals in electoral matters. They can contribute to the development of scientific research and electoral action research to improve the poorly managed polls in the DRC despite the existence of EFEAC which should have helped the DRC clean the entropy of the electoral space.

Besides the EFEAC, and in view of maintaining professionalism in dealing with electoral issues, the National Order of Electoral Experts (ONEE) was created in 2020.

It is an innovation by the master's degree graduates from the EFEAC who strive to promote professionalism of electoral processes and pragmatism in delivering their outcomes, and this is urgent not only in the DRC but in many African countries. ONEE could serve as a remedy for electoral megalomania that strikes many social and political actors, the worst consequences of which would be presenting oneself for election and power drunkenness which are serious viruses that erode democracy. Although participation in the polls is a matter of civic sovereignty, civil liberties and citizens' rights, there is unfortunately an imbroglio that would exceed the threshold of the controllable if it is not investigated by a scientific community such as the EFEAC which can suggest knowledge-based ways to ensure that the electoral environment is safe and democratic principles prevail and are respected by political actors. To achieve this perspective, the ONEE should be more active in providing electoral and civic education to both the grassroots and political leaders.

In the previous sections, we dealt with the semantic development of civil society and some key aspects of civil society in the DRC. In the following section, we discuss the two main activities of civil society pertaining to the activity of AETA, namely democracy and political transition. This section serves as the rationale for studying the relevance of AETA in the DRC.

11.5 AETA's Action for Democratisation

AETA is a platform of the Congolese civil society created in 2010 by a coalition of civil society organisations in accordance with Law 004/2001 of 20 July 2001 with general provisions applicable to non-profit organisations and public utility institutions. It is endowed with a legal personality under Ministerial Order No. 125 / CAB / MIN / J and DH / 2011 of 11 April 2011 (AETA 2012). AETA is therefore a level four structure made up of provincial and national networks and organisations. As such, it is based in seventeen of the country's twenty-six provinces where its work is carried out by organisational dynamics. Its headquarters are established in the city-province of Kinshasa, the capital of the Democratic Republic of Congo (AETA 2010, p. 2).

The AETA platform works primarily on issues of democratic governance, and it mainly focuses on voting processes. The number of its members approximate one hundred, and they develop programs of activities in the areas supported by its vision, namely of making elections a pillar of peace, political stability, and citizens' participation in the affairs of the country—in a way that builds national cohesion, inclusion, and promotes the rule of law. Thus, AETA has the mission of promoting democratic culture through developing democratic and electoral citizenry that places human rights at the center of all policies. AETA's intervention strives to ensure that the polls are credible, transparent, fair, legitimate, and stand as the pillar of durable peace, good governance, and sustainable development. Furthermore, these perspectives encourage inclusive citizens' participation in the life of the republic; it contributes to gender equality, improved security conditions, and the advent of the rule of law (AETA 2011, 2017b).

The main objectives pursued by AETA's interventions include the following (AETA 2011):

- Provide permanent electoral education of the Congolese population.
- Support citizens' control of the transparent and democratic management of electoral processes.
- Encourage the spirit of dialogue between leaders and key players during elections.
- Monitor the polls through the implementation of long-term and short-term elections' observation mission.
- Advocate peaceful, credible, reliable, fair, and regular electoral processes that respond to international standards.
- Work for the advent of participatory and democratic governance in the DRC.
- Work for lasting peace in the DRC.

The fulfilment of these aspirations requires an organisational competence which AETA has through its provincial and national representation. The structural organisation of AETA is described as follows: the national and provincial level, which represent statutory bodies of AETA.

- At the national level, a National Advisory Council operates as the supreme body, comprising all members. It works as the General Assembly and thus acts as a deliberative body, decision-making, orientation, and the principal component of the structure. In addition, at this degree, there also exists a Council of Wise Men, a Control Commission, a Permanent Secretariat serving as the Board of Directors, and a General Secretariat that manages the organisation daily (AETA 2017a).
- At the provincial level, the structure is covered by the Provincial Advisory Council (PAC)—a type of general assembly of all members of the structures at the provincial stage, and the Provincial Focal Point is a provincially based organisation that is responsible for keeping the organisational dynamics alive and members active. The PAC manages the structure daily in the provinces (AETA 2017a).

The aforementioned organisation follows a way that acknowledges the community's anchorage where citizens identify themselves with it and its action. The reliability of internal governance and the level of influence of AETA's structures in the context of its contributions to the democratisation of the DRC are sutained by the principle of autonomy and independence of this civil society organisation.

The idea of setting up AETA as a platform to play the framework of consultation of civil society organisations to support electoral democracy is originally exogenous. Although the approach to its implementation has been strategically masked, it was designed by the Open Society Initiative for Southern Africa (OSISA), an international organisation with leadership based in South Africa and the United States (AETA 2011). OSISA itself has offices in many African countries that are struggling with tumultuous and fragile democracies.

For OSISA, following the first democratic elections held by the DRC in 2006 in the aftermath of rebellions and armed conflicts that broke out in 1996, it was imperative to strengthen the role of civil society in supporting the process of consolidating the Congolese democracy. This was a strategy intended to get the country out of the political and socio-economic crisis into which the DRC was plunged under President Mobutu's dictatorship and armed insurrections led by Laurent Désiré Kabila that deposed him on 21 May 1997. The DRC's dark political, social, and economic crisis was exacerbated by seven years of two civil wars between 1996 and 2003, and the disastrous political transition that ended hostilities under the 1+4 formula—a political and peace settlement brokered in South Africa. This so-called 'Global and Inclusive Peace Accord' was signed in Pretoria on 17 December 2002 and was adopted in Sun City on 1 April 2003 between South Africa as the mediator, various fighting factions in the DRC, and CSOs. The Sun City 2002/2003 peace agreement showed several imperfections such as the lack of legal provision to trade justice by self-amnesty given to belligerents in violation of the provision of the rule of law (United Nations 2009) and the premature demise of the truth and reconciliation commission that could have served for transitional justice to support and maintain the framework of the agreement. The need for truth and justice emerges from the texts issued at Sun City that govern the transitional period (Centre d'Etudes et de Documentation et d'Animation Civique (CEDA 2003) Civil society's sitting at the negotiation table alongside belligerents from various rebel groups is tangible evidence of its ongoing commitment to democracy embodied by AETA's action to leverage the progress of the DRC's democratisation and peaceful political transition.

While AETA's engagement demonstrates that which CSOs are capable of in societies in transition, it is important to highlight some challenges with which AETA, as several other CSOs in the DRC, are confronted including self-sufficiency in terms of funding. In its first year of existence in 2011, AETA operated under the support of the OSISA Foundation in terms of administration, logistics and partly human resources. Thus, it was an OSISA project. This became acute as more financial resources were sought to support its activities for the 2016 electoral process, which was to enshrine the DRC's first democratic transition after the constitutional mandates of the incumbent President, Joseph Kabila Kabange, were exhausted. Another problem faced by AETA emerges from the influence and instrumentalisation of this platform by political actors. This situation has undermined the cohesion of its organs, which have been torn apart by conflicts of positioning and objectives.

To safeguard its structural power, and preserve its independence vis-à-vis politicians, the AETA platform has adopted a code of ethics for its members, and they are bound to observe this during the votes to ensure its neutrality and autonomy to monitor and facilitate electoral processes (AETA 2017a). This ethical rule stands as a credible, effective, and concerted contribution to the democratisation of the DRC. In the past, it has helped circumvent the negative effects of policies that contrast with AETA's standpoint insofar as impracticality and fair fulfillment of its statutory mission—to legitimise its political action—are concerned. Autonomy and independence are held as the *circumvallation* of AETA's credibility which can be vulnerable due to heavy dependency on international donors (South African, Belgian and the USA) for its running, since external sources of funds rely upon their respective governments who provide aid according to their rigorous criteria and the philosophy to which assistance is tied, and which may be at odds with CSOs' ethical principles, because of diverse ideologies inherent in development aids' providers.

11.5.1 Quadruple-Level Democratisation Monitoring Process

The discussion in this section refers to the period between 2011 and 2019. AETA's interventions were intended to accompany the successful outcome of the political transition and democratisation. However, it must be pointed out that during the period under consideration, the DRC experienced only one political transition due to the postponement of the elections by Joseph Kabila in his attempt to seek a run for the third term. This transition was, in its early days, characterised by violent conflicts, although it enjoyed a short lull because of the consensus produced by the comprehensive and inclusive political agreement of 31 December 2016 which was obtained through the mediation of the Catholic Church, one of the largest civil society components of the DRC. AETA's commitment to this framework has taken place at four levels: (1) reflexively, for the benefit of its members and other stakeholders; (2) Interventions for members and stakeholders in the democratic process; (3) Interventions for decision-makers and other key players; and (4) Raising awareness and mobilising the population or grassroots.

11.5.1.1 Reflective Level of Action

This is the strategic and pedagogical level of AETA's work in its role of supporting the DRC's democratisation process. It consists, above all, in strengthening the internal governance of AETA, by providing credible and transparent basic texts (statutes and internal regulations, procedural manuals, a code of conduct, gender policy, and institutional curriculum). This documentary framework serves to facilitate the operationality of statutory and strategic bodies within AETA and helps to establish mechanisms aimed for the transparency and accountability of its operation, and to consolidate the synergy between its members through permanent communication.

Strategic mechanisms of action are undertaken by thematic commissions relating to gender and voting, civic education, electoral observation, people with disabilities and elections, youth and voting, human rights and voting, media and voting, etc. There is also the *Electoral Hub* within AETA, which is a multidisciplinary technical think tank, charged with analysing occurring democratic and electoral issues to provide technical advice. The Electoral Hub is composed of personalities from various scientific disciplines: law, social anthropology, economics, demography, political science, philosophy, communicology, etc. In addition, AETA's other components include scientists, media professionals, students, social actors, private actors, trade unionists, citizen movements, etc.

The pedagogical aspect of AETA's intervention is the ongoing imparting and strengthening of skills needed by its members and partners in the electoral field. Three members of the platform had to train for a master's degree in elections at the Central African Electoral Training School (EFEAC). The principle guiding AETA's action is to make this platform the crucible of expertise that will solidify professionalism in accompanying electoral processes. The logic of AETA's pedagogical intervention is as follows: identification of the issue or problem surrounding it; deepening the narrow core issue of the Electoral Hub; the gathering of information through consultation with relevant stakeholders; technical analysis of the issue within the Electoral Hub for advisory purposes and making recommendations; and validation of the results of the technical analysis by the members (they mainly assess the relevance and veracity of the opinions, compliance with the statutory mission, their neutrality and impartiality but they also enrich them at the same time as it is the opportunity for their appropriation, etc.). The vision of AETA is further implemented through the analysis of technical opinions and the adoption of the recommendations developed with key players involved in electoral and transition processes, and broadcasting them through media coverage, and finally, by giving feedback to other actors involved to sensitise and mobilise the grassroot level around the ideals supported by AETA. This is necessary for advocacy to influence the decision-makers, as well as the monitoring and evaluation of the action.

Technical analysis of electoral issues has been one of AETA's innovative approaches since 2012 and part of its role in supporting the political transition and the democratisation of the DRC. It has imposed on itself a corrective strategy to tackle the imbroglio and confusion of that which is sometimes seen as partisan, detractive and obscurantist politicking around the issues that matter to people such as education, work opportunities, health, and people's socio-economic welfare. The need for objective, impartial, and relevant technical insight was thus claimed at all levels of AETA's structure—to put an end to the schemes of instrumentalisation and manipulation of citizens by politicians. For the period from 2012 to 2019, AETA has published twenty-three technical analyses that supply technical advice and recommendations mainly to the stakeholders of electoral processes that encompass various strata of society, namely: the CENI, the President of the Republic, the Parliament, the government, the Superior Council of Audiovisual and Communication (CSAC), the army and police, political parties, civil society organisations, Technical and Financial Partners (PTF), as well as the Congolese population.

Technical analyses have focused on issues related to electoral planning, constitutional contentions regarding the organisation of elections, the protective legal framework of electoral processes, their budget, the political and sociological challenges of the electoral environment, the constitution of electoral audits, electoral operations, transparency in the acquisition of electoral materials (electoral awareness supports, ballot boxes and voting booths as well as the voting machines), and security during and after elections (AETA 2017a).

11.5.1.2 Interventions for Members and Stakeholders in the Democratic Process

This level focuses on the capacity-building activities of the platform's members, civil society organisation leaders, political party leaders and members, and citizen movements on the current electoral and political challenges. The capacity of activities carried out by key players were sometimes based on the results of technical and scientific studies, monitoring and action research carried out in advance by AETA. These include activities such as training seminars, reflection meetings, restitution and validation sessions for field reports and technical analyses. These types of activities were also aimed at upgrading the players involved through the transfer of specific skills. They have also helped to foster synergy and cohesive action among stakeholders by sharing content, objectives, results, and action strategies (AETA 2012–2019). For example, AETA has organised the training for the members of political parties and civil society organisations on topics such as the Congolese electoral cycle (its issues, challenges, and burden), election monitoring and observation, electoral disputes, and the strategies of the electoral campaigns.

11.5.1.3 Interventions for Decision-Makers and Other Key Players

These interventions take the form of advocacy and lobbying—to influence the political will; to encourage the support and openness of key players to the democratic process; to decant the snags that hinder the transparency, credibility, and accountability of the process; and to create a socio-political atmosphere that is conducive to welcoming the results from polling stations by all parties, including the electorate. AETA ensures that the political environment is peaceful, and equally, that human rights are respected by all parties at all levels. This refers to actions such as advocacy, monitoring, organising political consultations, giving feedback about technical suggestions made by partners and stakeholders in the democratisation endeavor, and disseminating the results of action research on electoral and democratic issues. All these undertakings are assessed and evaluated by stakeholders, including the CENI, political leaders, Parliament, the Government, the President of the Republic, the Technical and Financial Partners involved in the democratisation process, the Audio-Visual/media committee, etc.

Institutions targeted by these actions are mainly the CENI's leadership, the President of the Republic, the offices of both houses of parliament, the government, the elected representatives of the people, political actors, the police and army authorities, media professionals, the Technical and Financial Partners, the United Nations Mission for the Stabilization of the Congo (MONUSCO), the United Nations Development Program (UNDP), the European Union (EU), the USA, Canada, Great Britain, international non-governmental organisations, etc.

Some important illustrations of the political role played by AETA under this advocacy heading comprise:

- AETA's participation as an expert in national consultations (2013)
- The political dialogue of the City of the Organisation of African Unity, OUA, (September-October 2016)
- The inclusive political dialogue organised by the National Episcopal Conference of Congo (CENCO) which resulted in the Global Inclusive Political Agreement of 31 December 2016 (December 2016)

- Accompaniment of the electoral process in 2018 and 2019, as an expert on the meetings of the tripartite evaluation advocacy missions for reporting to the international community, namely, in the USA, the European Union, Great Britain, Canada, the African Union, the Southern African Development Community (SADC), the CENI, and the National Agreement Monitoring Council (CNSA). The CNSA refers to the 31 December 2016's Accord.

11.5.1.4 Raising Awareness and Mobilising the Population

This level is crucial and inescapable by the fact that it has both anchored the actions of AETA, legitimised its intervention, and promoted the appropriation of the struggle for the democratisation of the country. Otherwise, elections would look like a simple, soulless mechanism—without giving true human and societal significance to all actions undertaken by AETA, which propagate information awareness through its members and partners with the aim of mobilising energies of the grassroot communities to being involved in advocacy and pressure campaigns aimed at influencing political and technical decisions related to democratisation. Pro-democracy movements have contributed greatly to the growth and effectiveness of this dimension of action. They were one of the main strategic areas of AETA's intervention.

However, many of these young people from citizens' movements have been victims of physical violence by the government to the point that many have lost their lives and had to be imprisoned. One can think of the martyrs Rossy Mukendi Tshimanga, a young member of the dynamic Collective 2016 and member of the network of Cocorico citizens, and Thérèse Kapangala, who were viciously shot dead during demonstrations of the citizens who demanded that elections take place and there be the peaceful transfer of power in February 2018. The Cocorico network of citizen movements, which integrates seventeen dynamic youth groups based in Kinshasa, the capital of the DRC, was set up under the educational support of AETA in 2016.

AETA has also acted as an awareness platform to awaken the socio-political consciousness of the DRC people regarding the cases of risks of human rights violations in the electoral context characterised by tensions, terror, and violence perpetrated by the ruling regime. For example, two young people from the citizen movement Cocorico who witnessed the murder of the pro-democracy activist Rossy Mukendi, during the February 2018 citizen protest march against President Joseph Kabila's third term in the Catholic parish of Saint Benoît in the commune of Lemba in Kinshasa, were subjected to intense threats of being arrested by the police and security service. These threats were intended to obstruct justice by dissipating their testimonies and concealing the evidence of that assassination. The tracking down of the youths in political revolution in the DRC, especially those involved in activism, went on, despite the risks that they faced, to be arrested or to be killed, and that led to political and social change, as one[7] of them commented: "With Mr Kabila's promise

[7] Statement from Alain Mulumba Kabeya, a 28-year-old Filimbi leader, Kinshasa, 2018.

to step down, 'We can say that we have won one battle,' he said. 'But our fight is not yet over'" (Chick 2018, n.p.).

It is worth mentioning that AETA took care of the two young people whose lives were under imminent danger, and exfiltrated them with the help of the technical and financial partnership and MONUSCO outside Kinshasa. It should be noted that the Secretary General of AETA himself was arrested on 30 June 2017 because of his remarkable commitment to holding a credible, transparent voting system and the peaceful transfer of power. These different levels of action reflect AETA's approaches to the democratisation of the DRC.

11.5.2 The Real Impact of AETA's Actions

Significant results that have impacted democratic change in the DRC stem from AETA's actions as part of its commitment to the consolidation of democracy. These demonstrate the leverage that civil society can provide in the democratisation of a nation. Six of them deemed essential are listed as follows:

- The "exoterisation"—making electoral professionalism suitable for people generally, not only for some people—has been carried through by encouraging civil society actors to develop interest in and be trained on the electoral issues. Because the administration and delivery of the polls' results have long been regarded as an esoteric philosophy and praxis—reserved to external experts—AETA has put structures to make voting activities a subject of study and specialisation within CSOs.
- The School of Electoral Training in Central Africa (EFEAC), which is an initiative of the late Father Apollinaire Malu Malu (a DRC Catholic priest and activist) who presided over the Independent Electoral Commission (CEI) (2003–2006) and the Independent National Electoral Commission (CENI); he was the first president of the European Center for Electoral Support (ECES) which he founded, until 2013, and in 2014, he was appointed secretary of the Electoral Knowledge Network of Central Africa. He resigned from his position of the CENI's president in the 2015 Independent Electoral Commission (IEC). While Fr Malu Malu excelled in his political career, there remain many unanswered questions about the true winner of the 2006 and 2008 elections which were allegedly characterised by massive fraud and irregularities. He represented the civil society as a priest. This indicates clearly how civil society has been active in the DRC's political struggle, especially regarding democratic processes and the transfer of power in transitional periods.
- AETA's approach to technical and scientific analyses of the way in which the polls are conducted has made this platform rise to the level of a key player in the search and pursuit of relevant electoral options that befit the political, legal, socio-anthropological, economic, and financial challenges of the DRC's context, simultaneously ensuring that the polls take place in accordance with the constitution of the country and remain consistent with international democratic standards,

namely, fairness, transparency, and freedom of the electorate to have all their votes counted.

- AETA has played an important role towards the implementation of two legal proposals regarding the texts related to the DRC's voting system. These entail the texts relating to the Organic Law 2013 on the Independent National Electoral Commission (CENI)[8] and the National Committee for the Monitoring of the Inclusive Political and Comprehensive Agreement of 31 December 2016. The two proposed laws initiated by AETA, and which were sent to the legislators on the responsibility of two national elected officials, are an immeasurable contribution to the democratisation of the DRC.
- The CNSA secured the release of political prisoners and activists who voiced their opinions, the consensus among political actors, facilitated the return of political exiles, promoted cohesion and peace between different national communities, and it carried through its monitoring and controlling mission over the holding of credible, transparent, just, soothed, and inclusive elections.
- Opinions and recommendations made by AETA provide relevant and objective insights that impact and guide electoral advocacy and political decisions in favor of improving the DRC's electoral processes. This consists of a study on electoral calendars, reviewing the electoral legal framework, monitoring the implementation of electoral operations, observing the audit of the electoral enrolment files and ballots, public monitoring of the budget of the electoral process, monitoring of the acquisition of electoral materials, etc.
- Various advocacy actions carried out by AETA, which involved competing political actors, paved the way for the transformation of the DRC's political environment, and this was supported by efforts produced by two electoral consultations and political agreements: the first that took place on October 2016 at the city of the Organisation of African Unity (OAU) which was mediated by the late Togolese politician Mr Édouard Kodjovi Kodjo (Edem Kodjo) who war the Secretary-General of the Organisation of African Unity from 1978 to 1983; the second refers to the 2016's New Year's Eve Global and Inclusive Political Agreement knowns the Saint Sylvester's Peaceful Transitory Deal that was mediated by the Catholic Church's leadership to salvage DRC's democratic process.
- The production and distribution of information materials that explained the circumstances and stakes involved in the extension of the date for holding general elections, and the implementation of the civic and electoral education program, have contributed to the electoral process's ownership by the Congolese population and its mobilisation through pressure actions in favor of holding the polls as required by the constitution.

[8] The Independent National Electoral Commission (CENI) is an institution supporting democracy, enshrined in the Constitution of the Democratic Republic of Congo (Article 211). It is governed by the Law n ° 13/12 of 19 April 2013 amending and supplementing the Law n ° 10/013 of 28 July 2010 on the organization and functioning of the National Independent Electoral Commission and by its Internal Regulations; at: https://www.devex.com/organizations/commission-electorale-nationale-independante-ceni-congo-135283. Accessed 22 December 2020.

These various interventions demonstrate the capacity of AETA to advance the DRC's democratic agenda and saw to the materialization of a successful outcome of the democratic process in 2011 and 2018. That conforms with any civil society organization's interventions at the spheres of control and influence; the results of which converge on sustainable alternatives to democratic change. At the center of the actions are the rights holders who are the population. The interactions of the three main actors in state governance (governance, and private and civil society) should primarily benefit rights holders.

11.6 Conclusion

Three decades after its launch, representative democracy is struggling to take off. Such a failure hinders any efforts deemed necessary to speed up institutional, political, economic, and socio-cultural reforms that have been lagging since the democratisation process started on 24 April 1990. Three decades later, the DRC's history and socio-political and economic life have remained synonymous with catastrophe and tragedy. Functional democracy is yet to be seen, touched, and experienced when people decide who should be given the mandate to represent them, and where the rule of law braces power separation in a real democratic state. Several trials of implementing democracy in the DRC have engendered non-democratic results and the use of political clientelism, human rights violations, and military muscle to quell protests and coerce people to accept that which they are forced to take. As a result, following the first three elections since the promising new dawn announced by Mobutu in 1990, the DRC's development and institutional reforms have stalled. The logic of the establishment to loot public treasuries; to use, permit, and condone bribery and corruption to leverage the status quo; and to institutionalise impunity in the management of the *res publica* have led to the demise of democracy and the perennation of dictatorship.

In this tumultuous journey towards the honest democratisation of institutions in the DRC, it was demonstrated throughout this work that civil society has painfully toiled to safeguard the narrow part through which real democracy remains possible and the Congolese people can live out this dream. In countries where democracy is emerging, especially in Sub-Saharan Africa, political development goes along with economic growth. Thus, social disasters that trigger the lack of faith in democratic ideologies are also described as the twilight of gnawed democracies, with the word *gnawed* revealing the syndrome of democratic fatigue (Van Reybrouck 2014). However, the DRC, through its CSOs, appears resilient and determined to overcome despondency and that rigged and manipulated polling results could get settled in the Congolese people's global consciousness.

This chapter had the task to support the relevance of civil society action, and particularly AETA, to strengthen democratic processes, facilitate political transition, and intervene for the peaceful transfer of power. We can infer that CSOs and AETA have a voice and place in the current political endeavor and transformation that should transfer power back to the Congolese people such that they are empowered and hold the prerogative to freely choose their representatives and hold them accountable and have the power to hire whom they wish in the parliament, the senate, and as the head of the State. This is an eloquent evidence of the inescapable role of civil society as it stands out from the simple theorisation based on the nomothetic approach (Gazibo/Mamoudou 2010). It follows that CSOs should continue to monitor the Government's action to create an environment that is conducive to peaceful electoral processes, trusted voting outcomes, and soothed transition which are critical to the economic development and social welfare. Economic growth, job creation, advancing vibrant infrastructure, health, education, housing, security, and sustainable development rely considerably upon a living democracy, peaceful transition processes, and the stability of institutions that seek to attain and promote the common good.

The turbulences and nightmares of past violence and civil wars resulting from the lack of power sharing have been quietened by the resiliency of CSOs. The last transfer of power between Joseph Kabila and Felix Antoine Tshisekedi seems to have inaugurated a new era and democratic culture, despite the claims by several observers that the results of the 2018 polls were tainted with fraud and that the victory was attributed to the candidate who did not truly win. The trials of the new era for the DRC under Felix Tshisekedi Tshilombo are still to produce desirable and durable economic out-turn, peace and security, social development and welfare, and provisions of more decent and deserved livelihoods and create better opportunities for the people, eradicating unfair poverty and bringing the country back to the path of global growth.

This path of development relies substantially upon a functional democracy that AETA has been promoting via its commitment to the Quadruple-level Democratization Monitoring Process, namely (1) reflexively, for the benefit of its members and other stakeholders; (2) for decision-makers and other key players; (3) with grassroots dynamics; and (4) for communities. AETA's role is vital to the democratization of the DRC, despite the hostile political environment, the prevailing insecurity and the perilous conditions under which CSO actors operate in this country. Thus, the institution of AETA, alongside the free activism of citizens' movements, remains a key factor for the advancement and fostering of democratic values, exercise and praxis in the DRC.

References

Agir pour les Elections Transparentes et Apaisées (AETA) (2011). Brève présentation de la plateforme AETA. Kinshasa: AETA [Unpublished].

Agir pour les Elections Transparentes et Apaisées (AETA) (2012). Rapport final du programme d'accompagnement du processus électoral crédible, transparent et apaisé en RDC. Kinshasa: AETA [Unpublished].

Agir pour les Elections Transparentes et Apaisées (AETA) (2012–2019). Consolidation des analyses techniques réalisées par l'AETA sur le processus électoral 2012–2019 (manuscript). Kinshasa: AETA [Unpublished].

Agir pour les Elections Transparentes et Apaisées (AETA) (2017a). Code de bonne conduite des membres en période des élections. Kinshasa: AETA [Unpublished].

Agir pour les Elections Transparentes et Apaisées (AETA) (2017b). Statuts modifiés portant création et fonctionnement de l'AETA. Kinshasa: AETA [Unpublished].

Ahmed, J.U. (2010). Documentary research method: New dimensions. *Indus Journal of Management & Social Sciences* 4(1), 1–14.

Aid to the Church in Need (ACN) (2018). *The Democratic Republic of the Congo: 12 million war casualties in the battle over natural resources*; at: https://acninternational.org/the-democratic-republic-of-the-congo-12-million-war-casualties-in-the-battle-over-natural-resources/. Accesed 20 Jan 2021.

Allen, J.A. (1994). The constructivist paradigm: Values and ethics. *Journal of Teaching in Social Work* 8(1–2), 31–54.

Andreevska, E. (2013). Transitional justice and democratic change: key concepts. *Lex et Scientia International Journal* 20(1), 54–64.

Arbour, L. (2007). Economic and social justice for societies in transition. *NYUJ Int'l L. & Pol., 40*(1), 1–27.

Babunga, B. (2018). *Le 23 février 1994, le Haut Conseil de la République – Parlement de transition (H.C.R.-P.T.) se réunit pour la première fois au Zaïre*; http://www.babunga.alobi.cd/2018/02/22/le-23-fevrier-1994-le-haut-conseil-de-la-republique-parlement-de-transition-h-c-r-p-t-se-reunit-pour-la-premiere-fois-au-zaire/. Accessed 21 Jan 2021.

Baines, E.K. (2007). The haunting of Alice: local approaches to justice and reconciliation in Northern Uganda. *International journal of transitional justice* 1(1), 91–114.

Belloni, R. (2001). Civil society and peacebuilding in Bosnia and Herzegovina. *Journal of peace Research* 38(2), 163–180.

Belloni, R. (2008). Civil society in war-to-democracy transitions. In A. K. Jarstad & T. D. Sisk (Eds.), *From war to democracy: Dilemmas of peacebuilding* (pp. 182–210). Cambridge: Cambridge University Press. https://doi.org/10.1017/CBO9780511755859.

Bentahar M (2014) Quelques rappels sur la société civile, sa place son rôle et son lien avec la démocratie. https://blogs.mediapart.fr/m-bentahar/blog/141114/quelques-rappels-sur-la-societe-civile-sa-place-son-role-et-son-lien-avec-la-democratie. 14 Nov 2014.

Bratton M (1994) Civil society and political transition in Africa. *IDR Reports* 11(6), 1–21.

Carothers T and Barndt W (1999) Civil society. *Foreign policy* Winter, 1999–2000(117), 18–29.

Centre d'Etudes et de Documentation et d'Animation Civique (CEDA) (2003) *Accord Global et Inclusif sur la Transition en République Démocratique du Congo [Global and Inclusive Agreement on the Transition in the Democratic Republic of Congo]. Signé à Pretoria (République d'Afrique du sud), le 16 décembre 2002 [Signed at Pretoria (Republic of South Africa), on December 16, 2002]* https://repositories.lib.utexas.edu/bitstream/handle/2152/4526/3626.pdf?sequence=1&isAllowed=y. Accessed 23 Jan 2021.

Chick, K. (2018). In Democratic Republic of Congo, activists face uphill struggle: Despite risk of arrest or death, many groups have sprung up to press for political and social change. *International Women's Media Foundation, December 17, 2018 10:00 PM*. Retrieved from https://www.thenationalnews.com/world/africa/in-democratic-republic-of-congo-activists-face-uphill-struggle-1.803860. Accessed 29 Jan 2021.

Clowes, W. (2016). *Police Thwart Youth Protest Against DRC's Kabila*. https://www.voanews.com/a/police-thwart-youth-protest-against-kabila/3613179.html. Accessed 16 March 2022.

Coghlan, B., Brennan, R.J., Ngoy, P. et al. (2006). Mortality in the Democratic Republic of Congo: a nationwide survey. *The Lancet* 367(9504), 44–51.

Colombo, A., d'Aoust O., Sterck, O. (2019). From rebellion to electoral violence: Evidence from Burundi. *Economic Development and Cultural Change* 67(2), 333–368.

Congo Forum (2009). *La société civile en R.D.Congo.* https://www.congoforum.be/fr/2009/03/la-societe-civile-en-r-d-congo/ Accessed 10 July 2020.

Correa, D.D.S., & Owens, W.R. (2010). *The handbook to literary research.* London: Routledge.

Curle, A. (1971). *Making peace.* London: Tavistock publications.

Cvetek, N., & Daiber, F. (2009). *Qu'est-ce que la societé civile?* https://www.issuelab.org/resources/20204/20204.pdf. Accessed 21 Dec 2020.

Democratic Republic of Congo (2001) *Loi N° 004/2001 Du 20 Juillet 2001 Portant Dispositions Générales Applicables aux Associations Sans But Lucratif et aux Etablissements d'Utilité Publique [Law N ° 004/2001 of July 20, 2001 on General Provisions Applicable to Non-Profit Associations and Public Utility Establishments].* http://www.leganet.cd/Legislation/Droit%20Public/loi0042001.20.07.2001.asbl.htm. Accessed 20 Dec 2021.

Gazibo, M. (2006). *Introduction à la politique africaine.* Québec: Les Presses de l'Université de Montréa.l.

Hall, A., & LaRocco, A. (2012). *Time works against justice: ending impunity in eastern Congo.* http://www.enoughproject.org/files/Ending%20Impunity%20in%20Eastern%20Congo.pdf. Accessed 12 Dec 2017.

Hamuli Kabarhuza, B., Mugumo, F.M., Shuku, N.Y. (2003). *La société civile congolaise: État des lieux et perspectives.* Brussels: Colophon.

Heaton, L. (2010). *Congo: A death toll rivaling the Holocaust.* www.enoughproject.org/blogs/congo-death-toll-rivaling-holocaust. Accessed 24 Nov 2020.

Hölkeskamp, K.-J. (2000). The Roman Republic: Government of the People, by the People, for the People? *Scripta Classica Israelica, 19*(2000), 203–223.

Human Rights Watch (2018). DR Congo: Security Forces Fire on Catholic Churchgoers. https://www.hrw.org/news/2018/01/19/dr-congo-security-forces-fire-catholic-churchgoers. Accessed 23 Nov 2020.

Jaysawal N (2013) Civil society, democratic space, and social work. *SAGE Open* 3(4), 1–2. https://doi.org/10.1177/2158244013504934.

Jeong, H-W. (2005). *Peacebuilding in postconflict societies: Strategy and process.* Boulder: Lynne Rienner.

Kouassi, C. (2012). *Société civile en RDC: sa naissance, son parcours, ses combats.* https://fr.africanews.com/2018/12/19/societe-civile-en-rdc-sa-naissance-son-parcours-ses-combats/. Accessed 12 Oct 2021.

Last, D.M. (2010). Civil Society and Peacebuilding. Routledge Encyclopedia of Peace. http://dx.doi.org/https://doi.org/10.2139/ssrn.2459768.

Laudani, R. (2012). Une démocratie a reconstruire: aux origines de la société civile. *Le Monde diplomatique, Septembre 2012.* https://www.monde-diplomatique.fr/2012/09/LAUDANI/48145. Accessed 25 Dec 2020.

Lauterbach, F. (2015). *Electoral violence in Africa.* https://www.academia.edu/38162543/Electoral_Violence_In_Africa_Political_Violence_And_Intimidation_As_An_Electoral_Strategy. Accessed 23 Jan 2021.

Lederach, J.P. (1997). *Building Peace: Sustainable Reconciliation in Divided Societies.* Washingston, D.C.: United States Institute of Peace Studies.

Leroux, S. (2018). Ce jour-là: le 16 février 1992, la « marche de l'espoir » est réprimée par Mobutu Sese Seko [That day: February 16, 1992, the "March of Hope" was suppressed by Mobutu Sese Seko]. https://www.jeuneafrique.com/531192/politique/ce-jour-la-le-16-fevrier-1992-la-marche-de-lespoir-est-reprimee-par-mobutu-sese-seko/. Accessed 23 Jan 2021.

Mayeda, D. (2012). DRC death toll exceeds five millin and almost no news and without coverage? http://www.sociologyinfocus.com/2012/11/26/aware-of-ongoing-violence-democratic-republic-of-the-congo/. Accessed 23 January 2021.

Mbombo, J-M.K. (2018). Congo: Nonviolent Struggle in the DRC. Making Sense of the Consent Theory of Power. *Conflict Studies Quarterly* October 2018(25), 49–67. https://doi.org/10.24193/csq.25.3.

McLaverty, P. (2002). Civil society and democracy. *Contemporary Politics* 8(4), 303–318. https://doi.org/10.1080/13569770216068.

Miles, M.B., & Huberman, A.M, (1994) *Qualitative data analysis: An expanded sourcebook.* Thousand Oaks: Sage Publications.

Moustakas, C. (1994) *Phenomenological research methods.* Thousand Oaks: Sage Publications.

Mudumbula, M.G. (2013). La problématique de la représentation unique de la société civile congolaise: enjeux et défis. Actes de la huitième synergie des ONGs de développement de Kinshasa, N° 8, Kinshasa, Novembre 2013. Kinshasa [Unpublished].

Mulegwa, P. (2018). RDC: l'Église catholique réclame de « continuer le combat » contre Kabila *Jeune Afrique*, 09 février 2018 à 16h09, Mis à jour le 27 février 2018 à 15h34. https://www.jeuneafrique.com/529236/politique/rdc-leglise-catholique-reclame-de-continuer-le-combat-contre-kabila/. Accessed 23 Jan 2021.

Mushi, M.F., & Kamavu, N. (1999). RDC, Bilan commun des pays: Organisation de la société civile. Bureau de Zone de Kinshasa: Organisation Internationale du Travail [Unpublished].

National Technical Secretariat of Civil Society (2013). Act of the first national civil society forum in DR Congo. Reportno [unpublished].

Nilsson, D. (2012). Anchoring the peace: Civil society actors in peace accords and durable peace. *International Interactions* 38(2), 243-266. https://doi.org/10.1080/17448689.2018.1484004.

Nsapu, P. (2018). Floribert Chebeya et Fidèle Bazana: les assassins courent toujours. https://www.jeuneafrique.com/565230/societe/tribune-floribert-chebeya-et-fidele-bazana-les-assassins-courent-toujours/Report Numberl, Accessed 23 Jan 2021.

Paffenholz, T. (2015). Civil society and peacebuilding. *Development Dialogue* 63, 108–118.

Paffenholz, T., & Spurk, C. (2006). Civil society, civic engagement, and peacebuilding. *Social Development Papers: Conflict Prevention and Reconstruction* 36(2006), 10–55.

Pelczynski, Z. (1984). *The State and Civil Society: Studies in Hegel's Political Philosophy.* CUP Archive.

Perera, S., Kambale, V. A., & Bussy, J. M. (2018). Youth Participation and Non-Violent Resistance in the Democratic Republic of the Congo: The Case of LUCHA. DLP Research Paper 55, Birmingham: Developmental Leadership Program / Dar-es-Salaam. https://media.africaportal.org/documents/Research-Paper-54.pdf. Accessed 30 Sept 2021.

Ritimo Network (2018). *Au Congo RDC: "La Lucha" et "Filimbi".* https://www.ritimo.org/Au-Congo-RDC-La-Lucha-et-Filimbi. Accessed 23 Jan 2021.

Tarif, M., & Vircoulon, T. (2016). *Transitions politiques: Les déboires du modèle de sortiede crise en Afrique.* Paris: Études de l'Ifri.

Towheed, S. (2009). Tools and techniques for literary research: Using online and printed sources. In D.dS Correa & Owens, W.R. (Eds.) *The Handbook to Literary Research* (pp. 9–35) 2nd ed. London: Routledge.

United Nations. (2004). The rule of law and transitional justice in conflict and post-conflict societies: Report of the Secretary-General. Retrieved from http://www.un.org/en/ga/search/view_doc.asp?symbol=S/2004/616.

United Nations (2009) Rule-of-Law Tool for Post-Conflict States. Amnesties. http://www.ohchr.org/Documents/Publications/Amnesties_en.pdf. 23 Jan 2021.

Van Reybrouck, D. (2014). *Contre les élections, traduit du néerlandais par Isabelle Rosselin et Philippe Noble.* Arles: Actes Sud.Warren, M.E. (2011). Civil society and democracy. In M. Edwards (Ed.) *The Oxford handbook of civil society.* Oxford Handbooks Online. https://doi.org/10.1093/oxfordhb/9780195398571.013.0030.

Yonekawa, M. (2012). Elections and democrazy in the Democratic Republic of the Congo: the dilemma of peace, justice and impunity. *Southern African Peace and Security Studies* 1(1): 77–83.

Zabondo, G.D. (2019). Démocratie et Ethique: Emancipation Politique et Sociale de L'Afrique [Democracy and Ethics: Political and Social Emancipation of Africa]. Paris: Editions L'Harmattan.

Zacharie, A. (2013). Mondialisation: qui gagne et qui perd. *Essai sur l'économie politique du développement, Bruxelles, Le bord de l'eau (coll.«La muette»).*

Zacharie, A. (2019). Mondialisation et national-populisme. La nouvelle grande transformation. Lormont, Bord de l'eau. https://www.cncd.be/Mondialisation-et-national. Accessed 1 Aug. 2021.
Zagzebski, L.T. (2015). *Epistemic authority: A theory of trust, authority, and autonomy in belief.* Oxford: Oxford University Press.

Chapter 12
Civil Society Organisations and School Peace Clubs in South Africa: An Outcome Evaluation

Dorothy Moyo

Abstract School peace clubs are a recent form of peace architecture. This chapter reports a qualitative outcome evaluation of a project involving 15 peace clubs set up by civil society organisations in the city of Pietermaritzburg, South Africa between 2017 and 2019. The evaluation examined the outcomes for learners and teachers – specifically, whether the peace clubs had resulted in them becoming peace agents and role models – and whether the schools in which the peace clubs were based had become less conflictual and violent. Learners live in violent environments and violence is the main way they use deal with conflicts. Peace clubs have educated learners to resolve their conflicts non-violently and to act as mediators at school and, at times, at home. This outcome seems to derive from both the subject matter from the peace club's curricula and the participative and experiential methods of learning which are little used elsewhere. While few teachers were willing to act as the peace clubs mentors, those who did so reported important personal benefits. Principals and teachers reported significantly lower levels of conflict and violence in their schools, which they attributed to the operation of the peace clubs. These results should encourage civil society organisations to engage in school peace club initiatives.

Keywords School violence · Peace clubs · South Africa · Outcome evaluation · Alternatives to Violence Project

12.1 Introduction

South Africa is one of the world's most violent countries with the homicide rate of 36.40 per 100k people in 2022—representing the eighth in the world, excluding accidental deaths and incidences of "self-inflicted murder" (suicide) (World Population Review 2022). South Africa ranks fourth in the world in crime index (76.06 per 100k people) in 2022 after Venezuela, Papua New Guinea and Afghanistan (Numbeo 2009–2022). A very high proportion of its violence is perpetrated by

---

Ms. Dorothy Moyo, PhD student in Peace Studies, Durban University of Technology; the author acknowledges the inputs of Professor Geoff Harris of Durban an University of Technology to this research project.

J. C. K. Kiyala and G. T. Harris (eds.), *Civil Society and Peacebuilding in Sub-Saharan Africa in the Anthropocene*, The Anthropocene: Politik—Economics—Society—Science 34,
https://doi.org/10.1007/978-3-030-95179-5_12

husbands, fathers and young men against women and girls, and against each other. The country is often referred to as the 'rape capital of the world', with 50 000 to 55 000 rapes and attempted rapes reported to police each year – and these are regarded as being just the tip of the iceberg. A respected estimate by Gender Links and the Medical Research Council (2010) is that only one in 25 victims report such crimes to the police.

South Africa's schools mirror the intense violence in wider society. A nationally representative survey carried out in 121 high schools in 2012 found that 22.2% of learners had been threatened with violence or had experienced assault, robbery and/or sexual assault in the preceding 12 months (Burton and Leoschut 2013). Another study (Mncube/Harber 2013) investigated violence in 24 schools across six provinces and found that 55% of learners had experienced violence at school, with 28% indicating that this violence was a daily occurrence. Not surprisingly, many learners regarded their schools with fear and apprehension.

School peace clubs are a relatively recent peace infrastructure, originating in Zambia around 10 years ago and now operating in a number of African countries. They are aimed at motivating learners to become peacebuilders in their schools and to equip them with the skills to do so effectively. The Mennonite Central Committee (MCC) – a faith based CSO – has been an important promoter and funder of peace clubs and have produced a range of curricula with a strong experiential and participatory emphasis on issues affecting learners in Zambian and South African schools, namely unwanted pregnancies, bullying, sexual acts between teachers and learners, etc. (Juma 2019). Peace clubs operate in a number of African countries and some of these are documented on NGO websites. However, there is very little by way of academic literature. A search of the *Google Scholar* and *Academic Search Complete* databases in mid-2019, using 'peace clubs' and 'Africa' as key words, produced less than a handful of relevant publications. Juma (2019) outlines the history of school peace clubs in Africa and provides an overview of their operation in a number of countries; Irene (2016) described an action research project with peace clubs in Nigerian schools; and Gulliksen (2015) examined the nature and extent to which children who had been involved in peace clubs in northern Uganda subsequently engaged in peacebuilding activities and the challenges they faced.

In 2012, the MCC began peace clubs in Pietermaritzburg, a city of some 600 000 (including adjacent townships) in KwaZulu-Natal (KZN) province. A second CSO involved in the project was the Alternatives to Violence Project (AVP), a training programme devised by Quakers in the United States in the 1970s to help prison inmates handle their conflicts non-violently. AVP has since expanded outside the prison context and into many other countries (John 2016). AVP-KZN had been conducting AVP workshops in secondary schools since 2003.

In 2016, the MCC funded AVP-KZN to set up school peace clubs, beginning with four clubs inherited from the earlier MCC initiative. The plan was to build up the number of schools with peace clubs to 34, with each facilitator handling between seven and nine schools. In March 2019, when the evaluation was carried out, there were 15 functioning peace clubs, 13 operating in schools and two with NGOs. Six other clubs had begun but had stopped, most because there was no teacher willing to assist. Of the 13, five were in primary schools and eight were in secondary schools.

The project employed four part-time facilitators working 15 hours during each school operating week, each of whom was assigned to several of the 13 school peace clubs, and a full-time project coordinator.

The evaluation reported in this article covered the three years 1 April 2016 to 31 March 2019 and was based on the intended outcomes of the project, which can be summarised as follows:

- Peace club learners will become peace agents and role models.
- The teachers supporting the peace clubs will become peace agents and role models. There will be a growth of peace identity in schools and in the wider community.

12.2 Research Methods

Three primary schools (PS1, PS2 and PS3) and four secondary schools (SS1, SS2, SS3 and SS4) were selected as 'typical' by the peace club's coordinator from the 13 schools for the evaluation; a summary of their characteristics is presented in Table 12.1. The four facilitators were asked to rate the performance of the selected clubs; one was rated as strong, four as mixed and one as weak. In the event, SS4 was not included in the study because of a learner's suicide the day before planned visit.

Table 12.1 Characteristics of the sampled schools. *Source* The authors

	Grades	Meeting frequency and time	Size of peace club	Size of school
Primary school 1 (PS1)	Grades 5–7	Once per week for an hour during cleaning period	20	590
Primary school 2 (PS2)	Grades 6 and 7	Class based, each class having a lesson once a month during the reading period	400	1837
Primary school 3 (PS3)	Grade 6	Once a week after school for 30 minutes	20	300
Secondary school 1 (SS1)	Grade 11	Break times and after school	30–40	900
Secondary school 2 (SS2)	Grades 10 and 11	Twice per week during sports period	14	325
Secondary school 3 (SS3)	Grades 8–12	Twice per week after school	50	1400

Data was collected from three sources:

- The AVP facilitators' files, supplemented with data provided by the coordinator.
- Interviews and focus group discussions with learners (76, in six school groups), principals (5), and interviews with teachers associated with peace clubs (7), the peace club facilitators (4) and the peace club coordinator.
- Observation and informal discussions.

The questions asked during the interviews were designed to assist in determining the extent to which the intended outcomes were met. Given that the interviews were conducted quite early in the school year, the responses largely referred to the experiences of 2018 and previous years. Very little quantitative data was available and no school kept an 'incident book' in which instances of indiscipline were recorded.

We utilised thematic content analysis (Braun/Clarke 2006; Nowell et al. 2017) to interpret the qualitative data we collected. That is, we identified themes which kept on emerging from the various data sources and focussed on trying to understand their various nuances. We found the use of direct quotations helpful in this process; unless otherwise stated, quotations reflect common experiences and widely held opinions.

12.3 Results and Discussion

Each peace club operates in its own distinctive way. Some only comprise learners with behavioural difficulties, others have volunteers only and some are mixed. Others have representatives from each class. As shown in Table 12.1, some meet during a formal class period while others meet during breaks or after school. Some have considerable facilitator and/or teacher involvement while others have little or none.

Primary and secondary learners are different. Secondary school learners can self-organise and can run the club with little outside support, whereas primary school learners need hands on assistance. Personal sharing plays a central role in all clubs, as does experiential learning based on the peace club curriculum booklets and extensive use of role plays.

12.3.1 Outcome 1: Peace Club Learners are Pace Agents and Role Models

Theme 1. Learners have a lot of anger and frustration, reflecting the intensity of violence in their communities

Learners have experienced and do experience violence in their homes and communities on a regular basis. At each of the sampled schools, learners said that they were affected by violence. Homes and communities do not model or provide training in peaceful methods of resolving conflicts. The violent ways of handling conflicts which are practiced teach learners that violence is normal and appropriate. Many learners bring this 'education', along with their anger and frustration, to school and vent it on others through fighting, bullying, insults and harassment.

A teacher at PS2 observed that learners come to school full of anger and aggression. At home, they are forced to do things they do not like, are spoken to harshly and feel disrespected by their parents and this is often exacerbated at school where they are teased and bullied. Learners from PS1 come from particularly disadvantaged homes and communities characterised by domestic violence, alcohol and substance abuse and child-headed households. The parents of most learners are young and are not involved at the school. A teacher from PS1 indicated that parents of most of their learners do not have the time or skills to support their children: 'There is just no parental involvement in the school. Parents don't come to the school when asked and if things get really bad [for their children], they change schools'.

Theme 2. Violence is the main way which learners use to deal with conflicts

As a result of growing up in violent communities, the majority of learners have been 'educated' to believe that conflicts can only be resolved through violence or threat of violence. A peace club learner from SS1 summed up this way: 'When it comes to handling violence, the only way most learners know is to use violence'. Another learner from PS2 summarised this understanding as follows:

> I used to be a fighter and whenever anyone at school wronged me, I would hit back. I would beat up people who irritated me with whatever my hands could get hold of, such as sticks ... [But now] I have learnt from the peace club that fighting does not help and conflicts can be resolved through dialogue.

A facilitator pointed out that before the introduction of peace clubs learners only knew that in conflict they must retaliate. However, through participation in peace clubs, learners now know that they can resolve conflicts non-violently and that there are alternatives to violence.

A learner at SS3 told how she was involved in an altercation with her friend a week back, during which her friend slapped her. Her natural response would have been to hit back but she reminded herself that she was a member of the peace club and that she should act differently. She calmed down, sent her friend a peacemaking message the following day and they later resolved their differences amicably.

It should be noted that the learner views reported here are those of 'the converted' and are their opinions about how other learners think and behave. They are most legitimate when speaking about their own experiences and (in theme 3) the differences which peace club learning has made in their own lives.

Theme 3. Peace clubs provide a safe space for learners to share their experiences, gain some healing and develop a sense of group belonging

At almost all schools, learners indicated that peace clubs provided them with a safe and confidential platform where they could share their experiences without fear of being judged or censured. They felt safe to tell their stories of anger, violence, drugs and alcohol that are rife in their homes and communities. For most learners, there was no alternative place for such sharing. Even if they told their parents that they were bullied or harassed at school, the parents did not pay attention. In the peace club, such issues are listened to and action is taken where necessary.

A learner from SS1 testified about the safety aspect of peace clubs:

> When I joined the peace club, I had a lot of issues around my life. I had internal conflict. I have now learnt to resolve my internal conflicts. The peace club enabled me to open up in a safe space and share what was troubling me. I realised that I was not alone and that other members of the peace club actually had more problems than what I had. The peace club helped me to heal from the hurt that I had.

The Principal of PS1 observed that the facilitator was regarded as a father figure or uncle and learners felt free to open up to him and tell their stories. This is against the background that homes and families do not give their children any psycho-social support. Peace clubs, it seems, fit the 'supportive group psychotherapy' model identified by Kaminer/Eagle (2010), where people suffering the same kinds of trauma receive support from each other by sharing common experiences and reactions. As a result, they 'gain authority over traumatic material so that it no longer becomes a dominant factor in their lives' (Kaminer/Eagle 2010, p. 107).

The group identity aspect is a strong factor in peace club success. The need for identity and belonging was mentioned by teachers and learners at all schools. In the words of a SS1 learner, 'A peace club is different to a class. If you belong to a peace club, you have to behave in a certain way and uphold what the peace club stands for. So you become accountable for your behaviour'. Many learners and teachers spoke of the importance of group membership for learners and the mutual support which the peace clubs offered. Several learners expressed a wish to have T shirts and badges with the peace club logo.

Theme 4. Peace club members have learnt to resolve conflicts non-violently and practice this at school and at home

There were some components of peace club curricula which many learners identified as major insights for them. Three which stood out were the distinction between conflict and violence, managing their anger and dealing with their conflicts using nonviolent, dialogue-based methods.

Learners at five of the six peace clubs had had lessons on anger management, which encouraged them not to bottle up their anger but to address the issues that made them angry through nonviolent means. The importance of dialogue in resolving conflicts was mentioned by learners at all the schools. They stressed the fact that they had realised that violence such as fighting did not solve anything and that it bred resentment and more anger.

A strong commitment to tolerance and respect ran like a thread through group discussions with learners, as evidenced by the following stories:

> Before I joined the peace club, I did not respect other people. I looked down on the disabled, the deaf and people leaving with albinism. I used to laugh at them and did not consider them as human beings. Now I respect all people (Learner, SS2).

> Growing up, I learnt to help people if they showed weaknesses to me. [But] I ended up having a lot of enemies who saw I was helping them because they were helpless and not because I really cared. Now I know that I must respect people because they are people and not because I am somehow more able than them (Learner, SS2)

A number of secondary school learners had attended AVP workshops and praised the development of listening skills and good communication. The following story demonstrates how using the 'I messages' which are taught at AVP workshops and peace clubs helped the learner resolve a conflict at home:

> At home, people I stayed with used to take my clothes and wear them leaving them dirty. This used to make me angry and I had a very short temper. After I learned about 'I messages', I was able to express my feeling to these members of the family and talk to them. The issue was amicably resolved.

Theme 5. Peace club members are peer mediators at school and mediate in disputes at home

There is evidence that peace clubs are positively impacting both the schools and families. At SS3, during break time, six peace club members perform monitoring duties to intervene in conflict situations and prevent them from becoming physically violent. Their interventions have reduced drug taking and physical violence.

At PS2, peace club members mediate and help resolve conflicts at the playground: A key feature of peace club mediation is that the learners do listen to their peers. In emphasising this, the Principals of SS1 and SS2 noted the value of peer mediation in reducing fighting and bullying in the playgrounds.

The practice of mediation also happens in the home:

> My mother and brother always fought over the time that my brother was required to be home in the evening – the curfew. My brother was of the view that he could do as he pleased because he was now over 18. My mother, on the other hand, was of the view that the house was hers. One day a fight between my mother and brother erupted and a lot of vulgar language was used. I intervened and got my brother and mother to dialogue over the issue. At the end, both parties agreed to review the curfew and reached a compromise (SS1).

A grade five learner at PS1 told how he stopped a fight between siblings aged 20 and 16 years who were fighting over the sharing of money they had earned from a cleaning job. After separating them, he facilitated a dialogue where the parents were also involved. The resolution was that the two siblings used the money to buy one item and to share it.

Learners in all the schools spoke of the challenges they faced in getting adults to resolve conflicts non-violently. A SS1 learner illustrated this challenge as follows:

> It is difficult for us as young people to talk to older people about violence because they say. What do you know when you were born in the 2000s and I was born in the 1940s? They will tell you they fought in the liberation struggle against apartheid where violence brought 'peace'. It becomes difficult then to reach out to such elderly violent people with the message of peaceful settlement of conflict.

12.3.2 Outcome 2: The Teachers Supporting the Peace Clubs are Peace Agents and Role Models

Theme 1. Teachers have benefited personally from peace clubs. They champion peace clubs and AVP

Some teachers have been positively influenced by the content of the peace club curricula. One remarked how 'It makes you ask why you are reacting to something in a particular way' (Teacher, SS1). She explained how she used to shout at students but has stopped doing this following an experience concerning a learner who always came to school late. A peace club member came and explained to her the duties the learner had (minding cows) before he could come to school which explained why he was always late. This positive change in a teacher's behaviour resulted from communication between learners and teachers.

A teacher from PS1 explained how her peace club involvement has changed her:

> When I was asked by principal to assist with the peace club, I was sceptical of what was supposed to be my role. [But] as I met the facilitators, something in me simply changed. It must be their manner of approach or the humble manner they portrayed. I also attended the first AVP workshop they were running for peace club learners and teachers together with my school principal. Dealing with learners is a headache and one needs to be patient and tolerant. Peace club has assisted me to build a heart for young ones and AVP expanded my mind and creativity to an extent that I have stopped shouting at learners. I issue basic instruction as well as talk about things before we do them and talk about things after we have done them. I find that everyday my class is always happy.

The principal at PS1 regards the peace club matron as 'just passionate about her learners'. She deals with cases of disruptive learners, referring those with health issues and learning difficulties to a clinic or social worker. She pays for things herself and has done counselling workshops e.g. on bullying. To relieve her of pressure, she is allocated a lower teaching load. Students from the past come back to the school to visit her in appreciation of what she did for them.

The Principal and teacher at PS3 have a very positive attitude towards AVP, whose philosophy on nonviolence accords with that of the school. In 2018, the peace club facilitator came every week, which greatly impressed the principal. In his words, 'the facilitators have been wonderful'. The impression given was that AVP stimulated thinking which had led to other nonviolent initiatives, like a 'buddy bench' where learners in conflict are made to sit and dialogue and resolve their differences nonviolently.

The Principal from SS2 explained that he believed in second chances and accepted difficult learners who had been expelled from other schools. Most of the learners at his school have behavioral challenges and, working together with the peace club, he seeks to win the difficult learners. His focus was on 'building relationships' as a pathway to a better school. He believed that quiet engagement improves matters ('You can win people to behave better by talking to them'), whereas yelling at learners only makes them misbehave more. The peace club and AVP philosophy tie up with his hence his full support of the project.

12.3.3 Outcome 3: Growth of Peace Identity in Schools and the Community

Theme 1. The school is now more harmonious

The new understandings of conflict, violence and nonviolence and the changed behaviour of peace club members have, in the opinions of Principals and teachers, positively influenced the whole school environment. The Principal at PS1 saw change begin with peace club members themselves as they changed their mindsets regarding conflict and violence. Improved behaviour is most noticeable when club members were previously learners with behavioural difficulties – 'Now they are more apologetic and remorseful' (Teacher, PS1). In addition, they encourage each another to avoid negative behaviour and model better behaviour to the rest of the school. Other learners envy what they see and also keep peace club members up to the mark. The improved behaviour happens both in the playground and in the classroom. 'The peace club supports better learner behaviour; it has made a lot of difference. We see the change ... (Principal, PS1). A learner expressed the link between her own behaviour and that of other learners as follows:

> I now understand what *umuntu ngumuntu ngabantu* (a person is a person through other people) means. By living a life free of violence my school mates and those outside school have come to realize that violence is not good... Now that I don't use vulgar language, I have realized that my friends also do not use vulgar language when they are around me (Learner, SS1).

The Principal at PS2 reported a dramatic reduction in fighting among grade seven learners because they are now aware there are alternatives. Learners are more able to say no to drugs and other temptations. A teacher at PS3 observed that there was a huge reduction in fighting in 2018 when the peace club operated and there have been no major discipline issues in 2019.

Theme 2. There is some networking between schools and with communities

AVP workshops have enabled the sharing of experiences by learners from different schools and networking. In April 2018, learners from five secondary schools were trained together as AVP facilitators, having previously undertaken basic and advanced workshops. During the three years of the project, AVP basic workshops were conducted for community members in seven communities with participating schools. No data was collected concerning the outcomes of these initiatives.

12.4 Conclusion

From our research, it is clear that these peace clubs are a significant peace infrastructure. They deal with issues which are not discussed or examined elsewhere and their use of experiential and participatory learning is very uncommon in schools, churches and other organisations. These differences are one reason why peace clubs are so attractive to learners.

In our opinion, there is no doubt that the project has met the first two intended outcomes i.e. learners and teachers have become committed and effective peace agents. Learners with no prior understanding of handling conflicts nonviolently now do so, with confidence and effectiveness, at school and even in their own homes.

We noted that our research involved talking to the converted – to peace club members rather than non-members – and in schools where peace clubs had continued. However, had we interviewed non-members and in schools where clubs had ceased operation, we firmly believe that our conclusions would be the same. In addition, we did not interview members of the community in which the schools were located. It is possible that more peaceful school communities may be a result of influences other than peace clubs. That said, our strong impression is that the clubs have acted as a catalyst for other changes, including attitudinal changes among staff and Principals, which have also contributed to more peaceful school communities.

Regarding the growth of a wider peace community, we cannot expect observable change in the short term, unless community-based peace clubs or equivalent were established with a similar participative and experiential learning approach. In the longer term, we can reasonably expect less violent communities than otherwise as peace club members move into adulthood and become parents. However, given the strength of the inter-personal, inter-group and structural drivers of community violence, the peace club influence at community level will be modest.

In quantitative terms, the project has fallen well short of its target of having peace clubs in 34 schools, each with a committed teacher as its guide and with each facilitator looking after seven to nine schools. However, given the challenges facing schools and any NGOs which work within them, the quantitative achievements of the project to date can be regarded as acceptable.

A final point relates to the limited involvement of teachers. While they were very positive about the beneficial effects of peace clubs, few were willing to take on the responsibility of mentoring them. The number of paid facilitators therefore sets a limit to the number of clubs which can operate. Can peace club goals be achieved without adding more work to already over-burdened Principals and teachers? Is there a way of compensating teachers who volunteer for peace club work via a reduced teaching load? What training can be given to teachers who run peace clubs? One answer may be the Learner Support Agents. These are young graduates appointed by the Department of Basic Education to schools with high levels of 'social skills' to support learners with issues like teenage pregnancy, gender-based violence and child-headed households. They could provide an important supplement to CSO inputs.

References

Braun, V. and Clark, V. (2006). Using Thematic Analysis in Psychology. *Qualitative Research in Psychology,* 3(2): 77–101. https://doi.org/10.1177/146879410100100307.

Burton, B., & Leoschut, L. (2013). *2012 National School Violence Survey.* Cape Town: Centre for the Study of Violence and Reconciliation.

Gender Links and the Medical Research Council. (2014). *The War @ Home: Preliminary Findings of the Gauteng Gender Violence Prevalence Study.* https://doi.org/10.1191/1478088706qp063oa. Accessed 20 Jan 2020.

Gulliksen, C. (2015). *A Study of Children's Participation in Peacebuilding in a Post-Conflict Society: A Case Study of Peace Clubs in Gulu, Uganda.* PhD thesis, Makerere University. Repositório do ISCTE-IUL: A study of children's participation in peacebuilding in a post- conflict society: a case study of peace clubs in Gulu, Uganda (iscte-iul.pt). Accessed 5 June 2019.

Irene, O. (2016). Building Infrastructure for Peace in Nigerian Schools. *African Research Review,* 10(2), 178–193. https://doi.org/10.4314/afrrev.v10i2.11.

John, V. (2016). Peace Profile: The Alternatives to Violence Project. *Peace Review,* 28(3), 369–375. https://www.tandfonline.com/doi/full/10.1080/10402659.2016.1201960.

Juma, M. (2019). Peace Clubs in Schools: African Experiences. In *Infrastructures for Peace in Sub-Saharan Africa* (pp. 165–183): Cham: Springer Nature Switzerland AG.

Kaminer, D. and Eagle, G. (2010). *Traumatic stress in South Africa.* Johannesburg: Wits University Press.

Mncube, V. and Harber, C. (2013). *Dynamics of Violence in South African Schools.* Pretoria University of South Africa. https://www.unisa.ac.za/sites/corporate/default/Colleges/Education/Research/The-dynamics-of-violence-in-schools. Accessed 27 Sept 2020.

Nowell, L., Norris, J., White, D. and Moules, N. (2017). Thematic Analysis: Striving to Meet the Trustworthiness Criteria. *International Journal of Qualitative Methods,* 16(1), 1–13. https://doi.org/10.1177/1609406917733847.

Numbeo. (2009–2022). Crime Index by Country 2022. https://www.numbeo.com/crime/rankings_by_country.jsp. Accessed 10 April 2022.

World Population Review. (2022). Murder Rate by Country 2022. https://worldpopulationreview.com/country-rankings/murder-rate-by-country. Accessed 10 April 2022.

Chapter 13
The Challenges and Opportunities for Civil Society in Peacebuilding: Lessons from the Matabeleland Collective, Zimbabwe

Dumisani Maqeda Ngwenya

Abstract *Gukurahundi*, (the rain that washes away the dirt the after harvest) is one of the darkest chapters of the history of violence in Zimbabwe. Between 1983 and 1987, the government of Zimbabwe carried out a pogrom against the people of Matabeleland and some parts of the Midlands provinces which left some 20 000 people dead, many more disappeared or maimed. For the duration of the late president Mugabe's rule, almost 37 years, no acknowledgement or healing programme was ever initiated. Any talk about this era was strongly suppressed by the government and the environment was generally hostile against any efforts to engage publicly on *Gukurahundi*. Although much was done by a number of CSOs and churches, it was always without much cooperation from the state. The advent of the so-called new dispensation, which saw President Mnangagwa come to power in November 2017, brought with it a slight glimmer of hope and a corresponding change of attitude by some CSOs in Matabeleland. A number of prominent Matabele organisations came together under the banner of Matabeleland Collective, to find an alternative way of addressing the violence and pain of *Gukurahundi* by seeking to engage directly with government and the president in particular, in a non-confrontational way. For the first time in the history of Zimbabwe, the president was able to come down to Bulawayo to engage openly on *Gukurahundi*. This was an historic occasion which resulted in government making certain commitments to address some of the concerns raised by the group. At the same time a considerable amount of opposition from people in the diaspora, politicians in the ruling party and opposition movements in Matabeleland, arose and there was serious backlash and accusations of "selling out". This chapter seeks to narrate and analyse the process that led to the formation of Matabeleland Collective, its achievements and challenges it faced, and the lessons that can be learnt from it.

Keywords Gukurahundi · Civil society · Matabeleland collective · Healing · Devolution · Mnangagwa

Dr. Dumisani Ngwenya is a peacebuilding practitioner and academic who has been working in the field of conflict transformation, trauma healing since 2003 in Zimbabwe.

J. C. K. Kiyala and G. T. Harris (eds.), *Civil Society and Peacebuilding in Sub-Saharan Africa in the Anthropocene*, The Anthropocene: Politik—Economics—Society—Science 34,
https://doi.org/10.1007/978-3-030-95179-5_13

13.1 Introduction

The history of the struggle for Zimbabwe's independence is a contested one. This is mainly because the Zimbabwe African National Union (ZANU), one of the two main nationalist parties (led by Robert Mugabe) which fought the struggle has consistently attempted to diminish the contribution of Zimbabwe People's Revolutionary Army (ZPRA) and the Zimbabwe African People's Union (ZAPU) out of the history of the struggle (Ndlovu-Gatsheni 2017). According to Ndlovu-Gatsheni (2017, p. 6) up until 1963, Joshua Nkomo had led the various big nationalist party formations starting with the Southern Rhodesia African National Congress (SRANC) in the 1950s to ZAPU. In 1963 there was a split in ZAPU which resulted in the formation of the Zimbabwe African National Union, led by the Reverend Ndabaningi Sithole, with Robert Mugabe as its Secretary General (Ngwenya 2018; Sibanda 2017). ZANU later deposed Sithole and eventually made Mugabe its leader.

It is not the purpose of this chapter to delve much into the reasons for the split save to say that it was a result of a lot of internal strife and differences – largely based on policy direction, but also on ethnicity – between Nkomo and some of the leaders in ZAPU. The end result of this split was that it widened the tribal rift between the Ndebele-speaking and Shona-speaking people in the country. While ZAPU's leadership was well representative of the country's ethnic demographics, it came to draw it's support mainly from the Matabeleland and Midlands regions, while ZANU drew theirs almost entirely from the Mashonaland regions. The split also led to the flaring of inter-tribal violence between the supporters of the two parties in the major cities, particularly Harare and Bulawayo.

After the 1965 Unilateral Declaration of Independence from Britain by the Rhodesian Prime Minister, Ian Smith, the race to recruit for the armed struggle increased the tensions between the two nationalist parties (Ngwenya 2018, p. 20). ZAPU's armed wing was the Zimbabwe People's Revolutionary Army (ZPRA) and ZANU's, the Zimbabwe African National Liberation Army (ZANLA).

The tensions between the parties naturally spilt into their fighting forces and there was no love lost between the two. The relationships were so bad that sometimes when the two met in Rhodesia, they would exchange gunfire (Abrams 2006; Alexander et al. 2000). Several efforts were made by the Frontline States[1] to unite the two fighting forces. One such attempt ended up very badly. In an effort to integrate the two armies into one army called the Zimbabwe People's Army (ZIPA), in June 1976 some ZPRA cadres were sent to the training camps of Mgagao and Morogoro (which were training camps for ZANLA) in Tanzania. Fighting broke out one morning, over the use of a kitchen, and ZANLA with the assistance of their Chinese instructors quickly armed themselves and fired at the unarmed ZPRA counterparts, some of whom managed to escape. It was a similar story with the ZPRA group sent to Mozambique; they were tortured and forced to denounce ZAPU. These incidents only served to harden the antagonistic attitudes between the two armies (Ngwenya 2018, p. 20).

[1] These included among others, Zambia, Mozambique, Tanzania, Angola, etc.

The political wings, ZAPU and ZANU, managed to unite, somewhat, for the 1979 Lancaster House negotiations. This unity, however, only lasted for the duration of the negotiations. The two parties went into the negotiations as the Patriotic Front. However, soon after the end of the conference, Mugabe announced that ZANU would contest the then upcoming elections on its own. As a sign to its commitment to the unitary approach, ZAPU incorporated Patriot Front into its name, becoming PF-ZAPU. The move, by Mugabe, contributed further to inter-party suspicions and antagonism that were already high. Mugabe won the February 1980 elections, garnering 57 of the 100 seats, while Nkomo won 20 (mostly in Matabeleland and the Midlands), three went to Bishop Abel Muzorewa's United African Congress Party (UANC) and the 20 other seats were reserved for the whites (this arrangement would fall away after 10 years). Although Nkomo's ZAPU were given some Cabinet posts, it was not long before they were kicked out by Mugabe. One of Mugabe's desires was for a one party state, and ZAPU stood in the way of that dream (Meredith 2008; Ngwenya 2018).

13.2 The Dawn of the *Gukurahundi* Era

During the war of liberation in Zimbabwe, South Africa's ANC worked very closely with ZAPU. Their armed wing, Umkonto WeSizwe (MK), fought side by side with ZPRA in then Rhodesia. So, when the ceasefire was effected in Rhodesia and the guerrilla armies sent to the assembly points for disarmament, demobilisation and creation of a new Zimbabwe National army (ZNA), some of the MK fighters went into the assembly point with their ZPRA counterparts (interviews with ex-ZPRA combatants). In addition, the ANC had hoped they could use Zimbabwe as a base to launch attacks into South Africa, so some of their arms had been brought into the country and cached in some of the ZAPU properties (Nkomo 1983). Furthermore, because of their distrust of the Smith regime, both ZANLA and ZPRA had left some of their fighters outside the assembly points and had also cached arms in case Smith reneged and they needed to resume the war.

This information was known to both parties; however, the cached arms on the ZAPU/ZPRA side later provided Mugabe with a perfect excuse to accuse his adversaries of being disgruntled with the outcome of the election and planning to topple his government. Furthermore, the mistreatment, humiliation and harassment of the ex-ZPRA fighters already integrated into the ZNA, led to disaffection of those still in assembly points. In addition, a series of provocative statements by ZANU politicians resulted in gun fights between ZPRA and ZANLA combatants who had been placed at camps close to each other in Entumbane suburb in Bulawayo. The fighting eventually spilt to other camps at Connemara in Gweru, Ntabazinduna about 45km outside Bulawayo and Chitungwiza township near Harare (see Ngwenya 2018 for a fuller history of this period).

The ex-ZPRA combatants now attested into the national army continued to face harassment and, in some instances, certain death. They were overlooked for high posts which were given predominantly to the former ZANLA cadres. This riled the ZPRAs who felt they were better trained and better fighters than their ZANLA compatriots. These frustrations and a sure fear for their lives drove some of the ZPRAs to dissert the army; some went back to their rural homes and families, while others sought refuge in Botswana, South Africa and overseas. A few decided to pick up arms and became a nuisance to the communities and farmlands owned by the white farmers. All these situations worked perfectly in Mugabe's favour, who by early 1982, felt he was in control enough to make his move to destroy Nkomo and ZAPU. In February of 1982, he announced that arms caches had been found in ZAPU properties and that this was clear proof that Nkomo and ZAPU had been planning to topple his government and that ZPRA 'dissidents' had become a menace which needed to be dealt with. ZAPU was expelled from Mugabe's government and some of the top ZPRA commanders arrested; Nkomo himself was fired as Minister of Home Affairs and had to flee to Britain. The most high-profile arrestees being general Lookout Masuku and Dumiso Dabengwa, who after a lengthy trial were found not guilty and acquitted only to be rearrested as they left court and detained, until 1986, under the Smith regime Law and Public Order Maintenance Act (LOMA), the very act that had been used to detain the nationalists arbitrarily, including Nkomo and Mugabe himself. Nkomo and ZAPU vehemently denied supporting the 'dissidents' and former 'dissidents' interviewed later confirmed they got no support from their party. In the midst of this confusion, South Africa, recruited some of these disgruntled ZPRA fighters and armed them, and were known as Super ZAPU and were better armed than the ordinary 'dissidents' (Alexander et al. 2000; Ngwenya 2018, pp. 23–24; Nkomo 1983).

In October 1980, Mugabe had apparently entered into a secret agreement with the North Koreans to train a new brigade which was to deal with 'internal security'. The deal, however, was only announced in 1982 upon the arrival of the North Korean instructors. This brigade was to be responsible directly to Mugabe and it operated outside the normal national army chain of command (Todd 2007, p. 37). The 5th Brigade officially known as *Gukurahundi*, a chiShona word meaning "the rain that drives away chaff before the spring rains," caused great havoc in Matabeleland and some parts of the Midlands provinces. It was responsible for the vast majority of the human rights violations and war crimes committed against the people in these provinces, who were perceived to be associated with Ndebele and ZAPU. It is estimated that about 20 000 civilians were murdered by the *Gukurahundi*, countless others were disappeared, tortured and raped in the most brutal manner. Villagers were exposed to the most inhumane treatment they had ever experienced. Pregnant women had their stomachs ripped open to show 'baby dissidents' (Overseas NGOs dossier), relatives were made to have sex with each other while everyone watched. Men would be made to climb trees, barking like baboons and told to jump down, many limbs were broken. People were burnt alive in huts, made to dig their own shallow graves and shot while piled up inside. Survivors were then made to sing and dance on top of these graves, and no one was allowed to cry.

Those killed were to be buried immediately, and as a result, many never received proper burials and some families still do not know where their relatives were buried up today. In Matabeleland South, at a place called Bhalagwe, a big concentration style camp was established by the 5th Brigade. In this camp, people were tortured daily and those who succumbed were thrown down a disused nearby Antelope mine. Smaller camps were found in other parts too (CCJP 2007, pp. 154–160; Ngwenya 2018, pp. 23–27; Scarnecchia 2011, pp. 91–94). Evidence suggests that the sole purpose of the 5th brigade was to deal solely with civilians, for they hardly ever engaged in any fights with the dissidents. Other military units, especially the police support unit dealt directly with the dissidents and committed less human rights abuses, but committed them nonetheless. Members of the 5th brigade, themselves, often told their victims that they had been sent to annihilate all Ndebele people (CCJP 2007, p. 82). On 22 December 1987, a Unity Accord was signed between the two parties after lengthy negotiations. The sum total of the agreement was the merger of the two parties resulting in the one-party state that Mugabe had always wanted. Nkomo became the second Vice President in the new Government of National Unity.

Apart from the cessation of the violence, there was no further peace dividend enjoyed by the ordinary citizens of these provinces. No efforts were made to address issues of justice, healing and reconciliation. Instead, a blanket amnesty was issued to both sides and the unwritten law was that this was never to be mentioned again as the two antagonists had dealt with it. In other words, only a limited political solution, meant to achieve a certain goal and not the needs of healing and nation building was important (Ngwenya 2018, pp. 26–28).

13.2.1 Early Efforts Made by Civil Society to Address Gukurahundi

Several efforts to bring to an end the harm being perpetrated by the 5th brigade were made by a number of groups but with little success. In March 1983, a group of international NGOs working in Matabeleland and calling themselves, the Overseas non-governmental organisations, wrote a letter to Prime Minister Mugabe and presented him with a comprehensive dossier detailing the brutal acts of the 5th Brigade. The dossier included reports and pictures of victims of the violence from doctors working in the rural hospitals. It also included a compilation of the people that had been killed thus far by the *Gukurahundi* and how they had died. The dossier would have been enough to cause Mugabe to want to act but it is not clear what his reaction to this was since this dossier was sent to him in confidence. However, judging from their complaint in the letter about the then Minister of Defence, Sidney Sekeramayi, it would appear that they were viewed as troublemakers. In the letter they stated the following:

> On another matter we wish to say that it was with great distress that we read the remarks attributed to Comrade Minister Sekeramayi in which the Herald (newspaper) of 10th March 1983, which reported that he accused some non-governmental organisations of becoming 'front organisations for Nkomo's anti-government, subversive and slanderous statements.' We wish to say that we know of no such non-government organisations. We believe we have acted in the best interests of all Zimbabweans in communicating our deep humanitarian

concerns privately and directly to you, and in responding to your request that we should do so (Overseas NGOs dossier 18th March 1983).

This dossier covers only 2 months but the extent of harm done right through Matabeleland is evident.

The Catholic Church has always been known for their social justice gospel and they have actively intervened or spoken out against injustices, even those perpetrated by states. The Zimbabwe Catholic Bishops' Conference was no different. Concerned with the reports coming in from their mission stations across Matabeleland, a delegation from the Catholic Bishops presented Mugabe with another comprehensive dossier of the evidence of the havoc caused by the *Gukurahundi* and a copy of their pastoral letter they intended to release to their folk. Mugabe did not take kindly to this and refuted these accusations. He accused the bishops of playing to 'the international gallery' and of being 'sanctimonious prelates'. Meredith (2009, pp. 62–63) has this to say about his reaction:

> He queried whether they were their own masters or "mere megaphonic agents of their external manipulative masters," adding: "In those circumstances, their allegiance and loyalty to Zimbabwe becomes extremely questionable." He continued: "The Church of Zimbabwe, whatever the denomination, must abandon forever the tendency or temptation to play marionette for foreign so-called parent churches whose interests and perspectives may, and often will be, at variance with the best interests of the people of our country." It should "attune itself to the realities of the new Zimbabwe."

Mugabe did appoint a Commission of Inquiry, led by Judge Simplicious Chihambakwe, which became known as the Chihambakwe Commission, in September 1983 and began gathering evidence in January 1984. However, in November 1986 it was announced that the Commission's report would not be made public (CCJP 1997, p. 9). Up today this report has never been made public and no one seems to know where it is – if it still exists. Even the constitutionally mandated National Peace and Reconciliation Commission (NPRC), set up by the 2013 Zimbabwe constitution, seems to have failed to access the report thus far.

Nevertheless, in 1997 the Catholic Commission for Justice and Peace and the Legal Resources foundation released to the public a comprehensive compilation of the atrocities committed by the 5th brigade and other armed groups. The reports were aptly titled "Breaking the Silence, Building True Peace: A Report on the Disturbances in Matabeleland and the Midlands, 1980 to 1988". It was later published and released internationally in 2007 under the title "Gukurahundi in Zimbabwe: a report on the disturbances in Matabeleland and the Midlands 1980–1988". It went further to quantify some of the losses suffered and some practical recommendations that could be taken to remedy the situation. Mugabe and his government ministers were given copies but there was no official comment forth coming. In fact this book was unofficially banned from Zimbabwe and having a copy in public exposed one to certain danger from state security operatives. In as far as public information is available these were the major efforts to engage directly with Mugabe and his government, and none were taken seriously. The closest he ever came to an acknowledgement was at the funeral of Joshua Nkomo, in 1999, where he is reported to have said about the *Gukurahundi* atrocities "it was a moment of madness".

There were other efforts from some pressure groups from the Matabeleland region who sought to keep the issue of *Gukurahundi* alive in the national social consciousness. One such group, formed in 1997, was made up of mostly young people and known as *Imbovane Yamahlabezulu*. It was a group of radical young people, some of them being students at university, while others were professionals. Their aim was to open up debate on issues pertinent to Matabeleland, chief among them, *Gukurahundi*. They managed to hold high profile public meetings with some of the ZANUPF and government ministers from Matabeleland, asking them to explain to the people who had given instructions for the atrocities. The group attracted attention from the state security agents who started hounding their leaders. The group eventually dissipated with some of the leaders leaving the country (ikhonaindaba 2013). Others included the Mthwakazi Action Group on Genocide and Ethnic Cleansing in Matabeleland and Midlands, based in the diaspora (Ndlovu-Gatsheni 2008, p. 188). These were followed later by current groups like *Ibhetshu Likazulu* (literally, the people's shield), formed in 2005 and has been at the forefront of advocating for government to address the issue of *Gukurahundi*. Every year in January they try to hold a commemoration for the victims of *Gukurahundi* and they have had several run-ins with the state security agents. Then there are organisations who chose to address *Gukurahundi* by working directly with the affected communities such as Ukuthula Trust and Grace To Heal among others.

13.2.2 The Matabeleland Collective (MC)

Sometime in 2017, two of the civil society organisations in Bulawayo came together to encourage people to participate in the biometric voter registration exercise that was going on at that moment. By October 2017, a few more organisations had joined in under the banner of the Western Region Envisioning Collective (which was later changed to Matabeleland Collective, at a strategic retreat). In their interactions with members of the public, a number of issues were raised with which people of Matabeleland were concerned. They then decided to convene a public meeting on 2nd December 2017, which was attended by over 500 delegates from civil society organisations (CSOs), community based organisations (CBOs) and church bodies. The key objective was towards promoting participatory citizenship within, healing from *Gukurahundi*; social inclusion and equalisation; compensatory devolution development, and sustainable economic growth while rebuilding the social fabric of Matabeleland (MC report 2019). At this meeting, the leaders were tasked with the mandate to "work on an issue-based construct, mobilising citizens through a consultative approach and national level engagement" (MC compendium n.d).

This was especially so because people felt that they had been accused of being cry-babies, always complaining but with nothing constructive to offer. People from Matabeleland feel that they lost 10 years to *Gukurahundi* and still continue to be marginalised socially, economically and politically (Ngwenya 2018, pp. 28–32). This was a shift in thinking, to present their grievances but to also offer possible solutions to the same. They were also tasked with making sure that their issues were brought to the highest offices in the land. The issues raised, numbering six key areas of concern, were later synthesised into four themes, namely; healing (the *Gukurahundi* genocide memorialisation and practical amends), economic freedom, social inclusion and equalisation, and devolution of power.

In February 2018, a two-day strategic retreat was held at a local hotel, in Bulawayo, with key church and civil society organisations from Matabeleland North and South and Bulawayo. Experts in the four thematic areas presented papers for each theme followed by intense discussions. The end result was a 111 page comprehensive document, the Matabeleland Collective Compendium, which set out clearly what the issues were and how these could be addressed by government, the international community and CSOs. The setting and tone of this document was *Gukurahundi*, its impact on the people of Matabeleland and how these four areas, if addressed, would result in a healed community and nation. Some of the issues raised within the four areas were as follows:

- **Healing** with two aspects, memorialisation and practical amends, and these were to include acknowledgment and apology, truth-telling, construction of memorials, identification and reburials, easy access to citizenship and identification documents, and reparations.
- **Economic freedom** which included infrastructure development, agrarian support, benefit by communities from their natural resources and minerals, and employment of locals first, especially for non-skilled jobs.
- **Social inclusion and equalisation**, to cover affirmative action capacitation, social services delivery, community level reparations and celebration of ethnic diversity, and
- **Devolution of power**, comprised of expediting implementation of devolution (as set out in the constitution), to be consulted on the devolution bill, compensatory development, and democratisation to encourage citizen self-agency (Compendium n.d. pp. 8–9).

Once the document was produced the idea was to have outreach programmes into as many community stakeholders as possible for buy-in. However, due to funding constraints, this didn't happen as planned. At the beginning of the process the membership of the Collective comprised of 66 community based organisations, civil society organisations and churches.

In May 2018, the group was able to meet the then Vice President, Kembo Mohadi, who was responsible for national healing in the Executive Branch of Government, and were able to discuss at length with him and he agreed with the contents and approach. The idea was to get him to organise a meeting between the group and the President. For a number of reasons, he seems to have been unable to organise this. Next, the group met with the commissioners of the National Peace and Reconciliation commission, who at that point and time were doing outreaches to gather views from the public about how they wanted to see the Commission functioning. They too were impressed and quite excited about the Compendium as it gave them a clear path to possible healing in Matabeleland. In fact, they pointed out that their work in the

Matabeleland region had been made easier since all the pertinent issues were laid out clearly, unlike the other regions whose issues had not yet been crystallised. Up to this point, members of the Collective seemed to be of one purpose and mind and the movement was gaining traction within the people in the region.

Sometime early in 2019, the group unexpectedly had a breakthrough in their quest to meet with President Emmerson Mnangagwa. Word had somehow got to him that civil society groups in Matabeleland were willing to engage him. The meeting was set for 21st March at Bulawayo State House (the Presidential Residence). During the group's preparatory meetings for its meeting with the President, they decided to adopt what they called a 'low hanging fruits' strategy. The idea was to look at all the issues in the compendium and start off the engagement with issues that required little effort from government and were non-threatening. The weightier ones were to be left for another time (it was later decided that issues of acknowledgment, apology, reparations, etc. would be handled by the chiefs who, by virtue of being the custodians of the communities and people affected by *Gukurahundi*, were better placed to speak authoritatively on these issues). Nevertheless, the issues to be presented were still very sensitive and the group did not know how the President would react or what might happen to them at or after the meeting, since the issue of *Gukurahundi* had been taboo under President Mugabe. Many had been arrested, harassed and hounded out of the country for being vocal about it. About 150 members of the group attended the open meeting.

In as far as meetings go, this was an extremely successful and historic one. In spite of all the heroic efforts of many men and women in the past, there had never been a meeting where citizens of the country had held a frank discussion with the President on issues of concern (outside their party structures), let alone the emotive and sensitive topic of *Gukurahundi*. Although all the four thematic areas were presented to the president, it was pointed out to him that *Gukurahundi* was the underlining factor in all of them. In its editorial, The Chronicle of 22nd March 2019 (a state newspaper), reported glowingly about the meeting, how it had broken the taboo and many years of fear and suspicion, and quoted the President as having said:

> I came here with an open mind to listen and I am satisfied that I made the correct decision because non-interaction creates fear, suspicion where there ought to be no fear, where there ought to be no suspicion at all. I go back today knowing that the social groups, non-governmental organisations, civic organisations have their country at heart.

The meeting resulted in an implementation matrix of the issues raised during the meeting, and a presidential directive to various ministries to attend to issues falling under them. It even had a timeframe for report-back. There was co-operation between the President's implementation team and leaders of the Collective in coming up with the final draft of the matrix. Several meetings were held to iron out issues and for feedback. Since not all members and leaders of the Collective could participate in these meetings, they soon became an added source of conflict, as members differed on how they should be held.

Since the meeting was private, the press was allowed to be present for the opening and closing remarks of the meeting. So their reporting was based on those public speeches. As soon as news of the meeting broke the whole situation went into a tailspin. It must be noted that Zimbabwe is a very polarised environment and almost every aspect of life is viewed in opposites. Both the public and private media push certain angles and agendas. The reporting of the event was very positive from the public media, as exemplified by the Chronicle editorial mentioned above. For a while, the private media was silent as if they were not sure how to report on the event. To make matters worse, one of the Collective members whose organisation deals with media and had been tasked with reporting for the organisation, had been arrested at the gate into State House by overzealous security personnel because he had a spent tear smoke cannister in his vehicle. This became the international story and focus for the private media.

Within a matter of hours, some of the prominent leaders of the Collective began receiving insults on social media, particularly Twitter and Facebook and accusations of having sold-out. Some of vicious comments came from people in the diaspora who spoke as if they had attended the meeting and knew more than those who had organised and attended the meeting. For the next few weeks, pressure from social media and private media escalated, and the attacks were unrelenting. Leaders of the Collective, who had impeccable records as human rights defenders were portrayed as sell-outs and the Collective as a 'Mnangagwa project'. Unfortunately comments from some of the leaders were taken out of context by the government press and used to show how the President was succeeding in his agenda as the 'listening president'.

As the pressure took its toll, cracks began to emerge in the Collective as people tried to distance themselves from being associated with the President's 'project' and to redeem their civil society struggle credentials. It became more and more difficult to reconcile the different views about direction, leadership styles and approach, which had existed previously. Details of business meetings and internal communication, began to be leaked to the private media and those opposed to the initiative. The leaks were so rampant and blatant. For instance, in a report by the Southern Eye newspaper, but carried by an online publication Zimeye, of 20 June 2019, titled "Mnangagwa Splits Matabeleland Collective", the reporter quotes liberally from minutes of the Collective's June 14th meeting. Communication made to members on their WhatsApp platform also made their rounds and would frequently find their way back to the leaders, coming from people who were not members of the Collective or on the platform.

In the meantime, the Collective continued to function somehow. They held a feedback meeting with a few key chiefs from Matabeleland and Midlands whose areas experienced the brunt of Gukurahundi. The meeting held on the 24th April 2019 was to appraise the chiefs on the meeting with the President and to encourage them to organise to meet with the President as chiefs to present the more serious issues. A meeting with chiefs, by the MC, had previously been held on September 1st 2018 to discuss the Compendium with them and to get their buy-in (MC report 2019, pp. 8–9). According to the same report, the meeting on the 24th April was well received by the chiefs, who supported the engagement efforts. However, they expressed reservations about the goings-on and stressed the need to handle the issue of *Gukurahund*i sensitively.

The chiefs were assisted by the Collective to organise their meeting with the President which was eventually held on 6th June 2019. Although the meeting apparently did not go as planned, due to the interference of the President of the Chiefs' Council (who has always been viewed as being against the idea of addressing *Gukurahundi* issues) they, nevertheless, presented their issues to him. The events leading to this meeting served to give those opposed to the process and doubting the President's sincerity, further ammunition to cast aspersions on the Collective and the engagement process. This was largely due to the insistence by the Chiefs' Council President, who is not from Matabeleland and so had not been invited, to attend and know what the chiefs wanted to talk to the President about. Part of the sensitivity of *Gukurahundi* lies in the fact that the 5th Brigade was Shona speaking and targeted Ndebele speaking areas. So for the local chiefs having a person from Masvingo province (a Shona speaking area) wanting to control proceedings was an insult to them and only confirmed their fears that the government was not serious about addressing the issue. On the day of the meeting, which was also held at Bulawayo State House, a group from the Mthwakazi Republic Party (which is agitating for a separate Ndebele State), gathered at the hotel where the chiefs were staying to demonstrate against the chiefs' president. They later issued a press statement condemning those they saw as not wanting to have the *Gukurahundi* issue addressed (Press statement 28th June 2019).

Activities on the implementation continued to be pursued in the meantime, but things seem to have been happening slowly. Almost a year later, the President returned for a feedback meeting on the 14th February 2020. Once again Matabeleland Collective was thrust into the fore and the noise erupted once again. This time the biggest bone of contention was that Mnangagwa was the wrong person to deal with the issue of *Gukurahundi* as he was one of its chief architects. In a NewsDay of 25th February, Silas Nkala interviewed a number of prominent people, including some who had been part of the Collective, and they all had negative views about any involvement by the President.

At about the same time, it came to light that some members of the Matabeleland Collective had broken away and formed another organisation, the Matabeleland Forum, in protest, accusing the organisers of the meeting of having a hidden agenda. According to an article in an online publication, NewZimbabwe.com of 13th February 2020, members of the new group accused the leader of the Collective of "unnecessarily showering Mnangagwa with accolades at the previous meeting" and further accused the Collective leadership of "doing things without consulting other members". Newspaper reports from this second meeting indicate that the Collective had not necessarily been compromised despite the accusations to the contrary. In a report carried by the Standard, a private weekly paper, under the heading "Apologise for *Gukurahundi*, ED told", the report states that the Collective told the President he needed to implement the matrix, since nothing had been done since the previous meeting a year ago. It also goes on to say that the group told the President that the people in the region were sceptical about his commitment to addressing *Gukurahundi* and that there was a need for him and his government to acknowledge and apologise for the atrocities. It further states that the group demanded the return of ZPRA/ZAPU properties confiscated by the government in 1983.

13.2.3 The Successes of the Matabeleland Collective

In spite of the vicious attacks and bad publicity the Collective received, it nevertheless achieved some historic mile stones as the narration above indicates.

Firstly, and perhaps most importantly, the Collective managed to unite, albeit briefly, the varying and often dissenting voices of Matabeleland. For the brief period that the organisation remained united, it commanded respect from many stakeholders, including government, foreign embassies and even among some sections of the security forces. Having 66 organisations agree on the fundamental issues and solutions, each with their own interests and agendas, was no small feat. Approaching the issues from a less emotional and more reflective and academic manner, assisted people to better comprehend the issues and even those who would normally have dismissed the *Gukurahundi* question were able to appreciate the issues. Being able to reduce the issues into a well written and professional document and capturing the essence of the question in a way that resonated with almost everyone concerned with issues from Matabeleland, was a great achievement. Although people disagreed with some issues in the approach, none faulted their Compendium.

Being able to bring the President, one of the chief drivers of *Gukurahundi*, to Matabeleland to listen to people talking about it in the manner in which the Collective did, is an historic achievement of great proportions. Telling the President to his face and publicly, that people were afraid to talk freely about what happened to them because of the security agents, that he needed to tell them that it was not unlawful to talk about *Gukurahundi*, the fact that people needed to heal and bring closure by knowing the truth of what happened and to be able to rebury their loved ones decently, had never been done before. Many people had desired it and fought for it over the years. The President's positive response to these issues, helped to unshackle the fear and oppressive environment of the previous regime (Ngwenya 2018, p. 29). A State owned radio station in Bulawayo was the first to host a discussion live on air soon after this meeting, thereafter, others followed suit. People were also able to feel a little more free to speak publicly about it. Nevertheless, many others, because of the many years of suppression, still felt unsafe talking about *Gukurahundi* in public spaces. The civil and intellectually sound presentations of the four thematic issues, in a non-hostile and respectful manner appears to have made it easier for the President to accept the demands.

This engagement helped to bring *Gukurahundi*, in particular, to the national debate. Whereas, in the past it was viewed as a regional and tribal issue, more and more people from other parts of the country began to embrace it as a national issue. It also helped that, at the same time, the NPRC had started its own processes and the topic of *Gukurahundi*, was at the top of their intentions.

Although the implementation process has moved slowly, there has been a few minor achievements. These centre mostly around issues of water provision. Information from the Collective also indicates that the issues of death, birth and identity documents for the communities affected by *Gukurahundi* were about to be actioned before the onset of COVID-19. Although the virus has slowed many aspects of life, consultations between the Collective and certain line ministries is on-going as they work towards the implementation of some of the issues in the matrix. This direct access to relevant government officials, even up to the President, is something that has not always been available to the ordinary people of Matabeleland.

13.2.4 Challenges Faced by the Collective

There are many challenges faced by the Collective and those by themselves could form a chapter of their own, however, I will try to focus on the most pertinent ones briefly.

The Collective's success in bringing so **many organisations together** became its Achilles heel. The Collective was an eclectic collection of organisations and individuals with wide ranging interests, agendas and allegiances. There were pressure groups, women's groups, special interest groups, peacebuilding groups, advocacy organisations, community based organisations, cultural groups, youth groups, etc. Although certain actions and approaches were agreed upon, it was never clear what philosophy of engagement undergirded the initiative. Rules of engagement for peacebuilding are different from those of advocacy, as they are different from those of pressure groups. In addition, some of the leaders of these groups had close political ties, while others had actually been political activists, party officials, and victims of violence at the hands of ZANUPF. So, right from the beginning, it was difficult to reconcile these varied interests. Although the Collective was supposed to be coalescing around the four thematic areas, and particularly *Gukurahundi*, it was difficult from the onset for some organisations to focus solely on the issue at hand without bringing in other issues of interest. The previous and perhaps, more abiding allegiances, it seems, were always going to prevail over what some might have seen as a short term goal in their scheme of things. Going through some of the WhatsApp discussions, one notices that people wanted the group to respond to all the other current political issues that kept arising. The group needed to have taken time to discuss and agree the parameters of their focus. This then raises the question of whether, in a country like Zimbabwe, where the tenets of democracy and rule of law are flagrantly flouted and human rights disregarded willy-nilly, it is possible to co-operate with the same government. Or better still, in pursuit of an important issue like the redress of past injustice, people should ignore other violations in order to achieve their ultimate goal?

Another big stumbling block, fostered by years of hostilities, was **lack of trust and the suspicions** that exists between government and civil society organisations. Since *Gukurahundi*, the people of Matabeleland have never trusted government. The deficit of trust has been compounded by years of perceived marginalisation and exclusion (Ngwenya 2018, p. 31). The people seem to trust the current President even less than the previous one. Perhaps he needed to have proactively participated in the trust building by implementing at least some of the more pertinent issues faster. As stated earlier, by his second meeting with the Collective, almost nothing had been attended to, except the provision of a borehole and discussions about the Matabeleland Zambezi Water Project which the people wanted returned to them, but these are peripheral issues. His pronouncements at the meeting and there after about people being free to 'debate' *Gukurahundi*, should have been quickly followed by tangible actions that supported his assertion. People have always doubted government's sincerity, when it comes to addressing *Gukurahundi*, the non-release of the previous commissions' reports is one major contributor. Even the NPRC has never really been fully accepted because people in Matabeleland see it as another attempt to go through the motions.

In negotiations, and indeed peacebuilding, the requirement to balance the need for **transparency and confidentiality** is a tricky one, especially when one is dealing with an entity like government. We noted, in the first section above, how some of the members of the breakaway group, Matabeleland Forum, complained that some leaders in the Collective were doing things without consultation. It was, perhaps, important for the members to at least know something about the behind the scenes negotiations. Modern political violence is designed to break the social fabric and community safety nets, the end result being that a community that has experienced traumatic events of this nature finds it difficult to trust each other (Ngwenya 2018, p. 36). So for some people a situation where lack of information about dealings with an entity they have never trusted in the first place, was unacceptable. On the other hand, we spoke earlier about the Collective being riddled with leaks of confidential information and perhaps the people at the forefront of the engagement feared that leaks would compromise progress.

Conflict mapping is one of the tools used in peacebuilding to analyse conflict. One aspect of this tool is the need to factor in the diaspora when mapping conflict. The diaspora compatriots have the potential to disrupt any peacebuilding efforts if they feel they don't serve their purpose, or are simply unhappy with being left out of the process. Diasporic studies indicate that people who have left or were forced out of their country, especially for political reasons, often maintain strong ties with their home countries, take a great interest in the political developments, and will often support the conflict back home financially or logistically (Missbach 2011). The Zimbabwean **diaspora** is no different. Those that were displaced by the government because of *Gukurahundi* and others forced out by the political events that led to the toppling of Mugabe, still maintain an active interest in political events back home and seem to have great influence among their followers. Twitter – and other social media platforms – has made cross border political activism much easier. They played a huge role in weakening the Collective, because they speak so authoritatively; and what they say is often taken as gospel truth by their constituencies, even though what they say is false or littered with half-truths. Some of them used to be prominent actors in the fight to have the government take responsibility for *Gukurahundi,* so they may feel as if they ought to be at the centre of any efforts in this regard. Perhaps the Collective should have engaged the diasporas, as they sought to do with other local stakeholders.

This leads us to the next stumbling block, the challenge of doing peacebuilding in the age of **social media**. Social media can contribute immensely to peace efforts but it can also destroy those efforts. The external pressure exerted by social media activists was complemented by the indiscipline within the users in the group itself. Netiquette is the set of acceptable online behaviour while, on the other hand, online ethics focuses on acceptable use of online resources in an online environment (webroot.com). Such rules assist to induce acceptable behaviour and helps members know what is, and is not appropriate to share publicly. It does not appear as if the group had any agreed way of how to treat confidential information, not only on social media but in general. However, as pointed out above, with such an eclectic collection and great number of people involved, these rules might still not have helped much. This does raise the need for peacebuilders, in this era, to think strategically about how to interact with social media. Apart from irresponsible members, hostile governments could eavesdrop into these conversations.

The final dilemma faced by the Collective is the concept of doing peacebuilding in the context of the perpetrator who is still in power and with the same instruments of violence still at their disposal– what Piccolino (2019) calls the 'victor's peace'. Several newspapers, some quoted above, and numerous social media postings all queried the logic of Mnangagwa's participation in the process of seeking closure. Their main argument was, how could he as one of the persons involved in perpetrating violence, be at the forefront of the healing process? At the time of *Gukurahundi*, he was Minister of State Security, under whom the notorious Central Intelligence Organisation fell. His utterances at the time, were crude and hurtful. There is a very real possibility that the accusations of trying to 'manage' the situation for his benefit might very well be true. On the other hand, how do people engage with the issue without involving the perpetrators? Some feel that the issue should be sent to the International Criminal Court, but who knows when that might be or whether it would actually happen? Besides, the government issued two sweeping amnesties for all involved in the violence, including the handful of dissidents who surrendered after the signing of the Unity Accord. At the same time the direct victims of *Gukurahundi* are dying due to old age, many are dying without having buried their loved ones properly, or knowing the truth about the disappeared, among other things. Unfortunately, peacebuilding can be a messy affair at times. It is important to mention here that the MC engagement with the government is on-going but, the initiative has been greatly weakened by the divisions in Matebeleland.

13.3 Drawing Peacebuilding Lessons from the Collective Intervention

Whatever the faults and weaknesses of the Collective and it's intervention, there is no doubt that valuable lessons for peacebuilders can be drawn from this.

The intervention has proved, to an extent, that it is possible to engage with perpetrators in a constructive manner and that such an approach is important. That is to say, people can engage on very emotional and sensitive issues in a civil manner and that this, in some cases, can yield the desired results without being aggressive or insulting. Timing is also of essence. In the early days of Mnangagwa's rule he portrayed himself and his government in democratic terms, and indeed there was a brief period in which it was easy to approach him. By taking advantage of this gap, the Collective was able to bring its agenda to the national platform.

Managing information flow is especially crucial in these times of social media. For coalition groups, agreeing on what can be shared with the large group and what to keep to those actively involved in negotiations makes sense. Diasporas are important in local peacebuilding interventions, and it is important, wherever possible, to engage with the influential ones in order to avoid efforts being torpedoed by them. Strategic thinking and prioritisation are necessary to achieve success. Leading leaders, with their own egos and visions of grandeur is challenging and perhaps the leadership approach needs to be different from leading the masses. The participation in the Collective was mostly at directors' level. These are people used to making decisions in their own right and therefore, are unlikely to want to be excluded from decision making processes. Much of leadership depends largely on the currency of trust and it is one of those things that leaders need to work hard to create.

The Collective broke the actual or perceived sense of regional lethargy and co-operation. The fact that, in the end, it did not work like clockwork cannot be an indictment of the effort.

Lastly, in a large grouping like the Collective, it might be helpful to take stock of organisational biases and strengths, and to agree that not all organisations or their people are suited for the sensitive work of peacebuilding at the level at which it sought to work.

13.4 Conclusion

Although the efforts of the Collective to try to deal with the injustices of the past with the perpetrators still in power might be a unique situation in Southern Africa, it nevertheless offers valuable lessons to all those interested in building peace after mass atrocities. Many questions will probably remain unanswered for now, perhaps some for ever. Was it a good idea to engage with the perpetrators? Should people have waited until they were out of power, and when might that be? Is it a good idea to have a group with such vast interests, allegiances, and areas of focus, involved in trying to deal with issues of healing and reconciliation? Whatever the answer to these questions might be, and whatever the final outcome of the Collective's initiative will be, there is no doubt that a historic milestone was achieved. Moving the *Gukurahundi* conversation from whispers in homes and small gatherings, always playing cat and mouse games with state security agents, to a national and inter-tribal one, conducted without fear of reprisals is in and of itself a life-time achievement. Whatever direction the issue takes now, it is unlikely to go back to where it has been all along.

References

Abrams, B.E. (2006). *Strategy of domination: Zanu-PF's use of ethnic conflict as a means of maintaining political control in Zimbabwe, 1982–2006*. Master's Thesis, Tufts University. https://dl.tufts.edu/concern/pdfs/9306t895p. Accessed 21 May 2009.

Alexander, J., McGregor, J., Ranger, T. (2000). *Violence and memory: one hundred years in the 'dark forests' of Matabeleland*. Oxford: James Curry.

Catholic Commission for Peace and Justice (CCJP. (1997). *Report on the 1980s disturbances in Matabeleland and the Midlands*. http://www.rhodesia.nl/Matabeleland%20Report.pdf Accessed 4 Aug 2020.

Catholic Commission for Justice and Peace (2007). *Gukurahundi in Zimbabwe: a report on the disturbances in Matabeleland and the Midlands 1980–1988*. Johannesburg: Jacana Media.

Ikhonaindaba blogspot (2013). *What happened to Imbovane pressure group?* https://ikhonaindaba.blogspot.com/2013/02/what-happened-to-imbovane-pressure-group.html. Accessed 5 Aug 2020.

Missbach, A. (2011). The Acehnese diaspora after the Helsinki Memorandum of Understanding: return challenges and diasporic post-conflict transformations. *Asian Ethnicity,* 12(2), 180–202.

Ndlovu-Gatsheni, S.J. (2008). For the nation to live, the tribe must die: the politics of identity and belonging in Zimbabwe. In B. Zewde (ed.) *Society, state, and identity in African History* (pp. 167–200). Addis Ababa: Forum for Social Studies.

Ndlovu-Gatsheni, S.J. (Ed.) (2017). *Joshua Mqabuko Nkomo of Zimbabwe: power politics and memory*. Cham, Switzerland: Palgrave Macmillan.

Ngwenya, D. (2018). *Healing the wounds of Gukurahundi in Zimbabwe: a participatory research project.*, Cham, Switzerland: Springer.

Nkala, S. (2020). Gukurahundi fast-track programme draws fire. The Newsday (Harare), 25 February 2020. https://www.newsday.co.zw/2020/02/gukurahundi-fast-track-programme-draws-fire/ Accessed 30 Sept 2021.

Nkomo, J.M. (1983). Informative letter to Prime Minister Mugabe. https://nehandaradio.com/2013/12/24/joshua-nkomo-letter-to-robert-mugabe-from-exile-in-the-uk/. Accessed 4 Aug 2020.

Overseas Non-governmental Organisations (1983). Dossier to Prime Minister Robert Mugabe [Classified document].

Piccolino, G. (2019). Local peacebuilding in a victor's peace. why local peace fails without national reconciliation. *International Peacekeeping*, 26(3), 354–379.

Scarnecchia, T. (2011). Rationalizing "Gukurahundi": Cold War and South African Foreign Relations with Zimbabwe, 1981–1983. *Kronos* 37(37), 87-10. https://www.researchgate.net/publication/261980049_Rationalizing_Gukurahundi_Cold_War_and_South_African_Foreign_Relations_with_Zimbabwe_1981-1983. Accessed 6 Aug 2020.

Sibanda, E.M. (2017). The contributions of Joshua Nkomo to the liberation of Zimbabwe. In S. J. Ndlovu-Gatsheni (Ed.), *Joshua Mqabuko Nkomo of Zimbabwe: power politics and memory*. Cham, Switzerland: Palgrave Macmillan.

Todd, J.G, (2007). *Through the darkness: a life in Zimbabwe*. Cape Town: Zebra.

Tshuma, N. (2019). *Milestone talks in Bulawayo…ED in frank discussions with Mat'land civic groups*. The Chronicle (Bulawayo) 22 March 2019. https://www.chronicle.co.zw/updated-milestone-talks-in-bulawayo-ed-in-frank-discussions-with-matland-civic-groups/. Accessed 23 June 2021.

Chapter 14
The Role of Civil Society Participation in Social Cohesion Processes in KwaZulu-Natal, South Africa

Paul Kariuki

Abstract Having played a crucial role in bringing down the apartheid regime, civil society remains a pillar of socio-political development and is an important player in holding government accountable in the democratic South Africa. Despite the remarkable progress made since the dawn of democracy, the social fabric of the nation has remained under severe strain due to unfulfilled promises and socio-economic and political challenges. Against this background and based on a desktop literature review, this chapter examines civil society's participation in social cohesion processes in KwaZulu-Natal, South Africa. Civil society organisations have employed a bottom-up approach to promote social cohesion, often acting as a mediator between the state and ordinary communities at the grassroots level. The chapter recommends that such participation should be intensified in order to deepen societal interrelationships that hold communities together across socio-economic and political strata.

Keywords Social cohesion · Civil society · Communities · Participation · Accountable · South Africa

14.1 Introduction

Civil society organisations have a long history. As pressure groups that keep governments on their toes as regards good governance, corruption, and implementation of policies and other vital issues (Banpasirichote 2004), they are seen by many as the last hope of the ordinary citizen. Some of these pressure groups are neutral while others may be compromised or sponsored by a political rival of the party in power. Whatever their stance, civil society organisations respond to the perceptions and yearnings of the masses in fighting bad leadership in order to improve the socio-economic status of the less privileged in society. Given that many political leaders ignore or pretend not to be aware of citizens' needs (Bhargava 2015), many of the developmental strides witnessed in South Africa are traceable to the influence of pressure groups.

Dr. Paul Kariuki, Research Fellow, School of Management, IT and Governance, University of KwaZulu-Natal.

J. C. K. Kiyala and G. T. Harris (eds.), *Civil Society and Peacebuilding in Sub-Saharan Africa in the Anthropocene*, The Anthropocene: Politik—Economics—Society—Science 34,
https://doi.org/10.1007/978-3-030-95179-5_14

As a maturing democracy, South Africa is experiencing a surge in the activities of interest groups, evident in the proliferation of civil society organisations with diverse ideologies aimed at improving citizens' welfare (Lehman 2008). The critical questions that informed this chapter were: Are civil society organisations able to foster peaceful co-existence of all in society without being compromised by rival political parties or victimised by the party in power? Is civil society able to promote societal cohesion now that the country is facing significant internal (in-country movement) and external migration (from neighboring African countries) to cities? Is civil society adopting a bottom-up approach in promoting social cohesion while acting as a mediator between the masses and government leadership?

The chapter is based on the premise that civil society still has a critical role to play in ensuring social cohesion, characterised by strengthened interrelationships among diverse population groups so that development can be achieved. The mediatory role that civil society plays is important, especially at present when the nation is still grappling with significant political shifts amidst the COVID-19 pandemic, whose devastating effects have eaten into the financial resources available for social protection to cushion the majority of the citizenry from socio-economic shocks. If the government and its social partners do not act decisively and speedily to address service delivery concerns, chronic poverty, and income inequality, among other challenges, civil protests may increase as citizens vent their frustration. If left unchecked, civil protests can quickly degenerate into violence and vulnerable populations are severely affected as they become easy targets for such frustration. It is in such moments that civil society plays a crucial role in easing the tensions between the state and citizenry by facilitating dialogue aimed at finding solutions to any challenges facing communities.

14.2 Problem Statement

The recent surge in civil protest in most cities in South Africa is an indication of unresolved issues emanating from failed policies that lead to poor service delivery. Scholars have identified structural crises such as high rates of unemployment, chronic poverty, income inequalities, and limited access to public social services such as housing and health as well as economic opportunities as key causes of ongoing tension between different population groups in society. Such tension is often characterised by violence and in extreme cases, internal displacement of vulnerable populations such as refugees and asylum seekers, most of whom live in townships and informal settlements. It is important to recognise that these structural crises are rooted in socio-economic policy and political leadership failures, and that the state cannot resolve them on its own. The state needs strategic partners to address these challenges that trigger xenophobic tendencies in South Africa. Civil society is one such strategic partner due to its proximity to local communities which are often the spaces in which xenophobic violence manifests.

This chapter examines civil society participation in social cohesion processes, focusing on building social cohesion between South Africans and African migrants, often referred to as "foreigners" (Tarisayi/Manik 2019:7). As a sector, civil society is well-placed to advocate peaceful co-existence between African migrants and their South African hosts, thereby promoting social cohesion.

A desktop literature review was conducted to gather secondary data on civil society and social cohesion in South Africa. Due to time constraints, no primary data was collected. Secondary data consisted of a wide range of publications, including research reports, and peer-reviewed journal articles that identify the role of civil society in social cohesion processes, as well as books and government policy documents that detail the legal frameworks underpinning South Africa's commitments to foreign nationals living and working in the country. Non-academic reports such as media reports were also reviewed to capture insightful opinions on the role of civil society in social cohesion processes in South Africa.

The chapter is organised as follows: The first section presents the problem statement, the objective of the chapter, the rationale, the methodology, and the assumptions made by the researcher. The second section presents a broad overview of civil society in South Africa as well as its participation in social cohesion. Section three discusses the methodology employed, followed by a conceptual analysis of key terminology in the fourth section. The fifth section discusses how social cohesion can be built and sustained given the changing socio-political and economic contexts in South Africa. This is followed by recommendations and a conclusion in the sixth and seventh sections.

14.3 Background: Civil Society in South Africa

Civil society groups are non-governmental societal organisations responsible for the defense of ordinary citizens against bad governance, corruption, failure, and poor service delivery by public representatives and elected figures. Civil society has become the last hope for the common person in promoting justice, ensuring responsive democracy, and peacebuilding (Foley/Edwards 1996; Ukase/Audu 2015).

The sustained pressure exerted by civil society groups in South Africa on the apartheid regime played a significant role in its collapse. Among many other activities, South Africa's civil society groups are known for their support for democratic principles, and for fighting corruption and holding the government accountable.

During the apartheid era, civil society organisations, which included professional groups, women's organisations, and youth groups and others, many of which were affiliated to movements such as the United Democratic Front, were considered as "radical groups". Indeed, some describe them as "opposition groups against the apartheid regime" rather than civil society organisations. Given that apartheid policies excluded and marginalised the majority of South Africans, the question arises as to whether such pressure groups fit the definition of civil society organisations. Had they been included in apartheid governance, would they have still constituted themselves as pressure groups? Rather, these groups constituted themselves as a parallel authority to the exclusionary state, thus existing as "liberation movements" rather than civil society organisations.

While civil society groups have recorded some successes in post-apartheid South Africa, they have also confronted many challenges. It is disheartening to note that undemocratic elements within the ruling political class have launched a war against pressure groups that they consider to be challenging their reckless and undemocratic practices. Civil society organisations have been victimised and their contracts with government agencies have been revoked. In extreme circumstances, major civil society actors have been murdered. For example, the chairperson of the Amadiba Crisis Committee and Sikhosiphi "Bazooka" Rhadebewas were assassinated due to their opposition to the mining of an environmentally sensitive area. Despite such intimidation, civil society groups and activists have continued to gain the support of the masses against the anti-democratic governance of political leaders in South Africa. This culminated in forcing two African National Congress (ANC) presidents (Thabo Mbeki and Jacob Zuma) out of office.

During Zuma's time in office, civil society groups mobilized against reckless governance, self-enrichment, nepotism, and corruption perpetrated by the then ANC president which led to his eventual prosecution. These organisations forced him to resign and also prevented him from installing any of his cronies as his successor. Civil society organisations have also continued to ensure that appointed or elected representatives abide by the Constitution and discharge their duties well. For example, the Council for the Advancement of the Constitution (CASAC) is well-known for taking up cases where such representatives have violated the Constitution.

Civil society groups have always been at the forefront of the campaign to promote good governance and a democratic ethos. Bad governance, which is a product of unilateral decisions has been a major focus of some pressure groups that have made consistent calls for government to consult widely with relevant stakeholders before making vital decisions that affect citizens. Organisations such as the Southern African Faith Communities Environment Institute (SAFCEI) and Earthlife Africa have been at the forefront of such campaigns. In some cases, they have taken the government to court. For example, a few years back, SAFCEI launched a court case when the government engaged in a secret procurement program to build new nuclear power stations without public consultation with relevant stakeholders (Le Cordeur 2016).

In 2016, Corruption Watch and the Right2Know Campaign instituted court proceedings against the findings of the Arms Procurement Commission (despite several protests by pressure groups aimed at disbanding the commission of inquiry due to perceived biases) which concluded that no corrupt practices were involved in the procedures undertaken by the government. These pressure groups have called for a review of the whole process by another commission of inquiry. Should they succeed, this would set a precedent for a review of several other verdicts by previous commissions of inquiry that violate citizens' rights.

14.4 Social Cohesion

The term "social cohesion" has not been fully understood as it is loosely used in both the academic literature and political discourse (AFD 2021). A socially cohesive society can be defined as one where citizens are able to live together despite their differences, exercising patience and tolerance of opinions, preferences, and choices regardless of their ethnicity, religion, gender, nationality, and class, among other forms of discrimination (Novy et al. 2012). More than 27 years since the dawn of democracy, South Africa has yet to emerge as a socially cohesive nation. Citizens are separated along spatial and social lines, as well as by class, race, and ethnicity (Todes/Turok 2018:20). The inequality that marks South African society has widened the gap between the poor and the rich, between racial groups, and between political leaders and their followers (Katumba et al. 2019:100).

The Agence Francaise De Developpement (AFD 2021) identifies five dimensions of social cohesion which are shown in Fig. 14.1.

These dimensions are interconnected, and they influence the extent to which a society becomes "socially cohesive and inclusive", where all ethnic groups respect one another and all citizens participate in civic life (Dickes et al. 2009:454). The various components are built upon *a set of attitudes and norms, that influence both vertical and horizontal interactions among members of a society* (Chan et al. 2006:290). According to Bernard (1999), these vertical and horizontal interactions intersect with other social spheres to ensure the inclusion of all citizens across the economic, political, and social-cultural spheres of society.

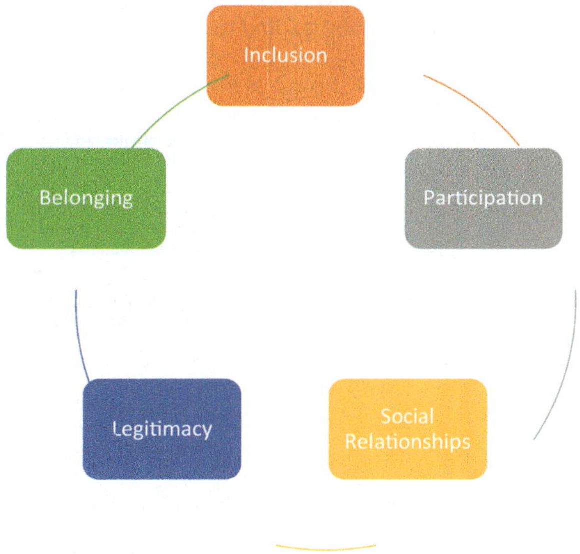

Fig. 14.1 Five dimensions of social cohesion. *Source* The authors

14.5 A Bottom-Up Approach to Building Social Cohesion in South Africa

Given South Africa's growing income inequality and unemployment, and persistent racial inequalities, social cohesion is a critical issue. Without a consensus-based working definition that reflects both intellectual discourse as well as citizens' lived experiences, it is impossible to determine the extent to which social cohesion exists in the country and how it can be sustained over time (Lefko-Everett et al. 2018). Civil society uses a bottom-up approach to build sustainable social cohesion (See Fig. 14.2).

The bottom-up approach emphasises joint consultation of all stakeholders involved in social cohesion processes. Civil society leads the consultations from below as it convenes a range of multi-stakeholder engagements at community level (Edinger-Schons et al. 2020). The process commences by identifying the key issues that need to be addressed. All stakeholders engage to develop a holistic community-driven process anchored by civil society.

After community consultations, civil society as a mandated structure engages the government and its agencies on key issues identified at the community consultation level. At this stage, civil society presents its views to the government, seeking a joint collaborative effort to address the identified issues. A joint goal/vision for the intervention is mutually identified and agreed upon. This step is important in ensuring all stakeholders' ownership and accountability (Caelesta/Madalina 2020).

The next phase involves government, civil society, and elected community representatives designing an action plan. This calls for clarification of the vision for the intervention, and the roles and responsibilities of the various stakeholders, timelines for the plan, accountability, and reporting as well as feedback mechanisms.

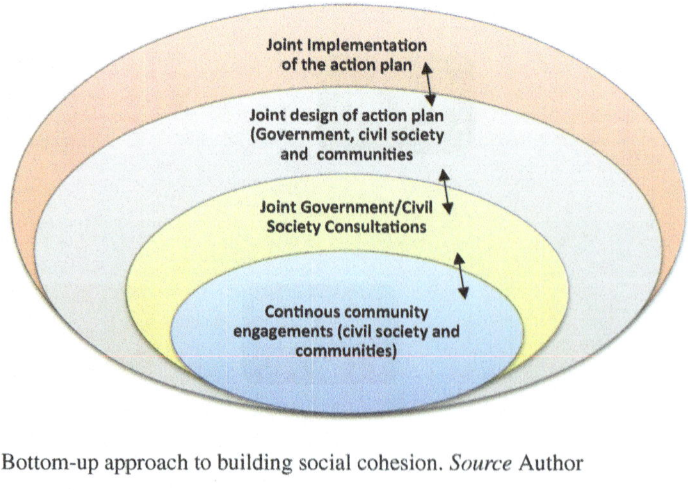

Fig. 14.2 Bottom-up approach to building social cohesion. *Source* Author

Once the action plan has been designed and aligned with the overall scope of the intervention, the agreements reached are sealed by way of a memorandum of understanding (MoU) that details the roles and responsibilities of each stakeholder. An implementation plan is then devised and mutually agreed upon. Implementation is tracked against a mutually agreed timeline. In between the various phases of the approach, the various stakeholders continue to engage, sharing insights, experiences, and ideas to ensure that the envisioned goal and its objectives are realised (Malan 2020).

The main characteristic of this approach is that it is driven by civil society, which receives a mandate from communities and acts as their voice when interacting with the government and its agencies. This ensures that the community's voice is heard and is part of decision-making processes in which community aspirations are considered and prioritised. As a joint process, the bottom-up approach ensures that civil society participates actively in social cohesion processes, proactively holding the process together by reaching out to key stakeholders on an ongoing basis as a trusted convener and advocate community interests. Moreover, this approach promotes ownership of the process by all stakeholders as they are involved in every phase (Malan 2020). Trust is built and reinforced throughout the engagement processes as the stakeholders share ideas and experiences bound by a mutually agreed common goal.

14.6 Case Study 1: The African Centre for The Constructive Resolution of Disputes (ACCORD)

The African Centre for the Constructive Resolution of Disputes (ACCORD) is a national not-for-profit non-governmental organisation based in Durban, whose work focuses on conflict resolution, peace building and research on migration, conflict mediation, negotiation, resolution, analysis as well as prevention in South Africa and throughout Africa, providing creative solutions to the challenges posed by conflict on the continent. The organisation was founded in 1992 and has remained focused on its vision of building African leaders' capacity in conflict resolution by addressing the underlying factors that contribute to under-development which often precipitates conflict on the continent.

The organisation's approach to social cohesion is a multi-pronged one that rests on three main pillars, namely:

- Strengthening local and national organisational and government capacities for peace to promote their proactive engagement in peace building and conflict resolution processes.
- Increasing the participation of women and youth (as demographics that are often affected when conflict occurs) in peace building and conflict resolution processes.
- Engaging in policy engagements that bring together all actors involved in peace building and conflict resolution. The actors include national governments, civil

society, academia, activists, and the media as well as research institutions, to name but a few.

Using this broad-based stakeholder approach to promote social cohesion, all actors pursue a common goal, with ACCORD facilitating interactions amongst them and developing action plans that are jointly developed and implemented. The model ensures that the voices of those most affected by conflict are heard and acknowledged and are used to develop mutually agreed upon actions plans that are binding on parties involved. The level of accountability amongst all actors is enhanced as the organisation convenes regular monitoring processes aimed at capturing the lessons learnt and using them to convene policy dialogues with governments, leaders of continental bodies such as the African Union (AU) and legislators so that appropriate decisions can be made to influence policy-making towards sustainable peace and development.

In 2015, following violent xenophobic attacks in KwaZulu-Natal, ACCORD led an inquiry, appointed by the then Premier, Hon. Senzo Mchunu, to determine the causal factors that led to the attacks on African migrants and what should be done to prevent similar occurrences in the future (ACCORD 2015). The organisation led a special reference group (SRG) that engaged affected communities, African migrants themselves, government departments, political parties, and business chambers, among others. Together with members of the SRG, ACCORD developed a report that detailed the causes of the violent attacks and mitigation strategies to prevent future conflict.

The report found causal factors that led xenophobic violence. The factors included (ACCORD 2015, pp. 59–109):

- Competition for resources that both African migrants and their local south African hosts need, and these resources are limited in supply.
- Limited opportunities for employment given the growing number of unemployment numbers in the country.
- Poor service delivery challenges – limited supply of housing, water and sanitation services among other public services.
- Limited public knowledge about migration, African migrants and the various reasons that promote migration of people into the province.
- Political commentary that is often inflammatory, especially in times leading up to an election, where politicians often use African migrants as an excuse of the challenges being experienced by local electorate meanwhile the failure to provide basic services is a governmental issue that has nothing to do with the migrants.
- Inaccurate media reporting on issues of migration and African migrants in particular and promoting a negative narrative and stereotypes about migrants as the cause of the many challenges that local South Africans are facing.
- Competition for resources and business opportunities between African migrant entrepreneurs and their local South African counterparts in communities where they reside.
- Insufficient responses to early alerts, especially by the law enforcement agencies.
- Limited efforts by government to facilitate integration of African migrants into communities and promote peaceful co-existence.

According to the report, most of these factors received minimum attention by government in both local, provincial, and national levels towards mitigating xenophobic conflicts. Some of the major consequences of the limited government intervention in addressing causal factors has been internal displacement of African migrants from their communities, strained relationships between African migrants and their local South African hosts due to mistrust promoted by stereotyping and negative media reporting, loss of livelihoods due to violence as well as loss of lives.

The report recommended the following (ACCORD 2015, pp. 167–175):

- Government at all levels of governance must strengthen intelligence and information gathering to ensure planned violence attacks against African migrants are prevented early.
- Policing must be improved throughout the province including the greater Durban Metro as well as all other geographical locations known to be "hot spots" for xenophobic violence such as townships.
- Address any institutional and structural gaps that impede the full implementation of all commitments (continental and global charters as well as treaties in reference to migration and community integration that South Africa has signed and committed to) as well as addressing border management and documenting processes for migrants, those in the country and those coming into the country.
- Develop innovative solutions aimed at reducing tensions in the small and informal trading sectors.
- Revisit the existing policies on migration to ensure that they speak to the lived realities of migrants and that they protect their rights as enshrined in the Constitution.
- Strengthen re-integration processes using a multi-stakeholder approach involving government departments, civil society, law enforcement agencies, organisations representing migrants and other relevant stakeholders and convene reintegration dialogues regularly towards promoting social cohesion in the province.
- Continue provincial-wide mass public education on the rights of African migrants including sensitisation campaigns for all civil servants including law enforcement and public health officials on the various categories of migrants and their rights.
- Establish provincial intelligence gathering structures and early warning systems to ensure information sharing across government and relevant structures towards effective collaboration and speedy response to any threats of violence against African migrants.

After the report was presented and handed over to the Premier of KwaZulu-Natal, several actions have followed in response to the report, led by the office of Premier in partnership with ACCORD and other relevant stakeholders. They are:

- The establishment of a broad-based stakeholder council, known as the KwaZulu-Natal Social Cohesion, Moral and Regeneration Council (KZNSCMRC). The council is chaired by the Premier and is composed of municipal mayors, religious leaders, civil society leaders, representatives from the South African Police

Services (SAPS) as well as government departments. The main aim of its establishment is to provide a platform for all stakeholders involved in promoting peace and social cohesion in the province. It formulates an action plan for mitigating any xenophobic conflict and the provincial government, led by the office of the Premier, providing the necessary political support to implement the action plan mutually agreed upon by all stakeholders in the Council.
- Civil society has organized itself by forming a network of organizations working on migration aspects in KwaZulu-Natal province. The organisations act as "early warning" systems as they are working in communities where xenophobic violence tend to erupt. ACCORD is part of this network alongside the United Nations High Commissioner for Refugees (UNHCR). The network meets regularly to appraise itself on matters affecting migrants and any other issues threatening peace and social cohesion in the province.

ACCORD continues to co-convene civil society and government dialogues in partnership with the office of Premier regularly reflecting on the state of social cohesion in the province and the progress made towards implementing recommendations is concerned. This approach ensures accountability on the part of the provincial government to its commitments as far as supporting and promoting social cohesion processes is concerned.

14.7 Case Study 2: The African Solidarity Network (ASONET)

The African Solidarity Network (ASONET), which was established in 2015, is a non-governmental organisation formed by African migrants living and working in KwaZulu-Natal. It represents their views on all aspects of migration, their lived challenges ranging from immigration issues to accessing public services, and other issues. Two pillars underpin ASONET's social cohesion processes:

- Building a strong African migrants' network in KwaZulu-Natal that engages the government at various levels to appraise it of the plight of African migrants in the province.
- Working with various government departments and civil society to promote the peaceful co-existence of African migrants and their South African hosts in the communities they live in.

Since its establishment, ASONET has successfully mobilised African migrants to join the organisation and work together to engage the government, its agencies and civil society as a legitimate entity that represents the plight of African migrants in KwaZulu-Natal. Using a bottom-up approach, it has been involved in many interventions to address tensions and conflict between African migrants and their South

African hosts. ASONET uses a combination of community dialogues and community radio stations to raise awareness of the plight of African migrants and migration. Over the years, this model has proven successful in educating the public about migrants and migration, and to some extent, reducing tensions between African migrant entrepreneurs and their South African counterparts in areas known for xenophobic violence in the greater Durban metro. For instance, when xenophobic violence broke out in 2015 in the greater Durban metro, ASONET mobilised other civil society organisations and African migrants in the affected communities (Isipingo, KwaMakhutha, Umlazi, Inanda) to engage with the political leadership at both local and provincial levels to address the challenges that led to the violence. This mediation process lasted for six months until normalcy was restored in the those affected areas. The mediation process involved convening community dialogues monthly, bringing together African migrants and local South African hosts as well as community political leaders, including representatives from civil society and organisations representing African migrants.

Whilst this mediatory intervention was important for building trust between African migrants and their local South African hosts, a relationship between ASONET and various government departments such as the Departments of Home Affairs and Arts and Culture as well as the Office of the Premier of KwaZulu-Natal was established, where ASONET alongside other civil society organi4sations such as the Democracy Development Program (DDP), KwaZulu-Natal Christian Council (KZNCC) and Zoelife to name a few, became one of the recognised civil society organisations representing the interests and the plight of African migrants in dialogues with the provincial government to ensure a long-term and sustainable peaceful co-existence between African migrants and their South African local hosts towards promoting social cohesion in the province.

Moreover, ASONET continues to convene reintegration dialogues with both city and provincial governments as well as with various associations representing migrants in its community dialogues. All these dialogues are convened regularly and serve as early warning systems to ensure any planned violence against African migrants is diffused before it even happens. They involve government departments, associations representing African migrants, law enforcement agencies, local community print and electronic media, as well as social security and public health officials to name a few. This multi-stakeholder approach is vital in promoting a broad-based societal understanding of the plight of migrants and in the process strengthen inter-agency relationships, a critical factor in promoting social cohesion in society. Additionally, these relationships also serve as useful interventions for mediating conflicts whenever they flare up. All these efforts are important in promoting social cohesion in the province.

As a member of the KwaZulu-Natal Civil Society Organisations Coalition (KZNCSOC), ASONET works together with other grassroots civil society organisations in ensuring the rights of African migrants are known and respected by ordinary citizens. This public education approach aims at enhancing community's understanding of migration, the reasons why people migrate, as well as the various categories of migrants, and in the process influence societal narrative on migrants, which is often laden with stereotypes that promote hatred and discriminatory behavior towards African migrants.

Over the years, ASONET, through its active engagements with city and provincial governments, has earned an opportunity to influence migration-related issues on behalf of African migrants in various government-led social cohesion interventions. The plight of ordinary migrants is known and heard by government and its departments because of ASONET's ongoing advocacy for integration of migrants into communities.

14.8 Case Study 3: The Democracy Development Program (DDP)

The Democracy Development Program (DDP), which was founded in 1993 to promote and consolidate democratic values and principles in South Africa is a non-partisan, non-governmental organisation based in Durban, KwaZulu-Natal. It advocates for an active citizenry that can hold those in power to account and a society in which the constitutionally enshrined Bill of Rights yields tangible benefits for all. Since its inception, the organisation has strived to build citizens' capacity to participate actively in building democratic institutions in society from grassroots to national levels, and mobilised civil society to speak with one voice as it engages with the state at various levels of governance whilst educating citizens on their rights and encouraging them to confidently participate in public life. It contributes to the nation's democratic development at community level by nurturing the lived values of democracy in individuals and communities. These values give rise to new norms and practices which support equality, active citizenship, and social justice. The DDP convenes community dialogues about shared humanity and building sustainable partnerships that influence social dynamics towards building a strong social fabric which supports social cohesion. Over time, it has facilitated mutual accountability amongst communities and actors involved in social cohesion building processes, working together towards a common vision of peaceful communities characterised by mutual respect, accountability, and transparency in all their dealings with one another, bound by common values.

For more than a decade, the DDP has been a convener of civil society in KwaZulu-Natal, and lately in Southern Africa. It has partnered with provincial and city governments together with other civil society organisations such as ACCORD, UNHCR, Africa Unite, Peace Builders Durban and ASONET, among others, to explore lasting solutions towards reduce xenophobic tendencies in the province. It mediated for peace in the 2008, 2015, 2017 and 2021 xenophobic conflict in the greater Durban metro. The mediation process included convening community dialogues between African migrants and the local South African hosts in various locations in the greater Durban Metro. The dialogues sought to explore solutions to persistent tensions between these two groups, often ending up in violence and increased mistrust among them. The DDP together organised African migrant associations engaged with relevant government departments, ward councilors, traditional leaders, local business leaders, religious leaders, and law enforcement agencies, to develop strategies to mitigate any xenophobic violence. The dialogues spanned over four years being held bi-monthly, with the DDP documenting the process all along to inform future social cohesion efforts in the Durban Metro and province.

The DDP's approach was premised on the agency of citizens and the power they hold as individuals to hold government accountable for its decisions. It used different mechanisms to build this citizen-centered, community-driven agency to co-create solutions to challenges. For instance, the organisation offered capacity building interventions to community-based organisations (CBOs) in those communities commonly known as "hot spots", where xenophobic violence often erupted with devastating effects. The capacity building efforts focused on educating communities both Africa migrants and South Africans on migration, migrants rights, human rights, advocacy, community building mechanisms, accountability, and community leadership skills empowerment among other aspects. The workshops were offered once a quarter for a period of four years. Combined with community dialogues, the interventions improved intercommunity relationships as citizens applied the knowledge gained from workshops. Overtime, the communities understood the power of individual agency in co-creating thriving, safe and peaceful communities, whilst holding each other accountable for commitments they make towards their community building efforts. African migrants' entrepreneurs in the different communities signed up Memorandum of Understanding (MoU) with their local South African counterparts began sharing trading skills and exchanging business ideas. Overtime, peace was restored in those communities, relationships were strengthened, and trust was rebuilt. Overtime, the tensions amongst the two groups began to lessen as they interacted regularly, developed business relationships, and embraced their cultural diversities. The DDP remained their convener and offered technical assistance by way of financial and non-financial resources to support joint regular community dialogues, targeted capacity building workshops including connecting them to other higher level societal networks that helped in strengthening business, self-leadership, emotional awareness, and community engagement skills among other aspects.

In summary, the bottom-up approach utilised by these organisations has been a catalyst in building socially cohesive communities in the greater Durban metro despite the socio-economic, political, and cultural challenges that continue to threaten peace in province. This approach premised on valuing citizens voices, ideas, aspirations, and dreams and actively engaging them in the planning, problem solving and decision making processes, provides the foundation for social cohesion processes. However, for these processes, they must be supported by political will as well as progressive, migrant-friendly policies that promote human rights for all including migrants whilst at the sametime strengthening inter-community relationships through progressive community reintegration processes aligned with the Constitution of the Republic of South Africa and the Bill of Rights. Moreover, ongoing multi-stakeholder dialogues played a critical role in supporting community-based social cohesion processes as African migrants and their local South African hosts continued to engage with each other. The dialogues suggested that social cohesion processes can produce lasting solutions towards peaceful co-existence when supported by all relevant stakeholders bounded by a common societal vision and are ready to hold each other accountable for mutually agreed upon commitments.

14.9 Looking to The Future: The Changing Role of Civil Society in Social Cohesion Processes in South Africa

As the last hope for the common citizen in South Africa, civil society organisations have adopted several strategies that benefit the public. According to Lewis (2018), civil society across South Africa has contributed "*immensely in social cohesion which has earned the trust of the masses*". Although some organisations no longer exist, many more progressive civil society organisations are actively addressing the most pressing challenges facing the country and threatening social cohesion such as inter-ethnic violence, racial discrimination, and violence against African migrants, to name but a few.

In the light of the changing socio-political and economic context, in which civil society's strength and resilience are tested, the sector needs to be more proactive in its advocacy efforts, holding government and societal leaders accountable whilst continuing its mobilisation and civic education efforts to ensure that communities remain connected, engaged, and informed with regard to various processes that promote society's resilience and capacity for collective solidarity (Cashdan 2000). At the same time, the sector should continue reaching out to government and political parties to promote social stability and uphold the democratic values enshrined in the Constitution (Kapundu 2017).

Furthermore, despite the pressing challenges experienced in the post-apartheid era, civil society organisations are committed to ensuring that the engagement space between the state and the public does not shrink over time. Civil society will remain a key actor, in complementing government's efforts to promote social cohesion in society, and spaces for engagement among all stakeholders involved in various social cohesion processes should be vigilantly safeguarded. This requires that civil society continues to strengthen its complementary and supplementary roles, working with all social partners and ordinary citizens.

The sharp decline in the public's trust in government highlights that trust is an essential, yet often overlooked, ingredient in successful social cohesion processes. Without trust in governments, markets and institutions, support for necessary reforms is difficult to mobilise, particularly when short-term sacrifices are involved, and long-term gains might be less tangible. If the current situation in the country is left unchecked, it could quickly descend into social unrest as citizens become increasingly despondent about the government's capacity to deliver on its promises. Civil society thus has a critical role to play as a complimentary partner to government efforts by mobilising itself to provide social services that ease the pressure on citizens at the grassroots level whilst at the same time engaging government at all levels to advocate speedy and better service delivery. This would reduce social instability which is highly likely to fuel xenophobic violence.

Moreover, current socio-political and economic circumstances call for a civil society that is connected and networked by establishing partnerships amongst itself and with multi-lateral organisations such the United Nations High Commission for Refugees (UNHCR) and international non-governmental organisations (INGO), to name but a few. The main aim of these joint partnerships is to pool their resources, and develop shared strategies and joint initiatives geared towards social cohesion processes, sealed by a memorandum of understanding (MoU) to guide their joint operations.

Civil society also needs to capitalise on digital platforms to continue its activism online, especially during the COVID-19 pandemic, where resources are under pressure and movement and in-person gatherings are restricted. Civil society needs to leverage various technology platforms to build networks and advocate peaceful co-existence among African migrants and their local South African hosts.

In summary, civil society should continue to play its multiple roles (facilitative, enabler, constructive challenger, and mediator) to ensure that citizens' voices are heard and that their interests are captured in social cohesion processes. Moreover, efforts should be geared towards engaging the state on key national development issues that can translate into better socio-economic wellbeing for all. The future of civil society has to be people-focused, adaptive, and creative. Using a bottom-up approach, civil society can mitigate the potential dangers that are likely to lead to social unrest in these uncertain times of COVID-19 and beyond.

14.10 Recommendations

- The following recommendations are made to ensure that civil society remains relevant in this era of significant socio-economic and political change that potentially threatens social cohesion and peaceful co-existence:
- **Reinforce the bottom-up approach in all its engagements with social cohesion processes.** This will enable all strategic social partners to work consultatively to identify the ever-changing societal needs and socio-economic challenges confronting the public and engage the government on them to find plausible solutions. As noted earlier, if these challenges are left unattended, they are likely to create a conducive environment for civil protests and thereby threaten social cohesion. It is incumbent upon civil society to continue playing this facilitative and mediatory role to ensure that the interests of the public, including vulnerable populations, are taken into consideration in all social cohesion processes.
- **Organise itself as a sector into a coordinated coalition so that it speaks with one voice when engaging with government and other stakeholders involved in social cohesion processes**. The sector is presently fragmented and characterised

by competing interests, partly due to dwindling funding as a result of COVID-19 which has affected the global funding environment, and partly to the shifting priorities of many civil society organisations as they adapt to the changing socio-political context. Fragmentation of the sector weakens its capability to participate meaningfully in social cohesion processes in the country.
- **Ongoing public education on social cohesion and its importance in building a caring society** through targeted civic education programs. These programs aim to educate citizens and communities at large about their role in promoting social cohesion in society. They are conceptualised, planned, and implemented by civil society organisations due to their proximity to citizens at grassroots level. In this way, civil society has an opportunity to transmit core values such as trust, commitment, and a sense of service, all of which promote peaceful co-existence in society.

Though challenged in many ways, civil society has potential to contribute to social cohesion in South Africa. Against the backdrop of a geopolitical and economic environment that is changing radically, further exacerbated by the COVID-19 pandemic, it is time for the sector to bolster its social cohesion-building efforts in South Africa. This chapter proposed a bottom-up approach as a model to strengthen civil society's participation in various collaborative partnerships and relationships involved in social cohesion processes in South Africa. The onus is on the sector to ensure that it remains vigilant and pursues inclusive engagements for improved outcomes of its social cohesion efforts. This will enable civil society to continue to play its multiple roles as a trusted societal structure that is essential to the success of the various cross-sectoral collaborations that are required to promote meaningful and sustainable social cohesion in South Africa.

References

African Centre for the Constructive Resolution of Disputes (ACCORD) (2015). Report of the Special Reference Group on Migration and Community Integration in KwaZulu-Natal. file:///C:/Users/Paul%20Kariuki/Downloads/ACCORD-Report-Special-Reference-Group-Migration-Community-Integration-KZN.pdf. Accessed 15 May 2021.

Agence Française de Développement (AFD) (2020). Conceptualizing and Analyzing Inequality and Social Cohesion in South Africa. https://www.afd.fr/en/carte-des-projets/conceptualizing-and-analyzing-inequality-and-social-cohesion-south-africa. Accessed 23 Apr 2021.

Banpasirichote, C. (2004). Civil Society and Good Governance: A New Chapter in Thailand's Political Reform? In F. Quadir, & J. Lele (Eds.) *Democracy and Civil Society in Asia*: (Vol. 1) (pp. 213–233). London: Palgrave Macmillan.

Bhargava, V. (2015). Engaging citizens and civil society to promote good governance and development effectiveness. https://www.adb.org/sites/default/files/publication/689831/governance-brief-042-civil-society-covid-19-asia-pacific.pdf. Accessed 24 Apr 2021.

Caelesta, B., & Madalina, B. (2020). Stakeholder engagement as a conduit for regulatory legitimacy? *Journal of European Public Policy* 27(11), 1599–1611. https://doi.org/10.1080/13501763.2020.1817133.

Cashdan, B. (2000). Local government and poverty in South Africa. Municipal Services Project. *Institute for Development Research* 113(4), 74–92.
Chan, J., To, H.-P., & Chan, E. (2006). Reconsidering social cohesion: Developing a definition and analytical framework for empirical research. *Social indicators research, 75*(2), 273–302.
Dickes, P., Valentova, M., Borsenberger, M. (2009) Construct validation and application of a common measure of social cohesion in 33 European countries. *Social Indicators Research* 98(3), 451–473. https://doi.org/10.1007/s11205-009-9551-5.
Edinger-Schons, L.M., Lengler-Graiff, L., Scheidler, S., Mende, G., Wieseke, J. (2020) Listen to the voice of the customer—first steps towards stakeholder democracy. *Business Ethics: A European Review* 29(3), 510–527.
Foley, M.W., & Edwards, B. (1996). The paradox of civil society. *Journal of Democracy* 7(3), 38–52.
Human Rights Watch (2020). 'They have robbed me of my life'. Xenophobic violence against non-nationals in South Africa. https://www.hrw.org/report/2020/09/17/they-have-robbed-me-my-life/xenophobic-violence-against-non-nationals-south. Accessed 2 July 2021.
Katumba, S., Cheruiyot, K., Mushongera, D. (2019). Spatial change in the concentration of multidimensional poverty in Gauteng, South Africa: evidence from the quality of life survey data. *Social Indicators Research,* 145(1), 95–115.
Kapundu, A. K. (2017). A critical evaluation of the roles and strategies of civil society organizations in development: a case study of Planact in Johannesburg. A project submitted in accordance with the requirements for the degree of Master of Arts, University of South Africa. https://uir.unisa.ac.za/bitstream/handle/10500/23561/dissertation_kapundu_ak.pdf?sequence=1&isAllowed=y. Accessed 30 Sept 2021.
Le Cordeur, M. (2016). "Safcei prepared to go to ConCourt over 'secret' nuclear deal", Fin24, May 23. https://www.news24.com/Fin24/energy-dept-finally-responds-to-nuclear-court-challenge-20160612. Accessed 16 Apr 2021.
Lefko-Everett, K., Burns, J., Nontshokweni, S., and Njozela, L. (2018). Exploring South Africans' understanding of social cohesion (SALDRU Working Paper Number 218). Cape Town: SALDRU, UCT. http://www.opensaldru.uct.ac.za/bitstream/handle/11090/905/2018_218_Saldruwp.pdf?sequence=1. Accessed 1 Oct 2021.
Lehman, H.P. (2008). The emergence of civil society organizations in South Africa. *Journal of Public Affairs: An International Journal, 8*(1-2), 115–127.
Lewis, P. M. (1992). Political transition and the dilemma of civil society in Africa. *Journal of International Affairs, 46*(1), 31–54.
Malan, N. (2020). Service Learning and Stakeholder Action: Technology and Education for Urban Agriculture in Johannesburg, South Africa. In A. Thornton (Ed.), *Urban Food Democracy and Governance in North and South* (pp. 177–191). Cham: Springer.
Novy, A., Swiatek, D.C., Moulaert, F. (2012). Social cohesion: A conceptual and political elucidation. *Urban Studies* 49(9), 1873–1889.
Ukase, P., & Audu, B. (2015). The role of civil society in the fight against corruption in Nigeria's fourth republic: Problems, prospects and the way forward. *European Scientific Journal, 11*(2), 171–195.
Xenowatch Report (2019) Xenophobic violence in South Africa: 1994-2018. An overview. Prepared by S. Mlilo and J Misago for the African Centre for Migration & Society. https://www.xenowatch.ac.za/wp-content/uploads/2019/03/Xenophobic-Violence-in-South-Africa-1994-2018_An-Overview.pdf. Accessed 2 July 2021.

Chapter 15
Building the Capacity of Civil Society Organisations in Nonviolent Campaigning: A Case Study from South Sudan

Moses Monday John

Abstract Civil society organisations (CSOs) have increasingly become vital forces in discourses, initiatives, and programmes that foster peace, democratic reforms, and service delivery across the world. Despite their important role, the civil society sector in Sub-Saharan Africa continues to face capacity deficits in its efforts to efficiently prevent violence, manage conflicts, and build peace in the region. In South Sudan, most CSOs have limited knowledge of nonviolent campaigning; its principles, methods, and strategies; and the specific skills needed to use them. This chapter focuses on building the capacity of CSOs in nonviolent campaigning with reference to a PhD action research case study from South Sudan. Drawing on data gathered through focus group discussions, interviews, questionnaires, and structured observations, this chapter highlights how primary data from a targeted action research team can help design training contents and methods. The chapter concludes with a discussion of the most significant short-term outcomes realised by 24 action research participants in their efforts to build peace through nonviolent campaigns and how the author envisages the experience can be replicated in other contexts of Sub-Saharan Africa and beyond.

Keywords Civil society · Nonviolent campaign · Peacebuilding · Qualitative research · Action research

15.1 Introduction

South Sudan's history of resistance dates from the early sixteenth-century Islamic sultanates (1504–1821) and the nineteenth-century Turko-Egyptian occupation of Sudan (1821–1885), the Mahdist State (1885–1898), and the Anglo-Egyptian Condominium colonial rule (1898–1956) to the independence of Sudan in 1956 (Sudan People's Liberation Movement 2016). The Islamic rule of Mahdiyya worsened conditions, as it raided and forcefully conquered some parts of South Sudan (Johnson 2016,

---

Moses Monday John is a PhD student in Peace Studies at Durban University of Technology. His doctoral thesis focused on building the capacity of the civil society organizations in nonviolent campaigning in South Sudan.

J. C. K. Kiyala and G. T. Harris (eds.), *Civil Society and Peacebuilding in Sub-Saharan Africa in the Anthropocene*, The Anthropocene: Politik—Economics—Society—Science 34,
https://doi.org/10.1007/978-3-030-95179-5_15

p. 2). In August 1955, merely a few months before Sudan would gain its independence, a civil war erupted in Torit, which lasted for 17 years (Nyaba 2000, p. 14–21). In 1972, the Addis Ababa Peace Agreement was signed between the Anyanya One Movement and the Khartoum government. Ten years later, the agreement was dishonoured, yet another example of an agreement being dishonoured by the Khartoum regime (Alier 1990, p. 2). The violation of the agreement, coupled with continued economic, social, and political marginalisation, led to the second civil war in May 1983.

On 9 January 2005, the Comprehensive Peace Agreement (CPA) was signed between the National Congress Party (NCP)-led government of Sudan and the Sudan People's Liberation Movement/Army (SPLM/A) and, among other agreements, gave the people of Southern Sudan the right to self-determination (Nyaba 2013, p. 10). The country attained its independence from Sudan on 9 July 2011, following a referendum held in January of the same year (Mulla 2018, p. 6). South Sudan became a sovereign state after more than 50 years of struggling for emancipation and freedom from social, economic, and political domination by Sudan.The political history of many countries is often associated with an armed struggle for freedom and independence. Freedom narratives typically support the common belief that violence is indispensable for obtaining freedom from foreign subjugation, and such a belief ignores the power of civilian-led nonviolent resistance and the historical role that it has played in many national quests for liberation (Bartkowski 2013, p. 1).

Unfortunately, in December 2013, tensions within the ruling Sudan People's Liberation Movement (SPLM) party led to the outbreak of armed conflict between the SPLM and the opposition movement, which became known as the SPLM-In Opposition (SPLM-IO). The conflict continued in July 2016 due to clashes in the presidential palace in Juba between the government and the opposition SPLM-IO forces, which led to the collapse of the peace deal signed in August 2015. The violent conflict has created one of the worst humanitarian crises in modern times. In September 2018, a Revitalised Agreement on the Resolution of Conflict in the Republic of South Sudan (RARCSS) was signed in Addis Ababa, Ethiopia. However, the implementation of the agreement is faced with some challenges, including a lack of funding, the limited political will to implement the agreement, and regular clashes with the holdout armed opposition groups.

People suffering from social, economic, and political emancipation can use a variety of means ranging from violent to nonviolent means or a combination of the two. Thus, a campaign can be violent or nonviolent depending on the means being used. A nonviolent campaign does not involve physical violence against human beings, their properties, and the environment. Individuals or groups take actions in pursuit of social, economic, and political objectives without violence. A nonviolent campaign, as the name implies, refers to a series of organised and observable tactics that are repetitively directed to prevent, transform, or address a specific injustice to achieve the desired goal. A campaign is not an event but a series of actions employed to put pressure on the opponent to achieve specific results (Mandikwaza 2016, p. 24). Civil society often constitutes a large part of pressure groups not only in Sub-Saharan Africa but also across the globe. There is much contention in scholarly circles regarding the meaning and definition of the term *civil society*. I define it as organic, diverse formal and informal groups or associations of people pursuing common, interrelated, or complex interests to improve their societal welfare. Thus, civil society organisations encompass more than merely formally registered non-governmental organisations. They include civic, faith-based, and community-based associations as well as professional unions and networks.

While stories of violence dominate the public discourse in South Sudan, student unions, faith-based groups, and traditional authorities have long had an active role in resisting colonialism, preventing conflicts from becoming violent, and building peace. For example, in the 1920s, throughout the southern region of Sudan, chiefs refused to cooperate with and pay taxes to the colonial administration, facilitate the free movement of their expedition forces, and assist the administrators in settling (Kacuol 2008, p. 21). An anecdote from the 1980s revealed that the South Sudanese rural women of Kachipo and Murle succeeded in stopping the communal fighting by refusing to have sex with their husbands until the men of the two rival tribes had made peace. At an individual level, a female peace activist refused to milk the cow that her husband had stolen (Japan International Cooperation Agency 2017, p. 8). Furthermore, ecumenical church clerics from the south travelled internationally to spur boycotts of oil companies doing business with the Khartoum regime. They campaigned against oil companies because of their roles in fuelling war in Sudan (Ashworth et al. 2014, p. 178). Churches and the New Sudan indigenous non-governmental organisations (NGOs) in the early 1990s facilitated the Wunlit people-to-people peace dialogue, which reconciled the Dinka and Nuer communities as well as the SPLM/A leaders (Ashworth et al. 2014, p. 158). In January 2014, the women of the South Sudan Council of Churches (SSCC) organised a prayer to denounce the war, which had broken out in late 2013. Monthly peace and anti-war marches of the civil society have increased pressure on the government and opposition groups to accommodate the will of the people (John et al. 2018, p. 9). Mass participation increases the effectiveness and longevity of nonviolent political struggles, particularly when women are involved (Principe 2016, p. 4). The New Tribe, a nonviolence social movement, which incorporates all 64 tribes of South Sudan, has released several press statements, petitioned the office of the president regarding prisoners who are being held unlawfully, and held peaceful demonstrations and online campaigns to end war and early and forced marriages (New Tribe 2019, p. 4).

Despite the relative importance of South Sudan's civil society sector in peacebuilding, democratisation, and service delivery, civil society organisations (CSOs) have struggled to make significant improvements to the socioeconomic and political conditions in the country. This is because they have been faced with significant internal and external capacity challenges. Internally, most CSOs working on peacebuilding lack capacity or are not even aware of the potentials of nonviolent techniques in peacebuilding. General knowledge of the principles, methods, and strategies of nonviolent action and its relationship to peacebuilding is limited among the South Sudanese. Recent examples of nonviolent action campaigns launched by CSOs have lacked strategic focus (John et al. 2018, p. 2). In addition, there is the challenge of poor documentation and coordination of activities between CSOs, particularly regarding those activities relating to campaigning to reduce people's vulnerability to different forms of violence and building peace. The gaps in the documentation of CSO's achievements and challenges in implementing nonviolent campaigns in the country are yet another challenge. Externally, CSOs face deliberate harassment, intimidation, and sometimes the detention of activists by state security apparatus. Thus, CSOs in South Sudan operate within a fragile, risky, and limited civic space in discharging their legitimate roles and responsibilities in society.

The main aim of this chapter is to discuss how my PhD action research project, with a focus on building the capacity of CSOs through training in nonviolent campaigning, was able to contribute to transforming the socioeconomic and political conditions in South Sudan. Training, whether for CSOs, diplomats, conflict parties, the military, the police and security forces, or other stakeholders in the field, is a key need in peacebuilding and development contexts. This is because many practitioners still lack some of the core competencies for engaging effectively in peacebuilding (Hallward/Hoger 2019, p. 237). As emphasised by Wilson (2014, p. 23), local actors should be empowered through training to take a lead role in local peacebuilding processes and their sustainability. Peacebuilding is any activity that aims to tackle different forms of violence, injustice, and unbalanced power relationships that prevent people from enjoying their rights and livelihoods and achieving their potential. Meaningful peacebuilding empowers individuals, groups, and their constituencies with the knowledge and skills to undertake nonviolent actions to balance power relationships and reinforce dialogue and meaningful negotiation to realise mutually acceptable solutions that satisfy the interests of the parties in a conflict.

This chapter contributes to the relevant body of knowledge by unveiling the experiences of members of the CSOs who applied the theories of nonviolence in peacebuilding, as reported in the project's short-term outcomes.

The specific objectives of the study were:

- To assess the effectiveness and capacity of the South Sudanese CSOs in the development and implementation of nonviolent campaigns between 2011 and 2017
- To explore examples of nonviolent campaigns that CSOs have implemented in selected countries in their efforts to transform socioeconomic and political conflicts
- To design, implement, and evaluate the outcomes of a nonviolent action training programme with a sample of 25 NGO workers, university students, members of faith-based groups, and other members of civil society

Research on nonviolence has never dominated the academic field of peace research, particularly in Sub-Saharan Africa and South Sudan. People's armed conflicts have been studied in depth, but very little attention has been paid to the contributions of unarmed resisters. Evidently, many library shelves are filled with books on wars in people's history, but almost no documentation on the nonviolent resistance of the ordinary people in South Sudan exists. Even in modern forms of media, a story is not worth reporting if it does not involve violence. The results are meagre coverage and documentation and the nonviolence potential being sidelined (Johansen 2009, p. 69).

To my knowledge, no previous research has been carried out on building the capacity of CSOs to develop and implement nonviolent campaigns in South Sudan. My study fills this gap and contributes to the body of knowledge in this under-researched area by documenting the short-term outcomes of a nonviolent action training programme with members of the country's civil society.

15.2 Methodology

I employed an action research approach and gathered qualitative data using focus group discussions, interviews, questionnaires, and structured observations. A cardinal prerequisite of successful research lies in the careful selection of the research methodology to achieve the study's aim and objectives, given the available time and financial resources. I preferred a qualitative design because I was interested in assessing the knowledge, attitudes, and behaviour of the action research participants towards nonviolence and nonviolent campaigning. Qualitative research is any data-gathering technique that generates narrative data rather than numerical data (Monsen 1992, p. 73). As the name implies, quantitative research deals with elements that can be counted, and it often uses the statistical manipulation of numbers to process data and summarise results (Locke et al. 1998, p. 123). Quantitative research is weak in understanding complex social interactions or settings in which people interact, and the voices of the participants are not directly heard or captured. Qualitative research compensates for this weakness (Creswell 2009, p. 9). Mixed methods research to utilises both qualitative and quantitative data within the same study. Creswell/Clark (2007, p. 5) noted that purposeful data integration enables researchers to seek a more panoramic view of their research landscape, viewing the social phenomena under investigation from different viewpoints and through diverse research lenses (Allison/Joanna 2017, p. 74). A mixed methods design is appropriate for fulfilling research objectives that neither quantitative nor qualitative methods could achieve alone (Ivankova et al. 2006, p. 18). Despite the benefits of using mixed methods in scientific inquiries, no research method is without its weaknesses. For instance, Wisdom/Creswell (2013) argued that the process of mixing methods within one study can add to the complexity of conducting research. It often requires more resources in terms of time, personnel, and technical skills in sample selection, data collection, data analysis, and data integration. Taking into consideration the challenges of a mixed methods study, I chose a qualitative design to gather and analyse data.

In the study, I followed an action research design, which encompasses four components: exploration, devising and implementing an intervention, and evaluating its outcomes. Action research is one of the few research approaches that embraces the principles of the participation, reflection, empowerment, and emancipation of people and groups interested in improving their social situation or condition (Berg 2004, p. 195). It aims to not only identify social problems but engage society in proactively contributing to resolving the prevailing problems. The primary reason for engaging in action research was to connect the theories of nonviolence in peacebuilding to the practices and test them. The approach helped me to shift away from the traditional methods of scientific inquiry that produce theoretical solutions and focus on those methods that are more practical and effective for building peace in the community. As Kaye and Harris have argued:

Research has to produce more than just a book. Without people's participation in developing a solution, the solution is more likely to fail as whatever is proposed has to be done with, not to, the people involved. People are capable of understanding and participating in change. Action research is intended to focus this understanding and then to expand it with the aim of change and improvement (Kaye/Harris 2017, p. 11).

The concept of action research can be traced to the works by John Collier (1890–1947) in the 1930s. However, the more systematic and methodological work on action research is linked to Kurt Lewin, who first used the term *action research* in 1944 and later published a paper entitled *Action Research and minority* in 1946. Lewin believed that it was possible to conduct an experiment in a real-life situation with the aim of achieving a specific goal that had a bearing on problem solving and social change (Bloor/Wood 2006, p. 10). The action research approach has "a cyclical inquiry process", which incorporates diagnosing a problem, planning action steps, implementing, and evaluating outcomes (Babbie/Mouton, 2001, p. 53). It allows for the collection of in-depth data from the small action team and the evaluation of both human actions and their perspectives (Babbie/Mouton 2002, p. 270). Data collection represents the key point of any scientific inquiry (Bryman 2016, p. 5). Figure 15.1 depicts an overview of the action research plan that I designed and followed.

I used a purposive sampling technique to select 24 action research participants from 32 respondents who participated in the focus group discussions (FGDs). They represented NGO workers, university students, faith-based groups, and other members of the CSOs working in peacebuilding. To identify and contrast the gaps in the South Sudanese peace movement, it was imperative to explore the literature to gain insights into how non-state actors in other parts of the world have employed the tactics and strategies of nonviolence to tackle their socioeconomic and political conditions. Hence, I explored literature from selected countries, which addressed the first and second objectives of the study. The insights obtained were used to develop a nonviolent action training curriculum and implement the subsequent training of 24 participants.

15.2.1 The Plan, Action Training Contents, and Methods

When preparing for any training, it is often crucial for the trainer or researcher to plan in advance to ensure that the training goal and objectives are achieved. In the case of this study, the preparations included socioeconomic, cultural, legal, and political context analyses to better understand the research context. This was particularly

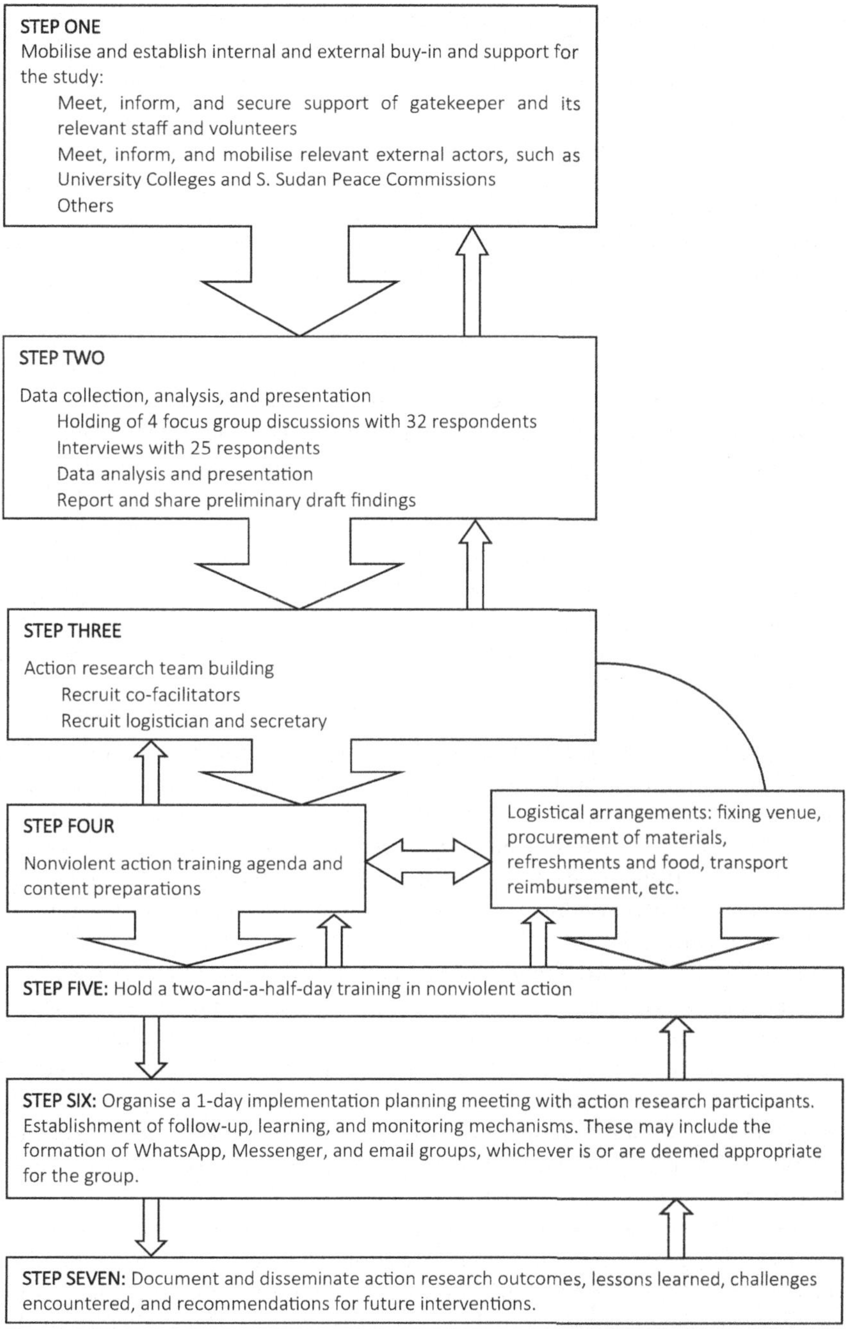

Fig. 15.1 An overview of the action research plan. *Source* Author's design 2019

important because nonviolent action training can be more effective when it is adapted to the sociocultural and political context and needs of the people in a given society. Connecting the training with the felt needs of the people will undoubtedly make it more relevant and sustainable.

I drafted a plan to guide me when to gather data, when to recruit co-trainers, support staff and participants and when to develop the training agenda and curriculum. Table 15.1 presents this plan.

A plan is one matter but its implementation can be another experience. Although I had planned to gather the data within 1 month, but it took four 4 months. This is because the timing for data collection is dependent on the availability of the respondents rather than the researcher. In addition, establishing the initial contacts and buy-in of relevant stakeholders is more vital than rushing to collect data as their support is needed. In my experience, FGDs require much time to group participants according to the constituencies that they represent and the preferred language. While the original plan was to target approximately six to eight individuals for a one-hour conversation, in practice, one FGD was attended by nine participants for obvious reasons. However, the participants did not have much time to emphasise what they wanted to discuss. Thus, the lesson that I learned was that the fewer the number of participants in the FGDs, the more time they would have to share their experiences. Accordingly, I reduced the number of participants in the subsequent three FGDs to five to six participants; this allowed more time for discussions, and I still reached the target of 32 respondents in the FGDs.

The number of training days was also adjusted from 2 to 2.5 days, which allowed the participants to cover the planned content. The adjustment was carried out in consultation between the co-facilitators, the research support team, and the participants. The extension enabled covering the planned agenda and responding to the participants' queries. WhatsApp and Messenger groups were created to share information and support the implementation of the participants' plan. This engagement proved to be useful in reminding the participants of their commitments and reporting the short-term outcomes.

All but two of the 26 participants who attended the FGDs and interviews went on to become action research participants. More than half of the participants were women. The selection of trainers was carried out after a careful analysis of a variety of factors, such as the power dynamics and balance between the male and female trainers. The identification of one experienced female nonviolent action co-facilitator to work side-by-side with the two male counterparts was crucial in the male-dominated society of South Sudan. It was meant to inspire gender mainstreaming and encourage women's participation in the peace movement.

To follow the ethical requirements of the Durban University of Technology and clarify and manage the participants' expectations, I requested each participant to sign a consent letter acknowledging that they were aware that no participant would be paid for participation or requested to pay co-researchers. The study topic and the training aim and objectives were also communicated in advance, as were the training days. I administered pretraining questionnaires before the actual training to assess the extent to which the participants understood violence, conflict, nonviolence, and

Table 15.1 Tentative overview of the data collection and training plan. *Source* Author

Activity	Why?	When?	How long?	Response achieved
Preliminary meetings with the key stakeholders	To inform the stakeholders, learn from them, secure their buy-in, and ensure the needed support for the success of the study.	The meetings were planned for 2 weeks (20 February to 5 March 2018). This lasted until 9 March 2018	14 days	Space, internet, access to libraries, and volunteers were secured at no cost. The support was received from the gatekeeper, the University of Juba, and other partners
Pretesting of research data collection tools	To validate the soundness, relevance, and correctness of the instruments.	6–11 March 2018. This lasted until 21 March 2018	5 days	The tools were revised to ensure the simplicity and clarity of the questions. The same tools were translated into simple Juba Arabic
Data collection and analysis	To obtain information about the research topic from the respondents' point of view	22 March to 21 April 2018. This lasted until 1 June 2018	30 days	Participants' understanding and perceptions of conflict, violence, and nonviolent campaigns and their effectiveness, with reference to the work of CSOs, were assessed and the capacity gaps were identified
Recruitment of co-facilitators and support staff	To recruit two experienced nonviolence trainers, two support staff for logistics and secretariat to co-plan the training	26 March– 2 April 2018. The recruitment occurred in April 2018	7 days	Two nonviolence trainers and support staff were recruited and engaged in the planning of the action project
Nonviolence training curriculum development	To ensure training content, methodology, and learning objectives are planned in advance	22 March– 5 April 2018. Agenda and training curriculum development were completed on 2 June 2018	14 days	The research team of three co-facilitators reviewed and adapted the training agenda and curriculum

(continued)

Table 15.1 (continued)

Activity	Why?	When?	How long?	Response achieved
Invitation of action research participants	To officially invite the participants for the training	26 March–2 April 2018. This did not occur until 3–12 June 2018	7 days	26 participants who attended the FGDs were invited, and only 24 attended the training
Sharing of training agenda with the researcher's supervisor and other trainers for comments	To seek feedback, clarify, and ensure that the methods to be used are relevant to impart the skills and knowledge in non-violent campaigning	5–7 April 2018. However, this took more than 2 months	2 days	Feedback was received from the researcher's direct supervisor, two co-facilitators, USIP nonviolent action trainer, and ONAD experienced trainers. Feedback was incorporated in the training agenda
Purchase and mobilisation of stationery, prepare the training venue, etc.	To ensure training materials and venue are ready and appropriate in time	8–10 April 2018. This was practically handled in July 2018	3 days	Stationery was secured, materials photocopied, and the training venue prepared in advance
Hold a 2-day nonviolent action training	To introduce the concept, principles, methods, and strategies of non-violent action and how they work in practice	12–13 April 2018. The training was convened between 26 and 27 July and on part of 31 July 2018	2 days	24 participants attended the training. Participants were assured that they could leave the training at any time they wished
Hold a 1-day implementation planning meeting	To develop participants' implementation plans	14 April 2018. The 1-day activity was held on 31 July 2018 for half a day	1 day	A total of 21 participants developed their plans. Three participants did not make it for health reasons
Regular follow-up and hold a 1-day evaluation meeting	To gather data for monitoring and evaluation and reporting purposes	15 May 2018– 15 January 2019. This took place on 12 June 2019	9 months	Short-term research outcomes were reported and documented to fulfil the third study objective

nonviolent actions. Five questionnaires were translated from English into Arabic for the Arabic speakers.

The data collected from the 24 completed questionnaires indicated that:

- All respondents were aware of conflict and violence. Close to half of the respondents argued that violent means could be more effective in South Sudan than nonviolent tactics, given the military background of the ruling elites and their harsh responses to peaceful protests.
- Almost all the respondents defined nonviolence as the absence of violence and the peaceful way of handling conflicts without harm. Most respondents said that they had a limited understanding and practice of nonviolent methods and strategies to shift power. Five participants acknowledged that they did not know about nonviolent action and its methods and strategies.
- All the respondents expressed having limited experience in planning, organising, coordinating, and implementing nonviolent campaigns.
- Three-quarters of the participants doubted the effectiveness and potentials of nonviolent campaigns to address socioeconomic and political conflicts, citing repressive responses and a lack of democracy. One participant commented that "Nonviolent action can best work in a democratic and less repressive context than in a very violent and repressive context of South Sudan".
- Almost all the participants were of the view that power lies with leaders who have political, military, and financial resources rather than with the citizens. Most respondents viewed power as the privilege of leaders. However, two respondents had different views: one respondent argued that both leaders and citizens have power, and that power means the ability to voluntarily influence decisions without coercion, while the other respondent emphasised that power lies in education for liberation and critical self-awareness and empowerment.[1]

In order to meet the third objective of the study, a two-and-a-half-day training session in nonviolent action for 24 action research participants was planned and implemented in Juba. The training was designed based on preliminary findings of the capacity gaps or weaknesses identified in the South Sudanese CSOs in their experience in planning and launching nonviolent campaigns, as reported in this chapter. Historical records underscored the importance of training in equipping and preparing the participants to carry out nonviolent campaigns and contribute to building peace. For example, the success of the civil rights movement in the United States was greatly attributed to training in the theories and practice of nonviolent action (Bloch 2016: 3). Sporting teams practice their skills before facing the competing team; the same is true with regard to undertaking nonviolent campaigns. Training is an integral part of human development and capacity building, which aims to ensure that individuals' knowledge base, skills, and abilities are strengthened.

The necessity of training was echoed by more than three-quarters of those reached through the FGDs and interviews as a need to bridge the capacity gap within civil

---

[1] Agnes Batuel and Wilson Saturlino, on 26 July 2018, during the pre-nonviolent action training assessment.

society to develop and implement nonviolent campaigns. Some respondents argued that the understanding of the concept and popular practice of nonviolent methods and strategies are still limited in South Sudan. This insight was expressed by almost all prominent nonviolent activists. One activist noted:

> Training in nonviolent action is critical to build movement's internal human resource capacity to train others who reach out to the grassroots population, inspire frontline nonviolent and human rights activists to connect their local peace activism to national peace movement and consolidate people's power.[2]

Training is important in educating participants on how to sequence and use diverse nonviolent methods and strategies to bring about a political change. It is necessary. There are needs to develop relevant campaigns to change unjust power relations between men and women as well as between the government and the governed. The training is also critical to developing context-specific tactics and supporting individuals in overcoming their fears and retaining strong self-discipline in resisting the temptation to respond to provocations or to fight back (Bloch 2016, p. 14).

Some respondents suggested alternative training methods, such as mentorship, seminars, reading literature related to nonviolent campaigns, watching documentary films, using creative arts and theatre, community and political dialogue, movement-building conferences, and exchange programmes as equally important potential avenues, tools, and opportunities to build the capacity of civil society to implement nonviolent campaigns. For instance, one respondent commented:

> Most training is often conducted with insufficient preparation and with no or less follow-up strategy and is not connected to specific ultimate change objectives. Training is often perceived as an end by itself rather than a means to an end. Thus, many community practitioners and researchers rush to organise training, to keep attendance sheet and training photos as evidence to justify training has taken place. This attitude and abuse of training approach have disappointed some community members to dislike training events.[3]

This constructive criticism informed me to critically prepare the training intervention in such a way as to respond to the participants' needs, aspirations at the community level, and beyond. I used different participatory training methodologies, such as brainstorming, role plays, exercises, case studies, group discussions, documentary films, questions, answers, etc. The available time, uniqueness, and relevance determined the use of each method. The training was conducted with members of CSOs based in Juba but engaged with wider civic and faith-based communities in the rural parts of South Sudan.

---

[2] Victoria Lawrence, on 14 June 2018, in Juba, during the one-on-one interview with the researcher to validate the preliminary findings from the focus group discussions.

[3] Deng Chol, in Juba, on 19 June 2018, during an interview with the researcher to validate the preliminary findings from the focus group discussions.

15.2.2 Day One: What is Nonviolent Action and Why Use It?

Prior to addressing the main theme of understanding nonviolent action, the training commenced with prayers, introductory remarks from the co-facilitators and support staff, and an official opening from the Organisation for Nonviolence and Development (ONAD) administrator. This was followed by a participants' introduction and affirmation exercise, where every participant shared one unique aspect (attitude, behaviour or talent) that they loved about themselves. The exercise helped to create a sense of trust as everyone felt that they had something unique of which they could be proud.

The participants then completed the pretraining questionnaire in either English or Arabic. This was because five participants were Arabic pattern students; they could follow the discussions in English but could best express themselves or write in Arabic.

Subsequently, the research team took the participants through a buzz groups process to establish the training community golden rules and outline what they expected to gain from and share with the training participants. This preliminary session was covered in 2 hours to allow the participants to interact, build rapport and fill in the pretraining questionnaires. This was followed by a 20-minute tea break. The break was necessary as some participants came from the outskirts of Juba City, and some might have missed their morning tea in order to arrive on time.

To introduce the participants to the concepts of nonviolence and nonviolent action, the researcher took the participants through an ideal village game. The participants were divided into three groups of eight. Each group was named after a fruit (Orange, Apple, and Banana). I instructed each group to draw (on a large flip chart paper), in not more than 15 minutes, a picture of an ideal village that they dreamed of seeing in South Sudan.

When the groups' depictions of their villages took shape with rivers, trees, schools, marketplaces, recreational centres, etc. on the paper, the facilitators transformed into the chief executive officers (CEOs) of a multinational corporation interested in extracting resources from the community (water, fossil fuels, land, etc.). Each of the three facilitators made several visits to each community, admiring the village and offering to use the land in return for money, jobs, electricity, etc. On the third visit, each facilitator escalated the conflict by taking some parts of the village, that is, by tearing off part of the community map for a factory, coffee plantation, mall, or other entity. This action angered the Orange and Apple village members and led them to protect their village violently using firearms and traditional weapons against the investors. The Banana village was unfortunately destroyed by the so-called investors and the game ended in chaos.

To unpack the game, the researcher involved the participants in a debriefing session where the researcher questioned the participants on what their feelings were when the investors visited the village, what actually happened (facts) and whether the communities won or lost, whether they had tried any tactics, if there had been a strategy, what had worked and what had not, and what lessons or experiences could one learn from this game for the future.

In response to these questions, the participants were able to comprehend that, in conflict situations, people respond either violently or non-violently or a combination of the two. One of the researcher's co-facilitators then introduced (on previously prepared placards) the definitions of nonviolence and nonviolent action, as well as their principles and methods, to the class. These definitions, principles, and methods have been developed and experimented with by peace movements and researchers over time, some of which are presented in Table 15.2.

Building on Sharp's 198 methods of nonviolent action, the researcher introduced the participants to the four broad categories of nonviolent protest and persuasion, non-cooperation (social, economic, and political), direct intervention, and constructive programmes. These methods and more had already been practised by people and

Table 15.2 The principles of nonviolence and nonviolent action

Principles of nonviolence	Principles of nonviolent action
Search for the centre of conflicts – nonviolence calls for action, not for passivity. We are called to act where people are suffering under violence, oppression, and injustice (King 1999: 101) • Show respect to your adversaries as much as you respect yourself and others – treat all people in a conflict as equals. Remember, nobody can degrade you without your permission. Gandhi argued that it was possible to hate the sin, not the sinner (Nagler 2012: 1) • Remember that everyone can change and that there is something good in all people. In nonviolence, we do not seek to be winners or rise over others; we seek to learn and to make things better for all. • Ends and means have to be compatible; for example, if we want peace, we have to pursue it in a peaceful way. Gandhi stated that peace is the road and not only the goal. • If suffering is unavoidable, take it on yourself rather than harming someone else. The cycle of violence stops with me.Nonviolence involves risk-taking.Thus, resisters must be willing to take risks when necessary and prepare to suffer any consequences to transform injustice (King 1999: 101) • No one has a monopoly on the truth – the challenge is to bring our own and our adversaries' truths together (John et al. 2006: 17)	• Support mass participation: successful nonviolent campaigns choose tactics that enable more people to participate. Larger numbers of participants strengthen the power and legitimacy of a nonviolent campaign and make it more difficult for the campaign to be crushed. According to Stephen/Chenoweth (2008) no campaign that attracted around 3.5% of the population has ever failed to achieve its objectives • Maintain nonviolent discipline: successful nonviolent campaigns use tactics that leverage power while maintaining nonviolent discipline. Keeping actions nonviolent increases participation and reduces the potential for repression by authorities. If repression does happen, it often backfires in favour of nonviolent activists in the form of loyalty shifts or defections from the authority's supporters • Invest in planning: Successful nonviolent campaigns and movements engage in ongoing strategic planning that harnesses assessment, sequencing, escalation, and innovation of tactics to lead to a successful end game. Planning encourages creativity, imagination, connectivity, and sustainability (Bloch/Schirch 2018: 9). • Unity of protesters and campaign leadership

civil society actors across the globe (Engler 2017, p. 14). I connected the methods of nonviolent action to the village game as potential options that the participants could have used or could use in future violent conflicts that they were likely to experience in their daily lives. To connect the principles to the methods of nonviolence, I made a brief presentation in which I explained the importance of using nonviolent methods, with reference to nonviolence principles, to prevent violence and realise success.

After a 40-min lunch break, the participants embarked on a context and power analysis. Common forms of violence and injustices in South Sudan were identified. To address the injustices, a co-facilitator took the training through the "Almighty finger" power exercise. To energise the participants after lunch, this exercise was performed in a circle outside the hall. The facilitator gave the following instructions: form a circle and then one person gets the "almighty finger" to silently or non-verbally give orders by pointing at any person in the circle to act – the orders could be "sit down", "go out", "look up", etc. After 3 minutes, the participant with the almighty finger would hand it over to another participant who would then use it. In the exercise, most participants obeyed the orders, while a few disobeyed. To understand the exercise, I debriefed the participants by asking them the following: What happened? What did the almighty finger do/ask? Why did you obey? Should we be disobedient sometimes? Why? Were Jesus and the Prophet Mohammed disobedient sometimes? Why? What effect did it have?

The responses from the participants regarding authority and power in relation to obedience revealed that most participants attributed an obedient response to a reaction to a fear of sanctions: the subjects obeyed because they felt threatened by the power holder. Two participants disobeyed because they were not happy with the way the authority was used, which they described as unfriendly. When the authority or the person with power was challenged, he (the person who played the role of a power holder in the exercise) asked for more volunteers to support the authority, but this again failed as more volunteers were unwilling to cooperate with the authority when they were ordered to do as such.

This exercise clearly communicated to the participants that power is never the property of an individual. It lies with the people, and once the authorities have been disobeyed or challenged, power vanishes.

After a 15-minute tea break, a 30-minute question-and-answer session was held. This allowed the participants to share their experiences and feelings about what they had learned and what they would like the co-facilitators and the support team to do differently on the days that followed. Suggestions for improvement were also provided. These included arriving on time, being precise, and avoiding repetition to allow for more participation and to save time. The participants requested handouts, which were provided.

Day one was concluded with prayers, and the participants were paid their daily transport reimbursements worth SSP 500 (a mere 500 South Sudanese Pounds), the equivalent of approximately USD 2.5.

15.2.3 Day Two: How Can We Plan and Implement Nonviolent Campaigns?

The second day began with prayers and a summary of what had been covered or learned the previous day. The participants cited the village game and the "almighty finger" exercise as the most creative methods that introduced the nonviolence methods and principles as well as the understanding of power. Most participants acknowledged that they then knew the different methods of nonviolent action, which they could use to manage and confront injustices when they occurred.

To explore how individuals and groups plan and implement nonviolent campaigns, an episode of the documentary entitled A force most powerful on South Africa's economic boycotts was screened and debriefed. The questions used included the following: Which methods of nonviolent action were used? Did it work? Did the organisers stick to nonviolent discipline? What else can we learn from the South African experience?

The researcher then facilitated a session where he divided the class into three groups: the national (South Sudan), regional (Africa), and international (Asia, Europe, and the Americas) teams. Each group was tasked to briefly (in 20 minutes) write down the nonviolent action experiences of which they knew in their specific context. After the groups had completed most of their assignment, the researcher directed each group to hang their flip chart paper on the wall and assigned one representative to report back. In a world café style, the participants toured the three different groups and were introduced to different experiences, and individuals from the groups were allowed to add unreported nonviolent action experiences to each other's groups. After the session, a summarised handout of contemporary stories of nonviolent action in South Sudan was handed over to the participants as a reference. The researcher also encouraged the participants to search for online sources to gain more knowledge on how nonviolent campaigns were being planned and executed in other countries of the world.

To practise skills in creating nonviolent campaigns, the researcher's co-facilitator divided the class into three groups. She directed each group to engage in a brief conversation and identify a violent scenario or injustice at home, in the workplace, or in the community that had not been challenged or resolved. By using creative role-playing activities, each group would then plan, prepare, and act using nonviolent methods to address or transform the injustice. Scenarios that were identified encompassed early and forced marriages, armed robbery, and corruption. The role plays demonstrated the participants' ability to plan and carry out nonviolent campaigns. This session lasted for an hour and 45 minutes, as more time was required to plan, rephrase, and present the different role plays and debriefings.

After a 45-minute lunch break, a co-facilitator introduced the class to the pyramid of strategic planning in developing and implementing nonviolent campaigns. The pyramid includes the vision and core values, mission, assessment, setting SMARTT goals, taking strategic steps, and implementing plans and tactics (Bloch/Schirch 2018, p. 51). The session was complemented with a brainstorming session on how strategic planning supports activists in building just and sustainable peace. As evident in Yogi Berra's well-known saying, "If you don't know where you're going, any road will take you." It was clarified that much as the participants may work individually to plan and tackle specific injustices, it was advisable that they work in groups and coordinate their actions with credible syndicate organisations, peace movements, and networks to have a greater impact. The class was also divided into two groups and was taken through a lion, hunter, rabbit, and mountain exercise. This exercise enabled the group to practise how to strategise for nonviolent action. In the session, the participants also brainstormed on how to deal with fear and state-engineered repression. Day two was concluded with the distribution of training and New Tribe background handouts.

15.2.4 Day Three: What Next?

Day three focused on the practice of nonviolent action. To gauge whether the participants were able to put their knowledge and skills acquired within the previous 3 days into practice, the participants were requested to share their most recent experience where they had used nonviolent action to resolve conflict and build peace. From the reported experiences, a female participant revealed that the training had empowered her, and, as a result, she had approached her husband and resolved family grievances. She reported:

> It all started with coffee taking over the weekend. I chatted with my husband, remembering the old good days following our marriage. I then requested if we talk about a concern that has been bothering me which my husband quickly accepted to listen to. I narrated to him how I felt irritated and bad when he insults me before the children and abuses me when drunk. I spend most of my time crying and regretting our marriage. Before I finished, my husband broke into silent tears. He responded, 'I did not know that I have offended you that much'; he then apologised for the bad behaviour and promised to love and respect me. He too appreciated the way I approached him. Since that time, our relationship improved, and love to each other increased.[4]

She added that persuasion is an effective method of non-violent action, which she had learned 3 days earlier. Other participants shared stories of intervening to resolve conflicts in public transport and the workplace and initiating mediation to resolve a domestic conflict. Although the stories demonstrated the immediate result of this action research project, more outcomes are yet to be harvested. This experiment revealed that the participants were capable of applying the methods of nonviolent action to change their lives and those of their communities.

The second part of day three revolved around planning the application of nonviolent campaigns, networking, and sustainability. Joint training and application planning enhances effective campaigns and ensures a higher possibility of success and informs one how to deal with obstacles as they occur (Bloch 2016, p. 14). The

[4] Grace Linda, on 31 July 2018, in Juba, during the 1-day implementation planning workshop.

researcher led the class through a simple planning session where the planning and the plan were defined. As Simon Sinek agreed with the former US President Dwight Eisenhower that, “Planning is everything and the plan is nothing” (Mrics 2021, p. 2).

To help each participant with that which they wanted or intended to work on to contribute to tangible peace, the researcher introduced simple planning questions, which included the following: what? (for the activity), why? (for the objectives), where? (for the location), when? (for time),with whom? (to establish who else would be involved) and how? (to establish which methods and strategies would be employed).

Each participant was given 35 minutes to prepare and present the post-training implementation planning not more than 5 minutes. The plans were subjected to questions for clarification and further improvement. In researcher’s view, planning constitutes an important aspect of life. Nevertheless, having a plan does not necessarily translate into achieving concrete results unless the plan is put into practice. So planning is crucial to guide the participants in implementing the skills learned to employ nonviolent campaigns. Though the plan was developed in a participatory manner, the co-facilitators may have, at no cost, encouraged and technically supported the implementation process, but the major responsibility for implementation and the choice of methods and strategy fall solely on the shoulders of the trained participants, as argued by Miller (2006, p. 42). In the development of effective strategies, activists must identify the issues and challenges at stake, consider where they come from, generate a dream of what the group wants, articulate its main objectives, and devise a strategy to collectively achieve these objectives (War Resisters’ International 2009, p. 35).

With regard to networking, the researcher’s co-facilitator posed questions to the class: Do you think it is necessary to stay connected after the training? Why and why not? How can we stay together and for what? In answering these questions, most participants voluntarily agreed to stay connected through WhatsApp and Messenger groups to share information, mobilise, organise, and support each other in implementing nonviolent actions and campaigns. Two social media groups were preferred because not all the participants had both Messenger and WhatsApp accounts. It was also agreed for the groups to be for closed members only, to be used to share related follow-up and application information, and to be named “PhD AR Parts” as a shortened version of “PhD action research participants 2018”. Two training participants voluntarily agreed to assist the researcher in establishing the network groups.

After this, the participants were given diaries in which to record their stories as they occurred. It was announced that there would be a follow-up meeting after 6 months, and 1 day would be allocated to an evaluation workshop that was to be convened in 2019.The exact dates would be communicated through the WhatsApp and Messenger groups.

15.3 Results

15.3.1 Measuring Short-Term Outcomes

At the end of the training, post-training questionnaires were administered. Each participant was assigned a unique code comprising the first three letters of their surname and a serial number. This code was only known to the researcher and was deliberately used to ensure that the researcher was able to assess each participant in terms of their level of knowledge and nonviolent action skills before and after the training. Thus, the same questionnaire that was used for the pretraining assessment was distributed to the participants according to their codes 3 days after the training programme in an effort to gauge the short-term outcomes of the training. The collected and analysed data indicated the following findings:

- Almost all the participants had understood that violence and conflict are two different concepts: they are similar but not identical.
- Almost all the participants had understood the concept of nonviolence and were able to clearly differentiate between nonviolent action and nonviolent campaigning as well as their principles, methods, and strategies.
- More than half of the participants had gained optimism and expressed faith in using nonviolent tactics to address injustices and conflict, citing empirical evidence as documented by Stephan/Chenoweth (2008). One young participant commented that he was 100% confident that nonviolent campaigns remained the only hope and means to defeat violence and restore just peace in South Sudan. He added that "we have used violent means for more than 55 years in Sudan and South Sudan, and it has not worked. The only means that we are yet to popularly try is nonviolent campaigns. If they have worked in other countries, I do not see why they cannot work in South Sudan. You never try, you never know."
- More than three-quarters of the participants were in agreement that power lies with the "group of people" or citizens at large and can only be enjoyed by a leader based on the voluntary consent of the governed. If a group of people or citizens choose, for any reason, to withdraw their power and disobey a leader, power vanishes.
- A handful of the respondents had appreciated the knowledge and importance of planning, organisation, and coordination in implementing nonviolent campaigns.
- Close to three-quarters of the participants had appreciated the different methodologies used by the co-facilitators to introduce different themes. Role plays; exercises; and documentary films, such as The Force Most Powerful, were graded as the most useful in internalising concepts and theories and using them in practice. Role plays on tackling specific real-life scenarios were liked by many.

Most participants rated the participatory approach as the best methodology used by the co-facilitators throughout the training. The facilitators adopted experiential learning techniques in which the participants reflected on and learned from their experiences and did not merely rely on the co-facilitators. The use of contemporary local examples of nonviolent actions connected theory to practice. The use of the village game; the "almighty finger"; role plays; and a film documentary, entitled The Force Most Powerful, were liked the most. The case studies of how nonviolent actions had been successfully employed in other countries of the world were stated as not having been exhausted by the facilitators.

The morning review of the previous day's discussions refreshed and reminded the participants to connect the different sessions. Evidently, it also informed the participants who had missed some sessions of the discussions. The translation of the questionnaires and some handouts from English to Arabic was viewed as helpful, as were some limited energisers used between the sessions, particularly in the afternoons and after lunch, which kept the training lively. The food and refreshments were rated as good, although one participant expressed environmental concerns over the use of bottled water and suggested using jugs instead.

The provided handouts were appreciated but fell short of the participants' expectations as some materials, such as success stories of nonviolent actions in other countries, had not been provided. Three participants expressed concerns over the training hall, which they generally rated as good but lacking air conditioning (the hall was equipped only with fans). The participants' transport reimbursements were rated as poor, as most participants were using motorcyclist transport services, which were charging twice as much as the daily transport reimbursements being paid.

15.3.2 Short-Term Outcomes

Short-term outcomes, as the term implies, are short-term changes that can be directly attributed to the action research project. They were determined by that which the participants and their institutions did differently as a result of the training and the consequences that would not have occurred without the contribution of the training. Building on the participants' testimonies and experiences, the training generally had a positive impact on most participants and their constituencies. The participants not only learned together but were able to undertake nonviolent campaigns more confidently after the training to change their conditions using various methods of nonviolent action, such as peaceful marches, stand-up and sit-in protests, petitioning, social media, and offline campaigns, aimed at bringing about change in their communities. The short-term outcomes are summarised in Table 15.3.

In addition to the short-term outcomes summarised in Table 15.3, the participants reported other benefits of training. For instance, almost three-quarters of the participants acknowledged that the training had positively shaped their worldview, attitudes, and behaviours towards nonviolence. It connected the participants to like-minded civil society partners, leading to the exchange of contacts and information, increased coordination, better organising skills and joint actions. The research project bridged the coordination and learning gaps between CSOs.

An elderly participant revealed that "the training was an eye-opener" for him. He added that:

Table 15.3 Outcomes and contributions of action research participants and their organisations in addressing the problems being faced

Title of nonviolent action	Which problem, violence, or injustice was addressed?	Contribution from action research participants and/or their organisations	Outcome story
Protesters marched against alleged mass rape in Bentiu	The systematic rape of women as a weapon of war has been practised with impunity in South Sudan.Most of these actions are being committed by men in uniform	Five action research (AR) participants were part of a women coalition tasked to prepare and commemorate 16 days of activism against gender-based violence (GBV).They seized this annual global event to protest the suspected mass rape	More than 1,000 protesters, drawn mostly from over 40 women and other civil society organisations, marched to the Ministry of Gender and the National Parliament on 10 December 2018 and expressed their disappointment in the alleged mass rape of 150 women in Bentiu (UNMISS 2018)
CSOs launched #MaMaraSakit (Juba Arabic for #NotjustAwoman) campaign	Negative perception that undermines women's potential in South Sudanese society	Eight AR participants and their organisations were involved in the planning and implementation of the campaign launch. They used their constituencies to defuse misperceptions of women	More than 60 women-and-youth-led civil society organisations launched the #MaMaraSakit campaign, Juba Arabic for #NotjustAwoman; she is everything.The launch took place at Logali House in Juba on 6 December 2019
Civil society petitioned the Minister of Information for poor telecommunication services	Poor quality and expensive telecommunication services in South Sudan	One AR participant mobilised other civil society organisations to sign the petition	On 9 January 2019, over 100 civil society organisations from across South Sudan petitioned the Minister of Information, Telecommunication, and Postal Services for poor and expensive telecommunication services (Radio Tarmazuj 2019).

(continued)

Table 15.3 (continued)

Title of nonviolent action	Which problem, violence, or injustice was addressed?	Contribution from action research participants and/or their organisations	Outcome story
Students peacefully protest tuition fee hike leading to its suspension	Taxes and tuition hikes are common responses from the governments to fix the collapsing economy in the country	Four AR participants, in collaboration with the students' union and other student associations, planned, organised, and implemented the campaign infront of the vice chancellor's (VC's) office, carrying placards demanding the revoking of the tuition hike	The president revoked the fees hike and returned the fees to the previous scale. This decision was celebrated by all the students at the University of Juba
Campaign lobby to prevent environmental pollution and its consequences in oil-rich Upper Nile region	Human beings and animals have reportedly died due todumping of expired chemicals in the bush, leakages of oil pumps, and smoke pollution of oil and gas industries	Three AR participants collaborated with CSOs, the University of Juba, and the Parliamentary Committee on Wildlife, Forestry, and the Environment and discussed the impacts of pollution in the oil-rich region	In August 2018, civil society activists peacefully lobbied the National Parliament to impose a penalty on oil companies that were not following environmental safety standards. They also called for immediate environmental audits to be conducted in the oil-rich Upper Nile region and the enactment and reinforcement of strong environmental policies

(continued)

Table 15.3 (continued)

Title of nonviolent action	Which problem, violence, or injustice was addressed?	Contribution from action research participants and/or their organisations	Outcome story
Women lobbied for a 35% women quota as representation ratio in all levels of the government in South Sudan	Women, like youth, have always been under-represented in decision-making processes at the family, community, and political decision-making levels	AR participants from women organisations lobbied throughout the year demanding an affirmative quotaof 50%. On 12 September 2018, the parties signed a peace deal granting a 35% women representation at all levels of the government	In 2018, women groups held grassroots, national, and regional meetings in Entebbe, Uganda,and Addis Ababa, Ethiopia,campaigning for an increase of 50% in women's representation at all levels of the government.They used media, workshops, radio talk shows, and lobbying in Ethiopia to influence the negotiating parties to consider their demand

> Nonviolent action is not actually new. Our ancestors have used its tactics severally without knowing it. The training helped me gained insights on how nonviolent action can be used strategically to shift power in conflict situation in order to set parties to dialogue and negotiate in good faith to address their grievances.[5]

Another respondent acknowledged that the training had helped him to become nonviolent in terms of disciplining his children:

> I learned that people sometimes use violence out of ignorance or lack of knowledge about alternatives to violence. I used to beat my children, thinking beating was a way of discipline. I was wrong and ignorant that it was actually violence. After the training, I changed to using nonviolent ways of discipline, such as cleaning the compound and dishes, writing a story about what had happened, and watering flowers. I also show my children love, care, and respect. I always explain to them, I want them to be good and responsible children to become successful in life. This way, I gained love and respect of my children. If one of them wrongs, he or she can report, seek forgiveness, and promise never to repeat the wrong.[6]

A few participants reported having incorporated the nonviolent action training components into their peacebuilding, human rights, and civic education interventions. Four participants underwent the advanced training of trainers in nonviolent action organised by local and international partners in Juba and Nairobi. Following the release of a report entitled Born to Be Married by the British charity Oxfam, stating that more than 70% of girls in Nyal, South Sudan, are married off before the age of 18 (Oxfam South Sudan 2019). An action research participant launched the #Back to school campaign targeting schoolgirl dropouts. She used community-based organisations and churches to educate the community, raised awareness of the short- and long-term impacts of early marriages and pregnancies, and encouraged dropout girls to re-enrol and continue with their schooling. The result was an increase in girls' enrolment in Magateein schools in Juba.

Not all the participants were optimistic about the power and potential of nonviolence. Some were pessimistic but, with time, learned to appreciate that nonviolent action is effective. A younger respondent reported:

> I did not believe that nonviolent action will work in South Sudan, but I tried it. I, together with more than 200 family members, blocked politicians from addressing funeral prayers held on 4 March 2019. This political non-cooperation succeeded. We insisted to just hear the word of God from the church leaders with no opportunity for political leaders to address the masses as usual. We did that as a protest for the government failure to pay salaries of civil servants for the last six months. Political non-cooperation with politicians who do not commit and fulfil their obligations and duties is the way to resist their injustices. The training helped me to think and act differently.[7]

Some participants reported that they had resolved disputes in their families and workplaces and offered free legal services to the victims of gender-based violence. Conflict management and peacebuilding approaches have often worked with nonviolent resistance to restore broken relationships and build harmonious communities.

---

[5] Arkangelo Tombe, 12 June 2019, in Juba, during the 1-day action research evaluation meeting.

[6] Solomon Gimba, 12 June 2019, in Juba, during the 1-day action research evaluation meeting.

[7] Lilian Patrick, 12 June 2019, in Juba, during the 1-day action research evaluation meeting.

15.4 Conclusion

People around the world have used nonviolent campaigns to build just peace and democratic societies. Despite the challenges of the limited capacity and shortage of literature on nonviolent campaigning, this chapter established that nonviolent activism has existed in South Sudan and that CSOs have on several occasions implemented the techniques of nonviolence in peacebuilding. The short-term outcomes of the action research project revealed that the participants not only learned together, shared information, and established sustainable networks but also used the tactics of peaceful marches, petitions, press statements, sit-ins, political non-cooperation, and other forms of nonviolent campaigns to advance social, political, and economic change. Supporting civil society with technical skills and knowledge on nonviolent campaigning as well as facilitating coherent peacebuilding coordination mechanisms and the attainment of financial resources are critical for building a safe, democratic, and prosperous country.

As Chigas/Woodrow (2018, p. 166) argued, significant progress towards sustainable peace can be achieved through more effective alliances, such as consortia, networks, platforms, etc. The establishment of the New Tribe coalition of nonviolence social movements is a practical example of efforts to strengthen and sustain CSOs' peacebuilding work. Close to half the action research participants have already joined the New Tribe coalition and other syndicate CSOs pursuing ongoing campaigns, using diverse tactics and approaches. The action research approach in peacebuilding facilitates such an experiment, which can be replicated in other contexts in Sub-Saharan Africa and beyond.

References

Alier, A. (1990). *Too many agreements dishonoured*. Reading: Ithaca Press.

Allison, S., & Joanna, S. (2017). Mixed methods research: Expanding the evidence base. *Evidence-based Nursing*, 20(3), 74–75.

Ashworth, J., Ruun, H., LoWilla, E., & Ryan, M. (2014). *The Voice of the Voiceless: The Role of the Church in the Sudanese Civil War 1983–2005*. Nairobi: Pauline Publications Africa.

Bartkowski, J. (2013). *Recovering nonviolent history. Civil resistance in liberation struggles*. Boulder: Lynne Rienner.

Berg, B. (2004). *Qualitative research methods for the social sciences*. Boston: Pearson Education Inc.

Bloch, N., &Schirch, L. (2018). *Synergizing nonviolent action and peacebuilding. An action guide*. Washington, D.C.: United States Institute of Peace Press.

Bloor, M., & Wood, F. (2006). *Keywords in qualitative methods*. London: SAGE.

Bryman, A. (2016). *Social research methods. The nature and process of social research*. 5th ed. Oxford: Oxford University Press.

Chigas, D., &Woodrow, P. (2018). *Adding up to peace. The cumulative impacts of peace programming*. CDA Collaborative Learning Projects. https://www.cdacollaborative.org/publication/adding-peace-cumulative-impacts-peace-initiatives/. Accessed 8 Nov 2019.

Hallward, M., & Hoger, T. (2019). Training and capacity building for peacebuilding and development. *Journal of Peacebuilding and Development*, 14(3), 237–241.

Ivankova, N. V., Creswell, J. W., & Stick, S. L. (2006). Using mixed-methods sequential explanatory design: From theory to practice. *Field Methods*, 18(1), 3–20.

Japan International Cooperation Agency. (2017). Country Gender Profile Republic of South Sudan. JICA, IC Net Limited, March 2017. https://www.jica.go.jp/english/our_work/thematic_issues/gender/background/c8h0vm0000anjqj6-att/south_sudan_2017.pdf. Accessed 12 Jan 2020.

Johansen, J. (2009). Waves of nonviolence and the new revolutionary movements. In M. Meyer and E. Ndura-Oudraogo (Eds.). *Seeds of new hope. Pan African Peace Studies for the Twenty first Century*. Library of Congress Cataloging-Publication Data.

John, M., Joseph, G., &Semjeback, M. (2006). *Report of three days basic nonviolence workshop organised by the Sudanese Organization for Nonviolence and Development*. 8–10 August. Khartoum, Sudan.

John, M., Wilmot, P., &Zaremba, N. (2018). *Resisting violence: Growing a culture of nonviolent action in South Sudan*. Washington, D.C.: United States Institute of Peace.https://www.usip.org/publications/2018/11/resisting-violence-growing-culture-nonviolent-action-south-sudan. Accessed 1 Dec 2018.

Kacuol, A. (2008). *Implementation offederalism in the Sudan*. Khartoum: Sudan Currency Printing Press.

Kaye, S.B., & Harris, G. (2017). *Building peace via action research*. Addis Ababa: UPEACE.

King, M. (1999). *Mahatma Gandhi and Martin Luther King Jr. The power of nonviolent action*. France: UNESCO. http://unesdoc.unesco.org/images/0011/001147/114773e.pdf. Accessed 13 May 2018.

Locke, F., Silverman, S., & Spirduso, W. (1998). *Reading and understanding research*. California: Sage Publications.

Mandikwaza, E. (2016). *Peaceful resistance and nonviolent actions for social change*. Review of Nonviolent Conflict and Civil Resistance: A Handbook. Edited by Gift Mwonzora. (pp. 1–42). https://www.nonviolent-conflict.org/wp-content/uploads/2017/05/Peaceful-Resistance-and-Nonviolent-Actions-for-Social-Change-Edknowledge-Mandikwaza.pdf. Accessed 19 Jan 2018.

Miller, C. (2006). *Nonviolent Transformation of Conflict-Africa*. Switzerland: University for Peace.

Mrics, P.B. (2021). *Planning is everything. The plan is nothing*. Spain: Worwath HTL Hotel, Tourism and Leisure industry report. https://cdn.horwathhtl.com/wp-content/uploads/sites/2/2021/02/IR_Planning-is-Everything.pdf. 30 Sept 2021.

Monsen, E. R., & Van Horn, L. (2007). *Research: Successful approaches* (3rd ed.). Chicago, Illinois: American Dietetic Association.

Mulla, R. (2018). *Federalism with particular emphasis to South Sudan*. Kampala: Fountain Publishers.

Nagler, M. (2012). *Six principles of nonviolence*. https://mettacenter.org/wp-content/uploads/2012/07/Nepal-articlePDF.pdf. Accessed 13 Dec 2017.

New Tribe. (2019). Strategic plan 2019-2021. Juba: New Tribe.

Nyaba, P. (2000). *The politics of liberation struggle in South Sudan*. An insider's view. Kampala: Fountain Publishers Ltd.

Nyaba, P. (2013). *South Sudan. The state we aspire to*. Cape Town: Africa World Books Pty Ltd.

Oxfam South Sudan. (2019). *Born to be married*. https://www.oxfamamerica.org/explore/research-publications/born-be-married/. Accessed 12 Dec 2020.

Principe, M. (2016). *Women in Nonviolent Movements*. Washington, D.C.: United States Institute of Peace. https://www.usip.org/sites/default/files/SR399-Women-in-Nonviolent-Movements.pdf. Accessed 30 Sept 2021.

Radio Tarmazuj. (2019). *More than 100 civil society petitioned the Minister of Information over poor mobile operators' services*. https://radiotamazuj.org/en. Accessed 31 Jan 2019.

Stephan, M. J., & Chenoweth, E. (2008). Why civil resistance works: The strategic logic of nonviolent conflict. *International security*, 33(1), 7–44.

Sudan People's Liberation Movement. (2016). *SPLM manifesto*. Juba: Sudan People's Liberation Movement.

United Nations Mission in South Sudan (UNMISS). (2018). Human rights investigators rush to South Sudan's Bentiu following spate of rapes. https://unmiss.unmissions.org/human-rights-investigators-rush-south-sudan%E2%80%99s-bentiu-following-spate-rapes. Accessed 7 Dec 2018.

War Resisters' International. (2009). *Handbook for nonviolent campaigns*. London: War Resisters' International.

Wilson, J. (2014). *Local peace processes in Sudan and South Sudan*. Washington, D.C.: United States Institute of Peace Press. https://www.usip.org/publications/2014/05/local-peace-processes-sudan-and-south-sudan. Accessed 13 Sept 2019.

Wisdom, J., & Creswell, J. W. (2013). *Mixed methods: integrating quantitative and qualitative data collection and analysis while studying patient-centered medical home models*. Rockville, MD: Agency for Healthcare Research and Quality.

Chapter 16
How Informal Peace Committees Complement Macro-infrastructures for Peace in Zimbabwe

Norman Chivasa

Abstract The prospects for achieving peace in any conflicted society hinge on its capacity to design and implement an inclusive infrastructure for peace in response to opportunities and challenges provided by conflict. Zimbabwe is one of the countries that has been held in stasis in the area of post-conflict peace building ever since independence in 1980. To contribute towards a strategy or approach to peace in Zimbabwe, this article reflects the present yet unaccounted for contribution to peace by informal peace committees in Zimbabwe. The emergence of informal peace committees in Zimbabwe affirms that addressing peace challenges is not only a technical issue requiring macro-infrastructures for peace, but requires the participation of all sections of the population affected by conflict. In contrast to macro-infrastructures for peace, which are structured, male-dominated and elitist, the features of informal peace committees include, but are not limited to, flexibility, gender sensitivity and inclusivity, the inclusion of different social groups in the community, the embracing of a participatory approach in which decision making is a shared process, and shared common interests and purpose. The article recommends that informal peace committees can by themselves hardly make an impact without the support of state institutions. Therefore the micro-macro synergy infrastructures for peace should be embraced in Zimbabwe if sustainability is to be achieved.

Keywords Informal peace committees · Micro-macro-infrastructure for peace · Zimbabwe

16.1 Introduction

This article is a reflection on the need to establish a strategy, mechanism or approach that contributes to sustainable peace in post-independence Zimbabwe. The search for a solution comes against the background that Zimbabwe represents a country in the global south that is held in stasis in the area of post-conflict peace building, primarily because infrastructures for peace (14P) in Zimbabwe have not adequately addressed opportunities and challenges presented by episodic conflicts and violence

---

Dr. Norman Chivasa is a Postdoctoral Fellow at the International Centre of Nonviolence, Durban University of Technology.

J. C. K. Kiyala and G. T. Harris (eds.), *Civil Society and Peacebuilding in Sub-Saharan Africa in the Anthropocene*, The Anthropocene: Politik—Economics—Society—Science 34,
https://doi.org/10.1007/978-3-030-95179-5_16

since independence (Tshuma 2019). In its classical sense, peacebuilding is about rebuilding the social, political, economic and cultural life of a society before or after a violent episode. Societies contribute to those aims either through macro or through micro peace initiatives. This article focuses on micro peace initiatives primarily because their peacebuilding interventions remain undocumented and limited to specific geographical locations (Noma et al. 2012).

This article, therefore, is a reflection of the present, yet unaccounted for, contribution to peace by informal peace committees (IPCs) by local people in post-independence Zimbabwe. IPCs are created as a response to a specific conflict with the main goal of contributing to conflict management and post-conflict peacebuilding. They can be formed as a preventative measure, with the goal of preventing the outbreak or escalation of nascent micro-level conflicts into violent and more widespread conflicts (Chivasa 2019).

There are two types of peace committee, namely, formal peace committees (FPCs) and IPCs. FPCs are composed of official members from both sides of a conflict and are created by a legislative framework. IPCs, on the other hand, are made up of individuals from all walks of life at community level and have no specific mandate or legislation (van Tongeren 2012). Members of IPCs are familiar with the community's everyday life and the conflicts in question and have as well an in-depth (albeit instinctive) understanding of the community's collective mindset, both of which capacities are factors that play a minor yet critical role in promoting peace and development. Similar structures have been created in the Wajir district in South Kordofan, Sudan; in Colombia and in certain districts in the DRC, Burundi, Uganda and Afghanistan, inter alia (Adan/Pkalya 2006; van Tongeren 2012), and in post-independence South Africa (Shearing et al. 2006). Accordingly, this article reflects on the contributions of IPCs to peace which macro I4P in Zimbabwe do not.

Even though IPCs are rapidly increasing in Zimbabwe, they have not been fully embraced in mainstream peacebuilding initiatives, in part owing to the elitist and exclusivist nature of mainstream I4P (Chivasa 2017). The problem that this article seeks to address is that the elitist I4P in Zimbabwe frown upon and ignore a novel surge of IPCs at community level which embraces both peacebuilding from below and within. The article acknowledges the value of macro-I4P, and does not seek to replace them, but rather, to bring IPCs into the mainstream peacebuilding efforts using local community structures as a vehicle to create a greater impact in addressing peace challenges in Zimbabwe.

To address the research problem, the chapter will address the following research objectives: to historicise IPCs and macro-infrastructures of peace in Zimbabwe (third section); and to discuss the comparative advantages of IPC's over macro-14P in Africa, and in Zimbabwe in particular (fourth section). The chapter concludes by arguing that the setting up of peace committees in a post-conflict situation to fight the legacy of the violent past by local people sets a pace worth emulating, as these peace committees could be the beginning of homemade solutions to peacebuilding challenges in Zimbabwe.

16.2 Research Approach

This article is a reflection of the present, yet unaccounted for, contribution to peace by IPCs in Zimbabwe. To systematically address the objectives of this article, the qualitative research approach was followed through the descriptive, explorative and normative tasks (Osmer 2008). The descriptive task led the researcher to historicise IPCs in Zimbabwe. The explorative task led the researcher to discuss the comparative advantages of IPC's over macro-14P in Zimbabwe. Sources of data were drawn through document analysis in the form of a literature review on IPCs in Zimbabwe and beyond. The works of Adan and Pkalya (2006); van Tongeren (2012); Shearing/Jenneker (2006) were useful in providing insights on IPCs and their contributions to peace.

16.3 Brief History of Macro-infrastructures for Peace in Zimbabwe

Regarding Zimbabwe, Maruta (2014) observes that from the mid-1800s to the 2000s, Zimbabwean communities have not realised sustainable peace in terms of addressing animosities and the healing of wounds involving a range of challenges: caused by racial and intertribal conflicts between blacks and whites; and the Shona and Ndebele peoples; the horrors of the colonial conflict; the subsequent civil war and electoral conflict in the 2000s. Machakanja (2010) contends that since independence, sustainable peace in Zimbabwe has remained a mirage. She argues that the impacts of violent conflict in June 2000, March 2002, May 2005 and June 2008 have been destructive to the extent that prospects of sustainable peace through pro-peace policies were obstructed and ultimately dashed. She concludes that the top-down approaches to peace since independence in 1980 have tended to neglect local communities as agents that can play significant roles in violence prevention and reconciliation processes.

Similarly, Muzavazi (2014, p. 100) notes that "the history of Zimbabwe is a continuum of political violence and social injustice perpetrated by one group of individuals against another." These problematic relations stemmed from tribal conflicts in the precolonial era, the racial conflict during colonialism and inter-communal tensions and hostilities that have all left negative imprints posing a threat to peace in modern communities (Tshuma 2019).

In response to a history of episodic violence, Zimbabwe has instituted three major I4P initiatives since 1980, namely, the national reconciliation policy of the 1980s, the organ on national healing, reintegration and reconciliation (ONHRI) of 2009, and the national peace and reconciliation commission (NPRC) of 2013. All these 14P have undergone academic scrutiny about the impact of such interventions. Machakanja (2010), Zhou /Hardlife (2012), Hapanyengwi-Chemhuru (2013), Muzavazi (2014) and Chiweshe (2016) examined the reconciliation policy of 1980 and established that it lacked political will and inclusivity. It also lacked participation by people at the grassroots; and failed to address structures of injustice. While these studies acknowledged that the reconciliation policy was noble and timely, they held that the policy left the Zimbabwean population divided.

Other studies that examined the ONHRI revealed that this peace process lacked political will, remained a paper tiger, and that like its predecessor, it left the fractured community unreconciled (Chinoputsa 2012; Muchemwa et al. 2013; Zembe 2013; Mhandara 2014; Muzavazi 2014; Chiweshe 2016).

Subsequently, the NPRC was instituted in 2013 to address the legacy of post-independence violence. As the current I4P initiative, the NPRC initially had a 10-year life span which was expected to expire in 2024, but developments in 2019 have shown that the life span of the NPRC has been extended. Theoretically, the NPRC is intended to carry on with the work of its forerunner, the ONHRI, which was also a precursor to Article 7 of the Global Political Agreement (GPA) signed between major political parties: the Zimbabwe African National Union Patriotic Front (ZANU PF) and two formations of the Movement for Democratic Change (MDC), in September 2008.

Like its predecessors, the NPRC is certainly not without challenges as the initiative is both elitist and informed by the liberal peacebuilding tradition, which is foreign to the culture of local communities in Zimbabwe. The liberal peacebuilding tradition is elitist, exclusivist and relies on actors external to the conflicted community (Mac Ginty 2008). Another gap within the NPRC is the non-recognition of the long-existing and firmly established customary courts of the Shona, Ndebele and other ethnic groups, and religious traditions, which has the potential to feed into the NPRC of Zimbabwe. There is extensive literature which confirms that mainstream I4P in Zimbabwe have predominantly been elitist without any regard to local resources for peace at the disposal of the community (Machakanja 2010; Muchemwa et al. 2013; Murambadoro/Wielenga 2015; Chiweshe 2016). This article acknowledges the contributions of IPCs to peace, which macro I4P in Zimbabwe do not.

16.4 Brief History of IPCs in Zimbabwe

Although many countries in Africa (such as the Democratic Republic of Congo, Kenya, Uganda, Burundi, South Sudan, South Africa and others) have established both FPCs and IPCs in the 1990s, in Zimbabwe IPCs are still new tracing their origins in early 2004, and not much is written about these peace formations.

The first attempt to establish IPCs in Zimbabwe was done by the Zimbabwe Civic Education Trust (ZIMCET), a grassroots organization founded in 2000 following electoral and farm-invasion-related violence. ZIMCET has facilitated the creation of IPCs throughout the ten Provinces in Zimbabwe. By the end of 2004, an estimated nine peace committees had been created in the Harare-Chitungwiza region, 11 in the Mashonaland region, comprising Mashonaland West and Central, 13 in the eastern region, comprising Masvingo, Manicaland and Mashonaland East, and 16 in the southern region, comprising Bulawayo, Matabeleland South and North and Midlands (ZIMCET 2014).

Before creating IPCs, ZIMCET facilitates conflict resolution workshops involving members of political divides and ordinary members. These workshops focus on sensitising communities about conflict management, gender issues, and violence against women and children. Approximately 72 workshops, which drew close to 3 804 participants, were held in Mashonaland West, while Mashonaland Central had 54 ZIMCET-driven workshops attended by an estimated 3 982 participants (ZIMCET 2014).

Another civic organisation that helped to set up peace committees was the Ecumenical Church Leaders Forum (ECLF), founded in 2008. ECLF was created in response to the upsurge of electoral violence in 2008 and was registered as a trust in 2010 (Cele 2013). Other civic organisations which have also advanced the IPCs framework in Zimbabwe are Heal Zimbabwe Trust, Envision Zimbabwe, which has facilitated the creation of peace clubs across the different provinces (Envision Zimbabwe 2014; Heal Zimbabwe Trust 2015). Not all civic organisations and communities involved in setting up IPCs in Zimbabwe are covered in this article, but insights from those mentioned provide a basis upon which the contributions of IPCs to peace in Zimbabwe can be understood and counted.

16.5 An Evaluative Discussion on the Comparative Advantages of IPCs over Macro-14P

This section discusses initiatives by ordinary people to avert violence amidst political turmoil. Examples include the street committees during apartheid and peace committees in post-independence South Africa, and IPCs in South Sudan.

16.5.1 The Case of Apartheid South Africa

South Africa is one of those countries that experienced violent conflict during the apartheid era. However, Lemanski (2008) praises informal community organisations in apartheid South Africa as game-changers when the country was experiencing political turmoil. One such grassroots initiative was the street committees. Street committees focused on bread-and-butter issues, "material needs", as Lemanski (2008, p. 396) puts it. Street committees emerged from townships to represent the aspirations of the poorest of the poor in South Africa. They became a notable challenge that subverted the political system of the day to give peace a chance.

Apart from street committees, another grassroots initiative called community development forums emerged, and it involved those people who were living in suburbs comprising ratepayers. The community development forum traced its roots from people at the grassroots, who reacted against what they perceived as peace challenges in their suburbs. They mobilised each other, and registered discontent through organised protests, marches and boycotts to lobby and force government to address issues that were affecting their wellbeing. These civic movements negotiated for the people's interests and needs, thus affirming the important role of grassroots in championing their shared interests and needs during a transition period. As Lemanski (2008) notes, these informal groups provided the poor and marginalised a window of opportunity to secure a voice for themselves when South Africa was undergoing a political transition. In the end, these grassroots efforts captured the attention of the perpetrators of the unjust system of apartheid and helped to mitigate the impact of violence in townships and suburbs.

16.5.2 The Case of Post-apartheid South Africa

In South Africa, foreign nationals, who include refugees, asylum seekers and migrants, among others, have been targets of xenophobic attacks by local South African citizens since the eruption of xenophobia in 2008. Peace committees have been instrumental in conducting peace education in xenophobic hot spots. The major tasks of IPCs in these communities have involved counteracting xenophobia. This was the case in places such as Alexander Township and Makause in Gauteng province, and Kwamashu and Isipingo in KwaZulu-Natal province. These formations help to reduce fear and promote community policing and improved community reporting of acts of violence (ASC 2015, 2016). As a result, these IPCs have helped to mitigate the spread of xenophobic violence against foreign nationals between 2019 and 2020 by alerting the police force and mobilising communities to quell potential attacks on foreign nationals in South Africa (Nganje 2021).

As the foregoing account depicts, IPCs "provide early warning before violence erupts, and because they are known and trusted by locals, communities turn to them as valuable mediators that prevent people from taking matters into their own hands when violence is imminent" (http://www.asc.org.za/2016/09/27/combating-drivers-of-xenophobia/). Typically, IPCs in South Africa are working hand in glove with formal justice systems in Alexandria, Johannesburg. This followed the involvement of IPCs in mediating conflict involving the death of one Mozambican national in the 2015 xenophobic attacks (Nganje 2021). Odendaal (2010) notes that IPCs act as a link between the state and local justice systems, administering both customary law and cosmopolitan norms. Adan /Pkalya (2006) note that IPCs are hybrid formations that combine both traditional and cosmopolitan values. In the Alexandrian case in Johannesburg, IPCs intervened in xenophobic conflict using the restorative justice (victim-offender centred) approach to complement the formal justice.

16.5.3 The Case of South Sudan

Sudan is one of those countries that have a long history of violent conflict. However, communities there were not negligent in that they came up with various alternatives, one of which involved the formation of IPCs to mitigate the impact of violence. A study on the transformative role of IPCs by van Tongeren (2012) revealed how the nonviolent resolution of conflicts has paid off in Sudan's local communities. He found that in South Kordofan (Sudan, where the Nuba Mountains are) and Unity State (South Sudan), a network of IPCs were aimed at responding quickly to conflicts, preventing smaller conflicts from escalating and helping communities resist any pressure to become involved in local conflicts, with the result that

54% of IPC interventions have resulted in communities "that previously fought alongside one of the parties now have chosen not to"

- [In] 80% of interventions where violence had occurred, no repeated violence has been reported.
- [In] 94% of interventions, the conflict appears to have been resolved or partially resolved" (van Tongeren 2012, p. 108).
- The foregoing review indicates that peace committees are tackling interpersonal violence, using conflict resolution and peacebuilding techniques such as non-violent resolution of conflict, dialogue, mediation and negotiation techniques to foster peace in Sudan.

16.6 Advantages of IPCs Over Macro-14P in Zimbabwe

16.6.1 Flexibility in the Creation of Peace Initiatives

In Zimbabwe, IPCs have been created using self-selection, where local people volunteer to join the committee, but with the community subsequently approving those individuals with qualities such as faithfulness, honesty, and trustworthiness, or who have abilities in resolving conflict (Sangqu 2014). Adan/Pkalya (2006) note that the self-selection process increases the chances of expanding the constituencies of peace committees because it is inclusive of all social groups, including the vulnerable and marginalised. The self-selection process is flexible in that it is based on the norms and values of the host community, which are largely dynamic in nature.

In contrast, macro-14P in Zimbabwe and elsewhere are created by structured and bureaucratic pieces of legislation. For instance, in South Africa, it was the National Peace Accord which brought about the establishment of 14P, which cascaded down to districts and villages (Nganje 2021). In Serbia, peace committees were established through the Committee on Inter-Community Relations in 2002, and in Sierra Leone, they were established through the District Code of Conduct Monitoring Committees (Odendaal 2010). In Zimbabwe, the ONHRI and NPRC were established by the GNU, all of which were elitist. By and large, these macro-I4P are noble, but not flexible and are not building on the local culture of the people affected by conflict. As a result, IPCs come out best because they are designed and implemented within the context where the conflict occurred, using the self-selection process because of iits flexibility and inclusivity. This kind of flexibility occurred during the formation of IPCs in the Wajir district in South Kordofan, Sudan; in Colombia and in certain districts in the DRC, Burundi, Uganda, and Afghanistan, among others (Adan/Pkalya 2006; van Tongeren 2012), and in post-apartheid South Africa (Shearing et al. 2006).

16.6.2 The Gender Sensitivities in the Informal Peace Committee Framework

One major important aspect that IPCs offer relates to their gender sensitiveness. This is so because any woman or man can chair a peace committee. For that reason, women can occupy strategic positions, such as that of chairperson, deputy or secretary, which are usually a preserve for men in traditional structures. This trend has to do with the gender sensitivities that underlie peace committees (Sangqu 2014). Of particular note is that there are no restrictions on women in the decision-making processes in peace committees, as is not the case in traditional structures such as customary courts.

In contrast, most macro-14P are male dominated, as was the case with the GNU in Zimbabwe. The GNU in Zimbabwe was male dominated from inception right up to its expiry in 2013. The NPRC is also accused of being maledominated by turning a blind eye to discouraging the participation of women and youth. As if that was not enough, in 2018, following the disputed presidential election results in which the MDC Alliance lost to ZANU PF, a quasi-peace process called political actors dialogue (POLAD) instituted by the ZANU PF government to resolve political stalemate in the country is also gender-sensitive (Tshuma 2019). The fact that IPCs are gender sensitive gives them a strong niche in peacebuilding because they take into account the experiences and aspirations of the most vulnerable groups in peace and conflict issues.

16.6.3 Inclusivity in the Peace Committee Framework

Peace committees comprise individual people representing different constituencies on the community level. For example, as Moyo (2014, p. 93) notes, peace committees comprise "civil servants, church leaders, traditional leaders, state security sector actors, political party leaders, women, youth and other stakeholders such as organisations operating at the community level." Sangqu (2014) adds that peace committees comprise different components of society, including youth, women, children and religious groups. She notes that these social groups represent different cultural, ethnic, political, religious, economic status and power dynamics existing in communities.

The merits of inclusivity and gender sensitivity of peace committees give them a very strong niche in peacebuilding initiatives. The inclusive nature of peace committees are building blocks for social harmony, cohesion and coexistence between group members at community level.

In contrast, macro-14P in Zimbabwe, in particular, the GPA, are comprised only of two major political parties. Other critical sectors such as civil society organisations, which include, among others, the Church, traditional leadership, women and youth, were not represented. For those women such as Priscilla Misiharambwi-Mushonga, it appears she was in the GPA representing a political party, not women in Zimbabwe, a specific social group. The youth were not represented in the GPA, thus again technically relegating critical stakeholders (women and youth) in peace issues. The inclusive nature of IPCs tells a story about their potential peacebuilding successes compared to macro-I4P.

16.6.4 The Non-elitist Diversity and Non-formality of Peace Committees

IPCs are rooted in the quest for non-elitist peacebuilding motivated by the realisation that ordinary people have the capacity to engage in peacebuilding (Young 2010). For that reason, it is not practical to overlook peace committees on the basis that they are not formal institutions, as to accept this is to allude to the fallacy that formality is superior to informality. What is critical about these formations is that they are diverse, non-formal models of peace promotion at the local community level.

The non-formal participatory approach is one of the major characteristic features of IPCs. A case in point involved the implementation by ordinary people of participatory approaches to tackle violence and rising crime rates, and the al-Shabaab terrorist attacks in Kenya. These locally initiated interventions produced relatively satisfactory positive results by lowering cases of violence and banditry (Anderson/McKnight 2014; Kioko 2017).

16.6.5 The Transformative Role of Peace Committees

There are reasons to believe in the transformative role of peace committees in some places where they operated, which resulted in reduced election violence as a result of bringing together stakeholders from conflicting parties to jointly explore ways of reducing electoral violence. In Zimbabwe, the establishment of inclusive peace committees comprising ZANU-PF and MDC members in Chivi, Nkai and Mudzi districts and other parts of the country helped to break down the polarisation that existed between members of different political parties (ZIMCET 2014). ZIMCET (2014) asserts that in some of Harare's high-density suburbs where peace committees were created, co-existence and tolerance was noticeable during the 2013 election by contrast with the two preceding elections. A similar outcome was attributed to the work of peace committees in the Mutasa district of Manicaland Province (ECLF 2014).

Another report by ECLF (2015) on the Chivi district of Masvingo Province showed that peace committees have contributed significantly by empowering local community members with skills on how to constructively handle conflicts. Many local people in the Chivi district seem to have changed the way they address conflict in their local traditional courts in which conciliation, as opposed to expulsion of offenders, is becoming almost the primary method of dealing with conflict (ECLF 2015). The capacity of IPCs to contribute to some reduction of violence and the embracing of participatory problem-solving are suggestive of the transformative power of these informal peace formations when compared to macro-14P.

16.6.6 Major Challenges of IPCs

The voluntary nature of IPCs is one of the challenges. This is so because peace committees depend upon individuals volunteering and serving, but if volunteers are not forthcoming, or do little after joining the committee, a peace committee can fail both in the short and long term (van Tongeren 2012).

Gender dynamics is another challenge in that within communities' discrimination against women can impede the participation and involvement of women in IPCs (Moyo 2014). If a community is male dominated, the composition of the IPCs will be predominantly male. In Nepal, for example, male domination of peace committees resulted in women losing confidence in the committees and subsequently avoiding participation, thus deterring their participation in local peace initiatives (Frogh et al. 2010). As Adan/Pklaya (2006) have pointed out, although IPCs draw most of their norms and values from both customary and cosmopolitan frameworks, they are faced with the reality of the exclusion of women and the youth, because traditional communities usually insist on maintaining the gender status quo.

16.7 Conclusion

The central argument of this article was to capture what IPCs are contributing to peacebuilding which mainstream I4P in Zimbabwe do not. The most important contribution is that of inclusivity. With respect to inclusivity, it is the calibre of people that make up the peace committee that matters; this relates to issues of personality, status in the community, level of education and degree of maturity, among others. The internal dynamics of the IPC hinge on the extent to which committee members are prepared to work together, and to which the committee is at peace with itself. This realisation often emanates from a clash of personalities, or a divergence in understanding the vision and mission of the peace committee and the needs and peace aspirations of the community at large. As a result, at the formation stage of the peace committee, a lot of attention and effort must be expended in selecting the right calibre of would-be members of the peace committee, and educating them on the vision and mission of the committee and the nature of its work before taking them in. For example, the members recruited must have the ability to perform the vertical link role, to work with all manner of people in the community, to function as a cohesive unit, and perhaps above all, to stay the course as the work ahead is not an overnight event.

Another important contribution of IPCs is that the community level social conflicts that the informal committees often deal with contribute to the political stability or otherwise of the community, as they are the fodder on which the political polarisation feeds. As a result, the more effective IPCs are at the community level of social interaction and relationships, the more relevant their work becomes at the political level in the community.

Overall, as evidence seems to suggest, creating more IPCs may mean helping local communities to have spaces to address conflict as a collective, as peace committees are indicative of local agency by ordinary people to take responsibility for their own peace and development, rather than looking to government to bring peace to their local communities.

References

Adan, M., & Pkalya, R. (2006). A snapshot analysis of the concept peace committee in relation to peacebuilding initiatives in Kenya: practical action. Nairobi: CORDAID.

Anderson, D. M., McKnight, J. (2014). Kenya at War: Al- Shabaab and Its Enemies in Eastern Africa. *African Affairs*, 114(454), 1–27.

Action Support Centre (ASC), (2015). Responses to violence and looting in Soweto. http://www.asc.org.za/2015/02/09/responses-to-violence-and-looting-in-soweto/. Accessed 20 April 2021.

Action Support Centre (2016). Combating Drivers of Xenophobia. http://www.asc.org.za/2016/09/27/combating-drivers-of-xenophobia/ Accessed 20 April 2020.

Kioko, E. M. (2017). Conflict resolution and crime surveillance in Kenya: local peace committees and Nyumba Kumi. *Africa Spectrum*, 52(1), 3–32.

Cele, S., (2013). Profile of the Ecumenical Church Leaders Forum (ECLF): Church and community working together for peace. In United Congregational Church of Southern Africa Newsletter, August, pp. 4–5.

Chinoputsa, L. (2012). National healing remains the most outstanding issue. *Crisis in Zimbabwe Coalition*, Zimbabwe Briefing, Issue 63, 1–7 March, 5.

Chivasa, N. (2017). Efficacy of informal peace committees to peacebuilding: Evidence from Seke district, Zimbabwe. *African Evaluation Journal*, 5(2), 1–11.

Chivasa, N. (2019). Lessons from community-based-driven peace initiatives. *The Sunday Mail*, 3 November, p-. 2. Harare, Zimbabwe. https://www.sundaymail.co.zw/lessons-from-community-driven-peace-initiatives. Accessed 23 June 2021.

Chiweshe, K. M. (2016). Efficacy of top-down approaches to post-conflict social co-existence and community building: experiences from Zimbabwe. *African Journal on Conflict Resolution*, 16(2), 11–34.

ECLF. (2015). Our achievements. http://www.eclf.org. Accessed 12 July 2015.

Envision Zimbabwe, (2015). Building bridges. www.envision.org.zw. Accessed: 23 June 2015.

Heal Zimbabwe Trust, (2015). Peacebuilding. http://www.fmcw.org/uploads/peacebuilding.pdf. Accessed 19 July 2020.

Frogh, W., Abdela, L., Okumu-Alya, F., (2010). From resolution to reality: lessons learned from Afghanistan, Nepal and Uganda on women's participation in peacebuilding and post-conflict governance. CARE International. https://www.carefrance.org/ressources/themas/1/709,CARE-FROM_RESOLUTION_TO_REALITY1.pdf. Accessed 20 July 2020.

Hapanyengwi-Chemhuru, O. (2013). Reconciliation, conciliation, integration, and national healing: possibilities and challenges in Zimbabwe. *African Journal on Conflict Resolution*, 13(1),79–99.

Lemanski, C., (2008). Houses without community: problems of community (in)capacity in Cape Town, South Africa. *Environmental and Urbanization*, 20(2), 393–410.

Machakanja, P, (2010). National healing and reconciliation in Zimbabwe: challenges and opportunities. Wynberg: Institute for Justice and reconciliation.

Mac Ginty, R. (2008). Indigenous Peacemaking Versus the Liberal Peace. *Cooperation and Conflict*, 43(2), 139–163.

Marimbe, D. (2019). A Review of Provincial Peace Committees: The NPRC's Step Towards Sustainable Peacebuilding in Zimbabwe. Focus on Peace Building, Issue 1, Heal Zimbabwe Trust. https://kubatana.net/2019/08/08/a-review-of-provincial-peace-committees-the-nprcs-step-towards-sustainable-peacebuilding-in-zimbabwe/. Accessed 20 Apr 2020.

Maruta, S. (2014). Re-thinking national healing and reconciliation in Zimbabwe. *Southern Peace Review Journal*, 3(1), 5–11.

Mhandara, L. (2014). Systemic transformation in Zimbabwe: towards true national reconciliation and socio-economic development. *Southern Peace Review Journal*, 3(1), 109–127.

Moyo, A. (2014). Community-based Healing and Reconciliation in Zimbabwe. In K. Wall, J. Aulin and Vogelaar, G (eds.), *Empowerment and Protection Stories of Human Security* (pp. 83-95). The Hague: The Global Partnership for the Prevention of Armed Conflict. https://www.gppac.net/resources/empowerment-and-protection-stories-human-security. Accessed 25 Oct 2020.

Muchemwa, C., Ngwerume, T. E, Hove, M. (2013). When will this long nightmare come to an end?, *African Security Review*, 22(3), 145–159.

Murambadoro, R., Wielenga, C. (2015). Reconciliation in Zimbabwe: The conflict between a state centred and people centered approach. *Strategic Review for Southern Africa*, 37(1), 31–52.

Muzavazi, F. (2014). Youth, Politics and Transitional Justice: Exploring the Role of the Youth in National Healing in Zimbabwe. *Southern Peace Review Journal*, 3(1), 92–108.

Nganje, F. (2021). Local Peace Committees and Grassroots Peacebuilding in Africa. In T. McNamee & M. Muyangwa (Eds.), *The State of Peacebuilding in Africa* (pp. 123–139). Palgrave Macmillan, Cham: Springer. https://doi.org/10.1007/978-3-030-46636-7_8.

Noma, E. Aker, D. Freeman, J. (2012). Heeding women's voices: Breaking cycles of conflict and deepening the concept of peacebuilding, *Journal of Peacebuilding & Development*, 7(1), 7–32. doi:https://doi.org/10.1080/15423166.2012.719384.

Odendaal, A. (2010). An architecture for building peace at local level: A comparative study of local peace committees. A discussion paper, The Bureau for Crisis Prevention of the United Nations Development Programme. https://www.un.org/en/land-natural-resources-conflict/pdfs/UNDP_Local%20Peace%20Committees_2011.pdf Accessed 19 Feb 2020.

Osmer, R. R. (2008). *Practical theology: An introduction*. Grand Rapids, Michigan, Cambridge: Wm. B. Eerdmans Publishing.

Paffenholz, T. (2014). International peacebuilding goes local: analysing Lederach's conflict transformation theory and its ambivalent encounter with 20 years of practice. *Peacebuilding*, 2(1), 11–27.

Sangqu, S. (2014). A crucial link: Local peace committees and national peacebuilding, *African Security Review*, 23(4), 422–424.

Shearing, C. Cartwright, & Jenneker, M. (2006). A grassroots governance model: African peace committees. In V. Luker, S. Dinnen & A. Patience (Eds.), *Law, order & HIV in Papua Guinea* (pp. 246–255). Canberra: Pandanus.

Van Tongeren, P. (2012). Creating infrastructures for peace–experiences at three continents. *Pensamiento Propio*, 36(37), 91–128.

Young, N., J., (Ed.), (2010). The Oxford International encyclopaedia of peace. Vol. 3, Oxford: Oxford University Press.

Zembe, S, (2013). Vision for peace and reconciliation in Zimbabwe. In D. Matyszak (ed.), *The Second international conference of transition justice in Zimbabwe.* Johannesburg, The institute for Justice and reconciliation and Zimbabwe Human Rights NGO Forum, 9–11 October, 33–40. https://www.ijr.org.za/portfolio-items/second-international-conference-on-transitional-justice-in-zimbabwe/ Accessed 21 April 2020.

Zhou, G., & Hardlife, Z, (2012). Public policy making in Zimbabwe: a three-decade perspective. *International Journal of Humanities and Social Science*, 2(8), Special Issue, 212–222.

Zimbabwe Civic Education Trust (ZIMCET), (2014). Peace committee concept. http://www.archive.kubatana.net/html/archive/edutra/040201zimcet.asp?sector=CACT&year=2004&range_start=61. Accessed 31 Oct 2014.

Chapter 17
Protecting and Nurturing Children Born from Rape in South Kivu Communities: A Challenge for Civil Society

Josephine Kimanu Mauwa, Sylvia Blanche Kaye, and Denis Mukengere Mukwege

Abstract Rape in wartime has been the multiplier factor causing stress and trauma, since violence against women often results in physical, moral, psychological, and physical wounds. Results include unwanted pregnancies, stigmas, and protracted rebellious attitudes of children born under such painful circumstances in the Eastern Democratic Republic of Congo (DR Congo) since the outbreak of the first war in 1996. Suspected of being potential criminals, like their genitors, and/or symbolising community war-crime victimisation, many children born from rape experience extreme violence and murder attempts. To protect themselves, such children join criminal groups. Traditional community leaders and local associations, as components of civil society organisations (CSOs), rely on cultural norms to nurture and protect this category of war-affected children from the systemic violence to which they turn as their protective mechanisms. This inquiry employs qualitative methods to document children's maltreatment and to assess the CSOs' efficacy in protecting these children. The results of this study are developed under the following concepts: the experiences of mothers as survivors and their responses; the experiences of children born from rape; extreme violence from persons and the community; children's reaction as a consequence of maltreatment; and the challenging functions of CSOs to protect children born from rape, justified by the predominance of culturally based norms in the Kalonge chieftaincy of South Kivu communities.

Keywords Children born from rape · Civil society organisations · Protecting · Nurturing · South Kivu

---

Josephine Kimanu Mauwa is a PhD student in Peace Durban University of Technology and junior lecturer Department of Peace and Conflict Transformation in Faculty of Social Sciences at the Evangelical University Bukavu, Democratic Republic of Congo.

Dr. Sylvia Blanche Kaye is a senior lecturer in the Peacebuilding Programme at the Durban University of Technology.

Dr. Denis Mukwege is a world-renowned gynaecologist, human rights activist, Nobel Peace Prize laureate from South Kivu Province of the Democratic Republic of Congo and Professor at school of medicine of the Evangelical University in Africa.

J. C. K. Kiyala and G. T. Harris (eds.), *Civil Society and Peacebuilding in Sub-Saharan Africa in the Anthropocene*, The Anthropocene: Politik—Economics—Society—Science 34,
https://doi.org/10.1007/978-3-030-95179-5_17

17.1 Introduction

Children born from rape and the protections needed for them remain an overlooked and unaddressed issue in many countries, from local to international levels. Civil society organisations (CSOs) have neglected protections, as have state institutions, in post-conflict societies (Rohwerder 2019). Although child maltreatment is a worldwide public health and social issue (Abbasi et al. 2014), with diverse and complex etiologies, the issue of child maltreatment in wartime as a consequence of war is more complex. The insufficient attention has attracted scholars and practitioners. Sexual violence, the systematic rape of girls and women in times of conflict (Lewis 2008), has caused a number of children to be born from rape in the Eastern Democratic Republic of Congo (DR Congo) since the eruption of the first conflict in 1996 (Douma/Hilhorst 2012; Houten 2018). Such children are engendered from rapists who belonged to various armed factions, such as rebel groups, militias, and governmental armies, or who are wartime opportunistic civilians (Human Rights Watch 2002, 2009; Stearns/Vogel 2015; Broache 2016; Geneva Academy 2019). Children born from rape encounter severe issues related to their identity, which exposes them to many types of maltreatment from their mothers' families, stepfathers, and the community simply due to the fact of being engendered by the so-called "enemies of the community". Nicknamed as "children of bad memories", "bad blood" (Rohwerder 2019, 4), or "cursed children" (Relief Web 2009, p. 2), this perception is a causal factor in their maltreatment, which starts from their birth. This challenging issue, in turn, affects social harmony and peace and destabilises relationships. As emphasised by the National Academy of Sciences (1993, p. 115), "Attitudes held before the birth of the child, such as negative maternal attitude toward an unwanted or unplanned pregnancy, have also been associated with later maltreatment". For some rape survivors, despite the children's innocence, the birth of children born from rape is a vivid image of the rape, a wound sustained by cultural norms (Mushagalusa 2014).

The protection of these children seems to be a great challenge in the communities where CSOs are still influenced by cultural norms, such as child integration into the clan, inheritance, a patrilineal system, respect for sacrality of cultural norms, etc., to engage in the protection of children born from rape. As with most post-conflict countries, interventions to provide protection for children born from rape remain a great challenge (Martin 2007; Rohwerder 2019). However, the basis of these challenges differs from one country to another and from one society to another. To protect themselves against community violence and to overcome the CSOs' challenges in protecting them, children born from rape have learned to use extreme violence. This is due to the fact that some local associations and traditional community leaders from CSOs, as discussed in this study, seem to be reluctant to protect children born from rape, thus avoiding transgressing cultural community norms. This neglect leads to maintaining systemic violence.

The overall objective of this study was to explore the challenges that CSOs face in protecting and nurturing children born from rape in South Kivu's communities.

The aim would be to restore a sense of self-worth that would limit rebellious and retaliatory attitudes demonstrated by these children as they grow in a volatile environment, most of the time being unaware of their fathers. The specific aims of this inquiry were the following:

- To document the treatment of children born as a result of rape in a sample of South Kivu communities
- To assess present efforts by CSOs to protect and nurture these children
- To propose more effective ways of carrying out these functions

The study used a qualitative methodology to collect and then analyse data. The data has been presented and discussed under four major sections: *the context of the study*, to understand the genesis of the dynamism of violence in the community; the *concept analysis* and conceptual framework, which present the philosophical foundation of the key concepts guiding this inquiry; the results of the study, providing details of the treatment of both rape survivors and their children born as a result of rape and their reaction to self-protection; and the discussion, based on the challenging functions of CSOs to protect and nurture children born from rape in the community folded up by the conclusive recommendations of CSOs' functions to nurture and protect children.

17.2 Context

War that has erupted in the eastern part of DR Congo, as mentioned above, has destroyed many aspects of the Kalonge chieftaincy due to the number of children born from rape. Such aspects include the relational, cultural, psychological, and social dynamics of the community. Composed of six sub-villages – Rambo, Fendula, Caminunu, Cibinda, Mule, and Cifunzi – the community has a significant number of children born from rape in the South Kivu province. Statistics show that approximately 10% of households, or 1,036 out of 10,587 households,[1] have at least one child born from rape, distributed as follows:

- Cifunzi, with at least 486 children
- Fendula, with at least 345 children
- Caminunu, with at least 181 children
- Cibinda, with at least 25 children
- The Rambo and Mule sub-villages were not involved in this study due to their minor numbers of children born from rape

These statistics are underestimates due to the fact that the incidents are not well documented in South Kivu. Statistics of such a population are almost unknown, apart

[1] Source: investigation done by the researcher, Josephine Kimanu Mauwa, and the traditional community leader, Kujirakwinja Bafunyembaka Sylvain, helped by the officer in charge of census (recenseur) of four out of six villages of the Kalonge chieftaincy, South Kivu, DR Congo, May–July 2019.

from some local organisations who hold their own statistics. When data is provided, it is under-estimated or nothing is mentioned (Panzi General Reference Hospital 2013). Therefore, the underestimated statistics of children born from rape in Kalonge village, as provided above, illustrates how critical this issue is as it constitutes a social problem in the community if nothing is done to accept these children, integrate them, and reverse the perceptions that they are cursed (Relief Web 2009).

Perceived collectively, children born from rape suffer from community stigmas, marginalisation, exclusion, rejection, extreme violence such as murder, and attempts at murder. They struggle to integrate into their mothers' families and their community due to implicit restrictions from traditional secret society groups, the traditional law-keepers of the community. This category of traditional leaders urges and influences other men's attitudes, encouraging the murder of those children born from rape. Secretive murders in a local community are known as one of the characteristics of traditional community law-keepers. This is demonstrated in Sierra Leone as well, where those social groups are accused of being actors of numerous ritual murders (Gumbu 2010). In other words, traditional community law-keepers determine the fate of children born from rape, and they are even at risk of infanticide. Because of this fact, children have learned violence as a defensive mechanism to equalise their pain and maltreatment in the community for their own safety. In this context of the reciprocal use of violence, where violence is used as a protective mechanism, whether against cursed children or against abusers from the community, a collective response is needed from CSOs, the intermediate sphere between the family, community, and the state (Meyer/Stacey 2010).

17.3 Research Design and Methods

The current study was designed as a case study that adopted a qualitative methodology, which utilised interviews, focus group discussions, and observations. Qualitative data was essential in terms of comprehending how children born from rape could be protected. The adoption of a qualitative methodology was also suitable because this research dealt with human feelings. This study applied a case study paradigm (Yin 2009; Baškarada 2014; Tight 2017) of collecting and analysing data. I used content analysis to gain meanings from the textual data (Hsieh/Shannon 2005), as discussed in the subsequent sections.

17.3.1 Sampling

The study used nonprobability sampling techniques (Ilker Etikan et al. 2016) based on purposive sampling (Teddlie/Yu 2007). It facilitated the selection of key respondents: children born from rape; their mothers, who are rape survivors; traditional leaders; and leaders of CSO associations. Twenty-one (21) children born from rape in the 12- to 18-year-old age bracket and their mothers participated in the research process. Ten local community leaders at the grassroots level were included, those who had witnessed firsthand the deep-rooted hatred and animosity of hostile groups in the society in their daily lives. They are also recognised as the third-level type of peacebuilders/actors in the grassroots leadership (Maiese/Lederach 2017).

Four associations and traditional/community leaders participated in this study. These included *Solidarité des personnes marginalisées dans la communauté* (SPMC) [Solidarity of marginalised people in the community], Action *d'Encadrement de Famille pour le Développement Intégrale* (AEFAD) [Family Support Action for Integral Development], *Comité locale Communautaire* (CLOC) [Local Community Committee], and *Comité de médiation et de conciliation* (CDMC) [Mediation and Conciliation Committee]. The sample consisted of participants representing all six villages of the Kalonge chieftaincy, notably Rambo, Fendula, Caminunu, Cibinda, Mule, and Cifunzi of the Kalonge chieftaincy in the South Kivu province located in the eastern part of DR Congo.

17.3.2 Data Collection

The data came from two categories of the respondent population: adults and children. Obtaining information from children is a complex process that needs careful attention. Proxy-reporting is no longer necessary to learn children's opinions and behavior. What was important was to collect accurate data directly from the children, even though this required skills, attention, caution, and suitable techniques (Finkelhora et al. 2014). The reason why children were involved directly in this study was because they have their own experiences, perspectives, and views on the way their parents, families, and communities treat them, and they are able to contribute toward breaking the cycle of violence. Children are seen as strong, capable, and knowledgeable experts on their own lives, possessing knowledge, perspective, and interest that is best gained from the children themselves (Einarsdóttir 2007). The second category is the one of adults: mothers who are rape survivors and traditional community leaders and leaders from CSO associations. Two types of questions were used to understand the issue of children born from the extreme violence of rape to gain insight on the intergenerational violence and the way CSOs manage intergenerational violence, aimed at breaking the cycle of the systemic violence. Data was collected over a period of 8 months, from December 2018 to July 2019.

17.3.3 Ethical Considerations

The children and adults who participated in this study were told about the concepts of anonymity and confidentiality, which have been applied in this study. They were informed of their rights and choice in having their names revealed – if they wanted to have their names used, this would be done as agreed. However, the concepts were simplified such that the children could understand them and to preserve their right to protection from any harm and inconvenience. For instance, the children were protected from the curiosity of their parents, peers, and other people who wanted to know what the children were saying. Thus, the children were aware of the confidentiality limit to stop someone from getting hurt (Thompson/Rudolph 2000). In addition, the choice of location was a requirement to ensure confidentiality of the children and avoid any influence on the children's answers. To be more effective, a social worker attended each meeting as he had the skills to prevent participants from suffering from secondary trauma and to work with children. The criteria for a child to attend an interview was also taken into consideration and carefully done. Only children born of rape and who were aware of that status were considered as interviewees.

The process of data collection from the children was challenging and complex. It first demanded the consent of their biological parents, adoptive parents, etc., and the consent of children under 16 years of age, which was renewed throughout the course of the data collection process. However, children who are 16 years old can give consent, with competence being defined as having enough knowledge to understand what is proposed and enough discretion to be able to make a wise decision in light of one's own interests (Fargas-Malet et al. 2010).

17.3.4 Qualitative Data Analysis

Kawulich (2004) has addressed the topics of ethnographic analysis, narrative analysis, phenomenological analysis, content analysis, discourse analysis, framework analysis, the constant comparative method, and the case study (Yin 2009; Tight 2017). Content analysis is the approach that was applied throughout the analysis of the collected data. To ensure the trustworthiness and the dependability of the analysis, it depended on the triangulation of information among the different sources of data collected. I have combined two processes to analyse the data, content and thematic analysis, with content analysis as the main one (Krippendorff 2004).

Three phases have been utilised: the preparation, management, and report phases. The preparation phase, which was based on the literature research, has guided the process toward an accurate sampling strategy and the selection of suitable units of analysis (Elo et al. 2014). The organisational phase included the identification of key points, codes, categories, and themes (Erlingsson/Brysiewicz 2017). The themes are the heart of qualitative data analysis (QDA; Creswell 2013); Sutton/Austin (2015).

The management of analysing data started with data collection, including texts, records, and images collected from the field. The data has been transferred from manuscripts to computer files to obtain an appropriate unified text in order to analyse the text manually and electronically. Collected in the mother-tongue language (Mashi) of the responders and in Swahili, the data was translated from those languages to English in order to facilitate the analysis. To identify themes, the following process was followed: moving from meaning units to condensed meaning units, codes, and finally, categories (Krippendorff 2013; Erlingsson/Brysiewicz 2017). Kawulich (2004) provided the guidelines on coding data, which should reflect the purpose of the research and be exhaustive, mutually exclusive, sensitive to the category content, and conceptually congruent – this is the start of the interpretation step (Krippendorff 2004; Hsieh/Shannon 2005). It was helpful to segment themes in order to obtain subthemes and report the frequency of occurrences. The report is the last step of content analysis.

17.4 Civil Society and Violence Protection

17.4.1 Civil society and Citizens' Protection

CSOs, functioning either as intermediate structures between the family, market, and state or as sectors (Meyer/Stacey 2010), have roles to play to prevent violence and bring peace back to the community. Globally, CSOs are known to be strong facilitators of democracy (Koselleck 1988; Bignami 2016; Klein/Lee 2019). The arena of their functions includes unconstrained collective action arranged around shared interests, tasks, and values (Antje/Dieter 2019). The functions of CSOs are still the same whether in the African or Western context, despite the criticisms of the fragmented African CSOs (Paffenholz/Spurk 2006; Asuelime 2017). CSOs' functions differ according to the dimension of their interventions at the international, regional, or local levels and specifically regarding the period of the intervention, peace, or post-conflict society (Fischer 2006). Apart from the democratisation and conflict transformation role (Fischer 2006), globally, seven basic functions of CSOs are the protection of citizens, monitoring for accountability, advocacy and public communication, socialisation, building community, intermediation and facilitation between citizens and the state, and service delivery that gains scholars' consent (Paffenholz/Spurk 2006; Wolfgang/Regina 2012). The CSOs' functions will be assessed to understand the challenges that the CSOs of South Kivu communities have in protecting children born from rape and breaking the cycle of violence.

17.4.2 Civil Society and Violence Prevention

CSOs' characteristics include closeness, legitimacy awarded by the local population and aggrieved people, knowledge of the root causes of violence and conflict, the capability and experience of working in programmes related to the prevention of violence, knowledge on violence dynamics, and positional capabilities of connecting to the larger group of CSOs around the world (Organization for Security and Co-operation in Europe 2018). Such capacities constitute functional advantages that grant CSOs the power to work against oppression and prevent violence, especially when they resort to the early warning mechanism of emerging threats, their spread, and protracted violence (Dörner/Regina 2012). However, despite the tribute given to CSOs on the ground, some limits have to be taken into consideration, which challenge CSOs' violence prevention functionality. This is due to the fact that many CSOs target symptoms instead of the root causes of violence (Organization for Security and Co-operation in Europe 2018). Concern for the root causes is important as they have the capability of addressing the sources of conflict (European Centre for Conflict Prevention 2006). They have the credibility to prevent violence, work with social groups in the community to identify and address grievances that weaken people and bring about peace and cohesiveness on the ground.

17.5 Results

The results of this study are conceptualised from the following sources: the experiences of mothers as survivors of rape and their responses, the experiences of children born from rape, the maltreatment received from other persons and the community, children's persecution, children's reactions as a consequence of maltreatment, and the challenging functions of CSOs to protect children born from rape.

Wartime has had profound effects on mothers who have given birth to children as a result of rape and has produced high numbers of victims. In addition to the family violence that rape survivors experience, they are also exposed to Intimate Partner Violence (IPV) and violence against their children born from rape. Narratives from rape survivors who submitted to abuse due to the presence of their children born from rape and the narratives of their children display violence ranging from minor cases to the extreme, such as murder.

17.5.1 Narratives from Rape Survivors

Rape survivors shared their experiences, as reported here. The first survivor, a mother, shared her story:

> We were gone in the forest, we met militiamen. They captured us, we were five women, and they were seven men. Their commander had taken me by force. I was his sex slave for two weeks. I found myself with pregnancy, I have fallen ill. I was brought to Panzi hospital. Arrived to Panzi hospital, their told me that I was pregnant. They asked me if I wished to abort the baby, I refused and decided to keep the baby, maybe that child will be important to me in the future. They told me to go back home and come back when I would be ready to give birth. One day as I gone to farm, unfortunately, I gave birth there. When my husband heard that I gave birth, he fled away to Hombo. After four months he was returning, saying that we had to kill the child. I refused and I told him that it was a girl, and I cannot kill her because she can be important. He told me that will end our marriage. He left. He came back later to tell me again to tell me that we must kill the child. I asked him how he wanted to

> kill her, will it be by the poison or by knife? He told me he will make her suck his penis like breasts. When she swallows the sperm, she will die, and we will bury her. We therefore divorced. This is my child who was about to be killed, her name is Shukrani.[2]

The second survivor told her story as follows:

> I was at home when I heard someone knocking on the door. I opened the door and see Militia Raiya Mutomboki who were knocking. They abducted me, my husband, and his brothers. They took us to jungle where they tied my legs and proceeded to rape me from behind. They were five of them who raped me in front of my husband and his brothers. As soon as they were done, they took us to another part of jungle called Nzovu where they killed my husband. I stayed there for four months, and they were constantly raping me. Eventually we managed to escape the militia camp. We were many in numbers. To my surprise, I discovered that I was pregnant. With that pregnancy I went home and found my in-laws. They asked me, "where are our brothers? You were taken together". I told them, "Your brothers were killed by militia, I'm the only survivor from the family among the escapees". They told me, "No, we have to kill you as well... you don't have right to stay here since your husband isn't alive". I managed to stay there through the misery for few months, until I gave birth. I gave birth to a boy. After I gave birth, my in-laws told me they had to kill the child because he was a product of rape. They kept pressuring me to give him up, but I refused. They are still insisting on killing the child, claiming that a Hutu seed cannot abide among them. They were following my child in everything he was doing and everywhere he was going. Every time I would wash and hang his clothes, they'd remove and throw them away. If I cook food for the child, they'd take it away, hoping to starve him to death. They said he is a bastard, they will kill him, but I refused the child to be killed. They don't have the child in their family. At times, they could take his schoolbooks from his back and throw them. They would beat him severely even when they found him with his age-mates. Up to present, the child does not have shelter. He started asking me sad question about his origins based on the maltreatments. I am unable to tell him the truth, no one can also tell him. I flew to Cifunzi. The second month, my brother's in-law followed me and started telling me how we were gaining weight while their brother was rotting. Thus, they claimed, they have to kill the child. They started beating him to the point he lost control, but by God's grace, he managed to escape.[3]

Rape survivors have experienced both family violence and IPV. They have been maltreated by their husbands or family in-laws for the simple fact of giving birth to children out of rape. A third survivor narrated:

> I have twelve children, eleven with my husband and one from being raped. However, my husband told me to kill the last one who was born from rape in order to lead life with him. I told him we are not witches to kill this child and he said he had a great witchcraft which will kill the child tonight. He said that the child must die. And I told him to show me that witchcraft. He told me that he will make the child suck his penis, and then the child will die tonight. I took my child, the whole night running from the village. For the moment, my child is 14 years old. If I had not taken this decision, the child would be dead. That's why they want to accept those children in this village. The lives of these children depend on their mothers. Even their half brothers and sisters discriminate on them. If we die, they will also die. If we live, they will also live.[4]

---

[2] Rape survivor, Cizara, 68 years old, interview, Cifunzi, Kalonge, 05 July 2019.

[3] Rape survivor, Mwana, interview, Cifunzi, Kalonge, 59 years old, 02 June 2019.

[4] Focus group discussion, rape survivors' focus group discussion, Cifunzi, Kalonge, 14 May 2019.

The fourth survivor expressed herself as follows:

> When we leave for the farm, his stepfather stays at home revealing his private part to the child. Then you go find the child's health degrading because of his stepfather's insane acts and the child may die. These men are pure evil.[5]

The fifth rape survivor stated:

> Custom kills. You know that custom is belief because if you believe in something could cause you wrong or misfortune. And all the Shi tribe people know that if a man reveals his private part to a child who isn't his biologically, the child must die. There is a term that he will use in Mashi tribe "look at your mother's friend". When the child hears these words in our custom, the child must die.[6]

The sixth survivor shared her ordeal in the following terms: "One day in my absence, my husband told my child: 'Look at your mother's partner [meaning look at that man's exposed genital organ, as a way of bewitching that child]'. When I came back from Ngweshe village, I found the child dead. It's been 15 years now".[7] To clarify the issue, as researcher, I had to ask a follow-up question: When you find the child dead, how do you know that the stepfather uses intimate parts (organs) on the child? How do you know that it is the forbidden custom that caused the death? A female community leader responded:

> The skin of the child will change the colour; he will become like he had Kwashiorkor. The child can die the same day or progressively die. There are some who die the same day. When we find foams or blood coming out of the child's mouth and nostrils, directly we know that it was his stepfather who did that. In our custom, if someone dies with blood in the nostrils and mouth, we know he has died because of a violation to our custom.[8]

17.5.2 Narratives from Children Born from Rape

Children born from rape shared their different stories, which are reported in this section. One of them started stating:

> My name is Shukrani. I'm the child of the pregnancy that my stepfather did not want. My stepfather wanted to kill me, but my mother did not want, and she fled with me. Following this, I wanted to end up my life, but my mother advised me not to kill myself.[9]

A second child reported:

> My name is Cuka, my mother told me that she had me in the jungle.[10] From the jungle we came to this village because we were chased from our family village. Even after we got here

---

[5] Focus group discussion, Cifunzi, Kalonge, 14 May 2019.

[6] Focus group discussion, Cifunzi, Kalonge, 14 May 2019.

[7] Rape survivors' focus group discussion, Cifunzi, Kalonge, 14 May 2019.

[8] Female community leader, Mwira, focus group discussion, almost 65 years old, 14 May 2019.

[9] Child born from rape testimony, interview, Cifunzi, Kalonge, 14 May 2019.

[10] In the children language, coming from jungle means that the mother got his/her pregnancy from rape in the jungle.

> in Cifunzi, they are still hunting us this side to kill us. They followed to kill me, but my mother refused. I was once followed on my way to the river, I ran to report to my mom. My mother then decided for us to move here. Before yesterday, they came to accuse me claiming that I curse their family. They asked my mother to offer me to killing, but she refused. So, they started to beat us. I fled to Cifunzi village.[11]

Children, themselves, have not only escaped from their own murder, whether at their early age or later in their lives, but also witnessed the murder of their age-mates. Some of these children shared the following stories:

> In my neighborhood, there are people who hate me because only of my good behavior, my obedience, and use to say, why this child does not refuse any service to people that need his help. I feel like that is not pleasing them and comes back against me because... Yesterday for example, there was a woman who gave us food me and my friend, I did not eat as I know that the lady hates me. On our way back home, my friend who ate that food had stomach-ache, then his belly started inflating. I took him to a prayer room. The pastor met on duty told me that he ate a poisoning food, the poison that can kill directly. Then I rushed to the hospital with him
>
> At the hospital, they told me that it was late, I have to go back home with him quickly. We went home. He didn't take long, he passed away. He told me, if he dies, they must bury him in the presence of his mother and friends. Then his clothes must be given to his brothers.[12]
>
> In my family, they don't bear with me (my uncle, aunts, and neighbors). When I greet them, they keep quiet. They have ravished my mum's farm... My grandmother is frightening me as she empoisoned me previously. She gave me human flesh to eat. She gave to my sister yomi the flesh of dog. It's when my mother went to consult a soothsayer (clairvoyant) that they told her all these stories... I feel unsafe when I'm alongside with my mother's co-spouses. My heart is grieved too much.[13]

Children born from rape experience hardships and rejection, and some are innocently killed. This is a serious social issue that needs redress. The following section deals with the aspect of the adverse consequences of being born in those circumstances of rape and unwanted pregnancies.

17.6 The Challenge to Civil Society Organisations

During the data collection process, local associations (SPMC, AEFAD, CLOC, and CDMC) and traditional community leaders were involved as CSO components in protecting children born from rape. The starting point to bring peace back and protect children is to break the cycle of systemic violence that has already been established in the community and strengthened by the respect for cultural norms. The cultural norms

---

[11] Child born from rape, interview, Cifunzi, Kalonge, 13 May 2019.

[12] Yani, child born from rape, interview, Cifunzi, Kalonge, 16 years old, 14 June 2019.

[13] This is a religious practice in the area where Protestants used to gather to pray. Most of the time, in the prayer room, there is at least one person on duty to assist the needy. Rayo, child born from rape, interview, 13 years old, 14 June 2019.

are what shape a community's attitude toward children born from rape. They determine the children's acceptance or rejection, integration or exclusion, and protection or the reluctance to protect them.

In answering the question concerning the cultural considerations of children born from rape in families and communities, traditional community leaders and the leaders of local associations did not seem to have positive perceptions of this category of children. Culturally, children born from rape do not have any particular value in the community, as community leaders reported:

> Culturally, they mean nothing. They are considered as incapable, powerless persons, without any value. They are those who have been hated. Called Mushana Ngozi,[14] a picked-up children... A picked-up children perception has affected, provoked stepfathers' behavior and way of reacting against them. That's the main justification of stepfathers' maltreatment, extreme violence, mistreating them and don't grant any value to them.[15]

To realise how challenging the issue of protecting children born from rape is and how it disturbs cultural norms, which is the main reason of avoiding the acceptance and integration of children, we look at the traditional leaders' narratives. These expressed deep emotions of disapproval of the idea. A wise man, traditional community leaders, and leaders of the community screamed and exclaimed loudly, followed by others:

> He! ... (strongly exclaimed) that is not possible! Introducing those children in our clans is to weaken our custom! ... What? Mixing modern law and custom to integrate those children astonishes me and gives me fright. I'm worried! ... The thing that a father said, remains yes and unchangeable. Clan, it's an issue of blood; it's an abomination to mix blood. We can't mix our blood with the one of those children! ... Law and legislative law is an issue of White people, Western civilization. For us, custom first! ... Law comes later and does not have same value as customs. For example, alliance by blood (kunyanana),[16] law can't reach the energy of that alliance in terms of respect, etc. ... As long as a person is not yet accepted into a clan, the integrative process to get inheritance does not exist to him/her... from Mobutu Sese Seko (former President of the DR Congo), cultural norms, custom have deep influence on the psychology of people more than legislative law... We never think about mixing our blood, you (researcher) are bringing new perception of those children![17]

Exclamative language expressed by the most known people such as wise men, local association leaders, traditional community leaders, and the traditional lawkeepers of the community show how outrageous, scandalous, and shocking the integration of children born from rape into the clan is. It is also expressed that CSOs never think about protecting children born from rape from maltreatment, whether it is from minor or extreme violence. In addition, they seem to be reluctant to protect children

---

[14] From Mashi tribe.

[15] Focus group of 10 traditional community leaders and leaders of local associations, Cifunzi, Kalonge, 05 July 2019.

[16] *Kunyana*: (in Shi local language) refers to a tradition or cultural practice of binding friends, families through the suck of blood between people.

[17] Focus group of 10 traditional community leaders and leaders of local associations, Cifunzi, Kalonge, 05 July 2019.

from extreme violence, which implicitly contributes to and justifies the maintenance of a culture of systemic violence, as participants' narrations showed:

> We are worried about the mix of culture, Rwando–Congolese culture. The reinstallation of those children born here in Kalonge long time ago around 1960 still a very sad experience that we still worried about. It was those children who went to support Kagame, Rwandan president, to fight against Hutu to freed Rwanda. We are worried to go back again to that cycle of war... Those children are a bomb that will burst later. Previously, it was Tutsi, but tomorrow, it will be Hutu. It's those children who will accomplish the balkanization project of the Congo.[18]

The core of the discussion lies in the challenges that the traditional community leaders and associations (SPMC, AEFAD, CLOC, and CDMC) that participated in this research claimed to have encountered in protecting children from extreme violence. The role played by some other participants, such as mothers who are rape survivors as well as children themselves, are not to be neglected.

Rape survivors have been subjected to many sanctions in terms of psychological, social, and economic factors to the extent of obliging them to participate in the murders of their children. I argue that children born from rape symbolize a traumatic image to abusers, reminding them of their genitors' crimes committed against family members or against their community. To take revenge in order to relieve themselves of their pain, the extreme displaced aggression has been directed to children born from rape. As noticed in both cases, the gender of children born from rape does not affect the extreme violence attempts. Being a child born from rape, especially from war criminals such as rebels and militias, has been enough to justify the murder of those children. To avoid bearing any responsibility for the crime of murder or being prosecuted by the judiciary system, many techniques of murder have been applied.

Extreme family violence against children born from rape is practiced by male perpetrators using superstition, which is a practice that has been placed at the forefront and trusted for the extermination of children born from rape. This practice seems to be a soft cold murder using mysterious approaches such as witchcraft, magical potions, or practices such as revealing a male private part to children mixed with superstitious powers. These beliefs are strongly spread and culturally accepted in the community, as survivors narrated when I asked how children born from rape were being murdered. Despite the fact that this practice is known and practiced by men in the community, no action to stop this criminal practice, whether from state judiciary institutions or CSOs, has been taken thus far. In addition, mothers do not have the power to stop the practice. The prevalent approach to which they resort is a preventative method to counteract the perpetrator's murder plan against their children, which involves displacement, escapement, ceasing contact with the perpetrator, or divorce to protect their children.

Children born from rape are trapped, caught, and squeezed into a complex configuration of violence, family violence, witnessing IPV, and community violence perpetrated against them, which have heavy physical and psychological effects. Their

[18] Focus group of 10 traditional community leaders and leaders of local associations, Cifunzi, Kalonge, 05 July 2019.

self-strategic attitude to protect themselves against other people's maltreatment as well as against murder attempts made against them does not seem to proceed as they have wished. For instance, as narrated above, some have lost their lives by trying to behave and having a positive attitude toward abusers. Children are victims of many psychological, physic, and relational forms of violence, with the incidents ranging from within their families to the outside community, an overwhelming and complex maltreatment situation including the murder practices mentioned above.

Children born from rape are aware of the weak quality of their relationships with their relatives and family members, causing them to live in great danger and at risk of losing their lives. Despite their efforts to maintain attachment to their relatives, they are conscious of being hated to death where their lives are in jeopardy due to the nefarious attitudes that act against them, such as a lack of love and tolerance, non-communication, non-acceptance, food deprivation, bullying, social distancing, family splits, and other forms of abuse. They are living in a traumatic environment.

Furthermore, not only do other people within the family and those outside the family submit children to xenophobic abuse, but the mothers who are rape survivors sometimes aggravate xenophobic sentiments while venting their anger and trauma upon their children.

Stepfathers are connectors of two sides of families, the mother's family and the mother's husband's family. This is the reason for classifying stepfathers as the first outer group of abusers of children born from rape. Within this category, stepfathers' family members are the most abusive, followed by the stepfathers themselves who disturb children's lives to the point of causing their deaths. Male adults from within the community constitute the second important external group of child abusers who make children feel unsafe and uncomfortable due to diverse forms of maltreatment, abusive attitudes, murder attempts on the children, a lack of communication, dislike, hate, depicted savagery, hypocrisy, and physical violence.

Such children's age-mates who are not their friends constitute the third critical group of child abusers due to the transfer of adult xenophobic attitudes to children. Age-mates apply the same attitude as adults against children born from rape through insults. As such, this becomes a reminder of a previous traumatic situation that has disturbed the children's mind – bringing back bad memories of being fatherless children born from rape survivors who are perceived as prostitutes. Age-mates, as the third important abusive group, make children feel unprotected as the children are bullied, disliked, humiliated, unappreciated, etc. Thus, children born from rape would define their level of safety through certain concepts and realities such as love, communication, and care which would allow them to feel protected. However, the contrary becomes proof of their unsafety. Abusers applying xenophobic attitudes based on the hated identity of children born from rape remains a significant traumatic issue where at least 87% of children[19] have acknowledged that it makes them feel unprotected and in jeopardy.

Children are depicted as outsiders, pariahs of the community who do not belong to the local community due to their identity. They are identified by their background

[19] This statistic came from the interview responses of children born from rape.

story, circumstances in which they were born, the situation of war, insurrection, rebellion, rape against their mothers, etc. As such, they are portrayed as Hutu children due to their morphology and according to their genitors' identity. They are directly told of their lack of resemblance to local people, maternal family members, or their stepfathers' families, which subjects them to psychological violence. Furthermore, this depiction means that they are wild and savage and from a jungle where their mothers have been taken for rape, determining their identity according to their genitors' criminal actions and places of crime. This causes children to feel guilty of being engendered by evil people, bandits, wild people, immoral individuals, etc., turning the genitors' identity into an insult. Thus, due to their morphology resembling that of Hutu people, the children acknowledge that they are not from the community and do not belong to their mothers' families nor completely to their stepfathers' families based on the historical background of their birth and their genitors' origin.

This demarcation between the *us* and *themp* expresses a xenophobic attitude, excluding children from the community. Xenophobic sentiments toward children's identity fold into a negative historical background of the genitors of children born from rape based on a criminal context. This, in turn, justifies the attitudes of male traditional law-keepers and other males, which influences the behavior of individuals, family members, and community members against these children. Their presence, viewed as a manifestation of a curse, leads to massive destruction and the mourning of the families and community due to the crimes of their genitors. Consequently, this allows for the accusation that they are responsible for community disharmony, the deterioration of relationships, and the misery of the family and community. This is the justification for venting all the negative emotions upon those children.

The persecution of children born from rape essentially comes from the protection of cultural norms and background of the previous and current consequences of foreign rebels' crimes in the community, which drive the community's behavior and attitudes to become hostile toward the children. Hence, the heart of the persecution lies in the children's identity as a risk factor motivating extreme maltreatment attached to the criminal background of their genitors. This has been justified by the cultural norms of the community, which, according to traditional community leaders, are transgressed. Children's narratives provide more information on the reasons for their maltreatment.

The challenges faced by CSOs to nurture and protect children born from rape include the issue of bloodlines, the previous historical background of crimes committed by foreign people from Rwanda, and the blame placed on the presence of children born from rape.

Notably, the extreme violence exhibited never attracts their attention. The children's integration into families is first an issue of blood and not an issue of human rights nor a legal issue. As such, the issue of blood becomes primordial and the base of decision making determining the protection, acceptance, and integration of those children. Traditions and customs take over all other systems to protect and integrate children into their families or communities. The concern of traditional law-keepers regarding mixing bloodlines is based on the protection of cultural norms. They disregard the application of *Western law* for fear that it could worsen or weaken their customs. Consequently, this attitude becomes a real challenge for social groups, leaders of associations from CSOs, and traditional community leaders who should be able to advocate, intervene in, and be involved in children's protection processes. Therefore, there seems to be a disconnection between the perceptions of CSO leaders and the values and functions of CSOs.

The communities' experiences of previous wars around the independence period of Africa and that of DR Congo (Christensen/Laitin 2019) have contributed to the societies' reluctance to integrate children born to these rape victims into society. During the independence period, Rwandan children who were residing within DR Congo were granted Congolese citizenship. According to traditional leaders, these same children, now grown up, went back to their native land and recently supported a rebellion in Rwanda, in 1994 (Lubunga 2016). Because of this, male traditional community leaders as well as association leaders of CSOs in the community argue that integrating children born of rape will produce the same effect. They consider these children to be threats to their communities and those of their fathers. This is the main reason of rejecting recent children born from rape of Hutu Rwandan rebels, their genitors. For them, this is a reasonable mechanism to break the transfer of violence in the local community and contribute to bringing peace back. Their presence in the community does not convince the elders, association leaders, and traditional leaders to protect them and accept and integrate them into the local community. The concern of a mixed culture has consequently become a strong feeling, a justification to harshly reject them and accuse them of being criminals coming from criminal genitors or of being a further source of political balkanisation of the DR Congo. There is a mix of cultural and political issues that aggravate the rejection and increase the reluctance to protect these children. This attitude blocks any further initiatives for protecting children.

The children are accused of causing disharmony and inciting violence against themselves from other people, especially from their stepfathers. However, community and association leaders, despite their reluctant attitude, have developed mechanisms to protect the marriages of their daughters who are survivors of rape, which is an indirect approach of protecting children from their stepfathers' extreme violence, murder attempts, or other forms of maltreatment. This system has been applied by experiences not led by cultural norms, a practice drawn from the result of the extreme violence against children from men, especially their stepfathers.

To nurture and secure the future of their children born from rape, mothers have adopted an inheritance technique which has caused more harm than good. This has divided households and families and destroyed marriages when the husbands discovered that their wives had been secretly investing to build a legacy for their children or to protect them from maltreatment. The children are mostly left with their mothers' families when the mothers get married to protect them from being killed by their stepfathers, as narrated in a focus group: "it's because of the experience that people have adopted that system in the community".[20]

A stepfather's capacity for murder has made the community become more protective of the children due to the threat of murder. For the sake of protection, they have

[20] Focus group of traditional community leaders and local associations

made a decision to keep the children in their mothers' families. This is because the practice of murder has not yet attracted and motivated CSOs or the judiciary system to initiate a dissuasive approach to stop the murder or maltreatment of children born from rape.

Children born from rape are considered worthless and lacking in identity. As such, no value is granted to them; in contrast, this perception triggers their maltreatment. They are accused of disturbing individual and relational harmonies and being the cause of their own maltreatment. In other words, as they mean nothing to other people, their presence is enough to provoke other people's emotions and a rise in negative attitudes, abuse, and maltreatment. Stepfathers, for instance, take advantage of this perception as children lack a clear identity and are not protected from the abuse of their genitors, which keeps them in abusive relationships. If stepfathers have this perception of a lack of identity, community leaders, traditional leaders, and leaders of local community associations also keep the same perception of dealing with unvalued children. Therefore, prior to the time of data collection, nothing concrete had been initiated by CSOs in thinking about their involvement to protect children from ill-treatment.

17.7 Explaining the Failure of CSOs

The failure of CSOs to nurture and protect children born from rape from extreme violence and persecution is justified by the cultural norms, which could be understood as having a double function, being constructive and destructive, depending on their use. Cultural norms have been seen as the most challenging key factor, complicating the interventions of South Kivu CSOs to protect such children in order to reduce violence and bring peace back to the community. As mentioned above, cultural norms can play a double function: causing violence or building peace. If cultural norms are the source of rising violence and the marginalisation of children born from rape (Neenan 2017), culture could also be used to protect children from violence and break the cycle of violence. Culture, then, becomes a useful resource to create harmony and build peace in the community (Lederach 1998). In the case of South Kivu, in the community of the Kalonge chieftaincy, cultural norms have caused more violence than peace; as argued by the World Health Organization (2009), the influence of norms within cultural groups has driven violence.

The responsibilities of traditional community leaders and local associations (SPMC, AEFAD, CLOC, and CDMC) are to protect citizens, socialise people, build communities, and advocate and monitor accountability. Such functions would serve as the basis of managing children born from rape. The issue of children's acceptance and integration into the families and community is widely rejected by the belief that clan blood relationships need to be preserved against any external threats. The belief in the conservation and protection of the bloodline of traditional community leaders and local association leaders could provide a foundation for building communities with unity, cohesiveness, solidarity, and togetherness (Ibuot 2013). In other words, there is the possibility of integrating others into the clan bloodline but under the conditions of a blood covenant with the benefit of togetherness. Kenyon (1969) asserted that because of the fragility of friendship, the concept of a blood covenant had been used in African societies to guarantee the continuity and strength of friendships and relationships, an old practice applied by many primitive societies, which also include the society of the Kalonge chieftaincy. This could be used as an open pathway to adapt the covenant to integrate children into the clan bloodline in order to limit the violence against them and their maltreatment. This inclusion mechanism functions under the principle of "do no harm" to each other in order to protect, build, and strengthen social relationships (Ibuot 2013). Hence, cultural norms allow for inclusion either by blood or by the creation of bonding from social relationships (Beidelman 2012).

However, the complexity of children's inclusion in the community and clans, in order to benefit from the protection of traditional community leaders and local leaders' associations, lies in the attitude of displaced aggression, as argued by Woollett and Thomson (2016: 1069): "displaced aggression can occur when someone cannot aggress towards the source of incitements or provocation, and instead takes it out on some-indicated". Children are paying the price of being rejected and excluded due to their genitors' criminality, and, in turn, the community members take revenge using cultural norms to justify their attitudes. As noted, however, there is the possibility of adapting their inclusion through the use of social relationship bonds if a blood bond does not justify their inclusion in the clan. The attitude of rejecting and excluding the children is reinforced by the patrilinear system which determines the group membership, considering the linear from the male population (Gupta 2009) and, therefore, excluding female children whose genitors are ignored.

Given that wartime rape, pregnancy, and children born from rape have been used as a way of disrupting ethnic bloodlines (Martin 2007), the likelihood that traditional law-keepers and the community's leadership will be reluctant to protect the children is increased. Reluctant attitudes of traditional community leaders as well as local associations (SPMC, AEFAD, CLOC, and CDMC) to offer protection are justified by cultural norms and produce many disadvantages. They encourage maltreatment of the children as findings provide evidence of violence to the level of extreme violence (murder) from families to the community level, consequently maintaining a culture of systemic violence and destroying and weakening community safety and peace, where advocacy is ignored to initiate, for instance, judiciary actions against abusers to hold them accountable. Therefore, the only source of protection remaining for children born from rape is their mothers' affection and the children themselves, given the failure of local CSOs.

17.8 Recommendations

Merrick/Guinn (2018: 1118) stated that "preventing child maltreatment requires... in developing long-term, sustainable solutions that address structural barriers". Structural norms have to be reviewed in order to challenge traditional norms and practices as tools to boost change regarding the attitude toward nurturing children. Considering the results of this inquiry, we have suggested recommendations as further structural approaches to address the mismanagement of the dynamism of violence, which has led to the maintenance of systemic violence in the community centered on the maltreatment of children born from rape in the post-conflict society of Kalonge. Recommendations include the following:

1. At the sociocultural level, two of the most relevant mechanisms that should be applied to nurture and protect children born from rape are community resocialisation and the bilateral kinship system.
   - Community resocialisation on the integration of children born from rape should use an adapted social covenant which has equal power to the blood covenant to influence the attitudes and behavior of people in the Kalonge community. This is to prevent both tiers, the people in the community and children born from rape, from using violence as an approach to manage the extreme abuse of children as the covenant will allow them to function under the same principle of "do no harm" to each other. The social covenant should be set as an entire system to achieve this goal for the cohesiveness of the community.
   - Based on the customary system, in a society where the number of children born from rape is very large and the cycle of violence is maintained, the bilateral kinship system is useful, where descents from both the male and female sides have rights to equal esteem in each lineage since the right of inclusion by blood reflects the mother's lineage as well. This flexibility would significantly encourage inclusivity, mutual protection, cohesiveness, unity, togetherness, harmony, and the building of peace in the community.
2. At the judiciary level, dissuasive measures, such as constraining criminals to hold them accountable for their crimes and extreme maltreatment in the murder of children, are used. On the other side, legislatively, laws guarantee the protection of the bilateral kinship system for an inclusive society.
3. At the human rights level, the creation and promotion of new rights that protect specific children in a post-conflict society, such as children born from rape, adapted to the context of each society should be guaranteed in order to avoid the situation of children being disenfranchised by their communities. Advocacy to mobilise local, national, and international organisations to obtain adequate responses from the government for the protection of the abovementioned legislative measures; to ensure capacity building of local leaderships at the grassroots; etc. is a necessity to concomitantly protect children and significantly reduce the maintenance of the culture of systemic violence.

17.9 Conclusion

The aim of this study was to explain the challenges that CSOs face in protecting and nurturing children born from rape in South Kivu communities, with the objectives of documenting the treatment of children born as a result of rape in a sample of South Kivu communities, assessing present efforts by CSOs to protect and nurture these children and proposing more effective ways of carrying out these functions.

The results of this study have shown the interpersonal and structural violence that have deeply affected rape survivors as well as their children born in post-conflict society. Rape survivors have been seriously psychologically, emotionally, physically, and financially violated because of having children from rapists who were considered enemies of the community. Children, on their side, have been extremely violated involving simple to extreme forms of violence in terms of murder, using superstitious approaches to kill them, and implicitly excluding them from families and the community. These criminal attitudes and practices against children have been reinforced by cultural norms and, somehow, irreproachably encouraged by traditional secret groups excluding those children from families, clans, and the community using their customs' justification of blood descendance privileging the patrilineal system.

The response to managing the maltreatment of children has been dynamically violent, where some children, especially boys, have learned to use violence in order to equalise their extreme pain and avoid the risk of being murdered by integrating into criminal groups, using guns to dissuade their abusers and take revenge. In fact, children born from rape do not expect much in the way of positive contributions from their mothers as role models who can stop their maltreatment or who can promote their integration and the improvement of their relationships with abusers. Mothers seem to be powerless, undermined by abusers, and limited by their miserable conditions to influence any positive change. Regarding the children, only those who have strong personalities and the power to confront and subjugate abusers can impact change and stop the extreme violence against them. Hence, they try to integrate into criminal groups, such as rebel groups and militias, to secure their protection, which is ensured by themselves or by their network crews. Those who do not perceive their protection through criminal networks turn against their mothers, blaming them for their unsafety.

On the other side, traditional community leaders and associations who participated throughout the inquiry acknowledged being challenged by cultural norms, which prevent CSOs from fulfilling their functions accurately. They are not able to protect and nurture those children due to the fact that they are applying, practicing, and sharing common cultural perceptions rooted in the belief of blood bonds and its justification, which has hindered their engagement in a protective role. Consequently, the mismanagement of children's integration into their mothers' families, clan, and community has led to systemic violence, preventing any initiative to build a cohesive community, burdening the peace process concomitantly, and maintaining systemic violence in the Kalonge community. Hence, sociocultural resocialisation, the judiciary, and human rights advocacy mechanisms to protect and nurture children and boost peace should lead scholars' studies and practitioners' interventions in the long-term perspective.

References

Abbasi, M. A., Saeidi, M., Khademi, G. and Hoseini, B. L. (2014). Child Maltreatment in the Worldwide: A Review Article. *International Journal of Pediatrics*, 3(1), 353–365.

Antje, D., & Dieter, N. (2019). Civil society and social movements: conceptual insights and challenges in African contexts. *Critical African Studies*, *11*(2), 176–192.

Asuelime, L. E. (2017). Civil society and the South African developmental state: An appraisal. *Journal of Social Development in Africa*, *32*(1), 45–68.

Asuelime, L. E. *Journal of Social Development in Africa*, 32(1), 45–68.

Baškarada, S. (2014). Qualitative Case Study Guidelines. *The Qualitative Report*, 19(24), 1–18.

Beidelman, T. O. (2012). *The Blood Covenant and the Concept of Blood in Ukaguru*. Cambridge: Cambridge University Press.

Benoit, D. (2004). Infant-parent attachment: Definition, types, antecedents, measurement and outcome. *Paediatr Child Health*, 9(8), 541–545.

Bignami, F. (Ed.) (2016). *Theories of Civil Society and Global Administrative Law: The Case of the World Bank and International Development*. Cheltenham, UK: Edward Elgar.

Broache, M. (2016). International prosecutions and atrocities in the Democratic Republic of the Congo: A case study of the FDLR. *The Journal of the Middle East and Africa*, 7(1), 19–38.

Child Welfare Information Gateway (2016). *Intergenerational patterns of child maltreatment: What the evidence shows*. Washington, D.C: U.S. Department of Health and Human Services, Children's Bureau. https://www.childwelfare.gov/pubs/issue-briefs/intergenerational/. Accessed 20 Apr 2021.

Christensen, D., & Laitin, D. D. (2019). *African States since independence: Order, development, and democracy*. New Haven: Yale University Press.

Creswell, J. W. (2013). *Qualitative inquiry and research design choosing among five approaches (Third edition ed.)*. Washington, D.C.: SAGE.

Dörner, W., & Regina, A. L. (2012). *Civil Society, Conflict and Violence: "An introduction"* London: Bloomsbury Academic. https://doi.org/10.5040/9781780931036.ch-001.

Douma, N., & Hilhorst, D. (2012). Fond de commerce? Sexual violence assistance in the Democratic Republic of Congo. *Disaster studies*, 2, 1–79.

Einarsdóttir, J. (2007). Research with children: methodological and ethical challenges. *European Early Childhood Education Research Journal, Routledge*, 15(2), 197–213.

Elo, S., Kääriäinen, M., Kanste, O., Pölkki, T., Utriainen, K., & Kyngäs, H. (2014). Qualitative content analysis: A focus on trustworthiness. *SAGE open*, January-March 2014, 4(1), 1–10. https://doi.org/10.1177/2158244014522633.

Erlingsson, C., & Brysiewicz, P. (2017). A hands-on guide to doing content analysis. *African Journal of Emergency Medicine*, 7, 93–99.

European Centre for Conflict Prevention (2006). *Agents for change: Civil Society roles in preventing war & building peace*. Amsterdam: European Centre for Conflict Prevention.

Fargas-Malet, M., McSherry, D., Larkin, E., & Robinson, C. (2010). Research with children: Methodological issues and innovative techniques. *Journal of early childhood research*, 8(2), 175–192.

Finkelhora, D., Vandermindena, J., Turnera, H., Hambyb, S. and Shattucka, A. (2014). Upset among youth in response to questions about exposure to violence, sexual assault and family maltreatment *Child Abuse & Neglect, the international journal*, 38(2), 2017–2225.

Fischer, M. (2006). *Civil Society in conflict transformation: Ambivalence, potentials and challenges*. Geneva: Berghof Research Center for Constructive Conflict Management.

Geneva Academy (2019). *The war report 2018, Democratic Republic of the Congo: Conflict in the eastern regions*. Geneva: Université de Genève.

Gumbu, M. (2010). *Secret societies in Sierra Leone*. http://www.thepatrioticvanguard.com/secret-societies-in-sierra-leone. Accessed 3 June 2021.

Gupta, M. D. (2009). *Family systems, political systems, and Asia's 'Missing Girls', The construction of son Preference and Its Unraveling*. Washington, D.C.: The World Bank, Development Research Group.

Houten, K. V. (2018). The in-between: Local civil society organizations and their community-based structures in South Kivu, the Democratic Republic of the Congo. *Journal of Civil Society*, 14(2), 116–134.

Hsieh, H.-F., & Shannon, S. E. (2005). Three Approaches to Qualitative Content Analysis. *Qualitative Health Research*, 15(9), 1277–1288.

Human Rights Watch (2002). *The war within the war: Sexual Violence Against Women and Girls in Eastern Congo*. London: Human Rights Watch.

Human Rights Watch (2009). *Soldiers Who Rape, Commanders Who Condone Sexual Violence and Military Reform in the Democratic Republic of Congo*. London: Human Rights Watch.

Ibuot, E. J. (2013). The impact of blood covenant on development in Nigeria. *Journal of African Studies*, 10, 97–115.

Ilker Etikan, I., Sulaiman Abubakar, M. and Sunusi Alkassim, R. (2016). Comparison of Convenience Sampling and Purposive Sampling. *American Journal of Theoretical and Applied Statistics*, Science Puplishing Group, 5(1), 1–4.

Kawulich, B. B. (2004). Data Analysis Techniques in Qualitative Research. *Journal of Research in Education*, 14(1), 96–113.

Kenyon, E. W. 1969. *The blood covenant*. Lynnwood: Kenyon's Gospel Publishing Society.

Klein, S., & Lee, C.-S. (2019). Towards a Dynamic Theory of Civil Society: The Politics of Forward and Backward Infiltration. *American Sociological Association*, 37(1), 62–88.

Koselleck, R. (1988). *Critique and critics, Enlightenment and the pathogenesis of modern society*. Cambridge, Massachusetts: The MIT Press.

Krippendorff, K. (2004). Content Analysis: An Introduction to Its Methodology (2nd ed.). *Organizational Research Methods*, 13(2), 392–394.

Krippendorff, K. (2013). *Content analysis: an introduction to its methodology (third edition)*. London: SAGE Publication, Inc.

Lederach, J. P. (1998). *Sustainable reconciliation in divided societies*. Washington, D.C: United States Institute of Peace Press.

Lewis, S. (2008). Sexual violence and the HIV and aids, turning the tide of HIV/AIDS in Africa. Paper presented at the United Nations Human Rights Council on Violence Against Women. Sussex, 2007. Stephen Lewis Foundation, 1–13.

Lubunga, E. (2016). The impact of conflict in the Democratic Republic of Congo on women and their response to peace-building. *Stellenbosch Theological Journal*, 2(2), 347–364.

Maiese, M., & Lederach, J. P. (Eds.) (2017). *Levels of Action (Lederach's Pyramid) Posted: July 2003*. Boulder: University of Colorado. http://www.beyondintractability.org/essay/hierarchical. Accessed 14 May 2021.

Martin, S. (2007). *Ending sexual violence in Darfur: an advocacy agenda*. Washington, D.C.: Refugee International.

Merrick, M. T., & Guinn, A. S. (2018). Child Abuse and Neglect: Breaking the Intergenerational Link. *American Journal of Public Health*, 108(9), 1117–1118.

Mushagalusa, A. K. (2014). The symbolic dimensions of wartime rape, a case study of Kamanyola community, Bukabu/South-Kivu province (Democratic Republic of Congo). Master, University of Witwatersrand.

National Academy of Sciences. (1993). *Understanding Child Abuse and Neglect*. Washington, D.C: National Academy Press.

Neenan, J. (2017). Closing the protection gap for children born of war, Addressing stigmatisation and the intergenerational impact of sexual violence in conflict. London: London School of Economics and Political Science.

Organization for Security and Co-operation in Europe (2018). *The role of Civil Society in preventing and countering violent extremism and radicalization that lead to terrorism*. Vienna: Organization for Security and Co-operation in Europe.

Paffenholz, T. (2015). Civil society and peacebuilding. *Development Dialogue*, 63(3), 108–118.

Paffenholz, T., & Spurk, C. (2006). Civil Society, Civic Engagement, and Peacebuilding. *Conflict Prevention & Reconstruction*, 36, 1–54.

Panzi General Reference Hospital (2013). *Annual Activity Report Panzi Hospital 2013*. Bukavu: Panzi General Reference Hospital. www.panzihospital.org Accessed 14 May 2021.

Relief Web. (2009). *DR Congo: Children born of rape face dismal future*. https://reliefweb.int/report/democratic-republic-congo/dr-congo-children-born-rape-face-dismal-future. Accessed 28 April 2021.

Rohwerder, B. (2019). *Reintegration of children born of wartime rape*. Brighton: Institute of Development Studies.

Santoso, A. (2010). Civil society and peace processes in Kivu. *Forced Migration Review*, 36, 35–36.

Saunders, D. G. (1986). When battered women use violence: Husband-abuse or self-defense? *Victims and Violence Journal*, 1(1), 47–60.

Stearns, J. K., & Vogel, C. (2015). *The landscape of armed Groups in the Eastern Congo*. New York: Center on International Cooperation.

Sutton, J., & Austin, Z. (2015). Qualitative Research: Data Collection, Analysis, and Management. *The Canadian Journal of Hospital Pharmacy*, 68(3), 226–231.

Teddlie, C., & Yu, F. (2007). Mixed Methods Sampling: A Typology With Examples. *Journal of Mixed Methods Research*, 1(1), 77–100.

Thompson, C., & Rudolph, L. (2000). *Counseling Children*. Belmont, CA: Wadsworth Brooks/Cole.

Tight, M. 2017. *Understanding case study research, small-scale research with meaning*. New Delhi: SAGE.

Verkoren, W., & Leeuwen, M. V. (2013). Civil society in peacebuilding: Global discourse, local reality. *International Peacekeeping*, 20(2), 159–172.

Wolfgang, D., & Regina, A. L. (Eds.) (2012). *Civil Society, Conflict and Violence: An introduction*. London: Bloomsbury Academic. https://doi.org/10.5040/9781780931036.ch-001 Accessed 8 May 2021.

Woollett, N., & Thomson, K. (2016). Understanding the intergenerational transmission of violence. *South African Medical Journal*, 106(11), 1068–1070.

World Health Organization (2009). *Changing cultural and social norms that support violence*. Switzerland: World Health Organization Press.

Yin, R.K. (2009). *Case study research design and methods*. Fourth ed. New Delhi: SAGE.

Part III
Macro-level Case Studies

Chapter 18
Indigenous Peacebuilding Approaches and the Accountability of Former Child Soldiers: African Case Studies

Jean Chrysostome K. Kiyala

Abstract Studies on children associated with armed forces and armed groups reveal the vast limitations and insurmountable legal challenges with which the conventional prosecutorial criminal justice systems and popular courts are confronted. This conundrum ultimately stands as an obstacle to creating a peaceable environment where young soldiers can peacefully transition from soldiering to civilian and civilised life. In a world attracted to the logic of retribution and punishment, this chapter instead argues for the viability and sustainability of building peace with such young soldiers, based on indigenous mechanisms of dispute resolution, healing, and reconciliation inspired by African philosophy and ethics and the extent to which these approaches can enhance communal peace with former child soldiers. These mechanisms include *bushingantahe* in Burundi, *baraza* in the Democratic Republic of Congo, *ka pahla* in Mozambique, *fambul tok* in Sierra Leone, *gacaca* in Rwanda, and *mato oput* in Uganda. Such indigenous institutions work as components of the traditional civil society that are founded on the African humanistic and communitarian ethics (*ubuntu* and palaver).

Keywords African ethics · African palaver · Civil society · Child soldiers · Peacebuilding · Ubuntu

18.1 Introduction

The child soldiering phenomenon is an international and widespread phenomenon that can be traced back to the era of the World Wars, during which many children participated in hostilities. For instance, children known as the *Hitlerjugend* (the Hitler Youth) were conscripted by the Nazis (Singer 2005; Vautravers 2008), notably the 12th SS-Panzer-Division *Hitlerjugend* who were deployed in 1944 during the Battle of Normandy against the British and Canadian forces. The *Hitlerjugend* abode by the political, legal, ideological, and military structural discipline of the National

Dr. Jean Chrysostome K. Kiyala is a Senior Lecturer, International Centre of Violence, Durban University of Technology, South Africa; Associate Professor and visiting lecturer the University of Bandundu, Catholic University of Bukavu and Evangelical University in Africa based in Bukavu (DR Congo).

J. C. K. Kiyala and G. T. Harris (eds.), *Civil Society and Peacebuilding in Sub-Saharan Africa in the Anthropocene*, The Anthropocene: Politik—Economics—Society—Science 34,
https://doi.org/10.1007/978-3-030-95179-5_18

Socialist society with regard to the subject of the *Hitler Youth—an original Nazi mass organization*, as written by Kollmeier (Kollmeier 2007). On a similar note, JayakumarJayakumar (2011) ascertained that the juvenile conscription by the Nazis was also regarded as an act of heroism, as many of Germany's juvenile fighters were extolled as heroes in 1945.

A report by Child Soldiers International [CSI] (2018, p. 4) states: "around the world more than 240 million children are living in countries affected by conflict. Many of them face violence, displacement, hunger and exploitation by armed forces and groups". A plausible explanation of this reality is the new warfare mutations, as maintained by the Lima Declaration on Juvenile Restorative Justice (UNICEF 2009), contending that while large-scale, international conflicts have decreased, new forms of warfare are emerging within countries. According to CSI's statement that follows: "56 armed groups and seven state forces [were] named by the UN Secretary-General as guilty of child recruitment in 2017" (Child Soldiers International [CSI] 2018).

In addition, it is estimated that 40 % of the global number of child-soldier effectives are found in Africa (Drumbl 2012, p. 5). The total population of children involved in hostilities varies between 300,000 and half a million children below 18 years of age (some as young as six years), with 40 % being girls, all of whom have been illegally recruited in more than 30 conflicts around the world (Popovski/Arts 2006; Singer 2005). While the practice of child soldiering has decreased in several parts of the world, it remains an entrenched practice in the Great Lakes Region, particularly in the Democratic Republic of Congo, or DRC (Child Soldiers International [CSI] 2018); in the Middle East, particularly those recruited by Da'esh (Morris/Dunning 2020); and in other parts of the world such as Myanmar (Chen 2014).

Efforts to prevent the ongoing child soldiering phenomenon around the world have extensively concentrated on the criminal justice system—by indicting warlords. However, such a strategy has had very limited deterrent effects as children continue to be recruited and used as soldiers. Recent studies have described the situation of boys and girls involved in armed conflict as abhorrent serial injustice and as human rights and humanitarian-law violations.

Based on experiments of indigenous jurisprudence and mechanisms of dispute settlement, healing, and reconciliation, this essay seeks to establish how such practices can help communities affected by the child soldiering phenomenon to build durable peace. It focuses on the cases of *bushingantahe* in Burundi, *baraza* in the DRC, *ka pahla* in Mozambique, *gacaca* in Rwanda, *fambul tok* in Sierra Leone, and *mato oput* in Uganda. All these models are rooted in the communitarian worldview of Sub-Saharan Africa which is sustained by the *ubuntu* and the palaver institutions. The main components comprise the introduction and the design and methods, which are discussed before reviewing the literature on child soldiering and peacebuilding. The literature review starts with explaining why carrying out peacebuilding activities with ex-child soldiers constitutes a moral and ethical imperative in African culture. This is further supported and elucidated by the pragmatic and philosophical assumptions of the humanistic philosophy of *ubuntu* and the African palaver (a dialogical institution). The subsequent section deals extensively with six indigenous approaches of addressing the consequences of war atrocities, namely those from

Burundi, the DRC, Mozambique, Rwanda, Sierra Leone, and Uganda. These countries have been selected because they have been worrisomely using child soldiers in protracted rebellions, civil wars, and other militarised hostilities.

18.2 Methods

This chapter is designed as an exploratory case study and uses six units of analysis (*bushingantahe*, *baraza*, *ka pahla*, *gacaca*, *fambul tok*, and *mato oput*), employing qualitative methods and the interpretivist paradigm; that is mainly based on literary research whose aim is: "to produce some critical, theoretical or historical writing. In most cases such work takes the form of a thesis or dissertation" (Correa/Owens 2010, p. 3). This design is conducted by utilising a dense and diverse range of data sources which include "databases, catalogues, bibliographies, dictionaries and multifarious other reference works – in printed and electronic versions" (Correa/Owens 2010, p. 4). The techniques and tools of inquiry adopted here essentially come from the exploration, analysis, and interpretation of online and printed sources. These approaches are useful in developing literary research (Towheed 2009).

The next sections examines issues pertaining to child soldiers and how that affect peace and are deemed as opportunities for peacebuilding.

18.3 Peace and Peacebuilding

Peace is necessarily a condition for achieving development. The sociological perspective sees peace as the absence of conflict and violence between individuals, communities, and states. The positive conception of peace sees peace as tranquility, harmony, justice. Peace is the creation of political, economic, and social conditions to support sustainable justice, security, and development (Ehrhart/Schnabel 2005). These scholars note: "In post-conflict situations internal and external actors must cooperate in mutually reinforcing the socio-economic, governance, and security dimensions of a highly fragile environment" (Ehrhart/Schnabel 2005, p. 7). Peacebuilding can also consist of preventing conflict, because conflict resolution and prevention are inclined to maintain peace. For example: UNDP uses both peacebuilding and conflict prevention because it has a mandate in both. It adopted the definition used in the Brahimi Report, and then observed how peacebuilding and conflict prevention are virtually synonymous (and uses the two concepts interchangeably) (Barnett et al. 2007, p. 42).

Similarly, the The World Bank (2007) report underlines that peacebuilding has multiple objectives, namely: prevention and management of armed conflict and sustenance of peace for at least a decade after violence has attenuated or ended; amelioration of the conditions for rebuilding the economy; accelerating development and democratisation, without being assimilated to these undertakings. To attain this vision, peacebuilding consists of three phases, notably: "prevention prior to the outbreak of violence, conflict management during armed conflict, and post-conflict peacebuilding for up to 10 years after the conflict ends" (The World Bank 2007, pp. 6–7). This very report explains:

> Peacebuilding is now understood more broadly. It often covers all activities related to preventing outbreaks of violence, transforming armed conflicts, finding peaceful ways to manage conflict, and creating the socio-economic and political pre-conditions for sustainable development and peace (p. 9).

Here, the concept of peacebuilding is broadened; however, its inclusion of almost every socioeconomic development, and efforts to reduce poverty and promote education or democratisation becomes problematic in defining the ends of peacebuilding activities, and where normal development activities start (The World Bank 2007). Similarly, "implicit causal linkages are constructed between general development activities and peacebuilding objectives, but there is little evidence to substantiate these links" (The World Bank 2007, p. 9). From this vantage point, provision of services is perceived as a means to enhance peacebuilding because this prompts the establishment of conditions propitious to peace (The World Bank 2007). The debates on the aspect of peacebuilding's improvement of "the political peacebuilding objectives seems to be tenuous and requires further exploration" (The World Bank 2007, p. 9).

Amidst several theories of peacebuilding, the works of Lederach (1997a), Paffenholz (2014b) and other pioneers in this field have underlined the extent to which local initiatives can leverage peace efforts and accelerate social cohesion which are vital for development. In divided communities, reconciliation, healing, and caring for affected persons rely upon local actors such as NGOs, religious and traditional leadership (Lederach 1997a). Even in a society in transition after large-scale conflict, diplomacy, advocacy, democratisation and human rights protection depends largely on CSOs (I-NGOs, L-NGOs, and transnational NGOs [T-NGOs]).

Previous debates in this book have largely reviewed the work of Paffenholz, among other scholars. The subsequent discussion turns to Lederach's (1997a) pyramid of peacebuilding that presents both the actors involved and the approaches that are adopted. It appears that CSOs are involved at all the degrees of action to build sustainable peace and reconciliation.

At Level 1, religious leaders with high reputation represent civil society; at Level 2, ethnic, religious leaders, academics or intellectuals, as well as humanitarian leaders from NGOs take part in peace processes; and finally, at Level 3 (the Grassroots) leaders emerge also from CSOs and feature prominently in the pyramid; these are: community/local leaders, leaders from indigenous NGOs, etc. The actions undertaken by or expected of these components can be summarised as follows: to unravel the drivers of conflict, deal with implosion and explosion of conflicts that go violent, and find realistic alternatives to contain these, and create infrastructure to maintain peace and social cohesion.

The pre-eminence of the Middle Range actors is explained by their positioning between the Top Range and Grassroots Levels, and Middle range leaders enjoy the trust of both top-level and grassroots actors (Lederach 1997a). He states: "They have more flexibility of thought and movement than top-level leaders and are far less vulnerable in terms of daily survival than those at the grassroots" (Lederach 1997b, p. 94). The pyramid clearly focuses on L-NGOs, whose role is crucial and indispensable in initiating, building and maintaining peace. Their actions may require financial aid, which would come from external partners, but their programmes for peace and reconciliation will spring forth from within the community. Here again, traditional, indigenous, or locally-based mechanisms of peacebuilding may prove effective to achieve the agenda of peace. They are culturally-based approaches and can be adapted to any context, particularly in the DRC where intrastate conflicts are entrenched. Peace is necessarily a condition for achieving development. Ehrhart and Ehrhart/Schnabel (2005, p. 1) assert: "Without a secure environment and a security system that ensures security even after the departure of international peace operations, political, economic, and cultural rebuilding are impossible". Both peace and development are part of the global processes of structural change within a social system.

Several actors participate in the construction of peace, which is moving towards a global vision as conflicts affect different levels of the social structure. Peacebuilding in South Kivu targets individuals (raped women, children from armed forces and groups, children in difficult situations) and communities at the local level (land conflicts, justice, inter-communal conflicts, etc.).

Peace-building programmes carried out by civil society organisations appear to focus on cyclical solutions rather than structural changes in recipient communities. Structural change would mean communities benefiting from peace-building programmes, changes in representations, practices, and structures to solve problems a result of past conflicts and violence, but also creating resilient practices that can maintain social cohesion and collective economic progress. However, what is observed is the fragility of the beneficiary communities even after the implementation of peace-building programmes, and hence the permanence of non-peace.

A comprehensive approach to build durable peace must not ignore indigenous practices that include local or cultural models of peace initiaves at Level 2 and Level 3 of peacebuilding pyramid of Lederach (Fig. 18.1) (Lederach 1997a, p.39).

18.3.1 Peacebuilding and Interventionist Research

Peacebuilding can utilise the action research model of engagement in trying to understand the challenges of post-war reconstruction, transforming and consolidating relations and restoring dignity and trust, and undertaking post-war development, in view of grasping all-inclusively the conflict, its consequences and promoting a democratic political culture (Stiefel 2001). One more instrument of peacebuilding can be "community mobilization", described by Erasmus (2001, p. 249) as "a means of tapping into the knowledge and resources of the local community and fostering a spirit of community ownership". He argues that community mobilisation is one of the most

Fig. 18.1 Lederach's three types/levels of peacebuilding's actors *Source* Author's adapted from Lederach's peacebuilding pyramid (1997, p. 39)

significant and efficient means of reducing conflict at the grassroots level, community mobilisation, and that needs to be sustained through dialogue, formally or informally, with all stakeholders (Erasmus 2001, p. 249).

This approach should be participatory in nature, in the sense that it should be involved in planning, implementing and evaluating the process via a team that designs the programme that should profit the community (Erasmus 2001). While community mobilisation can enhance peacebuilding, Erasmus observes that failure in community mobilisation could lead to disastrous consequence as in Somalia, namely: "(1) the development of a socio-politico-economic elite largely based in Mogadishu; and (2) the marginalisation of the traditional community and its leadership" (Erasmus 2001, p. 247). Furthermore, the experience of Sudan shows that the People's Liberation Movement/Army (SPLM/A) lead to a civil war 1983 during which the community was fragmented, disempowered and the military liberators turned to despots (Erasmus 2001).

It is important to broaden the understanding of peacebuilding by including the developmental aspect of this concept. Beside *direct intervention* by way of mediation, Reychler and Paffenholz (2001) note that peacebuilding initiative should include *indirect intervention* via development projects, relief aid, management of media and other issues that could trigger violent conflict (Reychler/Paffenholz 2001, p. xiii). This model of peacebuilding was tested through the *Aktion Africa Hilfe* (AAH) – German-based nongovernmental organisation operating projects in South Sudan, Somalia, and Northern Uganda; Even in this instance the AAH show the failure of community mobilisation not originating from the grassroots, which triggered conflict between the local community and an external NGO (Erasmus 2001). The next section starts linking peacebuilding to child soldiering phenomenon, and later, peacebuilding will be examined in specific cultural contexts.

18.4 Child Soldiering and Peacebuilding

This section overviews some features of the child soldiering phenomenon. This includes the definitional approach and the contexts in which children have been recruited and used as soldiers in the six countries of Sub-Saharan Africa. Other issues that are explored here encompass the linkages between peacebuilding and justice, and the justification for communal peacebuilding with former young soldiers.

According to the Rome Statute of the International Criminal Court (RSICC), Article 8(2)(b)(xxvi), a child soldier is any person under 15 years of age recruited into armed forces or used in hostilities: "Conscripting or enlisting children under the age of fifteen years into the national armed forces or using them to participate actively in hostilities" is a violation of children's rights (International Criminal Court 2002). According to the Optional Protocol to the Convention on the Rights of the Child on the Involvement of Children in Armed Conflict (OPAC), Article 2, coercive recruitment of persons under 18 into armed groups is completely outlawed: "States Parties shall ensure that persons who have not attained the age of 18 years are not compulsorily recruited into their armed forces" (United Nations 2000).

The Convention, in Art. 8(2)(b)(xxvi), permits those under 18 to be enrolled willingly to state armed forces under certain strict conditions to ensure that such recruitment is voluntary (United Nations 2000). It is evident that the conditions and circumstances of child recruitment do not reflect these legal requirements. The definition of child soldiers in this study was based on Article 2, paragraph 1, of the Paris Principles which states that:

> A child associated with an armed force or armed group refers to any person below 18 years of age who is or who has been recruited or used by an armed force or armed group in any capacity, including but not limited to children, boys and girls, used as fighters, cooks, porters, messengers, spies or for sexual purposes. It does not only refer to a child who is taking or has taken a direct part in hostilities (UNICEF 2007).

This description includes a range of activities in which children who participate in hostilities are involved, and which brings them under the category of child soldiers. This is the context in which the concept of child soldiers is used in this article. Persons called *child soldiers* who were involved in this inquiry were between the ages of 12 and 28, and they were conscripted or joined militias when they were below the age of 18. Although the term *children associated with armed forces and armed groups* is commonly used to designate this category of children, this study refers to them as *child soldiers*, since the definition provided by the Paris Principles covers a broad range of activities in which these children are involved. Under normal circumstances, a person who chooses a profession or career goes through a certain amount of training and is prepared to venture into this professional occupation once all the required conditions, including age as well as physical and mental aptitudes, have been met. Child soldiering is one case where unfit young persons are forced, or allowed, to perform onerous duties that are not permitted to them by law. Such children do not meet the professional requirements of military service (Honwana

Table 18.1 Situational repertory of child soldiers. *Source* The Author

Country	Duration of conflict	Nature of conflict	Estimates
Burundi	1993–2005	Civil war	14,000
DRC	1967–2002	Civil war	30,000
Mozambique	1976–1992	Civil war	8,000 to 10,000
Rwanda	1990–1994	Genocide	3,000
Sierra Leone	1991–2002	Civil war	10,000
Uganda	1980s–2010	Civil war	Over 25,000

2006). Neither forced nor voluntary recruitment changes justify the use of persons under 18 in armed groups and forces.

The crises emerging from globalisation include the use of child soldiers, inter-state conflicts, and the internationalisation or the regionalisation of conflicts which involve intra-state and inter-state actors and their proxies. The new warfare strategies have deepened and worsened children's vulnerability as they provide the means for children to easily operate in combat operations, namely in activities involving light ammunitions such as hand grenades, rocket launchers, AK-47 guns, as well as other activities in which children are involved to support direct military campaigns. The definition of child soldiers encompasses indirect military engagement that is inclusive of boys and girls taking a direct part in operations and being employed on the front line, and those employed in military logistics services such as being "cooks, porters, messengers, spies or being used for sexual purposes", according to Article 2, paragraph 1, of the Paris Principles (UNICEF 2007). The unlawful breaches of international law experienced in the brutal way children are enlisted and abducted demonstrate how the societal and cultural values to care for children and the elderly, solidarity, and healthy relationships with one's neighbor have disintegrated in post-colonial Africa. Table 18.1 presents the situational repertory of child soldiers in six African countries.

The overwhelming majority of child soldiers have been used in civil wars (Brett/Specht 2004; Honwana 2006) such as the Ugandan Civil War (Pike 2000-2021); the Rwandan Civil War and Genocide (1993–1994); the Burundian Civil War; the First Congo War and Second Congo War—also known as Africa's First World War and the African World War, respectively—occurring from 1996 to 2002 (Kiyala 2018a, 2018b); the Angolan Civil War occurring from the 1960s to 2002 (Pearce 2012); the Mozambican Civil War (Cahen et al. 2018); and the Sierra Leone Civil War (Utas/Jörgel 2008). External proxy forces play a significant role in instrumentalising ideologies and propaganda to sustain the monstruous wars breaking out in countries in conflict. The catastrophic consequences of a civil war are described in the following lines:

> And as human history has repeated time and time again, civil wars often leave deeper scars, are often more indiscriminate and more ruthless than are interstate conflicts short of either a world war or a nuclear war. The United States, for example, lost more people in its own civil

war in the 1860s than in any other single war in its more than 200-year history, including Vietnam and the two world wars (Mazrui 2004, p. 481).

With the elapse of time since colonialism, new leaders in the post-independence era extended their grip on power and falsified the democratic agenda upon which their countries' independence was founded, thus engendering new conflicts and novel warfare, as asserted as follows: "independence came to Africa, and most of the second half of the twentieth century revealed entirely new patterns of conflict, resistance, and violence" (Mazrui 2004, p. 479). In addition to these new trends of conflicts and violence, the impact of globalisation and the extent to which it has disrupted traditional societal norms of solidarity, peace among neighboring countries, and child protection; and prompted the use of child soldiers—a practice that desacralises childhood—is not to be underestimated (Kiyala 2018a). As a result, children accused of atrocities are called to account for their actions, and subsequently, this raises the question on how to build peace under these circumstances. The next section explores the conundrums ensued as prosecutorial justice mechanisms are employed to hold holding former child soldiers accountable; such enigmas would require a more realistic and fair strategy that will be discuses later, focusing on African philosophy and ethics.

18.5 Challenges of Accountability

While former child soldiers are primarily considered as victims and not only as perpetrators, pursuant to Article 3(6) of the Paris Principles (UNICEF 2007), there is a fundamental need for the victims of the crimes committed by child soldiers in transitional societies to have justice done for the atrocities that they have suffered. This consists of providing an opportunity for victims to meet their former tormentors who must be answerable for their alleged violations of international law, as provided by the United Nations document *The Rule of Law and Transitional Justice in Conflict and Post-Conflict Societies* (United Nations 2004). Although the Rule of Law encourages criminal trials in transitional contexts to denounce criminal behavior, establish the liability of alleged perpetrators, and ensure that adjudicative measures are pursued (United Nations 2004), the process of holding former child soldiers accountable needs to conform to international law pertaining to children in armed conflict (e.g., the Paris Principles and the Optional Protocol to the Convention on the Rights of the Child on the Involvement of Children in Armed Conflicts). A discussion on child soldiering and peacebuilding cannot be dismissive of justice issues which are often evoked as a prerequisite to progress with conflict settlements and peace deals.

Paragraphs 46, 47, and 48 of The UN Security Council Document, *The Rule of Law and transitional justice in conflict and post-conflict societies*, acknowledge the complexities and limitations of criminal prosecution, which can be evaded by perpetrators. This document suggests putting in place transitional justice mechanisms that can help to overcome the inherent limitations of criminal justice processes, help to satisfy the various needs of victims, and foster national reconciliation (United Nations 2004). Holding former child soldiers accountable responds to this necessity and the process of doing so may not be that which criminal courts do. It is a legal requirement of international law that children accused of violating human rights laws and humanitarian laws in armed conflict are called to account; secondly, it creates the opportunity to distinguish child soldiers as victims' identity from that of hero or perpetrator, thus avowing the stereotype of child soldiers as being "irreparable damaged goods", as discussed by Drumbl (2012, p. 7), or the prominent characterisation of such children as a "lost generation"—a warning raised by Singer (2010, p. 60). These narratives are reviewed and reconstructed by scholars. For instance, Drumbl (2012, p. 7) remarked that the depiction of child soldiers as presented above risks to generally underestimate the resilience of former young soldiers in the battlefield.

Rethinking the image and character of child soldiers is essential, beyond criminal accountability, in the process of peace efforts. I share this view with Drumbl (2012), and in the same perspective, Shepler (2005, p. 198) noted that "coming to terms with the participation of child soldiers, simultaneously perpetrators and innocents, is key to post-war reconciliation and peace building". Reconciliation and efforts to build peace with former young soldiers is multifaceted and may include healing rituals (Honwana 1997; Shepler 2005); reconciliation and peace search talks, such as is the case in Sierra Leone (Graybill 2010); a combination of traditional cleansing and enforcement of compliance by government officials to repair relations, as in the case of Uganda and the child soldiers of the Lord's Resistance Army (LRA) (Negara 2012); the use of the traditional concept of *gacaca*, which is integrated in the criminal justice system in Rwanda (Takeuchi 2011; Urusaro Karekezi 2012); and the process of public hearing and truth-telling of the Hunde of North Kivu in the DRC (Kiyala 2021). These processes facilitate the social reintegration of the child and ultimately discourage the phenomenon of voluntary re-enrolment, defections, and switching between different armed groups. Several questions arise when children are enlisted and used for military objectives, such as:

- What is their criminal responsibility in the violence that they are accused of carrying out?
- Why must child soldiers be held accountable for their actions or those that they executed under commanders' orders?
- How can peace and reconciliation be envisaged without justice?
- How effective is the conventional criminal justice system in adequately calling young soldiers to account?
- What are the alternative approaches holding historical recognition that have helped settle conflicts in local communities?
- How are the civil society organisations similar to local, social, and traditional groups responding to their communities' aspirations to peace?

To address these questions, we can argue from two perspectives: (1) the international legal framework; and (2) the moral and ethical imperative that underscores the communality and vitality of peace in African cultures. International law excludes punitive justice while encouraging restorative justice for children suspected of committing war-related atrocities when under 18 years of age, as stated in Article 3(6) of the Paris Principles:

> Children who are accused of crimes under international law allegedly committed while they were associated with armed forces or armed groups should be considered primarily as victims of offences against international law and not only as perpetrators. They must be treated in accordance with international law in a framework of restorative justice and social rehabilitation, consistent with international law which offers children special protection through numerous agreements and principles (UNICEF 2007).

Furthermore, Article 3(7) also establishes that "wherever possible, alternatives to judicial proceedings must be sought, in line with the Convention on the Rights of the Child and other international standards for juvenile justice" (UNICEF 2007). Unfortunately, in several countries, child soldiers exit hostilities without proper processes for accountability and reintegration into society. To support this point, several child soldiers in North Kivu Province are self-demobilised in an environment marked by a paralysed justice system and the absence of juvenile justice (Kiyala 2015). Taking into consideration available studies that argue on the limitations of the criminal justice system and its ineffective retributive sanctions (Noll 2003; Zehr 2013), former child soldiers need a forum where they can express themselves and tell their stories, the circumstances under which they joined fighting forces, the hardships they experienced in armed conflict, and how these have impacted their lives and relations with the members of their respective communities. Such a platform is necessary to change the stereotypes or labels placed on them, which are often associated with criminality, delinquency, and the rejection that they suffer in the community to which they return after exiting hostilities.

Article 3(6) of the Paris Principles envisages a restorative process as an alternative to retributive justice. Taking former child soldiers through a process of accountability is necessary from a humanitarian and legal perspective. On the humanitarian level, this would benefit children who are unjustly treated as delinquents and criminals and who, as a result, suffer marginalisation and exclusion and eventually return to armed groups (Kiyala 2015). From a legal perspective, one can argue for child soldiers' accountability on the grounds of the circumscribed actor model. According to Drumbl (2012, p. 17), "a circumscribed actor can do, has the ability not to do, and the ability to do otherwise than what he/she actually has done". This model corresponds to the condition of many child soldiers. It disillusions public opinion, according to which all acts carried out by child soldiers are remotely commanded and child soldiers do not always have choices to do otherwise. Some of their accounts reveal that they also take initiatives in their own capacities to carry out violent actions, though most of them are ordered to do so by their superiors (Kiyala 2015).

18.6 African Philosophy and Ethics

Africa experienced the erosion of its solidarity, sanctity of life, and its elites in the era when freedom meant using power as a tool to enrich oneself. A few selfless leaders such as Nyerere (with his ideologies of ujamaa, or African socialism; Kwame Nkrumah; and Panafricanism) and Nelson Mandela (who was responsible for reviving ubuntu, or African humanism) are remembered for moral regeneration which has remained a challenge with which despotic rules and power drunkenness in Zaire, Uganda, Zimbabwe, Sudan, Congo Brazza-Ville, Togo, etc. are confronted.

To remedy the erosion of societal ethos in the aftermath of large-scale human rights violations and human atrocities, some African countries have revived traditional approaches of conflict resolution for the healing of wounded communities. We will explore two concepts (ubuntu and the African palaver) that have fostered the way to peace, healing, and reconciliation. This section will attempt to show how the African social ethos and philosophy have shaped the indigenous mechanisms of dispute resolution and peacebuilding in societies in transition. The strength of the six models of accountability, healing, reconciliation, and peacebuilding discussed earlier (baraza, *gacaca*, fambul tok, ka pahla, Bushingantahe, and mato oput) is underscored by African communitarianism, the philosophy of ubuntu, the social ethics of ujamaa, and the dialogic institution of the African palaver. These are shaped as five overlapping layers of the peacebuilding graphic. The five levels of inter-relations in Fig. 18.1 represent the ontological characteristic of the human person as a social and rational being which features (1) physical agency that is exerted by the personhood and the community, and (2) the ethical and psychological principles (ubuntu, ujamaa, and the palaver) that mediate between the person and the community. The five layers overlap.

The person is the starting point, and the one who renders the remainder of the levels intelligible. At the second layer, we place ubuntu—the psychological and ethical force that defines the person's conduct vis-à-vis others and the community. Harmony, peace, sympathy, empathy, and mutual acceptance reside in the actuation of ubuntu. The third layer is similar to the second but it involves the principle of turning toward others, that is, accepting the other not as a stranger but rather as a brother or sister beyond the consanguinity and biological boundaries of human relationship. This is called ujamaa. The fourth level is identified as the palaver which is a powerful community disposition that encompasses attitudes; human ecology; and matters and meanings that spring forth from debates, discussions, or conflicts that need a settlement. The fifth level shows the community in which the human person practices the three virtues presented in the graph, namely, ubuntu, ujamaa, and the palaver.

How these layers of peacebuilding are based on the African social ethos and philosophy can be explained from the works of African philosophy writers such as Gyekye (1998) and Wiredu Wiredu (2005). For Gyekye (1992), a society is synonymous with an association. It is aptly described as an environment of sharing among individuals with regard to their common goals, interests, values, and etiquettes that fashion community life. This communitarian understanding of the human person implies that living in a community is not optional, personhood has a cultural dimension, isolationism is not a value, a person is naturally turned towards others, social relations are more a necessity than a contingency, and such relationships shape the person in his/her social environment (Gyekye 1992). The implications of this reasoning are such that reconciliation and reintegration of an offender are social imperatives. According to Wiredu (2004, p. 17), "right from the beginning of socialisation one is brought up to develop such strong bonds with large kinship units that one comes to see oneself as necessarily bound up with a community". From a normative point of view, the concept of personhood as the fulfilment of the progression of a human person to maturity denotes a moral dimension which Wiredu (2004, p. 18) explained as follows:

> A communalist ethic is one in which the interests of the individual are placed in a reciprocal adjustment with the interests of others in the community with reference to many specific circumstances of life and beyond the call of pure morality.

It is important to underline that in Gyekye's philosophy, based on the Akan culture, African morality is better interpreted as a form of humanism in the sense that the welfare or well-being of the community is the sole criterion that determines the goodness of a thought or action, and this is established by human beings themselves (Hallen 2005). This argument leads us to conclude that an action that intends to seek and build peace with former child soldiers both morally and ethically conforms to African morality and ethics, and it is an imperative, communalistic worldview. Considering that which morality encompasses, as explained by Wiredu (2004, p. 18), "communalistic rules of conduct are a clear extension of the imperatives of pure morality. Since both are defined in terms of human interests, the African ethics might be called humanistic, as opposed to supernaturalistic". Building peace and seeking reconciliation and societal harmony are moral and ethical obligations, as explained in the following paragraph.

As viewed by Masolo (2005), the dynamic interplay of morality and spirituality that characterises the human person creates the synergy that enhances humanism and socialism as a means of restoring wellness in wounded communities and their people. To emphasise the divine nature of a person, Masolo (2005, p. 54) stated that "on the question of the nature of a person too, although Origen believed in the Platonic trichotomy of personhood, his general theory has affinities with the Egyptian practice of approximating human nature to that of the divine". In other words, the African is by nature religious and repairs harm in the community—this involves rituals and sacrifices offered to the spiritual world (Healey/Sybertz 1996; Honwana 1997). This leads us to conclude that spirituality and morality are the two fundamental principles of traditional mechanisms of peacebuilding in Africa that are vehicles for the ubuntu spirit, ujamaa, and the palaver. We will focus on ubuntu and the palaver in this chapter. Ujamaa refers to the universality of the concept of brotherhood that stems from African socialism (Nyerere 1962; Stöger-Eising 2000). The third layer represents both the medium and the instrument that binds the community together and maintains peace and societal harmony. The last layer symbolises the locus of both ethics and the human person as an ethical agent of peacebuilding and reconciliation that has become a political framework to address the most egregious injustices and inhumanities of brutal post-World-Wars legacies globally, for example, in South Africa, Argentina, and Paraguay (Duvenage 2005). However, it should be honestly stated that truth-telling and efforts to take wounded communities to the path of reconciliation are laudable initiatives. However,

> Neither personal nor collective memory can undo a monstrous past. These difficulties, however, do not imply that our obligations to the past justify a policy of pardon and forgetting. The solution in this predicament may lie, as hinted above, in a policy of judgement and forgiving rather than punishment and forgetting (Duvenage 2005, p. 516)

In the same vein, Brounéus (2008, p. 57) supported the idea that peacebuilding literature and political rhetoric underscore the assumption according to which "truth-telling is cathartic or healing and will thereby advance reconciliation". A policy of healing memories should be considered and processes for such an undertaking should not disregard accused persons such as child soldiers, but on the contrary, include them in and treat them as partners of searching, finding, and maintaining durable peace. The importance of cultural support in peacebuilding processes is sustained by Brounéus (2008) as this can motivate post-violence conflict survivors to remain supportive of such procedures. It is to this that human communities affected by armed conflicts in which young soldiers are implicated aspire, and it demands context-based peace agenda due to the different cultural environments where conflict and violence erupt and develop. However, the principles that are evoked and applied to these multicultural and social contexts are consistent, namely, the spirit of ubuntu and the African palaver, because they are embedded in most African traditions, cultures, and consciousness. It is important to look into ubuntu and the palaver and examine the degree to which these ideals contribute to peacebuilding, particularly in war-torn communities where children have carried out human atrocities. Ubuntu has the potential to help overcome the intricacies of dealing with the aftermath of conflict when the demand of punishment overlaps with the urgency of pardon, reconciliation and peace.

18.6.1 Ubuntu and Peacebuilding

Before discussing the relationship between ubuntu and peacebuilding, it is necessary to define the concept of ubuntu and later establish how these two concepts could be linked. Ubuntu originates from the Nguni language, and it is often interpreted from the adage Umuntu ngumuntu ngabantu (translated as 'a person is a person through other persons' in isiZulu). This maxim has shaped the discourse regarding ubuntu since the 1990s (Gade 2013). Ubuntu is described as the heart of African ethics (Dolamo 2013); it is equated with restorative justice (Gade 2013) or an African jurisprudence (van Staden 2019); and finally, ubuntu is essentially and intrinsically peacebuilding for the fact that it aims to maintain harmonious interpersonal relationships, thus there is no antagonism in an ubuntu relation, family, community, and nation. The interconnected nature of ubuntu and peacebuilding is grounded in the communitarian ethical characteristic of society which implies a greater solidarity and equity among humans (Murithi 2006; Udo 2020). With ubuntu and the consciousness of our shared humanness and the responsibility to care for each other by avoiding and preventing instances of inhumanities that erode interpersonal ethical responsibility, post-conflict societies can rise from the hashes of hatred, bigotry, and human atrocities.

Reflecting on how Duvenage (2005) grappled with the way the community responds to violence while attempting to pursue justice and bring about reconciliation in the context of human rights violations perpetrated by the Apartheid regime, and the need to reconcile, bring peace, and move forward, Wiredu (2004, p. 20) noted that "the thinking underlying [the TRC] sought a mutual balancing of the imperatives of truth, justice, and reconciliation in circumstances in which a single-minded fixation on justice would predictably have generated violence of unpredictable consequences". To address the horrendous acts of human violence and past legacy of the South African Apartheid regime, the TRC embarked on a nationwide process of national reconciliation in which it did not seek to violently overthrow the minority white-led power and substitute it with the power of the formerly oppressed majority, but rather took the path of political compromise that helped avert a nationwide murderous revolution or civil war and paved the way for national reconciliation, a strong argument advanced by Duvenage (2004, p. 512) who wrote:

> The TRC was intended to bring this process of political compromise to a conclusion by, inter alia, granting amnesty to those who had committed gross violations of human rights for political objectives in the context of the conflicts of the past.

It is worth noting that the TRC stood as "a moral compromise, evading justice in a narrow sense by concentrating on truth and reconciliation. The objective was not to prosecute and punish the perpetrators of those atrocities so that justice can be done" (Du Toit 1996 cited in Duvenage 2004, p. 512). In the post-Apartheid era which was charged with emotions, bitterness, and a desire to retaliate, ubuntu prevailed against all odds. According to Duvenage (2004, p. 512), "[the TRC] was structured to operate through three specialized committees – one dealing with the violations of human rights, one on amnesty, and another on reconciliation and reparation".

While the South African TRC explicitly employed the constitutional framework that calls for ubuntu—an invitation for recognition and humane values and the nurturing of a reconciling spirit and opposing retaliation (Gade 2013)—the models of peacebuilding discussed above, with regard to reconciliation and healing for the wounds remaining after violent civil war in Africa, have held onto the instrument of the African palaver incorporating non-violent conflict resolution, African solidarity, and the communitarian social ethos of reconciliation and peacebuilding. This resonates with the Constitution of South Africa: "there is a need for understanding but not for vengeance, a need for reparation but not for retaliation, a need for ubuntu but not for victimization" (Constitution of the Republic of South Africa, Act 200 of 1993: Epilogue after Section 25) (Gade 2013, p. 9).

In a similar vein, Duvenage (2004) emphasised how ubuntu is opposed to vengeance, retaliation, and victimization while encouraging reparation. The justification for these attitudes is provided by that which Obenga (2004, p. 46) described as "consciousness", which enables people to distinguish what is correct from what is incorrect, and makes them capable of showing "commiseration, forgiveness and also of being ashamed of wrongdoing". Forgiveness, empathy, compassion, humility, honesty, caring, sharing, and brotherhood are the characteristics associated with ubuntu (More 2004). To illustrate this point, it is asserted: "accruing for moral anger are resentment, indignation, wrath, rage, hatred, and fury—reactive attitudes that might lead to non-forgiveness and the desire for revenge" (More 2004, p. 212).

Ubuntu is without boundary because it is the essence of an ethical intersubjective relation. For this reason, it has power as an African moral dynamic and worldview that has held primal societies together and in peace; furthermore, ubuntu can influence global peace through dialogue as the contemporary world becomes aware of the ethical dimension of this concept (Udo 2020). As a caution against a passive absorption of ubuntu, there were no guarantees that truth-telling would automatically signal the advent of truth and reconciliation that come after violent conflict, nor did it nurture the hope that the monstrosity that marked the Apartheid legacy would be undone (Duvenage 2004). However, the author underlines that:

> These difficulties, however, do not imply that our obligation to the past justifies a policy of pardon and forgetting. The solution in this predicament may lie, as hinted above, in a policy of judgment and forgiving rather than punishment and forgetting (Duvenage 2004, p. 516).

This concluding perspective suggests that even in the context of the monstruous past and memories of child soldiering brutalities, peace is attainable when the humanistic ubuntu ethos prevails as a persuasive obligation on society. The institution that awakens the human individual, stimulates the collective and communal consciousness, and opens the way for transformative action towards peace appears to be the African palaver, which is explored in the following section.

18.6.2 The African Palaver and Peacebuilding

The African palaver is rooted in the African custom of the peaceful resolution of disputes, and it takes place as an open forum where disputants have equal say in the discussion and who are partners in making decisions regarding the issue that has occasioned the palaver. According to Scheid (2011, p. 17):

> The palaver creates physical, social, and psychological space for open communication so that persons can be integrated into the life and expectations of their communities. Through the palaver African communities heal sickness, educate their members about moral standards, and reconcile enemies.

Bénézet Bujo is considered the promoter of the concept of Palaver (Scheid 2011). Other thinkers have also written extensively on the African palaver (Oborji 2020;

Scheid 2011). For instance, Maina (2008, p. 192) wrote the following: "African palaver is not useless talk but an inclusive moral discernment process in a community". The African Palaver emerged in the social ethical context that is not devoid of a religious dimension and that sustains African morality concentrated around the sanctity of life and the preponderant bonds between the human person and the community. This communitarian dimension of the African social ethos is that which validates this dialogical institution. From a theological and missiological perspective, Oborji (2020, p. 222) has ascertained that "Palaver is the interactive dialogue that animates many African communities' affairs, seeking holistic interventions on issues of life and maintaining relationships within the hierarchy of existence". The underlying teleological functionality of palaver is to enhance reconciliation, which is the key guiding principle of African communities, especially in the context where people have suffered harm of any kind. Coupled with the spirit of ubuntu and the palaver, peace dialogues and attempts to redress past legacies of politically motivated human rights violations have prevailed, such as that which was discernible in the activities of the TRC in South Africa and the 1990s national conference.

18.7 Indigenous Mechanisms of Conflict Resolution, Reconciliation and Healing

Discussing the various paradoxes of conflict patterns, violence, and resistance movements, Mazrui (2004) ascertained that post-colonial wars have been more destructive as compared to anti-colonial wars. These paradoxes encompass the paradoxes in racial deficit, fatal borders, religion, and ethnicity; the contradiction in identity and resources; the paradox in modern weapons and pre-modern armies; and finally, the internal contradiction between civil wars and interstate conflicts (Mazrui 2004, pp. 480–481). He ascertained further that "the seeds of the postcolonial wars themselves lie in the sociological mess which colonialism created in Africa by destroying old methods of conflict-resolution without creating effective substitute ones in their place" (Mazrui 2004, p. 480).

This section overviews the itinerary taken by post-conflict communities to revive and reimagine traditional mechanisms of conflict resolution and their contextualisation and adaptation to the demands created by the contemporary warfare mutations and their corollaries on human ecology and Sub-Saharan Africa Wiredu (2004) noted that while liberation struggles (for independence) have been put behind:

> [T]he philosophical problem of violence remains in Africa and everywhere else. Armed conflicts are raging in various parts of the world. In Africa one has to deal both emotionally and intellectually with the spate of military coups that have afflicted political life since the mid-1960s or so, not to talk of the variety of ethnic conflicts in which lives have been lost on an unspeakable scale (2005, p. 19).

Due to the varied nature and contexts of conflicts in Africa, six countries have used contextualised approaches to tackle the aftermath of past injustices marked by large-scale violations of human rights laws and international humanitarian laws to redress the past; prevent recurrences of such gross human rights abuses; forge peace, reconciliation, and the peaceful coexistence of different groups involved in hostilities; and move towards post-conflict reconstruction. Such community-rebuilding cannot solely rely on globalisation and external resources and technology. Local culture, in each context of post-conflict reconstruction, should not bypass local wisdom because such an approach will undermine peace efforts. This view is brought forth by Wessells (2007, p. 260) who wrote that "while Western approaches have much to offer, the silencing of local voices and the creation of dependency on external expertise create a sense of helplessness that is antithetical to healing and peacebuilding". His assertion leads to an understanding of peace as a form of comprehensive well-being that extends beyond the criminal verdicts which cannot provide healing.

Notwithstanding the efficiency of Western psychology and modern peacebuilding theories and praxes, these have shown some limitations in dealing with large-scale conflicts in Sub-Saharan Africa; in certain circumstances, a combination of Western psychology and indigenous models of dispute resolution and healing of wounded relations have been more productive. This has been the case in Mozambique after the civil war (Wamba 2004). The subsequent section explores the revitalisation and adaptation of alternative indigenous dispute resolution models to the context of child soldiering while establishing how successful and effective they are.

Many African countries have resorted to using locally based mechanisms of justice which have paved the way for fostering national reconciliation and nation-building. African Alternative Dispute Resolution (ADR) models have served multiple purposes as a jurisprudence, a conflict resolution tool, and a means of healing and transforming interpersonal and intercommunity relationships. The ultimate end of these approaches adopted by various communities in transition has consisted in preventing, and not only in attempting to prevent, the re-recruitment of child soldiers, but first and foremost, it has involved creating a peaceable environment to which these former young soldiers return and where they are accepted. The lack of peace and acceptance have been identified as push factors sustaining the remobilisation of young soldiers in armed forces and armed groups (Kiyala 2015; Wessells 2006). The fear of arrest and prosecution was singled out by child soldiers in the North Kivu Province of the DRC as one of the reasons that many children returned to armed groups and remained there—for their safety. This reality prompts us to investigate African dispute settlements which could encourage returnee child soldiers to be meaningfully reintegrated into society where they would participate in various projects of post-war reconstruction.

In pre-colonial Africa, there existed customary conflict resolution approaches where the elders of the family, clan, and village intervened to help resolve disputes peacefully. Where conflict broke out between villages or clans, the Chief at the level of chieftaincy would intervene to appease wrangling people or communities in disagreement. Traditional and religious leaders were actively involved in reaching settlements over contentious matters, namely, interpersonal squabbles or inter-clan and intercommunity fallouts. The role of religion was crucial because of the ambivalent dimension of offences—affecting the living and the dead, especially crimes, insults, and scandalous immoral behaviors. The practice of expiatory rituals was believed to carry ancestral power to heal wounded people, wounded land where grave offences had been carried out, and wounded relationships. With respect to these views, Shippee (2002) asserted that before the establishment of contemporary courts and the presence of judges, elders, and priests, and before the inception of legal councils, there existed clergymen, relatives, and neighbors who apparently helped resolve dissensions in society. To further expound this proposition, in the context of the USA, Shippee (2002) discussed the modern techniques and philosophies of conflict resolution based on three religions, namely, Christianity, Islam, and Judaism.

I will limit my exposé to six models, as featured in Fig. 18.2; they reveal that there is no one-size-fits-all response to the complexities of matters regarding child soldiers, and rather conflict resolution efforts are context-based, dynamic, and progressive. Cultural institutions, such as those presented in Fig. 18.2, reflect the extent to which civil society embraces peacebuilding agenda in the aftermath of violent conflicts. It is important to note that the mainstream Western-oriented criminal justice system is confronted with numerous constraints that make it ineffective to achieve durable peace in communities where children have allegedly carried out actions deemed as violations of domestic and international law. The models of accountability, the search for peace, and dispute settlement discussed here have yielded promising results, such as the vindication of victims, reparation of harm, apology, reconciliation, reintegration of offenders into society, peaceful coexistence, and the creation of environments for durable peace and interpersonal encounters, where the community members are not fighting rivals but rather partners in a new journey towards community reconstruction (Fig. 18.3).

The six models displayed in Fig. 18.2 promote reconciliation which is a fundamental aspect of peacebuilding. Reconciliation, in the context of Africa, goes beyond

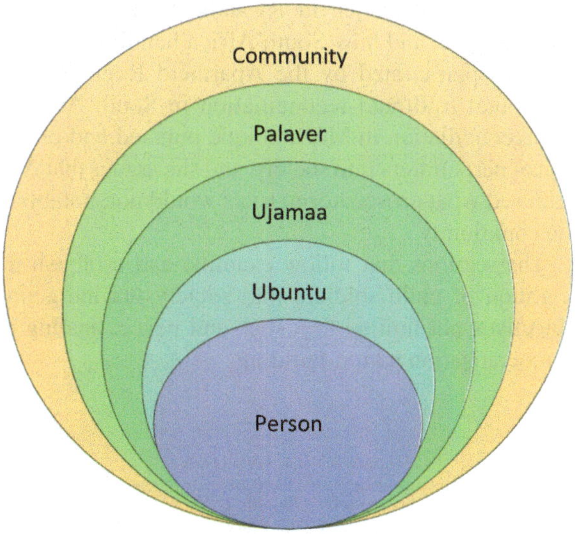

Fig. 18.2 Layers of peacebuilding based on the African social ethos and philosophy. *Source* The Author

Fig. 18.3 African local conflict resolution strategies. *Source* Autor

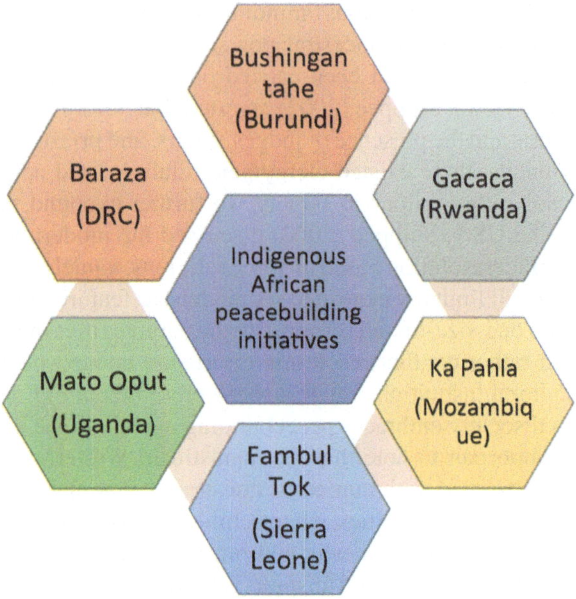

the socio-political arena to embrace the cosmological sphere of the visible and the invisible. For instance, Parker (2012) saw the necessity to reconcile people and heal them and their land, especially where the land had been affected by conflict-related violence and where death had occurred.

Indigenous approaches investigated in this chapter sustain the significance of three approaches dealing with large-scale violence in sub-Saharan Africa. These encompass pardon, punishment and amnesia (Graybill 2004). Studying post-genocide's mechanisms of justice in Rwanda, healing and peacebuilding in post-civil war Mozambique, and how South Africa handled justice to redress injustices and inhumanities perpetrated by the Apartheid Regime in South Africa, Graybill (2004) found that truth and reconciliation in South Africa, *Gacaca* and post-war haling and reconciliation in Mozambique pursued and prioritised the agenda of reintegration of perpetrators into society, and she argues that African accountability paradigms achieved what prosecutorial trials would not, notably reinsertion of perpetrators into the community.

The sections that follow examine and establish the perspectives of social reintegration of child soldiers into society via indigenous models that attain various objectives not limited to justice and peace, healing and reconciliation, and finally, democratisation nation-building.

18.7.1 Bushingantahe, Burundi's Indigenous Disputes Resolution Institution

In Burundi, the bushingantahe was an institution that existed in the pre-monarchical era and its role was to promote social integration (Reyntjens/Vandeginste 2001). The council of notables were composed of representatives of powerful lineages inhabiting a hill from both the Hutu and Tutsi ethnic groups; it was a component of civil society and an element of "counterbalance for the power of the central state"; and members of the councils (mushingantahe, or plural bashingantahe) were involved in the adjudication of local disputes, facilitating reconciliation of individual persons or families, representing the local population at a higher level, and sanctioning all forms of contracts such as marriages, gifts, and inheritance (Reyntjens/Vandeginste 2001, pp. 132–133). According to Kohlhagen (2009), the bashingantahe commonly represent a social group composed of wise men who are traditionally invested with the authority to settle local disputes in Burundian communities. However, the bashingantahe institutions have met some resistance insofar as their rehabilitation and integration into the normative legal system of Burundi is concerned, and despite several recommendations made by experts to effect legal reform, they have been barred from being incorporated into the Burundian statutory law (Kohlhagen 2009).

Reyntjens/Vandeginste (2001, p. 133) reported that the councils of bashingantahe were assimilated by colonial power as assessors of chieftaincy tribunals in 1943, and their proximity to the chief curtailed their independency, and also because of political ties with the chief. The council was established at the level of each colline de recenement (census hill) and received cases before they were submitted to the tribunal de résidence (the lowest court in the hierarchy of the judicial structure). It attempted to get people entangled in the conflict to settle their scores and proposed a solution; in the case that the disputing parties disagreed, they could take the matter to the tribunal which, in turn, would give a statute on the matter independently while keeping in mind the recommendations provided by the Council (Reyntjens/Vandeginste 2001, p. 133). The Unité et Progrès National (UPRONA) single party of Jean Baptist Bagaza and Pierre Buyoya, of the Tutsi ethnic group, politicised the functioning of the bushingantahe between 1987 and 1992 and maintained an upper hand upon it and its independency; this council revealed "the importance of traditional values (the common good) in the postconflict society" (Reyntjens/Vandeginste 2001, p. 134).

In Burundi, national and international voices have been calling for retributive justice to prosecute perpetrators of human rights violations carried out during civil wars. Unfortunately, such a call has been unsuccessful—to put in place truth-telling and accountability measures. Vandeginste (2009, p. 394) contends: "[...] in practice, there has so far been a complete failure to establish any kind of mechanism to deal with truth, accountability, reparation and/or reconciliation". It is worth noting here that there was a succession of peace agreements and attempts for reconciliation which could have utilised the traditional mechanism of community dispute resolution such as bushingantahe, where child soldiering issues could have been handled, but in practice, such an initiative was not attempted (Vandeginste 2009). Szpak (2017) explored transitional justice mechanisms and suggested that bushingantahe indigenous jurisprudence could be utilised as a transitional justice model for child soldiers in the context of Burundi in Africa and the Navajos' custom of naat'aani ture, the Navajos being a Native American people of the Southwestern United States.

A study conducted by Babatunde (2014), based on a documentary analysis of the literature on post-conflict social reinsertion of former child soldiers in Liberia and Burundi, found that traditional practices employed in the reintegration of former child soldiers are more promising and successful. This scholar argues:

> Traditional reconciliation rituals are a viable and community-based process that should form an integral aspect of the reintegration process, for former child soldiers as well as adult ex-combatants, and serves as a complementary mechanism to the contemporary peacebuilding process in post conflicts societies, considering its crucial role in addressing justice, psychosocial needs and promoting community reconciliation (Babatunde 2014, p. 388).

In this same perspective, Babatunde (2014) noted that the bashingantahe can utilise their social recognition and authority to facilitate the reintegration of ex-soldiers, child soldiers, and displaced persons into the Burundian post-conflict society; advance reconciliation agenda among them and their communities; and remake social interrelations and communal bonds, also noting that the bushingantahe institution is "a component for sustainable peace in Burundi" (Hofstede and Hofstede 2005 cited in Babatunde 2012, p. 386). In the context of violent conflict and interethnic confrontation in Burundi, from 1993, the bashingantahe intervened as a civil society group to apply ADR as a peace strategy to resolve those disputes that led to open and brutal clashes (Buszka 2019). It is worth nothing that the bashingantahe have been relevant as a traditional ADR institution and have demonstrated their ability to preserve peace while resolving conflict (Buszka 2019). Their actions have been successful, as observed by scholars such as Buszka (2019) and Nindorera (1998), and it is recommended that a link be established between these traditional practices and modern systems where ubuntu prevails in defining the dignity of the human person (Nahimana 2002). Some scholars have supported this ideal by suggesting the modernisation of the bashingantahe institution (Nahimana 2002) and others continue to sustain its relevance and efficacy in solving contemporary Burundian social matters (Muchiri et al. 2019). Based on its significance, both as indigenous jurisprudence and a traditional ADR institution, the bushingantahe/bashingantahe organisation is a viable platform to deal with child soldiering issues, including the social reinsertion of child soldiers and building communal peace with them.

There is a need for further research in appraising the challenges faced by the Burundian institution of bushingantahe and revitalising its multicultural originality and composition to resolve disputes peacefully and encourage the populations under its various jurisdictions to appreciate traditional moral norms and strive to promote the common good, reconciliation, interethnic social cohesion, and sustainable peace. This requires a revolutionary approach for child soldiers' accountability that is cognizant of child soldiers' individual and collective circumstances and that will remain aligned with international legal standards pertaining to justice for children involved in armed groups and armed forces.

18.7.2 Baraza: Democratic Republic of Congo's Conflict Resolution Institution

In the DRC, child soldiers are called the kadogo (little ones). According to the documented sources on the Congo war, many kadogo enlisted voluntarily to participate in the campaign that led to the late president Mobutu Sese Seko being ousted. They were not all coerced to fight and many of them joined after informed consent. The activities that they carry out do not separate them from other children's responsibilities in armed conflict. The concept of kadogo is broader and more inclusive, as provided by the definition of children associated with armed forces and groups from international law (UNICEF 2007).

In the first place, professional soldiering is supposed to be an adult job, not a venture for minors, that is, the term *kadogo* is understood from the definition provided by Article 2(1) of the Paris Principles cited above (UNICEF 2007). In addition, it is difficult to draw a line between the roles played by child soldiers, as described by international law, and those played by the *kadogo*. The kadogo include girls used for sexual pleasure, cooks, and porters in the DRC; all persons below the age of 18 with or without an active role in military operations; and individuals who have been mupori (in the bush, in the Kiswahili local language), synonymous with being in armed groups and militias. The kadogo do not choose to soldier as a career. Nonetheless, they become involved because of circumstantial constraints, without which they would not join.

The child soldiering phenomenon is ongoing. Child Soldiers International [CSI] (2018, p. 5) has reported that over 3,000 cases of recruitment took place in the DRC by armed groups in 2017. This is a problem that is not to be taken off the world peace agenda soon, especially in the DRC where armed conflicts have become endemic, partly because of the failed transitional justice mechanisms (Savage/Vanspauwen 2008), which include setting up a truth and reconciliation commission—as a corollary of this impasse, The United Nations Organisation Stabilization Mission in the Democratic Republic of the Congo (MONUSCO) has failed in its mission of ensuring that peace is achieved and conflicts are managed and resolved constructively is generally deemed more of a failure than a success. The recurring massacres of civilians and mass killings in the Ituri district, other territories of the DRC (Lalbahadur 2017), and where MONUSCO has spread its 15,000 troops could serve as evidence of this dereliction of duty by the UN Mission in the DRC.

African traditional societies have used various methods of resolving conflicts and lawsuits. This used to take place in various structures such as tribal/customary tribunals, at the royal palace, and at the chieftaincy's court. According to Rubbers/Gallez (2015), traditional leaders have presided over disputes, and in some instances, they have delegated a judge to preside over the palaver—proceedings/debates consisting of a dialogue during which the complainant, the accused, and any witnesses speak successively before the judges. The judges, in this instance, notables, deliberate on the matter with each other and advise reparatory fines to be paid by either party. Fines are no longer paid in kind with items such as arms, pearls, or elephant tusks, but mostly in cash (Rubbers/Gallez 2015). Zongwe et al. (2015) noted that although traditional leaders do not fall under the formal judicial system, they are empowered to pursue the arbitration and settlement of disputes in traditional communities.

In the DRC, the law permits customary courts to operate in various parts of the DRC until magistrate courts are established nationwide, and these customary jurisdictions help mediate and settle two-thirds of all disputes in the DRC (Zongwe et. 2015). Both customary tribunals and peace tribunals/courts represent the Tribunaux de Droit Ecrit (De Jure Tribunals), which include Appeal Courts, High Courts, and Peace Courts; and Tribunaux De Jure (Customary Tribunals), which function as District Courts, Chieftaincy/Collectivity Courts, and auxiliary courts (Rubbers/Gallez 2015). As the country undergoes several reforms, the customary judicial apparatus becomes affected. Replacing customary tribunals with the tribunaux de paix (peace courts) or magistrate courts has weakened the authority of several chieftaincies and tribal leaders in settling conflicts and imposing adequate reparatory fines.

Subsequently, baraza, which falls under the De Facto judicial order, which investigates both civil and criminal matters in the cities and districts of the DRC, has also been negatively impacted by those reforms.

The concept of baraza (in the Kiswahili language) means a place of meeting, a public forum, or a veranda where the elders meet to discuss matters pertaining to the life of the community. It also refers to a customary platform where issues triggering conflicts are discussed to prevent violence and the recurrence of unresolved disputes. Persons who violate social, ancestral, or customary norms appear before the baraza to account for their actions. Furthermore, this institution has existed as an organ of justice in local communities of the eastern DRC, and it is presided over by a Mwami (the Paramount Chief/King). The elders in baraza strive to instil moral values in children to dissuade them from joining armed groups and militias. This institution has facilitated the reintegration of former child soldiers in the baHunde community around Bweremana in both the South and North Kivu Provinces in the DRC. The mediation between victims and child-soldier offenders has led to the commitment by former child soldiers to become involved in security and the prevention of the recruitment of children by armed groups in that region (Kiyala 2021).

The local *baraza* differs from the provincial baraza, which consists of ethnic groups and foreign settlers in North Kivu Province (Kiyala 2016). The local baraza has jurisdiction over civil cases (sanctioning the engagement of couples preparing for civil marriage, settling payments of dowry, marriage, heritage) and criminal matters (murder, robbery/theft, physical injuries caused by aggression, incestuous relations in the family) only in a context where there is no competent court with jurisdiction over criminal matters. To some extent, the local baraza can refer serious criminal matters to peace tribunals that are available at the district level. It is also important to note that the baraza does not have jurisdiction over sexual offences, according to the new DRC Law of Sexual Violence (Congo 2006). Unlike the local baraza that serves a jurisprudence, the provincial baraza works on large-scale interethnic conflict resolution and prevention, because the use of child soldiers is partially sustained by interethnic hostilities and the emergence of community self-defence militias.

The outcomes of the baraza's deliberations include acknowledgement of guilt by the offender, recognition of the need to ask for forgiveness and to promise to refrain from re-offending, cleansing rituals, reconciliation between victim and offender, monetary reparations, admonition of the offender, and symbolic reparation. These results focus not on the perpetrators but rather on the source and causes of conflict and how these violate social and ancestral norms, what the consequences of such transgressions are, and the obligation to make reparation and to restore conviviality among stakeholders. The local baraza approach is in line with Rule 18(1) of the Beijing Rules (United Nations 1985); this encourages a community-based correctional approach.

As the lowest justice order that operates to ensure that each community, especially the eastern and north eastern DRC, complies with customary and ancestral norms, baraza continues to deal with civil lawsuits and disputes that include family-related matters such as marriage, divorce, land disputes, inheritance, etc. To some extent, baraza has handled cases of robbery, thefts, and murder (Kiyala 2016). In the backdrop of these explanations, it is fair to state that this indigenous jurisprudence has been effective in administering justice and resolving conflicts.

As far as child soldiers are concerned, after exiting armed groups and armed forces, they are required to undergo the DDR program at the end of which each child receives a document called kibali (certificate of exit from fighting forces) without which they remain vulnerable to harassment, arrests, and incarcerations by the police and security services. The local baraza established in communities to which children return from soldiering life could facilitate accountability processes and the reintegration of these children, especially where young soldiers have been unlawfully detained or prosecuted in contravention of international law. The insufficiency of local justice structures, such as the peace tribunal, suggest that the customary justice system must be reinvigorated and should regain its past legitimate authority through a constitutional mandate. This could possibly include jurisdiction on child-soldiering lawsuits, especially in the region of North Kivu where the child-soldiering phenomenon is deeply entrenched.

As components of civil society, traditional, tribal, or ethnic-leader leadership structures constitute important actors of peacebuilding. Through the baraza, the elders can influence forgiveness, restitution, and public commitment (not to re-offend) to redress the injustices authored by child soldiers, as part of a community-based response to their misdeeds. Alternative means for child soldiers' accountability should preclude retributive sanctions intended to cause suffering, attrition, and harsh treatment which are all contrary to international law, particularly Article 3(9) of the Paris Principles (UNICEF 2007). In instances where former child soldiers have been unlawfully incarcerated; the use of paralegals can be an additional strength to respond to injustices of the criminal justice system toward ex-child combatants. However, peace efforts initiated and supported by two civil-society institutions, namely, Caritas Goma and baraza, have substantially eased tensions and disputes between child soldiers and victims of their antisocial actions.

In the context of the state being fragilised by conflict, and where the launch of post-conflict justice and transitional justice reforms are delayed, conflict can be averted through the utilisation of locally based structures of conflict resolution to maintain peace. Baraza his been singled out as one of the traditional and cultural institutions that ethnic groups in the eastern Congo use, along with the Kyaghanda traditional model of settling conflicts between different social groups (Aertsen 2008). The Baraza institution maintains that criminal responsibility must be shared between families and armed group commanders for the crimes committed by child soldiers, since the baraza considers child soldiering as delinquency. Elders in traditional settings believe that children remain innocent and the responsibility for the armed violence of which they are accused is imputed to their parents, who fail to prevent their enrolment in fighting forces, and to armed group commanders, who recruit and use them as soldiers. The revival of the traditional baraza through the Intercommunity Council of Elders can be instrumental in achieving the desired transformation of the old (local) baraza into a useful instrument of child-soldiering accountability and social reintegration. This would be a step towards preventing the further enrolment and recruitment of children.

The use of child soldiers, being a novel phenomenon in the eastern DRC where armed conflicts are endemic, raises some concerns on how baraza should respond to this practice. The changing nature of contemporary conflicts and the large-scale destruction in which children are involved as soldiers implies the following:

- *Baraza* elders should find new peace strategies to respond to the challenges of neighboring communities entangled in rivalries leading them to take arms against each other, and thus precipitating the recruitment and use of child soldiers.
- It is important to address how the baraza should respond to allegations according to which community leaders are accused of being accomplices recruiting or sending children and adolescents to serve in auto-defense armed entities to protect their villages.
- How can the baraza deal with the violations of international law for which it lacks the jurisdiction? However, it is habilitated to operate as an ADR mechanism, which is competent in promoting peace.
- Due to the fact that local and tribal leaders have lost much of their prerogatives since the legislations on peace courts were adopted, traditional or tribal authority has subsequently diminished as their power has been weakened and restricted.
- Furthermore, the successful operation of the baraza may be impeded by the presence of armed groups, the militarisation of the communities, protracted conflicts, and finally, having to deal with matters pertaining to child soldiering, which is itself a new reality to be addressed by the baraza.
- When calling former child soldiers to account, baraza proceedings must conform to international criminal law in armed conflict which proscribes punitive sanctions for children found guilty of transgressing this law. Unfortunately, at the grassroots level, many traditional leaders may not be fully conversant with the international legal instrument related to international criminal justice.

In addition, in the context of the reform of the justice system in the DRC, the baraza would have some problems handling cases of sexual offences for which no one is exonerated. Article 42(bis) of the legislation on sexual violence, which proscribes any settlement of sexual violence outside of the court, states that "the official quality of the offender in no case can sexual violence exempt him from criminal liability nor constitute a cause for diminution of pain" (Congo 2006). Article 3(6) of the Paris Principles (UNICEF 2007) addresses how to balance this domestic legal provision with international legislation which considers child soldiers primarily as victims though they are also perpetrators of crimes in violation of international law.

To remedy these difficulties, the following approaches may be helpful: conscientise and sensitise people against ideologies that instil antipathy and hatred, enhance collaboration between peace tribunals and baraza for referral of child soldiers, give statutory and constitutional mandates to the local baraza to handle cases of human rights violations involving child soldiers, introduce new legislation to reinstate the power of customary tribunals and redefine the role of tribal authorities in holding perpetrators accountable, and finally, demilitarise the region and end conflict. This is necessary in the context where local tribal authorities have lost credibility because of their support for armed groups and auto-defense militarised organisations.

Finally, adopting the baraza as a model of justice for child ex-combatants would require adapting it to international legal standards on children in armed conflict. The Paris Principles recommends other systems of accountability for former child soldiers, namely, simultaneously mobilising informal and formal mechanisms (UNICEF 2007) of a communitarian nature. Thus, the baraza, as the lowest court, should have jurisdiction over actions of former child soldiers and can lawfully handle children accused of war-related crimes, including sexual violence. The elders constituting the baraza institution would examine all cases and make a decision on how to address each of them or refer them to the competent judicial structure for determination of their gravity and establishing criminal responsibility. Baraza is a viable forum to redress injustices authored by child soldiers; it is an alternative instrument of accountability that precludes retributive sanctions intended to cause suffering, attrition, and harsh treatment which are all contrary to international law. Its restorative sanctions are consistent with Article 3(9) of the Paris Principles (UNICEF 2007).

18.7.3 Ka Pahla: Mozambican Healing Rituals

Healey/Sybertz (1996, p. 324) explained the bond of communion between African people and the land. This is expressed through blessings and purification of the land. Such rituals are performed with the beliefs that the living owe their protection and socio-economic progress to ancestral spirits. However, there are spirits of the dead that can cause anathemas and destroy peace and societal harmony. The media used to communicate with the unseen world, spirits of the dead and ancestral spirits, through ku pahla (invocation and veneration) of the Supreme Being, as it is known in Southern Mozambique, and among the Tsonga as Xikwembu or Nkulukumba, the Creator (Honwana 1997, p. 296). The rituals to enter in spiritual conversation with the ancestors are performed by tinyanga (medicine men) who are believed to have the capacity to capture, cast out, or appease the dangerous spirits of the dead,

Mpfhukwa or ndau (Honwana 1997, p. 298). Ka pahla refers to the whole healing process which entails the invovation of ancestors, cleansing rituals, etc. (Honwana 1997).

Healing and reconciliation after the occurrence of civil war in Mozambique were conducted through ku pahla rituals. This occurred after soldiers, children, or refugees returned to society following 15 years of brutal civil war between the RENAMO and FRELIMO. The cleansing rituals were carried out under the assumption that the people involved in the war might have been contaminated through their exposure to dead bodies which may have made them unclean (Honwana 1997, p. 299). The role of the tinyanga is that of a secular mediator to play as an intermediary between the living and the ancestors in order to appease them for the wrongs of the soldiers and victims who might have suffered spiritual pollution due to coming into contact with dead bodies (Honwana 1997). This is expressed in the following text:

> [A]fter war, when soldiers, children, and refugees return home to their families, they are believed to be potential contaminators of the community and family [...] After war, men, women, and children need to go through a process of cleansing, which is a fundamental condition for their reintegration into society (Honwana 1997, p. 300).

In addition, reconciliation between the living and the hostile spirits is carried out through timhamba or ku pahla where the hostility is overcome and the ancestors are venerated and urged to stand against the malefic spirits in protection of the living by offering gandzelo (ritual offerings) in certain locations, such as under the tree of the ancestors—in the ancestral cemetery (Honwana 1997, p. 302). Other adequate rites include ku thlavela (to vaccinate with anti-malefic spirit products) which are performed on former child soldiers such that they are healed and cleansed from the consequences of participating in hostilities (Honwana 1997). To elucidate these rituals, she notes:

> Spirit possession is clearly a stabilising factor in social relations; it is by no means a static regulator of human endeavour and identity. On the contrary, it is precisely its capacity for allowing individual changes of identity that explains much of its ontological force and persistence (Honwana 1997, p. 304)

In African cosmology, restoring broken relationships and wounded persons takes precedence over other activities such as reconstruction, and it is crucial for healthy interrelations and societal harmony. Human atrocities, collective and other broad-scale violence pollute relationships between the living and ancestral spirits. Reconciliation in traditional societies, such as the Tsonga, is multi-dimensional and has cathartic effects, fostering peace and cosmic harmony. Unfortunately, the Western categories apply rationality while missing the fundamental psycho-social and emotional aspects of conflict resolution. The African worldview and the advancement of technology, as one of the consequences of globalisation, require balancing the context in which conflict occurs, drawing elements of ancestral wisdom and advancing the agenda of reconciliation rooted in culture and tradition, and at the same time, being in a dialogue with Western-inspired approaches. Ka pahla represents an approach that forges healing, restoration, and reintegration, and such an initiative can be combined with the Western psychology approach to attain a holistic healing mechanism (Wamba 2004).

Such healing, recognition, and benefits should not be selective with regard to the gender of the returnee young soldiers, such that young female soldiers are not disenfranchised and disadvantaged on the basis of gender—gender disparities are particularly evident in Africa where social disparities and economic inequalities affect mostly female citizens whose worth is undermined. Girl soldiers are not spared from this unfortunate and unfair reality; they become pregnant unwillingly and give birth to unwanted children. To illustrate this point, Wamba (2004, p. 15) writes that "in African traditional culture, women are viewed as unimportant and do not participate in decision-making processes. Not surprisingly, then, they were forgotten in the peace process, and national efforts to achieve social and economic reintegration privileged boys over girls". The challenge of ADR mechanisms rooted in African culture is to include girl soldiers in peace efforts and projects of post-conflict reconstruction; they should not face discrimination because of gender, but rather be given equal opportunities and attention to make them peacebuilders.

18.7.4 Gacaca, *Rwandan Post-Genocide's Justice Experiments*

Following the 1994 Rwandan Genocide, both domestic and international instruments have been employed to seek justice for all persons who participated in the genocide and other violations of international law, including war crimes and crimes against humanity. Former child soldiers have been included in the reintegration efforts undertaken at the Lake Muhazi Centre where these young soldiers underwent a three-month program that focuses on providing counseling, recreational activities, and job training aimed for social reinsertion of these children (Everman 2019). While Rwanda utilised child soldiers, it has similarly made efforts to fight this type of inhumanity—by demobilising these young soldiers in 2013, the Kigali Government acknowledged its role and responsibility in the use of child soldiers, though many of these young soldiers have remained in the DRC (Everman 2019). Besheer (2013) reported that close to 3,000 child soldiers were demobilised in 2013 amidst allegations made by the United Nations that Kigali had helped the M23 rebel groups recruit children in Rwanda who were sent to fight in the DRC.

Gacaca is a traditional model of a community meeting that was convened when there was a need for the members of one family of different families, or all the dwellers of one hill to participate, especially the males, except in the case that a female was part of the conflict (Reyntjens/Vandeginste 2001, p. 129). Usually, family elders preside over these procedures, and the family elders are endowed with wisdom to lead all the parties to reach a settlement acceptable by all participants, and the objective is "neither to determine guilt nor to apply legal regulations in a coherent and consistent manner (as one expects from state courts) but to restore harmony and social order in a given society" (Reyntjens/Vandeginste 2001, p. 129). The types of conflict with which the traditional gacaca deal are connected to land use and rights, cattle, marital issues, rights in inheritance, loans settlements, and property damages caused by individuals or cattle; generally, these conflicts are of civil lawsuits in the cases where they are brought before a court of law (Reyntjens/Vandeginste 2001, p. 129). Small criminal offences such as theft, and even some involving murder, though being of a criminal nature, could also be settled in *gacaca*, with no pressure of imprisonment (Reyntjens/Vandeginste 2001, p. 129).

Three strategies were exploited in Rwanda, namely, the International Criminal Tribunal for Rwanda, the Truth Commission—National Unity and Reconciliation Commission, and the Gacaca people's courts which will be covered in this paper. The International Criminal Tribunal for Rwanda (ICTR; *Tribunal pénal international pour le Rwanda, in French; and Urukiko Mpanabyaha Mpuzamahanga Rwashyiriweho u Rwanda*, in Kinyarwanda) was an international court established in November 1994 by the United Nations Security Council, in Resolution 955, to judge people responsible for the Rwandan genocide and other severe violations of international law in Rwanda or by Rwandan citizens in nearby states, between 1 January and 31 December 1994 (the Commission became a permanent body in 2002). The Truth Commission—National Unity and Reconciliation Commission—was recommended by the Peace Accord of Arusha in 1993 but started in 1999 and became permanent since 2002, under Charter: Law No. 03/99, led by 12 commissioners (United States Institute for Peace 1999).

The *Gacaca* consisted of a communal model of arbitrations and rendering of justice in Rwanda—also called "popular gacaca arbitration tribunals"—following the 1994 genocide, and it was conceived as a mechanism to "reduce the bottleneck of cases to be heard by the formal judicial structures and to include a component of reconciliation in the enormous postgenocide justice challenge" (Reyntjens/Vandeginste 2001, p. 130).

The impasse, caused by the inability of the justice system to deal with large-scale trials involving most of the population, who had to testify or admit to having killed or atrociously harmed fellow country men and women, gave rise to popular courts being established in different districts. "The Gacaca tribunals are special tribunals set up to handle specific categories of suspects, displace the jurisdiction of the ordinary courts and apply exceptional procedures" (Haile 2008, p. 23). It is reported that in 1996, close to 1,200 persons suspected of participating in the act of genocide were convicted or released under the national genocide legislation (Reyntjens/Vandeginste 2001, p. 130). The perceived danger of utilising a traditional mechanism and adapting and applying it to situations for which it was not perceived to be useful ran the risk of harming the traditional gacaca, which does not involve adjudication, and making it a hybrid mechanism that would utilise the prosecutorial retributive justice system to resolve disputes and attain reconciliation. As Reyntjens/Vandeginste (2001, p. 131) observed:

- Utilising "popular mechanisms to impose reconciliation by decree" as a top-down strategy does not ensure that it prevails.
- Applying the gacaca to prosecute genocide-related crimes and other wide-scale criminal offenses of a political nature goes beyond the purpose for which the gacaca was conceived and such extrapolation may result in popular revenge.
- The post-genocide gacaca tribunals bear the name of Gacaca but their operation and objectives are distant from the traditional approach which is not reflected in that which appears as a criminal justice mechanism.

As with any system which has weaknesses and strengths, gacaca helped accelerate criminal hearings, the sentencing of guilty persons, and the release of innocent people. After their inception 22 years ago, gacaca popular courts have required an assessment to identify the gaps between the traditional community-based conflict resolution mechanism and the present gacaca adjudication, to avert the danger of eroding the original institution by promoting a modernised mechanism that harms the earlier one. Are gacaca popular courts still relevant today in Rwanda where the judicial structures have evolved, after learning from their past shortcomings to take full responsibility for criminal prosecution and having revitalised the traditional/community-based approach of conflict resolution to carry on with its long tradition of dispute resolution and reconciliation? This is a civil society initiative that needs to be valued without contaminating it with emerging complex matters. There is a need for evaluation and the adoption of a new direction for the bottom-up reconciliation process which may result in healing, national reconciliation, and burying the hatchet to prevent acts of revenge from disenfranchised persons.

In Rwanda, an adequate transitional justice system has been operated within the context of gacaca popular tribunals and the International Criminal Tribunal for Rwanda. Its approaches to truth and reconciliation have been problematic and have raised questions on the perspective of genuine reconciliation (Haile 2008; Rettig 2008). It follows:

> Tackling the legacy of atrocities committed by all sides through a combination of prosecution, amnesty, and truth commissions, and recognizing the losses that Rwandans have suffered as a society can significantly contribute to the building of a new vision for a new Rwanda (Haile 2008, p. 50).

It is worth noting that the Rwanda Demobilisation and Reintegration Commission (RDRC), established in 1997, provided special assistance to children associated with armed groups, including girl soldiers, along with the disabled and chronically ill ex-combatants (Takeuchi 2011). Brounéus (2008, p. 57) noted that "in initiating the gacaca process, Rwanda follows a path like many countries emerging from internal conflict today. Truth and reconciliation commissions have become an important part of peacebuilding". Karbo/Mutisi (2008) averred that "the Gacaca process reflects the unison of hybrid approaches to peace and reconciliation by integrating culture and modern approaches in peacebuilding and trauma healing" (p. 16).

As argued throughout this essay, culture plays a significant role in peace processes because it embodies the principles that vehicle and sustain the peace itself. Such principles encompass the spirit of *ubuntu*, the dialogic institution known as palaver, and the communitarian dimension of interpersonal relations. It follows that any undertaking that seeks to construct a peaceable environment that welcomes former child soldiers to civilian life can be more effective and realistic if it considers local traditions and cultural values which are preserved and revived by an institution such as *gacaca*.

18.7.5 Fambul Tok, Sierra Leonean Healing and Reconciliation Mechanisms

Post-conflict mechanisms in Sierra Leone embraced a variety of approaches:

- There was adjudication via the Special Court for Sierra Leone, which was established as a judicial body by the government of Sierra Leone and the United Nations to hold accountable all persons who bore the greatest responsibility for serious violations of international humanitarian law and Sierra Leonean law for crimes committed after 30 November 1996 and during the Sierra Leonean Civil War.
- The Truth and Reconciliation Commission was set up, which was set as a condition of the Lomé Peace Accord with the aid of the international community in the aftermath of the 11 years of civil war between 1991 and 2002, during which the RUF, with support from the special forces of Charles Taylor's National Patriotic Front of Liberia (NPFL), confronted the government forces to overthrow the Joseph Momoh regime. President Ahmad Tejan Kabbah and RUF leader, Foday Sankoh, signed the Lomé Peace Accord on 7 July 1999. One of the objectives of this peace accord was to tackle impunity, address the needs of the victims, foster healing and reconciliation, and prevent the recurrence of violations and abuses suffered by Sierra Leoneans during the civil war. Joseph Christian Humper, the Bishop in the United Methodist Church, presided over the Sierra Leonean TRC.
- *Fambul tok* (family talk, from the Krio local language) was initiated as a three- to five-year program for reconciliation at the community level. Here, Sierra Leoneans resolve to take the route of old traditions, seeking to heal the wounds caused by civil wars by bringing together victims and offenders for ceremonies of truth-telling and forgiveness.

Hoffman (2008) averred that fambul tok constitutes a strategy of bringing all Sierra Leoneans—victims, offenders, or witnesses—back into the Sierra Leonean family for reconciliation and healing conversations; he remarks:

> [It] is a national initiative that addresses the need to foster a lasting peace from the village-level up [...] Fambul tok represents a way of drawing all members of Sierra Leone—whether victims, offenders, or witnesses—back into the Sierra Leonean family (Hoffman, p. 132).

War-affected communities are faced with the significant challenge to rebuild interpersonal trust and the social capital that have been destroyed by hostilities. Such societies are wounded and vulnerable to recurring violence if their members are not healed and the division between victims and perpetrators deepens. The sole way to bring wellness is through fostering healing, as noted by Hoffman (2008), for the case in Sierra Leone:

> The community healing processes are designed to prevent traumatic experiences from driving people in passiveness or renewed aggression; to encourage them to reflect on the past rather than withdraw; and to empower them to deal with past, present, and future conflicts (p. 132).

Post armed-conflict peacebuilding aims to repair and prevent further hostilities via healing; "this healing is necessary in order for individuals to contribute to sustainable peace and development" (Hoffman 2008, p. 135).

The slogan forgive and forget in Sierra Leone has become the leitmotif of post armed conflict and transition from despotic rule, where people are unlikely to experience justice and rather receive the suffering of past violence (Gibson 2002; Shaw 2005). This family-based justice approach and way of achieving justice and reconciliation and building peace diverges from the Western legal framing of justice and law concepts (Hoffman 2008). Fambul tok pursues justice with objectives such as reintegration instead of the alienation and retribution of perpetrators: "The predominant characteristic of these traditions and practices was an orientation toward reintegrating perpetrators into the community, instead of alienating them through punishment or retribution" (Hoffman 2008, p. 132). This mechanism was effective in enhancing forgiveness and social reintegration (Bangura 2008; L. Graybill 2010; Park 2010).

Reconciliation, in the context of post violent-conflict situations, presents multiple advantages, namely, healing wounded relations, cleansing perpetrators, holding wrongdoers accountable, attaining justice, and forging the path to durable peace.

18.7.6 Mato Oput, Ugandan Healing and Reconciliation Rituals

Any system of belief, philosophy, and human interventions shows limitations due to the nature of human imperfection, and African mechanisms that serve to build peace with former child soldiers reveal some shortcomings, either with regard to the conceptuality of the system or putting it into practice. The first difficulty relates to holding top political officials and high-ranking military officers accountable. In the context of Uganda, with the mato oput, Afako (2006) remarked that "while mato oput can address individual offences of non-state actors, it is not a suitable vehicle for probing the state's failures and violations". This would be the case with gacaca popular tribunals, baraza, and fambul tok. The second corresponds with the incapacity of the international community and national governments to compensate the multitude of victims of war, crimes against humanity, and genocide. I concur with Afako (2006) who regretted that protracted conflicts have hindered African hopes and victims have been helpless in protecting themselves against abusive army commanders and warlords. It is only by bringing war to an end and making this an absolute priority that today's victims can breathe a sigh of relief from the suffering of armed conflicts and hope can be brought to future generations, because the obstinate pursuit of criminal justice procedures could easily jeopardise peace efforts (Afako 2006). In a similar vein, Bangura (2008, p. 144) averred that "many Ugandans in the war-torn north who are weary of a conflict that has killed thousands of family members and forced two million of them into miserable camps, however, see the ICC as an obstacle to peace". This corroborates several scholars' work that show the limitations of the prosecutorial criminal justice system to advance peace agenda in post-conflict societies. However, scholars have debated on the possibility of achieving peace after prosecuting political actors accused of perpetrating egregious crimes, and if lasting peace can be achieved without justice. Johansen (2010) suggested that peacebuilding strategies should be cognizant of the varying conditions on the ground and the changing times in the environment affected by conflict and where the international judicial instruments are required to be applied. In addition, Johansen (2010) suggested that the peace strategy should also be informed by the three components of that strategy, namely, peace, justice, and accountability.

Following the civil war in Uganda, two strategies were undertaken by the government, namely, the Truth Commission and the Civil Society, notably the mato oput. The Truth Commission—Commission of Inquiry into Violations of Human Rights—lasting from 1986 to 1994 and instituted by the Charter of the Commissions of Inquiry Act Legal Notice No. 5 (conducted by six commissioners), was unfortunately interrupted in 1987 as a result of financial constraints (United States Institute for Peace 1986). The commission investigated cases of hundreds of thousands of civilians whose lives were lost under the military dictatorship of Idi Amin between 1971 and 1979 and later, as a result of hostilities endured under the subsequent government of Milton Obote from 1980 to 1985, who was ousted by Yoweri Museveni in 1986 (United States Institute for Peace 1986). During this time, several children were abducted and conscripted into soldiering by the LRA, and many of them became perpetrators of horrendous war crimes and crimes against humanity. Paradoxically, war-affected communities in Uganda are inclined to use indigenous justice mechanisms to hold perpetrators of war atrocities accountable. To elucidate this point of view, Bangura (2008, p. 143) wrote that:

> Traditional leaders from the LRA's Acholi ethnic group—who have been the main victims of attacks by both the government and LRA troops, including the abduction of their children as recruits—want LRA leader Kony and his fighters to undergo mato oput justice.

Mato oput is a ritual used as Acholi justice which literally means *drinking bitter drink* in the process for bringing about reconciliation in the aftermath of a homicide that has been carried out within the community (Afako 2006). After the massive escape of the LRA to the DRC and abandonment of the UN troops' mission to arrest war suspects, including the LRA leader, Joseph Kony, Ugandan President Yoweri Museveni asserted that the LRA had to confess their guilt for the crimes committed against victims and avail themselves for traditional justice within Uganda—this statement was made in the context of the Ugandan Amnesty Act (Afako 2006). In a similar vein, Bangura (2008, p. 143) also asserted that "the Ugandan government urged donors and human rights groups to accept traditional clan-based justice systems as an alternative to jail sentences for dealing with rebel war crimes". Museveni's comments on The Hague alternative were made on 30 May 2007 (Bangura 2008). The pronouncement made by President Yoweri Museveni was an act that demanded the rebels to sincerely repent for their crimes, and more than 7,000 followed the Amnesty dispensation and were reintegrated into their communities (Afako 2006).

During *mato oput* justice, the perpetrator takes responsibility for his/her actions and seeks forgiveness from the victims while promising to repair the harm; the offender and the survivors publicly share the bitter root drink from a calabash as a sign of an engagement to remember and bury the bitterness caused by the crimes which negatively affected their relationship (Afako 2006). Additionally, the *gomo tong* (bending of spears) ritual could also be utilised as it symbolises the termination of hostilities between groups following the process of truth-telling (Afako 2006). *Mato oput* has been revitalised by the traditional civil society of Uganda to pursue accountability and reconciliation following the Ugandan Civil War (or Bush War), also known as the Luwero War and War of Resistance, during which the Ugandan National Liberation Army and the Lord's Resistance Army (the LRA) confronted each other between 1981 and 1986.

The LRA committed heinous atrocities in violation of international human rights and international humanitarian law, and religious and traditional leaders among the Acholi people resorted to the old tradition consisting of the consumption of bitter root, known as mato oput in the Acholi language—this was adopted in the mid-1990s by local government officials and religious and traditional leaders as a model of justice that would bring about accountability for war crimes and crimes against humanity committed by the LRA's rebels, and pave the way to reconciliation and the reintegration of former belligerents.

Bangura (2008) explained the mato oput ritual in the following terms: "The ritual involves a murderer facing relatives of the victim and admitting his crime before the murderer and the relatives drink a bitter brew made from a tree root mixed with sheep's blood" (p. 143). The author also acknowledged the argument laid out by human rights groups expressing their lack of conviction toward the adequacy of this traditional approach (Bangura 2008). Preferring the cultural justice approach to the conventional justice approach demonstrates that African communities value reconciliation and peace that result from such a process more than prolonged hostility, enmity, and retaliation that is sustained by the conventional justice systems' court verdicts. The role of the traditional civil society actors in balancing the need for justice and trading justice among the Acholi communities of northern Uganda serves as a pathway to build peace between former belligerent children and their communities instead of adopting an intransigent approach capitalising on the ICC justice model to resolve innumerable violations of human rights and humanitarian law (Apuuli 2011; Nakayi 2008).

18.8 Conclusion

Several indigenous approaches to dispute resolution are conceived to deter antisocial behaviors and foster societal harmony and peace. It is believed that sustainable peace in the aftermath of violent conflicts cannot be attained without justice. Justice here is comprehended as a process of truth-telling and fair hearing from persons with a stake in an injustice for the purpose of establishing each party's responsibility, ensuring that reparations are made to the affected party, and ensuring that the offender commits himself/herself to refraining from reoffending. Justice is not equated to criminal retributive justice but rather emerges from mutual responsibility and the spirit of *ubuntu* which prompts each person to acknowledge the otherness of others in an ontological binding and interpersonal relation. Such justice is meant to be restorative in nature and foster peace-making. Even the agenda of holding former child soldiers accountable should be a just process that aims for peace. Building peace with young soldiers appears as a humanitarian necessity, a legal imperative, and a social need, and the utilisation of the baraza structure is aligned with international law pertaining to children in armed conflict. The *baraza* structure can serve both as a model of accountability and peacebuilding in which former child soldiers could be called to participate freely.

As with any value-laden system, particularly informal mechanisms of conflict settlement in transitional societies, the baraza structure and modus operandi has shortcomings and strengths. Further studies on the implementation of *baraza* are encouraged to be undertaken. Taking into consideration the difficulties inherent in finding qualified personnel for further praxes of systematic restorative justice programs, *baraza* appears more promising because it is an existing structure with a long tradition of handling civil and criminal matters at the level of villages where there is no competent adjudicative structure with jurisdiction to handle such procedures. Establishing the viability of baraza is instrumental in the process of giving to this structure a constitutional recognition and a statutory power to implement justice for former child soldiers accused of war atrocities.

A constant and engaged dialogue is vital in the local justice system such as the *baraza*, which has a long tradition of peaceful and non-adjudicative settlement of conflict. Justice here should be a context-based engagement that appreciates the synergy between traditional principles and children's interests, as laid out in the legal framework pertaining to children involved in armed conflict, and finally be oriented to achieving durable peace. This implies putting restorative justice in partnership with traditional justice in a context where child soldiers are encouraged to tell the truth about their actions and thereafter make amends.

Despite several hurdles being encountered by African ADR efforts (*baraza*, *gacaca*, *mato oput*, *fambul tok*, the *bushingantahe*, and *ka pahla*) in operating as a jurisprudence, their role in attaining justice, reconciliation, healing, and in remaining promising as peace strategies are in line with Rule 18(1) UN Standard Minimum Rules for the Administration of Juvenile Justice of the Beijing Rules (United Nations 1985). They encapsulate community-based correctional and reparative measures that are consistent with one dimension of the dimensions of Rule 18(1) of the Beijing Rules, as articulated in its commentaries: "Community-based correction is a traditional measure that has taken on many aspects. On that basis, relevant authorities should be encouraged to offer community-based services" (United Nations 1986). The generic term services does not preclude healing and cleansing rituals, seeking spiritual rehabilitation, and building peace.

On moral grounds, addressing injustices in which child soldiers are involved presents advantages that are twofold: it serves the interests of young soldiers—by forging the path of their well-being and acceptance in the community; and it is oriented toward the welfare of the community. As noted by Wiredu (2004), African morality seeks to balance the interests of the individual and those of the community from the vantage point of empathetic fairness. This is closely linked to the standard of justice by which child soldiers are expected to be handled fairly—without prejudice. From an ethical point of view, such justice should align with that which ethics entails, which according to Wiredu (2004), are systems of local norms or customs that regulate human interrelations. It follows that, as ethics satisfies the criteria and rules of life within a particular community and since morality is universal, justice can also be contextualised. Efforts to build peace can rely vastly on the way injustices are addressed, and the way of redress that is consistent with local wisdom and cultural precedents has shown promising effects such as forgiveness of child soldiers and their social reintegration, without which the prospects of lasting peace is very limited if not impossible. This is the case simply because a justice system that generates injustices cannot respond to the peace aspirations of a community. This has been tested with child soldiers, and the results have been the return of former child soldiers into armed groups and militias (Kiyala 2015).

It was demonstrated throughout this paper how all the mechanisms of peacebuilding that are grounded on non-adjudicative approaches, such as *ka pahla*, *fambul tok*, *mato oput*, *baraza*, *gacaca*, and *bushingantahe*, are able to quell *moral anger*, prevent revenge and the harboring of resentment, and manage hatred and rage; in addition, these culturally based models of peacebuilding encourage empathy, love, peace, forgiveness, etc. Based on the effects of *ubuntu* and palaver, as supported by the diversity of literature on peacebuilding activities post large-scale human rights violations, we can deduce that the process of building peace with former child soldiers that rely on indigenous jurisprudence and local wisdom remain the most productive perspectives of reconciliation that can lead to societal harmony. This is both the moral and ethical imperative and a contrarian and innovative response to the ongoing victimisation, rejection, ostracisation, exclusion, disenfranchisement, stigmatisation, and alienation to which several former child soldiers are exposed and with which they are confronted. However, a peace strategy should be more inclusive to provide equal treatment to both female and male soldiers, especially in Africa where girls are regarded as unimportant beings (Wamba 2004). The viability and effectiveness of traditional peacebuilding approaches lie in the established fact that criminal procedures may satisfy the desire of retribution but remain far from building durable peace among disputing parties and healing traumatised persons and wounded relations.

Protecting children and building peace are the most urgent human rights issues of our time, and any strategy to address these matters will remain counterproductive as long as communities and countries continue to undergo internal armed conflicts. It is imperative that the international community upholds its stance against the use of child soldiers and reinforces its protective legal standards to ensure that all children are removed from hostilities and have embarked on the global peacebuilding agenda, not as spectators but as partners, whether domestically or internationally.

References

Aertsen, I. (2008). *Restorative justice between the formal and the informal after large-scale violent conflicts*. Paper presented at the International Conference for Peace Studies, 17–19 September. CPS Working Papers No. 11, 2008, University of Tromsø, Norway.

Afako, B. (2006). Traditional drink unites Ugandans *BBC Focus on Africa. Friday, 29 September 2006, 07:50 GMT 08:50 UK*. http://news.bbc.co.uk/2/hi/africa/5382816.stm. Accessed 23 June 2020.

Apuuli, K. P. (2011). Peace over justice: The Acholi religious leaders peace initiative (ARLPI) vs. the International Criminal Court (ICC) in Northern Uganda. *Studies in Ethnicity and Nationalism, 11*(1), 116–129.

Babatunde, A. O. (2014). Harnessing traditional practices for use in the reintegration of child soldiers in Africa: examples from Liberia and Burundi. *INTERVENTION-AMSTELVEEN-, 12*(1), 379–392.

Bangura, A. K. (2008). The Politics of the struggle to resolve the conflict In Uganda: Westerners pushing their legal approach versus Ugandans insisting on their Mato Oput. *Journal of Pan African Studies, 2*(5), 142–178.

Barnett, M., Kim, H., O'Donnell, M., & Sitea, L. (2007). Peacebuilding: What Is in a Name? *Global Governance, 13*, 35–58.

Besheer, M. (2013). Rwanda Demobilizes Some Child Soldiers [Press release]. https://www.voanews.com/a/rwanda-demobilizes-child-soldiers-sends-others-to-fight-in-congo/1765082.html Accessed 16 March 2022.

Brett, R., & Specht, I. (2004). *Young Soldiers: Why They Choose to Fight* (Vol. null). Boulder: Lynne Rienner Publishers.

Brounéus, K. (2008). Truth-telling as talking cure? Insecurity and retraumatization in the Rwandan Gacaca courts. *Security dialogue, 39*(1), 55–76.

Buszka, A. J. (2019). When Alternative Dispute Resolution Works: Lessons Learned from the Bashingantahe. *Buff. L. Rev., 67*, 165.

Cahen, M., Morier-Genoud, E., & Do Rosário, D. (2018). *The War Within. New Perspectives on the Civil War in Mozambique, 1976–1992*. New York, Martlesham: James Currey/Boydell and Brewer.

Chen, K. (2014). *Comparative study of child soldiering on Myanmar-China border: Evolutions, challenges and countermeasures*. Singapore: Springer Science & Business Media.

Child Soldiers International [CSI]. (2018). Child Soldiers International Annual Reprot 2017–18. https://reliefweb.int/report/world/child-soldiers-international-annual-report-2017-18 Accessed 23 Nov 2020.

Correa, D. D. S., & Owens, W. R. (2010). *The handbook to literary research* (2nd ed.). London: Routledge.

Dolamo, R. (2013). Botho/Ubuntu: the heart of African ethics. *Scriptura: International Journal of Bible, Religion and Theology in Southern Africa, 112*(1), 1–10.

Drumbl, M. A. (2012). *Reimagining child soldiers in international law and policy*. Oxford: Oxford University Press.

Duvenage, P. (2005). The Politics of Memory and Forgetting after Apartheid In K. Wiredu (Ed.), *A Companion to African Philosophy* (pp. 509–518). Maldane: Blackwell Publishing.

Ehrhart, H.-G., & Schnabel, A. (2005). *Security sector reform and post-conflict peacebuilding*. Tokyo: United Nations University Press.

Erasmus, V. (2001). Community mobilization as a tool for peacebuilding. In L. Reychler & T. Paffenholz (Eds.), *Peace-building: a field guide* (pp. 246–257). Boulder: Lynne Rienner Publishers.

Everman, M. (2019). Post Genocide: Top 10 Facts About Rwanda Child Soldiers. https://borgenproject.org/top-10-facts-about-rwanda-child-soldiers/ Accessed 23 Oct 2020.

Gade, C. B. (2013). *Ubuntu and Restorative Justice: addressing the strife and divisions of the past in post-apartheid South Africa*. (PhD). Aarhus University, Department of Culture and Society.

Gibson, J. L. (2002). Truth, justice, and reconciliation: Judging the fairness of amnesty in South Africa. *American Journal of Political Science, 46*(3), 540–556. https://doi.org/10.2307/3088398.

Graybill, L. (2010). Traditional practices and reconciliation in Sierra Leone: the effectiveness of Fambul Tok. *Conflict Trends, 1*(3), 41–47.

Graybill, L. S. (2004). Pardon, punishment, and amnesia: three African post-conflict methods. *Third World Quarterly, 25*(6), 1117–1130.

Gyekye, K. (1992). Person and community in African thought. *Person and community: Ghanaian philosophical studies, 1*, 101–122.

Gyekye, K. (1998). Person and community in African thought. *Philosophy from Africa*, 317–336.

Haile, D. (2008). *Rwanda's experiment in people's courts (gacaca) and the tragedy of unexamined humanitarianism: a normative/ethical perspective*. online. Institute of Development Policy and Management. University of Antwerp. https://medialibrary.uantwerpen.be/oldcontent/container2143/files/Publications/DP/2008/01-Haile.pdf Accessed 16 March 2022.

Hallen, B. (2005). Contemporary Anglophone African Philosophy: A Survey. In K. Wiredu (Ed.), *A Companion to African Philosophy* (pp. 99–148). Blackwell Publishing: Malden.

Healey, J., & Sybertz, D. (1996). *Towards an African narrative theology*. Nairobi: Pauline Publications Africa.

Hoffman, E. (2008). Reconciliation in Sierra Leone: local processes yield global lessons. Fletcher Forum of World Affairs, 32(Suppl.), 129–141.

Honwana, A. (1997). Healing for peace: traditional healers and post-war reconstruction in Southern Mozambique. *Peace and Conflict: Journal of Peace Psychology, 3*(3), 293–305.

Honwana, A. (2006). *Child soldiers in Africa*. Philadelphia: University of Pennsylvania Press.

International Criminal Court. (2002). Text of the Rome Statute circulated as document A/CONF.183/9 of 17 July 1998 and corrected by procès-verbaux of 10 November. 1998, 12 July 1999, 30 November 1999, 8 May 2000, 17 January 2001 and 16 January 2002. The Statute entered into force on 1 July 2002. http://www.icc-cpi.int/nr/rdonlyres/ea9aeff7-5752-4f84-be94-0a655eb30e16/0/rome_statute_english.pdf Accessed 2 Feb 2021.

Jayakumar, K. (2011). Wars and the child-the unending saga of child soldiers. http://papers.ssrn.com/sol3/papers.cfm?abstract_id=1745284. Accessed 4 Dec 2020.

Johansen, R. C. (2010). Peace and Justice?: The Contribution of International Judicial Processes to Peacebuilding. In D. Philpott & G. F. Powers (Eds.), *Strategies of Peace* (pp. 189–230). Oxford: Oxford University Press.

Karbo, T., & Mutisi, M. (2008). *Psychological aspects of post-conflict reconstruction: Transforming mindsets: The case of the Gacaca in Rwanda*. Paper presented at the *Ad Hoc* Expert Group Meeting on Lessons Learned in Post-conflict State Capacity: Reconstructing Governance and Public Administration Capacities in Post-conflict Societies.

Kiyala, J. C. (2015). Challenges of reintegrating self-demobilised child soldiers in North Kivu Province: prospects for accountability and reconciliation via restorative justice peacemaking circles. *Human Rights Review, 16*(2), 99–122. https://doi.org/10.1007/s12142-015-0361-7.

Kiyala, J. C. (2016). Utilising a traditional approach to restorative justice in the reintegration of former child soldiers in the North Kivu Province, Democratic Republic of Congo. *Africa Insight, 46*(3), 33–50.

Kiyala, J. C. K. (2018a). *Child Soldiers and Restorative Justice: Participatory Action Research in the Eastern Democratic Republic of Congo*. Cham ZG: Springer International Publishing. https://doi.org/10.1007/978-3-319-90071-1.

Kiyala, J. C. K. (2018b). The impact of globalisation on child soldiering: challenges and opportunities in the context of the Democratic Republic of Congo. *International Social Science Journal, 66*(221–222), 271–284. https://doi.org/10.1111/issj.12132.

Kiyala, J. C. K. (2021). Responding to Epistemic Injustice against Child Soldiers. *The International Journal of Children's Rights, 29*(3), 662–700. https://doi.org/10.1163/15718182-29030006.

Kohlhagen, D. (2009). Les Bashingantahe écartés de la loi: la place de la justice traditionnelle burundaise après la loi communale de 2010. *Stefaan Marysse, Filip Reyntjens et Stef Vandeginste, L'Afrique des Grands Lacs. Annuaire, 2010.*

Kollmeier, K. (2007). *Ordnung und Ausgrenzung: die Disziplinarpolitik der Hitler-Jugend* (Vol. 180): Vandenhoeck & Ruprecht Göttingen.

Lalbahadur, A. (2017). Improving SADC Responses to Bring Peace to the DRC. *Policy Briefing 165, October 2017*. Accessed 24 Nov 2020.

Lederach, J. P. (1997a). *Building Peace: Sustainable Reconciliation in Divided Societies*. Washingston, D.C.: United States Institute of Peace Studies.

Lederach, J. P. (1997b). *Sustainable reconciliation in divided societies*. Washington, D.C.: USIP.

Les Lois sur Les Violences Sexuelles Loi N°06/018 du 20 Juillet 2006 Modifiant et Complétant le Décret du 30 Janvier 1940 Portant Code Pénal Congolais Loi N° 06/019 DU 20 Juillet 2006 Modifiant et Complétant le Décret du 06 Août 1959 Portant Code de Procédure Pénale Congolais [The Laws on Sexual Violence Law N°06 / 018 of July 20, 2006 Modifying and Supplementing the Decree of January 30, 1940 Bearing Congolese Penal Code Law No. 06/019 OF July 20, 2006 Modifying and Supplementing the Decree of August 06, 1959 Bearing Code of Criminal Procedure Congolese], (2006).

Maina, W. M. (2008). African communitarian ethics in the theological work of Bénézet Bujo. *Pacifica, 21*(2), 192–209. https://doi.org/10.1177/1030570X0802100205.

Masolo, D. A. (2005). African Philosophers in the Greco-Roman Era . In K. Wiredu (Ed.), *A Companion to African Philosophy* (pp. 50–65). Malden: Blackwell Publishing Ltd.

Mazrui, A. A. (2004). Nationalism, ethnicity, and violence. In K. Wiredu (Ed.), *A Companion to African Philosophy* (pp. 472–482). Malden: Blackwell Publishing.

More, M. P. (2004). Albert Luthuli, Steve Biko, and Nelson Mandela: The philosophical basis of their thought and practice. In K. Wiredu (Ed.), *A companion to African philosophy* (pp. 207–215). Maldane: Blackwell Publishing.

Morris, J., & Dunning, T. (2020). Rearing Cubs of the Caliphate: An examination of child soldier recruitment by Da'esh. *Terrorism and political violence, 32*(7), 1573–1591.

Muchiri, S., Murekasenge, J., Nzisabira, S. C. (2019). Exploring the Efficacy of the Bashingantahe Institution as an Indigenous Model of Problem Solving in Burundi. *Social Work Practice in Africa*, 197.

Murithi, T. (2006). African approaches to building peace and social solidarity. *African Journal on Conflict Resolution, 6*(2), 9–33.

Nahimana, T. (2002). Modernizing and integrating traditional judicial systems: the case of the Burundian bashingantahe institution. *East African Journal of Peace and Human Rights, 8*(1), 111–120.

Nakayi, R. (2008). Reaching Out For Justice and Reconciliation in Northern Uganda: The Proposed War Crimes Court and Traditional Justice. *University of Notre Dame, Centre for Human Rights, Working Paper* (5).

Negara, K. (2012). Bitter Root: two former Lord's Resistance Army commanders seek tribal justice in order to be granted atonement for their crimes [online video]: Al Jazeera 28 Jun 2012 06h36 GMT.

Nindorera, L.-M. (1998). Keepers of the peace: reviving the tradition of Bashingantahe in Burundi. *Track Two: Constructive Approaches to Community and Political Conflict, 7*(1), 0.
Noll, D. (2003). *Peacemaking: practicing at the intersection of law and human conflict*. Telford: Cascadia Pub House.
Nyerere, J. K. (1962). *" Ujamaa": the basis of African socialism*. Dar-es-Salaam: Tanganyika Standard, Limited.
Obenga, T. (2004). Egypt: Ancient history of African philosophy. In K. Wiredu (Ed.), *A companion to African philosophy* (Vol. 28, pp. 31–49). Maldane: Blackwell Publishing.
Oborji, F. A. (2020). The African Palaver Reconciliation Model and Mission. *International Review of Mission, 109*(2), 222–235.
Park, A. S. (2010). Community-based restorative transitional justice in Sierra Leone. *Contemporary Justice Review, 13*(1), 95–119.
Parker, R. (2012). *Healing wounded history: reconciling peoples and healing places*. London: Society for Promoting Christian Knowledge (SPCK).
Pearce, J. (2012). Control, politics and identity in the Angolan civil war. *African Affairs, 111*(444), 442–465.
Pike, J. (2000–2021). Uganda Civil War. https://www.globalsecurity.org/military/world/war/uganda.htm. Accessed 3 Feb 2020.
Popovski, V., & Arts, K. (2006). *International criminal accountability and children's rights*, (Vol. 4). Tokyo: United Nations University Press.
Rettig, M. (2008). Gacaca: truth, justice, and reconciliation in postconflict Rwanda? *African Studies Review*, 51(3), 25–50.
Reyntjens, F., & Vandeginste, S. (2001). Traditional approaches to negotiation and mediation: Examples from Africa. In L. Reychler & T. Paffenholz (Eds.), *Peacebuilding: A field guide* (pp. 128–138). Boulder: Lynne Rienner.
Rubbers, B., & Gallez, E. (2015). Réformer la «justice de proximité» en RD Congo. Une comparaison entre tribunaux coutumiers et tribunaux de paix à Lubumbashi. *Critique Internationale, 66*, 145–164.
Savage, T., & Vanspauwen, K. (2008). The conflict in the DR Congo: a story of failed transitions and interlocking conflicts. In I. Aertsen, J. Arsovska, H.-C. Rohne, M. Valiñas, & K. Vanspauwen (Eds.), *Restoring justice after large-scale violent conflicts: Kosovo, DR Congo and the Israeli-Palestinian case* (pp. 323–335). Culmcott House: Willan.
Scheid, A. F. (2011). Under the Palaver Tree: Community Ethics for Truth-Telling and Reconciliation. *Journal of the Society of Christian Ethics, 31*(1), 17–36.
Shaw, R. (2005). *Rethinking truth and reconciliation commissions: Lessons from Sierra Leone. Special Report 130, February 2005*. Washington, D.C.: United States Institute of Peace.
Shepler, S. (2005). The rites of the child: Global discourses of youth and reintegrating child soldiers in Sierra Leone. *Journal of human rights, 4*(2), 197–211.
Shippee, R. S. (2002). Blessed are the peacemakers: Faith-based approaches to dispute resolution. *ILSA J. Int'l & Comp. L., 9*, 237–255.
Singer, P. W. (2005). Western militaries confront child soldiers threat. *Jane's Intelligence Review, 1*(Suppl.), 1–7.
Singer, P. W. (2010). Caution: Children at war. *The US Army War College Quarterly: Parameters, 40*(4), 19.
Stöger-Eising, V. (2000). Ujamaa revisited: indigenous and european influences in Nyerere's social and political thought. Africa, 70(01), 118–143.
Szpak, A. (2017). Indigenous Mechanisms of Transitional Justice as Complementary Instruments to State Justice Systems. *Polish Political Science Yearbook, 46*(2), 55–70.
Takeuchi, S. (2011). Gacaca and DDR: the disputable record of state-building in Rwanda. *Japan International Cooperation Agency Research Institute, 32(July 2011)*. https://issat.dcaf.ch/mkd/layout/set/fullscreen/content/download/5951/50672/file/State%20Building%20Rwanda-JICA.pdf. Accessed 30 Sept 2021.

The World Bank. (2007). Civil Society and Peacebuilding: Potential, Limitations and Critical Factors. Washington, D.C.: The International Bank for Reconstruction and Development / The World Bank.

Towheed, S. (2009). Tools and techniques for literary research: Using online and printed sources. In D. d. S. Correa & W. R. Owens (Eds.), *The Handbook to Literary Research* (2nd ed., pp. 9–35). London: Routledge.

Udo, E. J. (2020). The Dialogic Dimension of Ubuntu for Global Peacebuilding. In E. Essien (Ed.), *Handbook of Research on the Impact of Culture in Conflict Prevention and Peacebuilding* (pp. 302–322). Pennsylvania: IGI Global.

UNICEF (2007). The Paris Principles, Principles and Guidelines on Children Associated with Armed Forces or Armed Groups, February 2007.

UNICEF (2009). Lima Declaration on Restorative Juvenile Justice. https://www.unicef.org/tdad/limadeclarationenglish(1).doc Accessed 23 Dec 2020.

United Nations (1985). Standard Minimum Rules for the Administration of Juvenile Justice ("The Beijing Rules"). Adopted by General Assembly resolution 40/33 of 29 November 1985. http://www.ohchr.org/Documents/ProfessionalInterest/beijingrules.pdf Accessed 19 Nov 2020.

United Nations (1986). United Nations Standard Minimum Rules for the Administration of Juvenile Justice. https://www.ojp.gov/pdffiles1/Digitization/145271NCJRS.pdf Accessed 25 May 2021.

United Nations (2000). Optional Protocol to the Convention on the Rights of the Child on the Involvement of Children in Armed Conflicts. UN General Assembly Resolution. Document A/RES/54/263. Adopted and opened for signature, ratification and accession on 25 May 2000 entered into force on 12 February 2002. http://www.refworld.org/pdfid/47fdfb180.pdf Accessed 16 June 2021.

United Nations (2004). The rule of law and transitional justice in conflict and post-conflict societies: Report of the Secretary-General. http://www.un.org/en/ga/search/view_doc.asp?symbol=S/171 02004/616. Accessed 18 June 2021.

United Nations (2004). The rule of law and transitional justice in conflict and post-conflict societies: Report of the Secretary-General. http://www.un.org/en/ga/search/view_doc.asp?symbol=S/200 4/616 Accessed 18 June 2021.

United States Institute for Peace (1999). Truth Commission: Rwanda 99. https://www.usip.org/pub lications/1999/03/truth-commission-rwanda-99. Accessed 23 May 2021.

Urusaro Karekezi, A. (2012). Transitional Justice and Gacaca in Post-Genocide Rwanda: theory and practice. *Peace and Conflict Management Review, 1*(1), Article 9, 12–13.

Utas, M., & Jörgel, M. (2008). The West Side Boys: military navigation in the Sierra Leone civil war. The *Journal of Modern African Studies, 46*(3), 487–511.

van Staden, M. (2019). Ubuntu: an African Jurisprudence, TW Bennett with AR Munro and PJ Jacobs. *Journal of South African Law/Tydskrif vir die Suid-Afrikaanse Reg, 2019*(1), 203–204.

Vandeginste, S. (2009). Transitional justice for Burundi: A long and winding road. In *Building a future on peace and justice* (pp. 393–422). Heidelberg: Springer.

Vautravers, A. J. (2008). Why child soldiers are such a complex issue. *Refugee Survey Quarterly, 27*(4), 96–107.

Wamba, L. (2004). *Complementary Strengths: Western Psychology & Traditional Healing. Rebuilding hope for child soldiers in post-war Mozambique.* Minneapolis: The Center for Victims of Torture. Wessells, M. (2006). *Child soldiers: from violence to protection.* Cambridge: Harvard University Press.

Wessells, M. G. (2007). Post-conflict healing and reconstruction for peace: The power of social mobilization. *Fear of persecution: Global human rights, international law, and human well-being,* 257–278.

Wiredu, K. (2004). *A Companion to African Philosophy.* Malden: Blackwell Publishing. Wiredu, K. (2005). A companion to African philosophy. Maldane: Blackwell Publishing.

Zehr, H. (2013). *Retributive justice, restorative justice.* In G. Johnstone (Ed.), A restorative justice reader (2nd ed., pp. 23–35). London: Routledge.

Zongwe, D. P., Butedi, F., & Clément, P. M. (2015). The Legal System of the Democratic Republic of the Congo (DRC): Overview and Research. http://www.nyulawglobal.org/globalex/Democr atic_Republic_Congo1.html Accessed 23 Sept 2021.

Chapter 19
Navigating Channels for Civil Society Participation in Conflict Prevention and Peace-Making in the African Peace and Security Architecture

Michael Aeby

Abstract The African Union (AU) and Regional Economic Communities (RECs) that are building blocks of the African Peace and Security Architecture (APSA) gradually elaborated structures for conflict prevention and peace-making and mediated negotiations in numerous conflicts in the last two decades. African civil society organisations (CSOs) with relevant expertise helped to build APSA structures, whilst CSOs in conflict-affected countries interacted with structures and mediators as stakeholders of interventions. However, the channels CSOs could use to participate in structures of APSA building blocks and their inclusion in mediations varied sharply. The chapter reviews channels for CSOs to participate in APSA structures and in mediations. It argues that the APSA saw a proliferation of policies for inclusive conflict prevention and peace-making, which political decision-makers and mediators would need to fully embrace. CSOs were routinely consulted in mediations, but the impact of their inputs on mediation agendas and agreements hinged on the volition of mediators and dominant conflict actors. CSOs had to constantly navigate channels for participation to identify pathways and dead ends.

Keywords APSA · AU · ECOWAS · SADC · Civil society inclusion · Mediation · Preventive diplomacy · Early warning

19.1 Introduction

The African Union (AU) and the Regional Economic Communities (RECs), which constitute the regional building blocks of the African Peace and Security Architecture (APSA), play a critical role in conflict prevention and peace-making in Sub-Saharan Africa, but their organisational structures and peace processes, which they mediate, provide few channels for the participation of civil society organisations (CSOs). This chapter critically reviews channels CSOs may use to interact with

Dr. Michael Aeby is a Visiting Academic at the University of Edinburgh and a Research Associate of the University of Cape Town's Institute for Democracy, Citizenship and Public Policy in Africa.

J. C. K. Kiyala and G. T. Harris (eds.), *Civil Society and Peacebuilding in Sub-Saharan Africa in the Anthropocene*, The Anthropocene: Politik—Economics—Society—Science 34,
https://doi.org/10.1007/978-3-030-95179-5_19

organisational structures for prevention and peace-making and be included in mediation processes that are mandated by the AU, Economic Community of West African States (ECOWAS) and Southern African Development Community (SADC).

The AU and RECs have gradually elaborated their institutions for conflict prevention and peace-making since the construction of the APSA was initiated with the adoption of the Protocol on the Establishment of the Peace and Security Council by the AU Assembly in 2002 (AU 2002; Karbo/Murithi 2018, pp. 1–5). The APSA pillars for decision-making, early warning, peace diplomacy, peacekeeping and the financing of peace operations of the AU and the equivalent institutional structures of RECs have, in recent years, been complemented with mediation support structures. African Non-Governmental Organisations (NGOs), which specialise in peace and security, contribute to the development of ASPA structures and local CSOs in conflict-affected countries interact with these structures as stakeholders of their conflict prevention and peace-making efforts. The AU and RECs maintain civil society interfaces to liaise with CSOs, but availability and viability of channels for civil society to participate in structures for conflict prevention and peace-making vary sharply (Aeby 2021, pp. 1–7).

Whilst mediation was already a conflict management strategy of the Organisation for African Unity (OAU), the AU and RECs facilitated many of the 294 peace agreements that were concluded in Sub-Saharan Africa between 2002, when the AU was founded, and 2015 (Coe/Nash 2020, p. 163). In 2018 alone, the APSA building blocks responded to 14 conflicts and facilitated 13 peace agreements. Most of these agreements, however, collapsed within a year (IPSS 2020, 52). The peace-making efforts of the AU and RECs have, moreover, traditionally relied on high-level mediations by sitting presidents, who often facilitate narrow power-sharing pacts between political and military elites (Nathan et al. 2015, p. 88; Sriram/Zahar 2009; Vandeginste 2013). Mediation guidelines, which the AU and RECs adopted in recent years, recommend the inclusion of civil society in peace negotiations and the implementation of agreements, and their mediators consult local civil society stakeholders. However, although CSOs play an important role in representing communities and in peacebuilding on the local level, civil society actors rarely get a seat at the negotiation table, where they would have a say on the content of agreements. Since the broad-based ownership of peace processes is, according to the inclusive peace paradigm, a condition to translate negotiated accords into stable political settlements and sustainable peace, a lack of civil society inclusion may undermine the mediation efforts of the AU and RECs (Aeby 2021, pp. 1–7).

This chapter serves to critically review and compare the policy and practice on civil society participation in conflict prevention and peace-making in three organisations that are part of the APSA: the AU, ECOWAS SADC. The objective of the discussion is to, firstly, assess the viability of channels the organisations offer to CSOs to engage relevant organisational structures, including decision-making organs, early warning systems, panels for peace diplomacy, and mediation support units. Secondly, the chapter serves to examine consultative mechanisms and other features of peace process designs that are commonly applied to include the voices of local society actors in negotiations that are mediated by the three organisations. The review of both organisational structures and mediation processes seeks to assess the benefits and limitations of channels civil society actors may choose to claim space for participation and represent their interests in prevention and peace-making efforts in the APSA.

The chapter argues that the AU and RECs saw a proliferation of policies and guidelines for inclusive conflict prevention and peace-making, which were drafted by consultants from NGOs and technical experts in the organisations, who diffused the inclusive peace paradigm. The extent to which political decision-makers and high-level mediators embraced and translated these norms into practice varied considerably. The organisations provided a range of access points for CSOs to engage APSA structures, but several civil society interfaces and relevant structures were either defunct or need reform. Whilst local civil society stakeholders were routinely consulted in mediations, the impact their inputs made on mediation agendas and agreements hinged on the volition of lead mediators and dominant conflict parties. Inclusive peace process designs were proposed in mediation guidelines but seldomly applied in practice. The channels for civil society participation in the APSA, thus, included viable pathways and dead ends, which civil society actors had to constantly navigate anew (Aeby 2021, pp. 54–59).

The discussion focuses on: (a) African NGOs that specialise in peace and security; (b) local civil society groups, which include national and subnational-level CSOs in conflict affected-countries that are stakeholders of interventions; and (c) regional NGO networks that act as intermediaries between local CSOs and the intergovernmental organisations (IGOs). It examines and compares the AU, ECOWAS and SADC as the organisations are key components of the APSA, include important subregion, and enable an insightful comparison owing to different policies and practices on civil society participation. The findings that are presented in this chapter were produced through a comparative content analysis of policy frameworks, analysis of organisational structures, and process tracing of selected mediations. The empirical sources include policy documents, communiqués, and 48 semi-structured interviews with officials of the organisations, civil society actors and experts, which were conducted between May 2019 and February 2020, which constitutes the cut-off point of the analysis. The interviewees were selected owing to their in-depth knowledge of APSA structures and peace processes, and their perspectives heavily inform the findings of the explorative study, which are preliminary owing to the limited representativity of the data. Since the SADC Secretariat did not assist our research, the discussion of the SADC case relies on external sources. The chapter summarises findings that were previously published in a report *on Civil Society Participation in Peace-Making and Mediation Support*, which the Institute for Justice and Reconciliation produced in collaboration with the German Corporation for Development Cooperation (GIZ).[1] The research was, moreover, funded by the Swiss National Science Foundation. The following sections include a brief overview of the role of civil society in conflict prevention and peace-making, the three case studies on

[1] For a detailed list of interviews and primary sources, see the original report: (Aeby 2021).

the AU, ECOWAS and SADC, a comparison of their policies and practices, and a conclusion.

19.2 Civil Society in Peacebuilding, Peace-making and Intergovernmental Organisations

Civil society has traditionally been ascribed the role of building peace and preventing conflict on the community level, whilst IGOs act as peacemakers by mediating peace negotiations between political and military elites. Unsuccessful state-centric peacebuilding initiatives, the collapse of elite pacts, and regional implications of local conflicts have led to paradigm shifts in the research and practice of conflict prevention and peace-making. The inclusive peace paradigm emphasises the need to involve civil society actors in international efforts to prevent and mitigate intrastate conflicts to foster broad-based ownership of peace initiatives, translate negotiated agreements into stable political settlements, and achieve lasting peace. The following section provides an overview of the peace research literature that underpins efforts foster civil society participation peace processes and IGOs.

19.2.1 Civil Society and Conflict Mitigation

Civil society plays a transformative role in different theoretical traditions, which ascribe intertwined democratic and peacebuilding functions to CSOs that are relevant in the context of peace negotiations and transitional governance processes. Besides articulating interests of local communities, overseeing government and diffusing pacifist and democratic norms among citizens, CSOs can build peace on the grassroots by conducting informal dialogue, Track 3 mediation and reconciliation processes, monitoring violence, offering peacebuilding training, and providing basic services (Carothers/Ottaway 2000; Belloni 2008, pp. 178–97; Barnes 2009). CSOs can transform conflict constructively and reduce violence by shifting public attitudes, addressing root-causes, defining agendas for peace, and mobilising communities for peace initiatives (Barnes 2009, pp. 131–47). Owing to these conflict-mitigating functions, civil society is equally thought to play a critical role in preventing the escalating of conflict. However, the peacebuilding role of civil society should not be romanticised as CSOs are not necessarily peace-loving. Civil society tends to consist of heterogeneous actors, who may be politically divided, rally around particularistic identities, oppose compromise, spread divisive messages, commit violence and form paramilitary groups (Barnes 2009, p. 144; Belloni 2008, pp. 178–97; Hellmüller 2020).

The inclusive peace paradigm is underpinned by the critique of state-centric liberal peacebuilding, the local turn of the 2000s that stressed the need for local ownership, and the conflict transformation school of peacebuilding, which sees local communities rather than international actors as peacebuilders (Chesterman 2007; Heathershaw 2013). The need promote public participation in peace-making to ensure stakeholders "owned the process" was also a major takeaway of national dialogues in South African and Northern Ireland (Barnes 2002, 2009, p. 143). More recent research has come to focus on the interactions between international and local peace actors, the relationship of peacebuilding on national and local levels, and different arenas of conflict. Internationally mediated peace-making among national elites and peacebuilding in communities are often regarded as complementary. But the sponsorship of local initiatives by international actors can undermine the agency of local actors, erode the legitimacy and relevance of interventions for communities, and foster the imposition of external agendas (Mitchell 2012, pp. 11–13; Mitchell/Hancock 2012, pp. 161–178). The interdependence of local and international interventions and conflict arenas commands research on civil society participation in the conflict prevention and mediation efforts of the APSA.

19.2.2 Civil Society Inclusion in Peace-Making

The inclusion of powerful political and military elites and armed groups, who can veto a peace process and act as spoilers, is a long-standing concern of research on international mediation and peace agreements (Nilsson/Söderberg Kovacs 2011). Normative critiques of narrow power-sharing pacts among violence-makers, who often lack democratic legitimacy, and the frequent collapse of such pacts from the background of research on inclusive peace-making, which focuses on the inclusion of non-dominant groups, including civil society actors, in negotiations, agreements and implementation processes (Sriram/Zahar 2009).

Civil society inclusion is, from a normative perspective, meant to empower communities to protect their rights and take leadership in peace processes, and give greater legitimacy to negotiations and agreements (Aulin 2019, p. 39; Hellmüller 2020; Zanker 2014, pp. 62–88). From a practical perspective, including CSOs is supposed to foster local stakeholders' confidence in peace processes, add alternative perspectives to talks to overcome deadlocks, and prevent groups from turning against the peace process (Barnes 2009, p. 145; Bell 2019, p. 12). The principal rationale for inclusion of civil society is that the representation of a broad range of societal actors is thought to translate into local ownership of peace processes and the institutions they produce, whilst elite deals are considered unlikely to achieve sustainable peace (Barnes 2002; Carl 2019, p. 6). The theory of change that underpins the inclusive peace paradigm assumes that inclusive negotiations, agreements and implementation processes will lead to more stable political settlements that are based on a more equitable distribution of power within society. These political settlements enable a mutually acceptable modus vivendi among competing elites and form the basis for sustainable peace (World Bank and UN 2018, p. 22; Bell 2019, p. 11).

The inclusive peace paradigm has come to reflect in policies of the United Nations, including the Sustainable Development Goal 16 on peace, justice and strong institutions, which aims at building "inclusive societies" (UN 2015, p. 16). Guidelines for conflict prevention and mediation of the AU and RECs have also established inclusivity as a policy norm (ECOWAS 2008; AU 2014a). In practice, however, civil society inclusion in peace-making encounters formidable obstacles. Mediators are reluctant to include additional delegations at the negotiation table as it complicates negotiations and causes selection problems. Powerful conflict parties oftentimes resist the inclusion of CSOs that lack coercive power (Paffenholz 2015, pp. 71–76).

Consultative mechanisms and other features of peace process designs can enable civil society stakeholders to participate in negotiations. National dialogue conferences permit a broad range of societal groups to directly participate in talks, but inclusive dialogues are often unfeasible in the political realities of violent conflicts (Berghof Foundation 2017). Consultations by mediators or the establishment of a Civil Society Room can permit CSOs to formally feed into negotiations, but consultative mechanisms may have little impact on the negotiation agenda (Hellmüller/Zahar 2019). Informal channels and relations with powerful negotiating parties that CSOs can use to include their demands on the agenda may outweigh their formal representation in separate delegations and consultations (Aeby 2016, pp. 717–24; Barnes 2009, p. 140; Paffenholz 2015, pp. 71–76).

Peace agreements can stipulate the participation of civil society in the implementation of provisions, and a large-N study by Nilsson suggests that peace agreements that include civil society are more sustainable (Nilsson 2012, p. 243). Whereas a 2007 study showed that few agreements mentioned civil society participation in implementation processes, (Bell/O'Rourke 2007, p. 293) IGO's adoption of inclusivity as a policy norm may render such provisions a more frequent feature of accords. The implementation of transitional mechanisms and reforms provides more opportunities for participation than negotiations and CSOs can play a vital role in monitoring the implementation of agreements (Paladini et al. 2019, p. 35). Whether opportunities for participation materialised, thus, not only depends on the conflict actors but on the promoting of inclusive peace process designs by mediators and IGOs that sponsor agreements.

19.2.3 Civil Society Participation in Intergovernmental Organisations

The UN and African IGOs committed in their constitutive treaties and protocols to engage civil society to promote development. Whereas some organisations have dedicated structures to liaise with CSOs, others rely on NGO networks that act as implementation partners and interface between intergovernmental and local civil society organisations. Common relational models of IGOs and CSOs include the facilitation of CSO activities by IGOs, dialogue between IGOs and CSOs in joint forums, and partnerships in which IGOs and undertake joint programmes based on mutual objectives. IGOs may outsource functions to NGOs that act as service providers or develop amalgamated structures (WANEP 2019, pp. 3–5). IGOs generally partner with NGOs that can advance their objectives and have the suitable competences, programming, leadership, governance structures, accountability procedures and verifiable impact (Rudo/Bronwen 2009; Morris/Rudy 2016; WANEP 2019, pp. 3–5). Whereas African IGOs partner with a variety of NGOs on development, the politically sensitive domain of peace and security leaves less room for co-operations (Söderbaum 2007; Morris/Rudy 2016).

The AU and RECs nonetheless partner with African NGOs that specialise in peace and security and provide expertise to develop policies, operationalise APSA structures, and analyse conflicts. The African partner NGOs, which have the necessary resources and expertise, for instance, includes ACCORD, whose embedded staff supported the secretariat of the AU Panel of the Wise. The Centre for the Study of Violence and Reconciliation was instrumental in developing the AU's Transitional Justice Policy. Femmes Africa Solidarité promotes women's leadership in AU conflict prevention and peace-making efforts. The Institute for Peace and Security Studies (IPSS) issues APSA assessment reports and trains AU officials. The Institute for Justice and Reconciliation (IJR) is on the AU roster for technical experts on reconciliation. (Murithi 2018, pp. 13–23). The Institute for Security Studies (ISS) monitors the PSC and supports AU programmes ranging from early warning to peacekeeping. This includes the Training for Peace (TfP) programme to strengthen peacekeeping capacity. The Life and Peace Institute partners with the AU and IGAD on conflict prevention and peacebuilding in North-East Africa. Oxfam, which is headquartered in Nairobi, has an AU Liaison Office and a Peace, Security and Humanitarian Affairs Programme. The West African Network for Peacebuilding (WANEP) is ECOWAS' principal partner for early warning and as entered a partnership with the AU (Aeby 2021, p. 15; Murithi 2018, pp. 13–23).

Whereas African NGOs and academics inform the development of APSA structures and their responses to conflicts, such partnerships involve great challenges. Since NGOs and IGOs are donor-dependent, their partnerships and the outsourcing of functions to NGOs are susceptible to changing donor agendas. Most importantly, some African states are suspicious of CSOs and their participation in the APSA (Aeby 2021, p. 15).

19.3 African Union

The AU has developed an elaborate policy and institutional framework for conflict prevention and peace-making, but the development of the corresponding infrastructure lags behind and the policies on civil society participation contain ambiguities. The Constitutive Act of the African Union established the Economic Social and Culture Council (ECOSOCC) to interface and partner with civil society (AU 2002, p. 22). Since this model has proven unviable to partner with expert African NGOs, the AU Commission (AUC) and PSC have adopted a more flexible approach. AU mediation guidelines declare inclusivity a mediation principle, but the application of these recommendations hinges on the will of mediators and conflict parties (AU 2014a, pp. 48–51; AU 2014b). The section outlines the AU's policy framework for civil society participation, involvement of NGOs in AU structures, and channels local CSOs can use to engage in AU-facilitated peace processes.

19.3.1 AU Policy Framework

The constitutive documents of the AU, strategic policy plans, as well as mediation guidelines entail provisions relating to both, the involvement of partner NGOs in structures for prevention and peace-making as well as the inclusion of local civil society stakeholders in peace processes.

The 2002 Constitutive Act and the 2003 Protocol on the Establishment of the Peace and Security Council envisage that NGOs participate actively in efforts to promote peace, security and stability, including early warning and research (AU 2002, p. 22). For this purpose, the 2004 Statute of the ECOSOCC establishes a Peace and Security Committee, which was meant to be the principal interlocutor for corresponding issues. To be eligible to ECOSOCC, however, CSOs must not only be registered in AU member states and share the AU's objectives but draw 50% of their resources from membership contributions (AU 2004). Whereas these criteria were meant to ensure African ownership, they effectively barred vast sections of African civil society from participating in the AU because well-capacitated NGOs, like the AUC itself, usually depend on donor assistance (Aeby 2021, p. 19).

The 2008 Livingstone Formula, which the PSC adopted to regulate its interactions with CSOs, envisaged civil society participation in early warning analysis, mediation support, training, civilian aspects of peacekeeping, local peacebuilding, humanitarian relief and post-conflict reconstruction. However, the PSC resolution underscored the primacy of the ECOSOCC and applicability of its eligibility criteria for NGOs to contribute to the work of the PSC and Peace and Security Department (PSD) (AU 2008). These rules would effectively prevent the PSD from cooperating with important African NGOs, which have relevant expertise on peace and security, and the PSC from interacting with CSOs from conflict-affected states. The 2013 Maseru Conclusions, therefore, introduced the principles of relevance and flexibility, which were henceforth applied to enable the PSD to work with suitable NGOs based on a Memorandum of Understanding (MoU). However, the Maseru Conclusions reaffirmed the ECOSOCC criteria, thus, perpetuating uncertainties and grey arease (AU 2014b).

Strategic policy plans, including the APSA Roadmap 2016-2020, Agenda 2063 and Silencing the Guns report, declare the objectives of enabling citizen participation, fostering collective ownership of a common vision for the AU, and developing plentiful partnerships with civil society and academia to prevent and mitigate violence. Yet, the plans do not give guidance on how these objectives can be actualised. The

2017 report on "The Imperative to Strengthen Our Union", which was produced under Paul Kagame's leadership to inform a comprehensive reform of the AUC, candidly states that the AU must become more relevant to citizens. But it omits reforms to the framework for civil society participation (Kagame 2017, 10). The ECOSOCC Statute, thus, remains in place although the experience since its enactment shows that it is impractical, inhibits rather than enables participation, and, thus, contradict the AU's strategic policy objectives (Aeby 2021, p. 19; ISS 2019).

A series of mediation guidelines and standard operating procedures for mediation support, which the AU developed in recent years with the assistance of consultants from African and international NGOs, detail that experts from NGOs may contribute to mediation training, advise mediators to design peace processes and agreements, facilitate multitrack dialogues, and assist the management of mediation knowledge by retaining lessons from missions and conducting research (AU 2012, 2014a, 2016a; Nathan 2009). Whilst such co-operations are possible thanks to the principles of flexibility and relevance introduced by the Maseru Conclusions, the AU could eliminate ambiguities in its policy framework by abandoning the ECOSOCC Statute (Aeby 2021, p. 19).

19.3.2 AU Policy Framework for Local CSO's Inclusion in Peace Processes

The organisational mandate of the Constitutive Act to engage civil society and the objective of policy plans to involve citizens command that local CSOs be included in peace processes. In terms of the PSC Protocol, the Council shall consult or invite CSOs that are involved in conflict situations to address meetings (AU 2002, p. 8). Whereas inclusivity constitutes an AU mediation principle in terms of the 2014 AU Mediation Support Handbook, the guidelines to include local CSOs in AU-facilitated dialogues constitute non-binding recommendations. CSOs are generally projected as stakeholders who must be consulted rather than included at the negotiation table (AU 2014a, pp. 7–10). The 2008 Plan of Action to strengthen the AU's mediation capacity advises mediators to consult CSOs because their direct representation in talks is often unfeasible (Nathan 2009, p. 16). The 2012 Standard Operating Procedures for Mediation Support instruct mediation teams to identify CSOs in conflict analyses and spell out mediation objectives for these stakeholders. The duty of consult CSOs lies with the lead mediator (AU 2012, pp. 6–37). An AU reader on Managing Peace Processes lists rationales, obstacles and modalities to include CSOs alongside political business actors in negotiations (AU 2013, 3:39–62). The Mediation Handbook advises mediators to strike a balance between keeping negotiations manageable by limiting the number of delegations and making them more legitimate by including many stakeholders. Whilst alluding to national dialogues, the Handbook equally treats CSOs as additional actors, who must be consulted, rather than delegations at the negotiating table (AU 2014a, pp. 63, 73).

19.3.3 AU-CSO Interface: ECOSOCC

The ECOSOCC, as Murithi shows, was included in the 2002 Constitutive Act of the AU owing to pressure by civic groups, who demanded that the AU architecture include a civil society interface (Murithi 2005, p. 112–36). Although the 2004 Statute established that CSOs would need to be ECOSOCC members to contribute to the work of the AUC, the operationalisation of the 150-member Council and its organs proved slow and its Peace and Security Committee was still not functional by 2015 (Amr 2012, p. 176). The Citizens and Diaspora Directorate (CIDO) in the Office of the AUC Chair served as ECOSOCC's secretariat and connection to AU organs until 2019, when an ECOSOCC Secretariat was set up in Lusaka. The latter is hoped to breathe life into the Council. In 2017, the ECOSOCC started consultations on Silencing the Guns and violent extremism (Aeby 2021, p. 21; Lwizi 2019). Yet, in 2019, an ISS report concluded that the ECOSOCC was "dysfunctional" and impeded civil society's access to the PSC (ISS 2019, p. 8).

The civil society interface remains hamstrung the fundamental design flaws of its Statute. It is widely acknowledged that the ECOSOCC's restrictive membership criteria exclude vast sections of African civil society and, thus, undermine rather than grow African ownership of AU institutions (Aeby 2021, p. 21; Rudo/Bronwen 2009, 27; Nathan et al. 2015, p. 152). Since donor-assisted expert NGOs are ineligible, the represented CSOs have lacked expertise to inform peace and security policy. The ECOSOCC has a passive advisory role and responds to request by AUC departments rather than to proactively initiate policy initiatives. The Councils' representativity was further limited by the fact that only 26 of 55 AU member states set up ECOSOCC chapters (Aeby 2021, 21; Amr 2012, pp. 172–83; Nathan et al. 2015, p. 150; ECOSOCC 2020). Authoritarian governments sought to deploy CSOs that echo their positions. Many African CSOs shun the ECOSOCC owing to these limitations, its marginal influence and the perception that states do not take the council seriously. Instead, they resort to other channels to influence decision-making in the AU. Whereas a review of the ECOSOCC Statute may ease these grievances, a platform that is independently managed by civil society would be better suited to ensure access and gain the trust of CSOs (Aeby 2021, p. 21).

19.3.4 AU Peace and Security Council

In principle, the PSC could invite CSOs that met the ECOSOCC criteria to address its sessions based on the 2008 Livingstone Formula, which it adopted owing to the recognition that its proceedings could benefit from CSOs input and sustained lobbying by NGOs (Aeby 2021, 21) But meetings between ECOSOCC and the PSC failed to take place because, as the ECOSOCC stated in 2018, the Livingstone Formula was never operationalised (ECOSOCC 2018). The application of the principle of flexibility from 2013 onwards enabled the PSC to consult relevant think tanks and local CSOs.

Such direct briefings henceforth became an important channel for CSOs to inform the AU's decision-making organ on peace and security. However, member states could prevent critical civil society voices from addressing the council, and a database for "relevant" CSOs had not materialised by early 2020 (Aeby 2021, pp. 21–23).

Submissions to the PSC Chair, who sets the agenda, the AUC Chair, who mandates mediators, the Commissioner for Peace and Security, who directs the PSD; and Permanent Representatives of member states can constitute an important channel to inform decision-makers. Whether such submissions stand a chance to inform deliberations depends on their quality and CSOs' reputation. A crucial way to inform PSC deliberations is to share analyses with embassies and representatives of member states. Well-capacitated NGOs with offices in Addis Ababa monitor and analyse the PSC agenda to proactively produce demand-oriented analyses, which can be easily absorbed by the relevant decision-makers (Aeby 2021, pp. 21–23).

19.3.5 AU Continental Early Warning System

Serving as a nerve centre that gathers information on conflict risks to alert the AUC Chair and PSC, (AU 2014a, p. 14) the situation room of the Continental Early Warning System (CEWS) analyses data from reports, news clippings and the early warning systems of RECs. The CEWS has in recent years increased its analytical capacity, but the early warning centres of some RECs that are not fully operational. Translating early warning information into prompt action by political decision-makers remains a formidable challenge. Early warning information is not widely shared in the AUC as states that deny conflict risks respond sensitively if they appear on its radar (Aeby 2021, p. 22; Arthur 2017, 10; ISS 2017; Noyes/Yarwood 2013, p. 251).

The CEWS was operationalised with the support of ISS. (Aeby 2021, p. 23) In line with the Livingstone Formula that envisages early warning collaborations, (AU 2008, para. 10) the CEWS leverages conflict analyses by think tanks (Aeby 2021, p. 23; ISS 2017). After entering an MOU in 2018, WANEP deployed a liaison officer to link its West African to the continental system (WANEP 2019, p. 43). The CEWS has since entered collaborations with further Africa NGOs and provides training to civil society actors (Aeby 2021, p. 23).

19.3.6 AU Panel of the Wise

In terms of the PSC Protocol, the five eminent members of the Panel of the Wise (PoW) have a mandate to advise the PSC and engage in preventive diplomacy (AU 2014a, p. 14; Gomes Porto/Ngandu 2014, p. 185). The PoW and equivalent panels of RECs are linked through the PanWise network, which has been inactive in recent years although having enabled fruitful collaborations in the past (Aeby 2021, p. 23; Nathan et al. 2015, p. 8). In 2017, the structures were completed with FemWise

Africa to strengthen the role of women in conflict prevention and mediation (Ngandu 2017). FemWise has gained considerable traction and serves to train and deploy women mediators, who are recruited from among local peacebuilders, to facilitate multitrack dialogues (Aeby 2021, p. 23).

The PoW was operationalised in 2007 with the assistance of peacebuilding NGOs. ACCORD seconded staff to its PoW secretariat until 2018, when the AU decided it should be self-reliant. ISS and TfP continued to provide technical assistance to the PoW to organise high-level dialogues on Silencing the Guns, whilst IPSS supported training for senior officials (Aeby 2021, p. 23). The International Peace Institute and ACCORD assisted the production of the PoW's thematic reports (AU 2014a, 14; AU 2019; de Carvalho 2017, p. 8).

When undertaking missions to a total of nine countries between 2007 and 2015 to advice the PSC and meet stakeholders, mostly in the context of elections, the PoW engaged in consultations with local CSOs, thereby providing a vital channel to civils society to inform conflict prevention efforts (Apuuli 2018, 160; Gomes Porto/Ngandu 2014, pp. 188–94) However, a 2014 APSA assessment found that the PoW did not sufficiently interact with CSOs and the PSC did not appropriately follow up on its recommendations. (Gomes Porto/Ngandu 2014, p. 197; Nathan et al. 2015, p. 8). The members of the PoW, whose term began in 2018, have been prevented from embarking on preventive missions as states turned down their good offices. Whereas the underutilisation of the PoW deprives local CSOs of an important channel, the locally recruited FemWise mediators provide new opportunities for communities to feed into multitrack dialogues (Aeby 2021, p. 23).

19.3.7 AU Mediation Support Unit

The small Mediation Support Unit (MSU) launched in April 2019 with a mandate to backstop mediators, build mediation capacity, manage mediation knowledge, and network with relevant actors. The development of mediation support structures in the AU started over a decade earlier with the support of international and African NGOs that specialise in mediation: The NGOs promoted the buy-in of stakeholders to create the MSU by presenting a vision and convincing AU officials, states and development partners of the added value; facilitated the transfer of research-based knowledge and best-practice standards in initial trainings; drafted instruments for procedures, mediation principles, job descriptions and resource requirements; and provided critical analyses of the operationalisation process. The involvement of NGOs also bore challenges: African leadership was limited as few African NGOs had expertise on mediation support, meaning that many instruments were drafted by European NGOs. Since expert NGOs must be entrepreneurial to secure contracts with IGOs and donors, they have an interest in carving a niche for themselves when designing APSA institutions. NGOs can also foster organisational consistencies by wooing different actors and divisions in the AU Commission. Norms that NGOs carry into guidelines and training documents they design, may not be shared and put into practice by mediators and support teams. Crucially, if the development of new structures is mainly driven by NGOs and donors, they risk lacking the necessary buy-in of political decision-makers to function properly (Aeby 2021, p. 24).

The planning of mediation missions by the MSU left little room for NGO participation, except for the provision of analytical inputs. The nascent MSU worked with expert consultants from African NGOs to develop its capacity-building programme and local civil society actors were among the recipients of trainings for FemWise mediators. A Knowledge Management Framework to retain and produce comparative mediation knowledge was produced with the assistance of the Centre for Humanitarian Dialogue (HD), but the MSU lacked resources to put a knowledge-management system into practice. A partnership with African research institutes would permit the outsourcing of tedious research in a hybrid knowledge-management system. Similarly, experts from African NGOs could populate a roster for technical experts who can support mediators to design peace processes and draft agreements as the AU lacks the resources to maintain a standby panel of experts as in the UN (Aeby 2021, pp. 24–30).

19.3.8 Inclusion of Local CSOs in AU Mediations

Whereas AU mediation guidelines propose inclusive peace process designs, the feasibility and application of inclusion mechanisms to permit a wider range of societal actors to participate in peace negotiations depends on the conflict situation, mediation mandate, personality of the lead mediator, and whether the AU or a REC leads a mediation effort. AU mediators nowadays routinely hold consultations with civic stakeholders during mediation missions and consider their position papers. Such direct consultations are, according to AU official and NGO practitioners, an effective channel to inform both the mediation agenda and mediators' reports to the AUC Chair and PSC. Whilst consultations are standard, the processing and integration of contributions by consulted civil society actors into the mediation agenda remains a challenge. For practical reasons, lead mediators can only consult a limited number of CSOs, and AU missions cannot liaise with civil society actors without the approval of governments that may regard such consultations as an interference in internal affairs. The MSU may prove instrumental in streamlining the processing of inputs and consulting local CSOs to analyse conflicts and map stakeholders ahead of future mediation missions (Aeby 2021, p. 29).

AU Liaison Offices and Special Representatives, who have a longer-term presence in conflict-affected countries, provide a key channel for local CSOs to inform mediations as AU mediators rely on their knowledge of conflicts and stakeholders. Whereas Special Representatives can give local CSOs credibility, peacebuilding NGOs share their analyses with Liaison Offices, using them as an access point to raise concerns within the AU system (Aeby 2021, p. 29). By 2016, the AU had established 17 Liaison Offices, whose mandate included reaching out "to people on the ground"

(AU 2016b). However, this vital access point was unavailable in states that opposed a longer-term AU presence (Aeby 2021, p. 29).

FemWise can provide additional access points for local CSOs to feed into AU mediations. FemWise aims at mulittrack mediations that enable communities to participate in dialogues. FemWise mediators are well-placed to serve as conduits between local CSOs and the AU as they are peacebuilders from communities. The impact such multitrack mediations make on the negotiations and the content of agreements is to be seen (Aeby 2021, p. 29).

Whilst CSOs may use the above channels, civil society inclusion in AU-facilitated peace processes encounters formidable technical and political obstacles. Local CSOs often lack resources, communication skills and knowledge on technical issues and the AU system to make an impact on peace processes. To participate in peace processes and inform mediations, they may depend on well-capacitated NGOs that liaise with AU organs. Civil society tends to be the weakest in war-torn and authoritarian states that curtail civic freedoms. The biggest obstacle to civil society inclusion in AU-facilitated peace processes is the distrust of governments that regard civil society activists as a threat, Western proxies and un-African (Aeby 2021, p. 29).

19.4 Economic Community of West African States

ECOWAS began to elaborate its peace and security institutions prior to the creation of the APSA in response to civil wars in the 1990s (Adetula et al. 2016, p. 21). ECOWAS' policies envisage a highly inclusive approach to conflict prevention and peace-making, but not all relevant structures are operational (ECOWAS 2008). The ECOWAS Commission provides a variety of access points for civil society and partners with regional networks that serve as interfaces for local CSOs and bolster its early warning capacity. Whereas ECOWAS mediators routinely consult local stakeholders, CSOs' access to ECOWAS structures and mediations is uneven (Aeby 2021, p. 33).

19.4.1 ECOWAS Policy Framework

The constitutive documents, policy plans and mediation guidelines of ECOWAS prioritise conflict prevention and enshrine a comprehensive framework for NGOs' involvement in ECOWAS structures and the inclusion of local CSOs in peace processes. The revised ECOWAS Treaty of 1993 added the prevention and resolution of intrastate conflicts to the mandate of the organisation, which was founded in 1975, but only envisaged civil society participation to advance economic integration (ECOWAS 1993). According to the 1999 Mechanism for Conflict Prevention, Management, Resolution, Peacekeeping and Security, which established ECOWAS' peace and security architecture, the Council of the Wise must include civic leaders, whilst special envoys and the ECOWAS Commission must coordinate interventions with relevant NGOs (ECOWAS 1999). The 2008 ECOWAS Conflict Prevention Framework (ECPF) introduced a bottom-up approach to conflict prevention and transformation and firmly established a human security paradigm. The ECPF envisages collaborations between ECOWAS, NGOs and states across the board, including early warning, preventive diplomacy, matters of democratic governance, human rights, the rule of law, natural resource governance, cross-border initiatives, security sector governance, disarmament, the ECOWAS Standby Force, humanitarian assistance, women peace and security, youth empowerment, and peace education. According to the comprehensive strategic plan, ECOWAS shall facilitate creative conflict transformation interventions that are led by civil society and states (ECOWAS 2008, pp. 9, 69, 72). The ECPF describes CSOs as bona fide partners and stipulates that ECOWAS shall partner with NGO networks based on MOUs, so they can contribute to policy-development and implementation, channel civil society concerns, and spearhead prevention and peacebuilding in member states (ECOWAS 2008, pp. 9, 25, 69, 72, 74). The role of NGOs in ECOWAS peace-making efforts is detailed in the 2019 ECOWAS Dialogue and Mediation Handbook, which emphasises multitrack mediations that may be led by NGOs and highlights examples of dialogues that were facilitated by WANEP and women's networks (ECOWAS 2017, pp. 25, 53, 59 ,72). Overall, ECOWAS policies and guidelines, thus, provide very strong foundations for NGO's involvement in prevention and peace-making.

The ECOWAS Mechanism also provides for the inclusion of civil society in peace processes by stipulating the Special Representatives of the ECOWAS President must liaise with relevant CSOse (ECOWAS 1999). Whilst focusing on civil society-led interventions, according to the ECPF, ECOWAS and states should mobilise local CSOs to assists mediations (ECOWAS 2008, p. 25). The 2018 ECOWAS Mediation Guidelines establish inclusive mediation as a mediation principle. The Guidelines recommend that not only primary conflict parties, but all relevant political, armed and social groups, including those who oppose a peace process, should be considered as participants of dialogues. Reflecting the inclusive peace paradigm, the Guidelines, which were drafted with the assistance of Finland's Conflict Management Initiative, suggest that inclusion contributes to effective negotiations by assuring the buy-in of stakeholders and the public, encouraging parties to make peace, enriching negotiation agendas, and increasing the legitimacy and sustainability of agreements. CSOs that may be represented in negotiations can hail from all segments of civil society, including trade unions, religious organisations and women's groups (ECOWAS 2018, p. 55). The Dialogue and Mediation Handbook projects local CSOs as participants in multitrack dialogues, stakeholders who may be consulted, recipients of training, and catalysts to grow public support for peace processes. Local CSOs are, thus, seen as participants in dialogues, workshops and consultations on Track 2 and 3 rather than delegations alongside the main conflict actors on Track 1 or in national dialogue conferences (ECOWAS 2017, pp. 53, 56, 57, 89).

19.4.2 ECOWAS' Civil Society Interface and Major Access Points

The West African Civil Society Forum (WACSOF) was launched in 2003 to serve as the official interface for CSOs to liaise with the ECOWAS Commission. Since then, the Commission has entered partnerships with a variety of NGOs and introduced additional access points, including the Human Security and Civil Society Division (HSCD) and ECOWAS Conflict Prevention Framework Secretariat.

Unlike the AU, ECOWAS has traditionally relied on an independent civil society network, WACSOF, as the primary interface for civil society. The vulnerability of the interface model was illustrated in 2014 when internal organisational problems decapacitated WACSOF. Whilst WACSOF underwent a change in leadership, ECOWAS renewed its commitment to the partnership. ECOWAS' financial support to the NGO network, however, raised concern among civil society actors over its ability to deal critique at ECOWAS. Despite its vulnerability, the interface model, which builds on an independent civil society network, is better suited to give CSOs a credible platform than an organ of an intergovernmental organisation. The ECOWAS Commission has, moreover, concluded MOUs with an array of West African NGOs research institutes on a needs-basis. In the domain of conflict prevention and peace-making, WANEP has become ECOWAS' primary partner. The combination of a membership of over 500 grassroot CSOs and well-capacitated structures of an expert peacebuilding NGO give WANEP a unique competitive advantage (Aeby 2021, p. 35).

In 2019, the ECOWAS Commission introduced the HSCSD, which oversees the engagement of civil society for the entire organisation. Whilst being located in the Department of Social Affairs and Gender rather than the Department of Political Affairs, Peace and Security, the HSCSD, is responsible for matters of human security, including women, peace and security, the protection of displaced persons, vulnerable children, and human trafficking. For this purpose, it collaborated with the ECOWAS Gender Development Centre and sectoral CSOs, The HSCSD devised a strategy to broaden ECOWAS' engagement with CSO and promoted the mainstreaming of gender in prevention and peace processes and was instrumental in launching FemWise ECOWAS. The modalities for the deployment of ECOWAS' FemWise Component, were, however, yet to be determined in 2020 (Aeby 2021, p. 35).

The ECPF Secretariat is a further important access point for CSOs and was established in 2015 to promote the implementation of the ECPF, review progress, coordinate stakeholders and mobilise resources. It has close ties to WANEP and holds meetings in the 15 member states to sensitise state, business and civil society actors to the plan. Starting in October 2019, the ECPF conducted a Youths for Peace Programme together with WANEP and states' youth ministries, which comprised training and national youth dialogues (Aeby 2021, p. 35).

19.4.3 ECOWAS Mediation and Security Council

The Mediation and Security Council (MSC), which comprises 10 ambassadors, is the standing decision-making organ for peace and security according to the Mechanism (ECOWAS 1999). Unlike the PSC or UN Security Council that respond to pressing crises, the MSC only meets twice a year alongside the ECOWAS Authority. Urgent decisions, such as the deployment of mediators, are made by the ECOWAS president in consultation with the ambassadors. CSOs cannot address the MSC, but WANEP started to provide quarterly briefing sessions to the ambassadors on the situation in the region in 2019. The objective of the briefings is to complement early warning reports with analyses that must not abide political imperatives. For this purpose, WANEP issues independent reports that the NGO shares with the ambassadors (Aeby 2021, p. 36).

19.4.4 ECOWAS Early Warning and Response Network

With a central situation room, five national centres, 77 field monitors, including 15 from WANEP, in 15 states, the ECOWAS Early Warning and Response Network is Africa's most sophisticated early warning system. The operationalisation of additional national centres to replace regional centres was underway in 2020. WANEP is an integral part of ECOWARN and an embedded Liaison Officer at the ECOWAS Early Warning Directorate coordinates its civil society component. National early warning system managers at each of WANEP's 15 national offices collect information according to a distinct set of indicators. They feed data from WANEP's independent West African Early Warning and Response Network (WARN), which comprises up to 20 community monitors per country, into ECOWARN. The combination of information that is gathered by ECOWARN and civil society permits the triangulation of data on issues such as the role of state actors in conflicts. Besides feeding into ECOWARN, WANEP issues independent opensource reports for CSOs and political actors (Aeby 2021, p. 36).

ECOWARN's outstanding qualities include its human security indicators, adaptation to sub-regional risks, mixed methods, and ability to respond to conflicts on the national level. By partnering with WANEP and leveraging WARN to operationalise ECOWARN, the ECOWAS Commission demonstrated the commitment to civil society participation. According to Amandine Gnanguênon, ECOWARN has fostered a culture of prevention, human security, transparency and participation in ECOWAS (Gnanguênon 2018).

ECOWARN encounters enduring challenges that relate to the complexity of indicators, data quality, new types of conflict risks and using data for policymaking. Most importantly, the political process of translating early warning into timely responses to conflict risks is opaque. The reliance on a singular civil society network that has the relevant early warning capacity in the region represents a considerable vulnerability as ECOWARN would be heavily affected if WANEP experienced organisational problems. Despite the partnership that is based on mutual trust and ECOWAS' commitment to civil society participation, continued state-centric security approaches by political actors and bureaucracy complicate the coordination of the governmental and civil society component of the system. The integration of WANEP structures into the ECOWAS system, moreover, implies a sensitive trade-off for the NGO's independence, as civil society monitors have no hand in the response to early warning signs by political actors (Aeby 2021, p. 36).

19.4.5 Council of the Wise

The ECOWAS Council of the Wise (CoW), unlike the AU PoW, has an unequivocal mandate for both mediation and preventive diplomacy (ECOWAS 1999). The CoW became operational in 2001, embarked on fact finding missions to raise conflict risks with the ECOWAS President, and undertook joint activities with the PoW (ICG 2016, p. 6). However, despite its achievements, in recent years, the council remained defunct and an effort to review its statute that started in 2016 was yet to be completed in 2020 (ECOWAS 2016). The reinstatement of this key structure of the Mechanism, which must comprise eminent civil society leaders, was hampered by a lack of political and financial support. In the absence of the CoW, ECOWAS relied on Special Representatives of the President as mediators. The omission to reconstitute the Council impeded the development of ECOWAS' mediation system and deprived civil society of an important pillar to contribute to conflict prevention and peace-making in ECOWAS (Aeby 2021, p. 37).

19.4.6 ECOWAS Mediation Facilitation Division

The development of the The Mediation Facilitation Division (MFD) was initiated in 2007 and it became operational in 2015. (Odigie 2016, p. 4) NGOs were involved in the operationalisation process from the onset, whereby WANEP participated in preliminary discussions, whilst the Centre for Humanitarian Dialogue drove the needs assessment through which its terms of reference were drafted. WACSOF, WANEP and the West African Civil Society Institute (WACSI), amongst others, gave input in workshops. Whereas the involvement of NGOs posed few challenges thanks to already existing partnerships, the initial set of mediation support instruments were drafted by European NGOs and not optimally adjusted to the West African context.

International NGOs also took the lead in drafting training instruments for the MFD's capacity-building programmes, and trainings for high-level officials were facilitated by the Clingendael Institute and Legon Centre. WANEP, meanwhile, assisted the rollout of a comprehensive training programme for over 470 actors from state and non-state entities across the region. The planning and backstopping of the numerous mediation missions which the MFD supported since 2015, was too political sensitive for NGOs to be involved, but the MFD used WANEP and ECOWARN reports. Whereas ambitious plans to establish a mediation resource centre to manage mediation knowledge did not materialise, the MSU collaborated with NGOs to exchange and disseminate knowledge, for instance, through ACCORD publications and by contributing to research by IJR. A roster for technical experts to backstop mediators was still the subject of discussions, in which the Kofi Anan Peacebuilding Training Centre participated. WANEP moreover managed the roster for FemWise mediators, who had been trained with WANEP's assistance (Aeby 2021, pp. 37–42).

19.4.7 Inclusion of Local Civil Society Actors in ECOWAS Mediations

ECOWAS-mandated mediators generally consulted a wide range of local stakeholders during missions. WANEP played an intermediary role by assisting mediation teams to identify stakeholders who needed to be consulted and facilitating consultations by giving local CSOs access to ECOWAS representatives. This included the prevention and mending of electoral conflicts, whereby WANEP set up Election Situation Rooms to monitor risks. Whereas the ECOWAS MFD consulted local women mediators, WANEP was facilitated dialogue on Track 2 and 3 together with its affiliates. In the case of Guinea Bissau, WANEP's director, who served as the advisor on mediation and dialogue to the Special Representative of the UN Secretary General, acted as contact point between CSOs, ECOWAS and the UN, and directly supported mediation efforts that were mandated by a regional contact group. In Côte d'Ivoire, WANEP assisted the monitoring of the negotiated accord (Aeby 2021, p. 40).

Consultations with local CSOs and WANEP's intermediary role bore challenges. The coordination, consistency and frequency of consultations and the efficient handling of inputs to inform mediation agendas left from for improvement. The inclusion of local CSOs' concerns on the mediation agenda and the ability of WANEP to facilitate consultations hinged on the personality and will of the ECOWAS mediator. Where sitting presidents were appointed mediators, the inclusion of civil society generally proved most difficult. Mediators' divergent objectives towards conflicts and state-centric imperatives impeded WANEP's ability to facilitate consultations. WANEP was well-placed to function as intermediary thanks to its network and local knowledge, but the NGO assumed an inadvertent gatekeeper role for local CSOs, making it easier for WANEP affiliates to access mediation teams (Aeby 2021, p. 40).

19.5 Southern African Development Community

SADC replaced the Southern African Development Coordination Conference in 1994. Its Organ on Politics Defence and Security emerged from the defensive Front-lines States Alliance in 1996. Both predecessors had served to coordinate resistance against the South African apartheid regime that destabilised the region and their major legacy consisted in an anti-imperialist ethos of the organisation that is dominated by former liberation party governments (Khadiagala 2012, pp. 26–35). The development of SADC's peace and security institutions, including the mediation infrastructure, faltered. Relations between some member states and civil society were characterised by suspicion, but SADC nevertheless partnered with selected NGO networks. Policies on conflict management require SADC to engage civil society, but corresponding mechanism were not operationalised. Whilst NGO participation in institutions for prevention and peace-making was minimal, the inclusion of local CSOs in SADC-facilitated peace processes varied sharply (Aeby 2021, p. 44).

19.5.1 SADC Policy Framework

The revised SADC Treaty of 2001 entails an organisational mandate to consolidate, defend and maintain democracy, peace and stability in the region, whereby SADC should involve the people of the region and key stakeholders, including civil society and the private sector. In the SADC Treaty, member states committed to establish SADC National Committees, which should comprise the key stakeholders to enable citizens to interact with SADC to oversee and initiate policymaking (SADC 2001).

The Strategic Indicative Plan of the Organ (SIPO) was first introduced in 2002 and developed by security officials with little input from think tanks and without meaningful consultations with CSOs (van Nieuwkerk 2013, p. 149; van Nieuwkerk 2014, p. 149). SIPO postulated that the Organ should encourage civil society to contribute to conflict prevention, management and resolution without outlining an action plan (SADC 2002, p. 19). When SIPO II was introduced in 2010, accredited civil society networks were invited to comment on the final draft that was produced without the assistance of security think tanks (Aeby 2021, p. 44). SIPO II remained in effect beyond its initial lifespan that expired in 2010 although being partly outdated at its adoption (van Nieuwkerk 2013, p. 150). Whilst lacking a business plan to achieve objectives in the domains of politics, defence, state security, public policy and police, SIPO II, whose implementation would include the development of SADC's APSA component, reflected a shift towards a human security paradigm (SADC 2010, 2; van Nieuwkerk 2013, 150; Aeby 2019, p. 35). According to SIPO II, SADC strategies to "prevent, contain and resolve intrastate conflict by peaceful means" should lead to "enhanced participation of civil society." The Organ should consider collaborations with regional research institutions to exchange experiences, undertake studies, and organise discussions on the involvement of civil society in Organ activities (SADC

2010, pp. 23 – 35). The modalities of such collaborations would need to be elaborated in a future strategic plan.

19.5.2 SADC's Civil Society Interface

SADC National Committees, which should be the main civil society interface in terms of the Treaty, were, either hard to access, unknown to stakeholders or defunct in most SADC member states. This was not only a missed opportunity to involve citizens in SADC's work, but according to Dimpho Motsamai, national committees could, in principle, provide a platform for SADC envoys to include local civil society actors in prevention and mediation efforts (Motsamai 2018, pp. 102–105).

Rather than the National Committees, three accredited networks acted as NGOs, labour unions and religious organisations respectively: The SADC Council of NGOs (SADC-CNGO), the Southern African Trade Union Coordination Council (SATUCC), and the Fellowship of Christian Councils in Southern Africa (FOCCISA) (Hulse et al. 2018). Thanks to sustained efforts spearheaded by SADC-CNGO, the SADC Secretariat became gradually more accessible to CSOs. CNGO, which maintained a peace and security programme, set up a mediation task team in 2012 to link it to the mediation structure SADC set out to operationalise, support local mediation initiatives and provide training. Thanks to an MOU with the SADC Secretariat and sustained relationship-building with SADC executives, SADC-CNGO could open channels to provide input to policymaking. But even the accredited networks could only engage the inaccessible Secretariat, which in itself has little authority within the SADC architecture, on a point-to-point basis. Access depended heavily on personal relations and contributions tended to make little impact on policies (Aeby 2021, p. 46).

SADC-CNGO saw a sharp decline in its organisational capacity and had to terminate the mediation programme following internal challenges and a cut in donor assistance. The SADC case, thus, illustrates the vulnerability of the interface model that depends on an independent network to organisational constraints of NGOs. A further limitation results from the fact that national NGO councils, who are the intermediary between SADC-CNGO and local CSOs, are only partially representative in countries like Zimbabwe where civil society is politically divided. Some CSOs attempt to engage the SADC Secretariat directly or via alternative regional networks. However, CSOs often lack the understanding of SADC's workings and resources to effectively inform conflict prevention and peace-making (Aeby 2016, 708; Aeby 2021, p. 46).

Since the mid-2000s. SADC-CNGO and its partners, who convene the annual SADC Civil Society Forum, advocated the creation of a SADC Non-State Actor Mechanism to enable a more structure engagement with the Secretariat. These efforts prompted SADC to commission the Southern Africa Trust (SAT) to draft a proposal for a Mechanism that resembles the AU ECOSOCC and was in principle approved by the Council of Ministers in 2016 (Southern Africa Trust 2018). Whereas the proposal was later amended to include the work of the Organ on Politics, Defence and Security in the thematic scope of the Mechanism, the Ministerial Committee of the Organ was yet to decide on the plan in 2020. A business actor platform was, meanwhile, promptly operationalised. The Non-State Actor Mechanism was hoped to enable a continuous and transparent engagement with SADC with clear feedback processes. In the worst case, the Mechanism would serve to feign civil society participation without meaningful impact on policies and to select CSOs that were deemed acceptable or exercised self-censorship (Aeby 2021, p. 47).

19.5.3 SADC Summit and Organ Troika

In the SADC architecture, the Summit of Head and State plays an immediate role in responding to conflicts, while the Organ Troika serves as primary decision-making body between Summit meetings. Both bodies are made up of heads of states, who hold closed meetings, and have no formal consultative mechanisms (Aeby 2019, p. 22). Whereas the annual Civil Society Forum and ordinary Summit take place at the same time, lobbying political decision-makers during the Summit was not a viable strategy. To inform SADC decisions, NGOs targeted receptive liberal democratic governments, South African in particular. During Namibia's tenure, NGOs that sought to raise concerns relating to the political crisis in Zimbabwe, were afforded a meeting with the SADC Chair. South African-based NGOs moreover cultivate relations with the Department of International Relations and Cooperation to inform SADC-mandated mediations. NGOs also seek to inform South Africa's policy on conflicts in SADC by advocating their positions through the government party, its trade union partners and the media (Aeby 2016, p. 707; Aeby 2021, p. 47).

19.5.4 Regional Early Warning Centre

The SADC Regional Early Warning Centre (REWC) should contribute to conflict prevention, management and resolution and, according to the APSA blueprint, feed into the CEWS (Hendricks/Musavengana 2010, 19; SADC 2020). However, the secretive REWC consisted of a small situation room, focused on state intelligence rather than human security, and lack resources. In practice, REWC was neither integrated into the CEWS, nor did it contribute information to support SADC mediations. Since the REWC was an intelligence organ, its staff included operatives who were recruited from central intelligence organisations of states. In the case of Zimbabwe, the Central Intelligence Organisation had a track record of abducting and killing civil society activists (ICG 2011, p. 3; Sachikonye 2012, 36; Human Rights Watch 2014, 6, p. 22). The REWC was, thus, neither fit to support prevention and peace-making nor to partner with civil society. It would need to be replaced with system that focuses on human security indicators, feeds data into the CEWS and produces analyses for SADC's mediation infrastructure.

The Common Market for Eastern and Southern Africa (COMESA), whose membership overlaps with SADC, has a human security-centred early warning system that feeds into the CEWS, but an independent civil society early warning system for Southern Africa does presently not exist (Porto Gomes 2013). Southern African NGOs would be well-placed to develop a civil society-based early warning system, such as West Africa's WARN, thanks to the technical expertise of think tanks that support the APSA, the relatively progressed regional integration of civil society, the existence of NGO networks that monitor issues pertaining to elections, gender and conflict, and the experiences gained by monitoring historic peace processes.

19.5.5 SADC Panel of Elders and Mediation Reference Group

The decision to strengthen the mediation capacity of SADC as part of its APSA component was taken by the Summit in 2004, but the creation of corresponding structures only returned to the agenda in 2008. By 2010, a drafting process that involved regional an UN experts produced a plan to create a mediation infrastructure consisting of a Panel of Elders (PoE), Mediation Reference Group (MRG) and Mediation Support Unit (MSU) (Hartman 2013, pp. 6–7). The members of the PoE, who must include civic leaders, were appointed in 2014, but nominations that needed states' approval continued in 2018. According to the planned architecture, the Elders would serve as stand-by panel for both preventive diplomacy and mediation, which would either lead or assist mediations in coordination with the Chair of the Organ and Executive Secretary of the SADC Secretariat. However, by 2020, there was no indication that the Panel had been deployed to engage in peace diplomacy, with the exception of one member, Joaquim Chissano, who continued his prior role as SADC mediator for Madagascar. Rather than deploying the Elders to Lesotho, SADC stuck to the old practice of mandating a sitting president, Cyril Ramaphosa, to lead the mediation. In Zimbabwe, SADC's crisis management efforts in response to the 2017 coup were led by the SADC Chair. Efforts by the erstwhile SADC facilitator, Thabo Mbeki, to facilitate dialogue in 2019, did not take place under an official SADC mandate. The failure to use the Elders for peace diplomacy indicates a reluctance by political decision-makers to abandon entrenched practices to institutionalise the mediation infrastructure that the Secretariat developed with the assistance of donors and technical experts. Whereas SADC-CNGO was not consulted on the establishment of the PoE, civil society stakeholders from conflict-affected SADC countries who contributed to our research were often unaware of its existence (Aeby 2021, p. 48).

The SADC mediation infrastructure comprises an MRG, whose ambiguous mandate includes advising the Elders and directly engaging in mediation. The nine members of the MRG first met in 2015 after being nominated in 2012. The advisory group comprises ambassadors, former government officials, whose background in mediation in unclear, and two civil society representatives, including the director of ACCORD (SADC 2015). Apart from the deployment of the MRG Chair, Leonardo Simão, to act as Chissano's adviser in Madagascar, there was no indication that the MRG had been used to assist peace diplomacy by 2020. Whereas the MRG and MSU organised meetings, the Secretariat, apparently, did not require its assistance (Aeby 2021, p. 48).

Apart from being represented in the MRG, ACCORD signed an MOU with SADC in 2015 to assist conflict analysis and mediation training for civilians, police and military (SADC 2016). Whilst the SADC mediation infrastructure was generally underutilised, training was the area where the MRG are most likely to become active.

19.5.6 SADC Mediation Support Unit

The MRG launched with a staff of three in 2014 in the Organ Directorate after the design of the structure had commenced in 2008. During a pilot phase that ended in 2018, the MSU was financed through a Regional Political Cooperation Programme with support from the European Development Fund (SADC 2019). The operations and team of the MSU were downscaled after the programme expired in 2018 as SADC did not mobilise the resources that were required to maintain the structure, leading to a loss of expertise and momentum in the development of the mediation infrastructure. NGOs involvement in operationalising the structure was minimal but significant. The director of the Centre for Mediation in Africa, who had assisted the development of the SADC Organ Protocol and plan of action to strengthen the AU's mediation capacity, assisted the design of the mediation infrastructure with the financial support of GIZ. Whereas SADC-CNGO had advocated the mediation infrastructure, it was invited to a workshop and had informal exchanges with SADC officials and the planned structures (Aeby 2021, pp. 49–53).

The SU supported several diplomatic missions and assisted the SADC envoy to identify civic stakeholders in Madagascar, but NGOs were not involved in its operational support activities. Before its operations were downscaled, the MSU made remarkable strides in capacity-building, providing mediation and dialogue training over 450 intergovernmental, state and non-state actors. The training curriculum was developed and evaluated with the assistance of consultants and ACCORD. By 2020, there were no indications that the MSU engaged in systematic knowledge-management activities, which would constitute a primary area where the SADC Organ could meet the SIPO II objective of working with regional research (Aeby 2019, pp. 49–53).

19.5.7 Civil Society Inclusion in SADC Mediations

Before the SADC mediation infrastructure was introduced, SADC facilitated the negotiation of agreements and implementation processes in Lesotho, Madagascar and Zimbabwe, whereby the space for civil society inclusion varied markedly and heavily depended on the will of mediators and political actors.

The SADC facilitation and negotiating parties shut civil society out from the negotiation of Zimbabwe's 2008 Global Political Agreement, although the accord included a comprehensive reform plan that would require broad-based societal support to be implemented. The facilitation team accepted submissions by CSOs, which the facilitators considered identical to those of the political parties. The SADC team, moreover, held hearings with religious leaders and women's groups (Aeby 2016, p. 712). The implementation of transitional mechanisms, such as the constitutional reform, implementation monitoring mechanism, and organ for national healing remained under the control of political parties, whilst CSOs sought to impact the GPA process from the margins. The constitutional reform involved a consultative mechanism that was required to receive UNDP support, but as the parties handpicked participants, many important CSOs boycotted the process. Most CSOs used the limited space and pragmatically engaged in flawed GPA mechanisms whilst undertaking independent activities to foster peacebuilding and a democratic transition (Aeby 2015, pp. 111–147; Aeby 2016, pp. 712–717). When the South African SADC Chair engaged in talks with political and military actors as the 2017 coup that ousted President Mugabe unfolded, there were no indications of consultations with civil society (Aeby 2021, pp. 51–53; Africa Confidential 2017).

Following the 2009 coup in Madagascar, SADC took the lead in the mediation, which that had been initiated by the AU and UN, after the Summit mandated Chissano to facilitate dialogue and a transition. When the mediation was still under the AU's aegis, it was controversially decided that the negotiations should only include four political formations representing the coup leader and three former presidents. The UN envoy had pledged for an inclusive dialogue (ICG 2010, 25; Witt 2016, p. 147). The four parties negotiated the SADC-facilitated Maputo Accords and the Addis Ababa Additional Act of 2009, which failed to end the crisis. (Joint Mediation Team 2009) In the dialogue was subsequently expanded to negotiate the SADC Roadmap among 11 political groups (Joint Mediation Team 2011; Nathan 2013, p. 6). As Antonia Witt shows, the Chissano team regularly consulted civil society groups, but the impact of CSO consultations on the content of the roadmap that resembled previous accords was doubtful (Witt 2017, p. 214). Whilst the SADC-facilitated negotiations faltered, Malagasy churches and NGOs organised a parallel Malagacho-Malagache national dialogue, which SADC embraced in 2010. The SADC Roadmap eventually enabled the installation of a recognised transitional government and envisaged a civil society participation in monitoring and reconciliation mechanisms, which the churches rejected. Instead, the churches organised a dialogue conference in 2013, whose participants issued a statement calling for the replacement of the SADC Roadmap with an inclusive transition (Witt 2017, p. 218; FFKM 2020).

Between 1998 and 2014, SADC repeatedly intervened in Lesotho and facilitated political dialogues to contain post-electoral conflicts and the politicised armed forces. Dimpho Motsamai shows that SADC was initially reluctant to engage civil society but came to praise the Christian Council of Lesotho (CCL) for facilitating a dialogue between 2009 and 2011 that produced an agreement on electoral reform. When the SADC mediation resumed in 2014, interparty talks were facilitated by the CCL and presided by the Namibian SADC Chair (Motsamai 2018, 150–179). After SADC mandated the South African Deputy President to mediate dialogue, the CCL and Lesotho Council of NGOs remained highly involved and SADC-CNGO served intermediary between SADC and CSOs (Aeby 2021, p. 52; Shale/Gerenge 2016).

The three SADC mediations largely took place before the new mediation infrastructure was introduced. The MSU interacted local CSOs and assisted the identification stakeholders for consultations in follow-up missions to Madagascar. To see whether the introduction of the PoE, MRG, and MSU benefits the inclusion of civil society in SADC mediations, the SADC mediation infrastructure would need to be rendered fully operational (Aeby 2021, p. 52).

19.6 Comparative Insights

The discussion of the policy frameworks of the three organisations, involvement of NGOs in structures for prevention and peace-making, and inclusion of local CSOs enables the following comparative insights.

19.6.1 Policy Frameworks for Civil Society Participation

The policy frameworks of the three organisations set different conditions for civil society participation as they are underpinned by security that range from state-centric conflict management to holistic human security and inclusive peace in ECOWAS. The conditions also differ as, in the cases of the AU and ECOWAS, detailed guidelines on mediation practices complement strong provisions for civil society participation in statutory documents and strategic policy plans, whereas SADC is yet to elaborate such guidelines.

ECOWAS elaborate policy framework envisaged civil society participation across the board and sees NGOs as leading actors in conflict prevention and peace-making, intermediaries between ECOWAS, states and communities, and facilitators of multi-track dialogues (ECOWAS 2017, pp. 57, 89). The AU's framework encourages collaborations for conflict prevention and peace-making, including early warning and mediation support, but prioritises ECOSOCC over African NGOs with suitable expertise (SADC 2010, pp. 23 – 35). SADC's policies propose collaborations with research institutions but omit further details (SADC 2010, pp. 23 – 35). Whereas the policy frameworks give legitimacy to NGOs' involvement in APSA structures, they include major grey areas. Closing these grey areas could either help guaranteeing spaces for participation or constrain otherwise informal interactions between NGOs and IGOs (Aeby 2021, p. 54).

In terms of civil society inclusion in peace processes, mediation guidelines of the AU and ECOWAS declare inclusivity a mediation principle, which mediators must balance against the practicability of negotiations. Rather than recommending civil society delegations be included at the negotiation table, the guidelines portray local CSOs as actors that must be consulted or included in dialogues on subordinate tracks (AU 2012, p. 9, 2014a, 73; ECOWAS 2017, 2018). Consultations, however, risk becoming a window-dressing exercise if they have no meaningful impact on the mediation agenda and content of agreements. Since the mediation guidelines for inclusion are recommendations rather than rules, their application not only depends on the will of the dominant conflict parties, but on the mediators and decision-making organs that define mediation mandates.

19.6.2 Participation in Organisational Structures

The access points the three organisations provided for CSOs to inform policymaking in the domain of peace and security vary sharply. The interfaces the organisations use to liaise with civil society illustrate the benefits and limitations of the respective models. The ECOSOCC had still not become the primary interface for CSOs to participate two decades after it was introduced by the AU's Constitutive Act. ECOSOCC shows that an interface prevents rather than enables meaningful participation if state actors define restrictive rules for CSOs to engage the organisation. It excluded vast sections of African civil society, deprived the AU of vital inputs from citizens, and risked undermining its credibility vis-à-vis civil society. ECOWAS and SADC, meanwhile, show that an interface model that depends on an independent civil society network is vulnerable to organisational challenges and funding constraints of NGOs. However, the model that builds on an independent civil society platform, which is owned and managed by its stakeholders, is better suited to provide an accessible and credible channel for CSOs to represent their interests vis-à-vis the IGO. Alternative funding models may prove instrumental in developing an interface that balances the independence and sustainability of the platform (Aeby 2021, p. 19).

The institutionalisation of channels for CSOs to inform decision-making bodies for peace and security differs between the AU PSC, which invited relevant organisations to address the Council, the ECOWAS MSC, whose members received quarterly briefings from WANEP, and the SADC Summit and Organ Troika, which consist of heads of state and had no formal consultative mechanism. Whereas only a few selected CSOs could present statements to the PSC and ECOWAS ambassadors, these platforms were an important channel to give visibility to the concerns of the invited NGOs and their constituents, including local civil society actors from conflict-affected countries. Informal channels to inform political decision-making in the organisations were equally important. For this purpose, NGOs conducted custom-made analysis and lobbied representatives, embassies and foreign affairs departments of influential member states as well as ruling parties and their civic allies. To stand a chance to make an impact, the NGOs had to foresee the agenda of decision-making organs, identify relevant states, and customise their inputs for recipients (Aeby 2021, p. 20).

The involvement of NGOs in early warning systems and the latter's suitability for such partnerships varied dramatically between the organisations. Whereas WANEP was an integral part of ECOWARN and tapped WARN into the system to enrich the data with reports by civil society monitors, the CEWS partnered with expert NGOs from across the continent. SADC's state intelligence-focused REWC would need to be replaced with a system that focused on human security and served to inform prevention and mediation. Southern African NGOs had great potential to establish an independent early warning system and share analytical outputs with stakeholders in states and the APSA (Aeby 2021, p. 56).

The panels for preventive diplomacy and mediation of the AU and ECOWAS, which include eminent civic leaders, collaborated with NGOs to compile reports, organise high-level dialogues, and to build bureaucratic capacity (Aeby 2021, 57; de Carvalho 2017, 8). Stakeholder consultations by the panels provided for an important channel for local CSOs to inform prevention, mediation and reports to decision-making organs (Gomes Porto/Ngandu 2014, p. 188; Nathan et al. 2015, p. 52). However, the AU PoW was underutilised, the ECOWAS CoW defunct and the SADC PoE yet to become fully operationalised in 2020. FemWise Africa, on the other hand, not only served to bring women to the forefront of peace diplomacy but aimed at promoting multitrack mediations and establishing links to local communities from where FemWise mediators were recruited (Aeby 2021, p. 56).

The mediation support structures of the three organisations were designed and operationalised with the support of consultants from expert NGOs. The NGOs helped to secure the buy-in of political decision-makers and donors, facilitated knowledge-transfer, drafted guidelines, critically reviewed drafts, and adjusted them to the regional context. Challenges related to a lack of African leadership as few NGOs had the relevant expertise; the entrepreneurial imperatives of NGOs; the adaptation of guidelines to regional realities; the compatibility of normative preferences which the NGOs transported into instruments; and an apparent lack of buy-in by political stakeholders for structures whose development was promoted by NGOs and donors (Aeby 2021, p. 56).

The politically sensitive backstopping of mediation missions left virtually no room for NGOs to be involved except for the use of analytical inputs from think tanks. Whereas the MSS interacted with local CSOs during missions, the most immediate contribution they could make to promote civil society inclusion was to propose inclusive peace process designs and streamline stakeholder consultations to ensure inputs made it on the mediation agenda. Civil society actors were involved in the capacity-building activities of the MSS as experts, trainers and trainees. Whilst African NGOs could assist the rollout of training on a broad scale, the involvement of multiple NGOs, who were keen to offer their services with the support of donors, sometimes led to a duplication of efforts. The development of knowledge management systems to retain and produce comparative mediation knowledge required great resources and, thus, saw little progress. African research institutions that already make up for the lack internal institutional memory in the APSA would be well-placed to become part of a hybrid system, in which mediation support officers debrief mediators and external researchers carry out tedious comparative analyses to inform future mediations. Since a standing team of experts as in the UN is too expensive, thematic experts from NGOs could be put on standby to assist mediators to design peace processes and draft agreement provisions (Aeby 2021, pp. 57–60).

19.6.3 Inclusion in Mediations

Local civil society actors could use several channels to engage mediation teams and inform the mediation agenda in peace processes, but the availably and viability of these channels varied between the three organisations and peace processes.

The channels, firstly, included stakeholder consultations by lead mediators, which were common in peace processes that were mediated in all cases, but depended on the specific political context as illustrated by the three SADC mediations. Secondly, owing to their long-term presence, local expertise and key role in supporting mediators, liaison offices, where available, provided an effective channel for local CSOs to inform mediations. Thirdly, FemWise mediators, who included local peacebuilders, could, in principle, serve as channel for communities to engage mediation teams. Fourthly, regional NGO networks with formalised ties to the organisations could act as intermediaries and assist local CSOs to access mediation teams. WANEP, in particular, could help ECOWAS mediation teams to identify civil society stakeholders for consultations and facilitated dialogues on subordinate tracks (Aeby 2021, p. 57).

The availability and viability of these channels to inform mediations varied owing to the way inputs from stakeholder consultations were processed and included in the mediation agenda. Unless consultations make an actual impact on negotiation processes and the content of agreements, consultative mechanisms amount to window-dressing. The holding of consultations could be rendered more systematic, frequent and consistent. Consultations and their relevance heavily depended on the volition of mediators and conflict parties as guidelines for inclusive mediation are not mandatory. Where sitting presidents lead mediations, civil society inclusion tended to be more difficult. The structures and mediators of the organisations preferred to work with trusted partner NGOs, who had an inadvertent gatekeeping role for local CSOs, who sought to engage mediation teams. Local CSOs oftentimes lacked the necessary resources, expertise, communication style and understanding of the IGOs' functioning to inform mediations, as civil society tends to be weak in war-torn and authoritarian states that curtail civic freedoms. Where the government deemed the presence of liaison offices, mediation missions and consultations an infringement of its sovereignty, an engagement with local CSOs was impractical (Aeby 2021, p. 57).

19.7 Conclusion

In sum, the three organisations that constitute key components of the APSA provide a range of channels for CSOs to inform conflict prevention and peace-making, whose viability depends on the respective IGO and peace process. The APSA has seen the proliferation of policy guidelines for civil society inclusion, which were drafted by technical experts and consultants from NGOs, who diffused the inclusive peace paradigm. But political decision-makers and high-level mediators are yet to fully embrace these norms to put them into practice. The civil society interfaces and key access points for CSOs, such as panels for preventive diplomacy and mediation, either needed reform or were underutilised. Inclusive peace process designs that were proposed in guidelines were yet to be standardised in practice. To prevent consultations with local CSOs from amounting to window-dressing, mediation teams needed to ensure that civic stakeholders' inputs were meaningfully included in mediation agendas and agreements. Civil society actors, who seek to participate in APSA structures and mediations, must, thus, constantly navigate the available channels anew to tell apart pathways and dead ends (Aeby 2021, 54–59).

References

Adetula, Victor, Redie Bereketeab, and Olugbemi Jaiyebo. (2016. *Regional Economic Communities and Peacebuilding in Africa.* Policy Dialogue. Uppsala: Nordiska Afrikainsitutet.

Aeby, Michael. 2015. Zimbabwe's Gruelling Transition: Interim Power-Sharing and Conflict Management in Southern Africa. Fribourg: University of Basel.

Aeby, Michael. 2016. Making an Impact from the Margins? Civil Society Groups in Zimbabwe's Interim Power-Sharing Process. *The Journal of Modern African Studies* 54: 703–28.

Aeby, Michael. 2019. *SADC – The Southern Arrested Development Community? Enduring Challenges to Peace & Security in Southern Africa.* 14. Policy Dialogue. Uppsala: Nordic Africa Institute.

Aeby, Michael. 2021. *Civil Society Participation in Peacemaking and Mediation Support in the APSA: Insights on the AU, ECOWAS and SADC.* Cape Town: Institute for Justice and Reconciliation.

Africa Confidential. 2017. The Crocodile Snaps Back. *Africa Confidential*, November 16.

Amr, Mohamed Sameh. 2012. The Economic, Social, and Cultural Council of the African Union. In *The African Union legal and institutional framework: a manual on the pan-African organization*, ed. Abdulqawi. Yusuf and Fatsah Ouguergouz, 157–183. Leiden: Martinus Nijhoff Publishers.

Apuuli, Kasaija Philip. 2018. The AU's Peace and Security Architecture: The African Standby Force. In *The African Union: Autocracy, Diplomacy and Peacebuilding in Africa*, ed. Tony Karbo and Timothy Murithi, 149–182. International Library of African Studies; 65. London: I.B. Tauris.

Arthur, Peter. 2017. Promoting Security in Africa through Regional Economic Communities (RECs) and the African Union's African Peace and Security Architecture (APSA). *Insight on Africa* 9: 1–21. https://doi.org/10.1177/0975087816674577.

AU. 2002. Constitutive Act of the African Union. African Union. Addis Ababa.

AU. 2004. Statutes of ECOSOCC. Addis Ababa.

AU. 2008. Livingstone Formula - Conclusions on a Mechanism for Interaction between the PSC and Civil Society Organizations in the Promotion of Peace, Security and Stability in Africa, PSC/PR/(CLX), Addis Ababa.

AU. 2012. Standard operating procedures for mediation support. Addis Ababa.
AU. 2013. *Managing Peace Processes, Towards More Inclusive Processes.* Vol. 3. Addis Ababa.
AU. 2014a. *African Union Mediation Support Handbook.* Umhlanga Rocks: ACCORD.
AU. 2014b. Maseru Conclusions on Enhancing the Implementation of the Livingstone Formula PSC/Retreat.4./6. Addis Ababa.
AU. 2016a. *Report of the Meeting on the Operationalization of the African Union Mediation Support Unit.* Addis Ababa: African Union.
AU. 2016b. AU Liaison Offices. *African Union.*
AU. 2019. *Report of the African Union Panel of the Wise on Improving the Mediation and Resolution of Natural Resource-Related Conflicts Across Africa.*
Aulin, Jenny. 2019. Civil Society Inclusion in Peacebuilding. *Accord*: 39–41.
Barnes, Catherine, ed. 2002. Introduction: - Owning the Process: Public participation in peace-making. *Accord* 13.
Barnes, Catherine. 2009. Civil Society and Peacebuilding: Mapping Functions in Working for Peace. *The International Spectator* 44: 131–147. https://doi.org/10.1080/03932720802693036.
Bell, Christine. 2019. The New Inclusion Project. *Accord*: 11–17.
Bell, Christine, and Catherine O'Rourke. 2007. *The People's Peace? Civil Society, Peace Agreements and Participatory Democracy.* SSRN Scholarly Paper ID 1359801. Rochester, NY: Social Science Research Network.
Belloni, Roberto. 2008. Civil Society in War-to-Democracy Transitions. In *From War to Democracy: Dilemmas of Peacebuilding*, ed. Anna Jarstad and Timothy Sisk, 177–89. Cambridge: Cambridge University Press.
Berghof Foundation. 2017. *National Dialogue Handbook: A Guide for Practitioners. Berghof Foundation: Berlin.* Berlin: Berghof Foundation.
Carl, Andy. 2019. Introduction: Navigating Inclusion in Peace Processes. *Accord*: 5–9.
Carothers, Thomas, and Marina Ottaway. 2000. Introduction: The Burgeoning World of Civil Society Aid. In *Funding Virtue : Civil Society Aid and Democracy Promotion*, ed. Marina Ottaway and Thomas Carothers, 3–20. Washington, D.C.: Carnegie Endowment for International Peace.
de Carvalho, Gustavo. 2017. *Looking for a Home: Mediation and the AU.* 1. Africa Report. Addis Ababa: ISS.
Chesterman, Simon. 2007. Ownership in Theory and in Practice: Transfer of Authority in UN Statebuilding Operations. *Journal of Intervention and Statebuilding* 1: 3–26. https://doi.org/10.1080/17502970601075873.
Coe, Brooke, and Kathryn Nash. 2020. Peace process protagonism: the role of regional organisations in Africa in conflict management. *Global Change, Peace & Security* 32. Routledge: 157–177. https://doi.org/10.1080/14781158.2020.1777094.
Diallo, Alimou. 2019. Abuja.
ECOSOCC. 2018. ECOSOCC Operationalises Livingstone Formula. April 6.
ECOSOCC. 2020. National Chapters Currently in Operations. *ECOSOCC.* May 12.
ECOWAS. 1993. Revised Treaty of the Economic Community of West African States. *Refworld.*
ECOWAS. 1999. Protocol Relating to the Mechanism for Conflict Prevention, Management, Resolution, Peace-keeping and Security.
ECOWAS. 2008. ECOWAS Conflict Prevention Framework. Regulation MSC/REG.1/01/08. Abuja.
ECOWAS. 2016. ECOWAS Develops New Statutes to Reposition 'Council of the Wise.' *ECOWAS.*
ECOWAS. 2017. *ECOWAS Dialogue and Mediation Handbook.* Abuja.
ECOWAS. 2018. *ECOWAS Mediation Guidelines.* Abuja: ECOWAS.
FFKM. 2020. Pour un gouvernement neutre et une autre transition - Résolutions de la réunion d'Ivato conduite par le FFKM. *Madagascar-Tribune.com.* August 17.
Gnanguênon, Amandine. 2018. *Afrique de l'Ouest : faire de la prévention des conflits la règle et non l'exception.* Bruxelles: l'Observatoire Boutros-Ghali.
Gomes Porto, João, and Kapinga Yvette Ngandu. 2014. The African Union, Preventive Diplomacy, Mediation, and the Panel of the Wise: Review and Reflection on the Panel's First Six Years. African Security 7: 181–206. https://doi.org/10.1080/19392206.2014.952582.

Hartman, Henrik. 2013. The Evolving Mediation Capacity of the Southern African Development Community. *Conflict Trends*: 3–10.

Heathershaw, John. 2013. Towards Better Theories of Peacebuilding: Beyond the Liberal Peace Debate. *Peacebuilding* 1: 275–282. https://doi.org/10.1080/21647259.2013.783260.

Hellmüller, Sara. 2020. The Changing Role of Civil Society Actors in Peacemaking and Peacebuilding. In *Routledge Handbook of Peace, Security and Development*, ed. Osler Hampson, Alpaslan Özerdem, and Jonathan Kent, 1st ed. Oxon: Routledge.

Hellmüller, Sara, and Marie-Joëlle Zahar. 2019. UN-led Mediation in Syria and Civil Society. *Accord*: 84–87.

Hendricks, Cheryl, and Takawira Musavengana. 2010. The security sector in Southern Africa. Pretoria: Insitute for Security Studies.

Hulse, Merran, Lisa Gürth, Helena Kavsek, Verena Stauber, Daniel Wegner, and Jan Weinreich. 2018. *Civil Society Engagement in Regional Governance: a Network Analysis in Southern Africa.*

Human Rights Watch. 2014. *The Elephant in the Room: Reforming Zimbabwe's Security Sector Ahead of Elections*. Johannesberg.

ICG. 2010. *Madagascar: sortir du cycle de crises*. 156. Rapport Afrique. Bruxelles: International Crisis Group.

ICG. 2011. Zimbabwe: Political and Security Challenges to the Transition. *Africa Briefing* 70.

ICG. 2016. *Implementing Peace and Security Architecture (III): West Africa*. 234. Africa Report. Brussels: International Crisis Group.

IPSS. 2020. *APSA Impact Report 2018*. Addis Ababa: Institute for Peace and Security Studies.

ISS. 2017. Denialism plagues Africa's early warning system. *ISS Africa*. April 19.

ISS. 2019. *The Horn of Africa should improve citizen engagement*. 113. PSC Report. Pretoria.

Joint Mediation Team. 2009. Transition Charter (translated version). PA-X Database. Maputo.

Joint Mediation Team. 2011. Roadmap for Ending the Crisis in Madagascar: Commitments by Malagasy Political Actors (Translated version). PA-X Database.

Kagame, H E Paul. 2017. *The Imperative to Strengthen Our Union*. Addis Ababa: AU.

Karbo, Tony, and Timothy Murithi, ed. 2018. Introduction. In *The African Union: autocracy, diplomacy and peacebuilding in Africa*. International Library of African Studies 65. London: I.B. Tauris.

Khadiagala, Gilbert M. 2012. The SADCC and its Approaches to African Regionalism. In *Region-Building in Southern Africa: Progress, Problems and Prospects*, ed. Saunders Chris, Dzinesa Gwinyayi A, and Nagar Dawn, 25–38. London / New York: Zed Books.

Lwizi, Gladys. 2019. Zambia honored to host AU economic, social and cultural council HQ. *Zambian Business Times*, June 11.

Mitchell, C. R. 2012. Introduction: Linking National-level Peace-making with Grassroots Peacebuilding. In *Local peacebuilding and national peace: interaction between grassroots and elite processes*, ed. C. R. Mitchell and Landon E. Hancock, 11–13. London ; Continuum.

Mitchell, C. R., and Landon E. Hancock. 2012. Between Local and National Peace: Complementarity or Conflict? In *Local peacebuilding and national peace: interaction between grassroots and elite processes*, ed. C. R. Mitchell and Landon E. Hancock, 161–178. London; Continuum.

Morris, Odhiambo, and Chitiga Rudy. 2016. *The Civil Society Guide to Regional Economic Communities in Africa*. African Minds.

Motsamai, Dimpho. 2018. Evaluating the Peacemaking Effectiveness of SADC. Thesis.

Murithi, Tim. 2018. *Tackling Africa's Peace and Security Challenges: The Role of Non-State Actors*. Cape Town: Institute for Justice and Reconciliation.

Murithi, Timothy. 2005. *The African Union: Pan-Africanism, peacebuilding and development*. Aldershot, Hampshire, England ; Burlington, VT: Ashgate Pub.

Nathan, Laurie. 2009. *Plan of Action to Build the AU's Mediation Capacity*.

Nathan, Laurie. 2013. *A Clash of Norms and Strategies in Madagascar and teh AU Policy on Unconstitutional Change of Government*. 4. Mediation Arguments. Pretoria: Centre for Mediation in Africa.

Nathan, Laurie. 2019Skype.

Nathan, Laurie, Michelle Ndiaye, and Yahia Zoubir. 2015. *APSA Assessment Study 2014*. Addis Ababa: African Union.

Ngandu, Yvette. 2017. *Operationalisation of "FemWise-Africa."* Addis Ababa: African Union.

van Nieuwkerk, Anthoni. 2013. SIPO II: Too little, too late? *Strategic Review for Southern Africa* 35: 146–153.

van Nieuwkerk, Anthoni. 2014. The Strategic Culture of Foreign and Security Policymaking: Examining the Southern African Development Community. *African security* 7: 45–69. https://doi.org/10.1080/19392206.2014.880031.

Nilsson, Desirée. 2012. Anchoring the Peace: Civil Society Actors in Peace Accords and Durable Peace. *International Interactions* 38: 243–266. https://doi.org/10.1080/03050629.2012.659139.

Nilsson, Desirée, and Mimmi Söderberg Kovacs. 2011. Revisiting an Elusive Concept: A Review of the Debate on Spoilers in Peace Processes. *International Studies Review* 13: 606–626. https://doi.org/10.1111/j.1468-2486.2011.01080.x.

Noyes, Alexander, and Janette Yarwood. 2013. The AU Continental Early Warning System: From Conceptual to Operational? *International Peacekeeping* 20: 249–262. https://doi.org/10.1080/13533312.2013.838393.

Odigie, Brown. 2016. *The Institutionalisation of Mediation Support within the ECOWAS Commission*. 42. Policy & Practice Brief. Umhlanga Rocks: ACCORD.

Paffenholz, Thania. 2015. Civil Society and Peacebuilding. *Development Dialogue* 63: 108–117.

Paladini Adell, Borja, and Sean Molloy. 2019. More Inclusive Monitoring of Peace Agreement Implementation: Barometer Initiative in Colombia. *Conciliation Resources*. February 21.

Porto Gomes, João. 2013. COMWARN: COMESA's Distinctive Contribution to the Early Warning System. In *Towards an African peace and security regime : continental embeddedness, transnational linkages, strategic relevance*, ed. Ulf. Engel and João Porto Gomes, 125–151. Global Security in a Changing World. Farnham, Surrey, England: Ashgate Pub. Company.

Respondent 7. 2018Email.

Respondent 9. 2020. Addis Ababa.

Respondent 10. 2020. Gaborone.

Rudo, Chitiga, and Manby Bronwen. 2009. *Strengthening Popular Participation in the African Union: A Guide to AU Structures and processes*. Johannesburg: OSISA & Oxfam.

Sachikonye, Lloyd. 2012. *Zimbabwe's Lost Decade: Politics: Development & Society*. Harare: Weaver Press.

SADC. 2001. Consolidated Treaty of the Southern African Development Community. Gaborone.

SADC. 2002. Strategic Indicative Plan for the Organ on Politics, Defence and Security Cooperation, Gaborone. SADC.

SADC. 2010. Revised Edition of the Strategic Indicative Plan for the Organ, Maputo. SADC. Maputo.

SADC. 2015. Constitutive Meeting of the Mediation Reference Group.

SADC. 2016. Conflict Analysis and Mediation Refresher Training for the Mediation Reference Group. June 10.

SADC. 2019. SADC, EU Jointly launch €15 million Support to Peace and Security in the SADC Region (SPSS) Programme. April 5.

SADC. 2020. Regional Early Warning Centre. *SADC*. https://www.sadc.int/sadc-secretariat/services-centres/regional-early-warning-centre/. Accessed June 19.

Shale, Victor, and Robert Gerenge. 2016. Electoral Mediation in the Democratic Republic of the Congo, Lesotho and Kenya. *Conflict Trends*.

Söderbaum, Fredrik. 2007. Regionalisation and Civil Society: The Case of Southern Africa. *New Political Economy* 12: 319–337. https://doi.org/10.1080/13563460701485276.

Southern Africa Trust. 2018. *Taking SADC to the People: The Proposed SADC Mechanism for Engagement with Non- State Actors*. Johannesburg.

Sriram, Chandra Lekha, and Marie-Joëlle Zahar. 2009. The Perils of Power-Sharing: Africa and Beyond. *Africa Spectrum* 44: 11–39.

UN. 2015. Sustainable Development Goal 16. *Department of Economic and Social Affairs*.

Van Nieuwkerk, Anthoni. 2020Skype.
Vandeginste, Stef. 2013. The African Union, Constitutionalism and Power-Sharing. *Journal of African Law* 57: 1–28. https://doi.org/10.1017/s0021855312000149.
WANEP. 2019. *Practice Guide on CSOs Engagement with Intergovernmental Organizations*. Accra: WANEP.
Witt, Antonia. 2016. Ordering by Default: The Politics and Power of Post-Coup Interventions in Africa. PhD, Leipzig: Universität Leipzig.
Witt, Antonia. 2017. Mandate impossible: Mediation and the return to constitutional order in Madagascar (2009–13). *African security*: 205–222. https://doi.org/10.1080/19392206.2017.1352397.
World Bank, and UN. 2018. *Pathways for Peace. Inclusive Approaches for Preventing Violent Conflict*. Washington, DC: The World Bank.
Zanker, Franzisca. 2014. Legitimate Representation: Civil Society Actors in Peace Negotiations Revisited. *International Negotiation* 19: 62–88. https://doi.org/10.1163/15718069-12341270.

Part IV
Enhancing the Role of Civil Society in Peacebuilding

Chapter 20
Centre-Periphery Relations Between Civil Society Organisations and External Funding Partners: A Case Study of South Kivu's Peace and Development Agenda

Philippe Mulume-Oderhwa Kaganda and Jean Chrysostome K. Kiyala

Abstract Ideological control of peacebuilding efforts and sustainable development by International Non-Governmental Organiaations (I-NGOs) and funding agencies at the expense of Local Non-Governmental Organisations (L-NGOs) or indigenous initiatives has gained momentum in contemporary research. While acknowledging global peacebuilding efforts and the progress achieved in development projects, the "center-periphery" model of relations between donors, International-Civil Society Organisations (I-CSOs), I-NGOs and L-NGOs, and between these organisations and target populations is perceived as unbalanced and unfair. This study suggests a power-balance whereby CSOs can mitigate adverse effects of ideologies and orientations conceived and controlled by external partners with insufficient consultation and decision of L-CSOs and local communities. The findings of this study are largely drawn from the literature reviewed and from semi-directive interviews with 18 members of 13 NGOs based in South Kivu.

Keywords Civil society · Development theory · Center-periphery theory · Peacebuilding

20.1 Introduction

Peacebuilding processes as well as contemporary conflicts fuel political and scientific debates and encourage unearthed activism actions in the so-called post-conflict zones. Internal and external actors under the umbrella of civil society worldwide have been involved in various projects and agendas aiming to either prevent broad-scale conflicts or find peaceful resolution mechanisms for them. Where violent conflicts have destroyed infrastructure and the social fabric that is necessary for integral and

Dr. Kaganda O. Philippe, Dean, Faculty of Social Sciences. Evangelical University in Africa, Bukavu, South Kivu Province, Democratic Republic of Congo; Official University of Bukavu.

Dr. Jean Chrysostome K. Kiyala is a Senior Lecturer, International Centre of Violence, Durban University of Technology, South Africa; Associate Professor and visiting lecturer the University of Bandundu, Catholic University of Bukavu and Evangelical University in Africa based in Bukavu (DR Congo).

J. C. K. Kiyala and G. T. Harris (eds.), *Civil Society and Peacebuilding in Sub-Saharan Africa in the Anthropocene*, The Anthropocene: Politik—Economics—Society—Science 34,
https://doi.org/10.1007/978-3-030-95179-5_20

durable development, international non-governmental organisations (I-NGOs) and locally born initiatives strive to save conflict-affected areas through rescue plans, coordinated development projects, and putting in place peace infrastructure to prevent the recurrence of violence and human atrocities.

It is important note that I-NGOs strive to network with local/domestic organisations—to advance development agendas and, similarly, promote peace issues and offer options for positions in official governments (The World Bank 2007). In their variety and diversity, NGOs generally work under stringent funding policies and their interventions are limited to areas pre-determined by their policies. I-NGOs that operate to promote development projects have different interests that are not beneficial to areas not supported by those I-NGOs' modus operandi. Hence, participation in localised peace and development projects is quite limited.

Several hurdles to peace and development efforts need to be cleared to comprehensively assist the populations in need of aid. While individual community ownership is now an accepted principle in development cooperation, in conflict settings the need for a participatory approach in the geographies and spatialisation of I-NGOs is stronger, and yet harder to achieve. The lack of input from intended beneficiaries and environments to leverage access to funds calls for a greater involvement of civil society to ensure that communities in need can call for funding despite their positioning vis-à-vis the predefined criteria and policies that govern I-NGOs' decisions to fund projects (or not) in areas that require assistance. Civil society is instrumental in post-conflict peacebuinding situation; however, it happens that certain norms and policies meant to address conflicts are inposed by I-NGOs that intervene to intervening I-NGOs (Verkoren/Van Leeuwen 2013). As a result, locally based NGOs' members feel ostracised from funding processes. For this reason, donors also apply rigorous criteria in selecting beneficiaries of peacebuilding projects. For instance, The World Bank (2007, p. 31) warns: "Without a thorough analysis of civil society, donors may inadvertently support spoilers and actors that are not working for peace and social cohesion".

The prospects for effective civil society action in South Kivu should be cognizant of the contradictions in the design of peacebuilding and development projects, the selection of beneficiaries, the nature of intervention, the choice of priorities and of internal and external partners to engage in the field. To address those difficulties raised above, this inquiry is guided by the following questions:

- What is the nature and extent of LNGO involvement in peacebuilding in South Kivu?
- What are the limitations of their interventions?

This study is conducted under the following assumptions:

- The participation of NGOs in peacebuilding and development projects is done through the execution of medium-range projects in various fields, which are grafted onto the philosophies, geographies, philosophies, and programs dictated by external donors.

- The assistance of external funding institutions depends on their organisational culture, philosophy, geographic areas of interventions and ideology, which often disqualify local initiatives that should be considered as genuine beneficiaries.
- The "center-periphery" relations between CSOs, comprised of I-NGOs and L-NGOs, and between these organisations and the needy populations are too unbalanced and unfair to accelerate peacebuilding and development actions.

To remedy the negative impact of this "center-periphery" relationship among peace and development interventions in South Kivu and globally, this study suggests power-balanced and fair relations between CSOs and their funding partners to mitigate the adverse effects of ideologies and orientations conceived and controlled by outsiders through which L-CSOs and local communities appear to be marginalised.

This inquiry focuses on the organisational approach to peacebuilding as a condition for development, while empirically linking this to the contradictions between I-NGOs' intervention policies, the credibility and resolve of L-NGOs to successfully undertake funded programs, and the relation between I-NGOs and L-NGOs in developing and implementing development and peacebuilding programs in a spirit of complementarity and partnership while respecting the principle of subsidiarity. This model of cooperation is expected to be overseen by civil society advocacy groups.

In the context of South Kivu's multifaceted conflicts, ranging from ethnic, identity, land, power, intra-state armed conflicts to cross-border armed hostilities (Vlassenroot 2013), peacebuilding is illusive without rebuilding the social capital and engaging civil society in initiating and strengthening efforts to grow and sustain social cohesion. According to the World Bank's 2007 report, the focus of actions that enhance social cohesion should seek cooperation among rival groups, and could encompass delivery of services, which is achievable via "mixed user committees or joint development committees" (The World Bank 2007, p. 21).

Peacebuilding starts from healing wounded relations, repairing the broken social fabric, building social capital, and creating further accountability measures and conflict prevention mechanisms. Civil society can be instrumental in this process.

This chapter is divided as follows: following the introduction, the literature on the links between civil society, peacebuilding initiatives, sustainable development, and funding processes is reviewed. A conceptual analysis of civil society, including its relation to social capital and peacebuilding, is discussed, and this opens up the conceptual framework, which explores development theory, center-periphery theory, and the "Local First" and "In-Between" models of engagement by civil society in peacebuilding. These paradigmatic frameworks are employed to guide this study and the linkages amongst them are sought and explained, and the extent to which they can slow or advance peacebuilding and development is also discussed. The next section deals with funding and program impact on peacebuilding and development in South Kivu. Study design and methods are explained; the results are tabled; and a discussion section follows, leading to the conclusion.

20.2 CSOs, Peacebuilding and Sustainable Development in South Kivu

20.2.1 Historical and Geographical Background

The geographical field of this study is South Kivu. Spread over 69,130 square kilometers, South Kivu is one of the 26 provinces of the Democratic Republic of Congo. This province is in the east of that country, and borders Rwanda and Burundi. It is divided into 8 territories: Fizi, Idjwi, Kabare, Kalehe, Mwenga, Shabunda, Uvira and Walungu. Its population, divided among several ethnic groups, is estimated at 5,772,000 (DRC National Institute of Statistics 2015). Rich in mining and forestry resources, the province has a low level of industrialisation and its economic sector is dominated by agriculture and informal trade and the tertiary sector. Its rural areas are plagued by insecurity orchestrated by national (Mayi-Mayi) and foreign (the Democratic Forces for the Liberation of Rwanda [FDLR] and the National Liberation Forces [FNL]) armed groups. A study published in 2015 revealed the existence of 81 armed groups in North and South Kivu. Two years later, in 2017, a new study indicated an increase to 120 well-identified armed groups (Stearns/Vogel 2015). The conflictual background of the South Kivu poses serious concerns for development which requires a great many donor interventions. However, protracted conflict can paradoxically restrain the flow of aid from external partners. Thus, understanding the goals of funding partners is crucial to strategising the way to access funds from donors.

20.2.2 Program Goals and Funding

Peace projects carried out by several organisations worldwide follow the opportunities and vision of donors whose funding mechanisms, according to The World Bank (2007, p. 13), encompass "multi-donor trust funds for specific countries or single donor funds. Dedicated funds can be established at headquarters", such as the World Bank's Post-Conflict Fund or GTZ's Peace Fund, or at field level such as the UNDP Peace and Development Trust Fund in Nepal.

The objectives of donors are thus defined according to the general framework proposed by the so-called external partners, which are international organisations, vectors of the ideologies and cultures of others. However, the objectives of the programs refer to the degree of violence and social deconstruction and their consequences on individuals or communities. Unfortunately, prospective recipient populations for international aid are not involved in defining the objectives of the programs for which they are beneficiaries. Thus, the perceived needs that justify granting development subsidies are transformed into expressed needs, which unfortunately are not the needs of the community, and such an approach results in internal inappropriateness of both funding and the programs for which aid is granted. Organisations deployed to implement received aid, act according to areas chosen by donors; that explains the downstream subventions to communities, and the course of action to obtain community members who commit themselves to the execution of the said programs. Organizational objectives hide the individual objectives of project designers and the ideological goals of external organisational and institutional partners.

20.2.3 Impact of Peace-Building Projects and Programmes

Measuring the impact of the actions of local civil society organisations in the construction of peace remains difficult to establish, given the resurgent dynamics of conflict, but also and above all because of the lack of evidenced variables to determine peace. For this reason, the organisations concerned face the challenge of establishing the impact of their peace interventions, and in defining the necessary qualitative and quantitative indicators. They are content to identify achievements against the objectives set in advance.

The interventions of non-governmental development organisations in South Kivu mainly address the cyclical problems which are a result of conflict and violence in order to alleviate their psychological, physiological, socio-economic and environmental effects such as trauma and diseases related to rape, the breakdown of social ties, crime, the social rejection of children from armed forces and groups, children without families, disinherited individuals, killings and massacres between communities, natural disasters, the schooling of children, under-disclosure or misinformation, etc. Considering some of the achievements made by these organisations, they play an essential but partial and insufficient role in the return of social peace to South Kivu.

Nevertheless, their strong? interest in the above-mentioned cyclical problems limits the impact of the programs implemented. The short duration of these programs, designed and implemented according to the rationales of the donors alone, cannot bring about expected structural changes such as improving attitudes and collective representations in relation to peaceful coexistence between different communities; rationalisation of productive capacity (industrialisation and large-scale production, large-scale processing); institutional reforms and improved governance (fair justice, reliable and competent political structures); etc. The development of peace is based on structural changes born of internal dynamics supported by international mechanisms integrated with local logics in a horizontal partnership horizontality.

20.2.4 Funding Peacebuilding and Civil Society

The seventy-fourth session of the UN General Assembly Peacebuilding Fund, (see Document A/74/688) made strides in its support of peacebuilding program, and it participated in efforts that aim at prevention of…, to which increasing funds have been directed through "cross -pillar strategies" (United Nations 2020, p. 3). The same documents states:

> The Fund contributed to further advancing the implementation of the United Nations-World Bank Partnership Framework for Crisis-Affected Situations with a $4.4 million investment in the establishment of the Humanitarian-Development-Peacebuilding and Partnership Facility (United Nations 2020, p. 11).

In 2019, the DRC received USD7 971 886 for an immediate response facility towards peacebuilding (United Nations 2020). Funding was aimed at empowerment of women and young people via "innovative and bottom-up approaches", and this was held to be central to the Fund's portfolio: this sanctioned "the direct funding of $12.9 million to civil society organsations, including five national organisations in Colombia, the Democratic Republic of the Congo, Madagascar and Sierra Leone" (United Nations 2020, p. 3).

- A report by Broadbent (2020) explored the literature on global mechanisms by which donors financially support civil society actors. These are: "(a) Direct support to individual or umbrella organisations; (b) Via Southern government; (c) Via Intermediaries – largely Northern NGOs" (p. 1), and these could be facilitated through the following four fund-dispatching structures: "Core funding, 'Basket' funding, Umbrella funds, [and] Multi-Donor Trust Funds (MDTFs)" (Broadbent 2020, p. 2). In addition, financing capacity building of NGOs was an essential concern that a 2009 study singled out, as it is mentioned among two major themes that emerged from that study: Firstly, there exists the need for ongoing or scaling-up of civil society actors, and for recognition of the fundamental role that civil society plays in development. The concept of "partnership" is also widespread in the literature.
- Secondly, capacity building was identified as a crucial component of any support to civil society, whether that is "direct" (such as providing subsidies for organisational development) or "indirect" (like providing financial backing to any CSO to help it build the capacity for participation of more local actors) (Broadbent 2009).

Moreover, each of the four mechanisms of disbursing donations involves risks, not only for donors, but for recipients as well. Broadbent (2020) alludes to the following general risks:

- Marginalisation of smaller organizsation because of the bias of the centralised subsidising system towards larger or more professionalised NGOs.
- Dependency on donated resources becomes probable amongst NGOs who receive funds on regular basis.
- High risks of "funding delays and short-termism".

- Interference of Southern governments in allocations and channeling of subsidies to civil society actors are potentially risky to adequate financial management of funds, "as well as increasing the potential for governmental 'co-option' of civil society actors – thereby weakening their claims to autonomy and objectivity" (Broadbent 2020, p. 2).

According to Tembo et al. (2007) cited in Broadbent (2020, p. 5), the leading ways of backing civil society via multi-donor interventions (MDIs) encompass:

- Umbrella funds to support a variety of actors (e.g. Common Fund for Supporting Civil Society in Nicaragua)
- Sector programs (e.g. Multi-stakeholder forestry program in Indonesia)
- Multi-donor trust funds (MDTFs)
- Basket funds for specific actors (e.g. Tanzania Media Fund)
- Core funding to individual CSOs (e.g. Ghana Research and Advocacy Program (Broadbent 2020).

It is evident that CSOs are well-supported financially by various structures at international level. Sad to say, the partnering between CSOs and institutions that fund their intervention programs is based on a "dependency theory" that sustains a dependent relation between the "periphery" representing CSOs in the South, and the "core", constituted of wealthy funding institutions in the North. For instance, the European Union (EU) is biased in promoting CSOs; it provides aid based on their funding ideology, as observed by Mahoney/Beckstrand (2009, p. 30): "It favors EU-level groups and groups that promote a European identity through pro-EU activities, EU integration promotion, democracy and civic engagement promotion and intercultural exchange and youth education and engagement". CSOs in the South are excluded in decision-making processes about funding. This situation needs redress through advocacy because many CSOs are dependent on subsidies from the EU, which represents >60% of their annual income (Haynes et al. 2019, p. 11). It is necessary to review the power balance between the North or the "center" that retains the monopoly of funding, and the South or the "periphery" whose operations and survival depend heavily on the "center". Approaches that may contribute to advance new relations between the "core" and the "periphery" would include, as mentioned earlier, empowering CSOs to build capacities that will boost self-reliance, and making them partakers in all strategic planning and funds disbursement policies before getting projects underway.

Through testimonies and examples, whether they are programs for raped women or child soldiers, the reader discovers with amazement the dilemmas and perverse effects of this outsourcing strategy. The proliferation of players fuels rivalries, dilutes responsibilities, leads to greater financial opacity, higher coordination costs and risks of corruption, loss of decision-making control and field knowledge, etc. Through this case study, international aid is an "ill-ordered charity" that urgently needs to be called into question.

As a result, local peace-building organisations are not consulted enough, that is, instrumental use in the service of 'outside' ideologies whose origin of resources and real purposes they do not know. This imbalanced rapport between donors and L-NGOs stems from donors' philosophies and actions on the ideological and axiological determinants of the funding partners. The direction given by funders is reinforced by the non-subsidisation of the pacification actions of local NGOs by the Congolese state. Indeed, for lack of state support these local NGOs are exposed to the ideologies of the "outside" through aid in the construction of peace.

The relationships between donors and beneficiaries of peacebuilding and development projects are defined as a partnership from which two positions emerge: The donor position is occupied by international organisations and institutions such as La Benevolencya, CORDAID, International Rescue Committee (IRC), USAID, the World Bank, etc. They design projects, provide funds, direct activities within a specified time frame, and monitor the use of the resources granted. The position of executor concerns local organisations (local civil society). Their mission is to develop the projects and implement them. Respondents are unanimous on the need for flexibility of their donors to changing projects depending on the context, as long as prior notice is given. Funding is provided through the presentation of a project duly accepted by the donors. To some extent, local organisations are "beneficiaries' and donors are 'donors.'?

Funds destined for the implementation of peacebuilding programs are provided by international organisations as part of multinational cooperation. Some donors operate directly from their headquarters in the West, others indirectly through agencies seconded to South Kivu Province (IRC, CORDAID, UNDP, UNICEF, Women's Plus Foundation, International Alert, the European Union, the World Bank, etc.). Instead of playing its hegemonic role in civil society, the Congolese state aligns itself among the beneficiaries of the actions of organisations through programs of institutional capacity-building, material support, etc. given to certain state institutions such as the prosecutors, the provincial divisions of social affairs and justice, the Territories, etc. As a result, through uncontrolled external financing from within, the Congolese state and national peace-building organisations are embarked on an alienating system defined by "rigid external determinism," in the words of Touraine (1976, p. 59).

20.2.5 Relationships Between International Donors and LNGOs

Earlier in this study, we examined the seven functions of civil society in local peacebuilding (Paffenholz 2015). The 2007 Report by the World Bank's Social Development Department acknowledged the unique and distinctive contributions that civil society can make to peacebuilding; it recommended that the support for civil society should be broadened its conception and embrace structures outside NGOs and organisations that are formally constituted (The World Bank 2007). The same report underlines how direct external support can strengthen civil society at various levels and states: “Local ownership and partner-led program identification are key, as are a solid understanding of the “intermediary chains” and “insider-outsider” partnerships” (The World Bank 2007, p. v). Alluding to the seven core functions of civil society (Paffenholz 2015), the World Bank’s report can be very helpful in designing objective-driven initiatives that can make peacebuilding more productive; the programming should be grounded in a “rigorous analysis, including the conflict and political setting, civil society itself, its enabling environment; and its peacebuilding experience and constraints” (The World Bank 2007, p. v).

Reflecting on the functions of civil society as envisioned by Paffenholz (2015), Hayman (2013) argues that other dimensions such as mediation and the combatants involved in armed hostilities, should be considered and she emphasises the advantages that locally-led peacebuilding initiatives offer over internally-led actions aimed to foster self-help, sustainability, and relevance. She observes:

> A Local First approach looks primarily for the capacity within the country or society and only brings in external assistance where no local capacity can be found. But, Local First does not mean local only—very often the most effective solution involves a partnership between internal and external organizations (Hayman 2013, p. 17).

Besides Local First, there has been a model that Van Houten (2018) names as The In-between where local civil society organisations engage in partnership with community-based structures to develop locally based peacebuilding interventions that would open opportunities to access international funding aimed at peace efforts in South Kivu. Local First and The In-between do not preclude community agency and the principle of subsidiarity (communities or individuals achieve what they can by their proper means without totally relying on external agents to solve their problems).

According to Evans (2013, p. 47),

> [s]ubsidiarity advocates a social order for the more efficient functioning of society. Specifically, if individuals or “subsidiary organisations” are left to resolve the matters closest to them, larger organisations, such as the State, are better able to carry out their allocated functions.

International CSOs and networks have placed global issues on the international agenda, successfully launched international campaigns (e.g. to ban landmines and blood diamonds, publish-what-you-pay) and partnered in key international conferences and consultative processes. It is reported: “I-NGOs can provide valuable support to domestic CSOs, but in many cases are not considered as part of that country’s civil society” (The World Bank 2007, p. 6).

Notwithstanding the contribution of external/international civil society actors in domestic affairs, the principle of subsidiarity is not antithetical to a globalised peacebuilding vision. It rather serves to empower local actors and awaken their conscience to take ownership of the vital issue of peace at the level of each community. Evans (2013, p. 54) notes: “Subsidiarity provides for the empowerment and moral enrichment of the individual through allowing the individual to help themselves without interference from a higher association”. The emerging global consciousness about local ownership of peacebuilding by either L-NGOs or community-based structures has the potential to lead each community or society to sustainable development and peace. Locals understand better the contours of their challenges and they should empower themselves to be the catalyst of solutions, the reason being that local conflicts are born of and supported by local actors who can ipso facto be partakers of reforms and transformative action. The intervention of international actors would then back up locally – generated peace and development agendas with supportive funding. This has been the vision of civil society organisation in South Kivu (Aembe/Jordhus-Lier 2017; Van Houten 2018).

The interaction between locals and outsiders is a salient premise of sustainable development in the sense that, as pointed out by Hayman (2012), development initiatives such as delivery of services and goods that are spearheaded by CSOs, private sectors and governments should be a continuous process, and such projects should be appreciated, since they leave local organisations stronger than before. Thus,

> Local First implies that outsiders engage with local perceptions of problems and solutions and seek out and build on pockets of effectiveness wherever they are found. Doing so can make maximum use of local knowledge, reinforce self-help and self-reliance, and offer encouragement and self-confidence to local organizations (Hayman 2012, p. 13).

However, when dealing with intercommunity conflict, the Local first approach can be limited in initiating and achieving durable peace when stakeholders in a particular conflict are entangled in an adversarial relation and hostilities. In such circumstances, local initiatives would yield limited results. This view does not undermine the action of I-CSOs to network with domestic organisations, as they strive to advocate development and offer alternatives to individual states' officials; they carry out their actions under the institutionalised UN system, and their influence continues to expand and grow (The World Bank 2007).

Domestic civil society tends to have little involvement in direct facilitation between parties in conflict, especially when it involves actual peace negotiations, as this role is primarily played by external parties, especially governments (e. g. Norway in Sri Lanka) or multilateral agencies (e.g. the UN in Guatemala). In some instances, this role can be taken up by international CSOs as in the case of Comunita di Sant'Egidio in Mozambique or the Geneva-based NGO Center for Humanitarian Dialogue which facilitated the first negotiations in Aceh (The World Bank 2007). Donor engagement with CSOs is often fragmented and short-sighted. External funding and support are often limited to a small sub-set of CSOs (particularly development-oriented NGOs), while many local-level and membership-based organisations are bypassed.

These sections dealt with the role of local civil society in North Kivu based on the referential framework of *Local first* and The *In-between* and established how such approaches enhance peacebuilding strategy and development, that is followed now by a discussion of the theoretical framework that guides this chapter, notably civil, society, development and peacebuilding, and center-periphery theory.

20.3 Conceptual Framework

Two theories are utilised to frame and guide this chapter, namely development and peacebuilding theory and center-periphery theory. These concepts are relevant to this chapter for several reasons. Firstly, most contemporary hegemonic conflicts occur in sub-Saharan Africa where development and democracy remain the key challenges of peacebuilding. Secondly, most peacebuilding projects in sub-Saharan Africa and around the Globe depends on foreign aid and capitals that are provided by donors in the North, also perceived as rich or prosperous countries and institutions. The connection between these two and civil society is obvious for the fact that NGOs that make one category of civil society (Paffenholz 2015).

20.3.1 Development and Peacebuilding

The domains of development and peacebuilding are embraced by CSOs; however, the accessibility of funds for projects related to these fields is stonewalled by funders' policies. In this regard, the The World Bank (2007, p. v) considers: "Civil society contributions to development and peacebuilding can be categorised in a variety of ways, but donors largely employ actor-oriented perspectives." It is worthwhile noting that the World Bank recommends a move toward "a functional perspective, centered on the roles that different actors can play in conflict situations" (The World Bank 2007, p. v). The same report states:

> Such a functional perspective would enable donors to better analyze existing and potential forms of civil society engagement in peacebuilding. In particular, it would help clarify policy and programming objectives, select civil society partners, and help to set outcome indicators to improve monitoring and evaluation (p. v).

The stance taken by the World Bank opens up a more realistic and effective approach to be adopted by funding institutions to make L-NGOs more productive as they make up one of the components of civil society.

However, in development studies, several scholars have produced the following arguments:

> The foreign penetration, diffusion, and acculturation of modern values, techniques, and ideas from the centers to the periphery does not necessarily produce development. In most of the cases, this process contributes to the subordination of the underdeveloped countries to the centers (Grosfoguel 2000, p. 360).

Militarised conflicts in the Kivu Province and the immediacy of the search for peacebuilding and sustainable development is well understood in the context of globalisation and democratisation, in which civil society reemerged with its multi-functional agenda. In the context of unstable sub-Saharan Africa, the Balkans, in the regions of Central and Eastern Europe, ethnic tensions, civil wars and xenophobia have assuaged the enchantment with democracy (Roniger 1994). The great mutations and changes in the public and political sphere, the shattered ideological perceptions of the of the world in the twentieth century, have consolidated the "democratic and neoliberal 'readings of reality'" (Roniger 1994, p. 1). As far as power-balance is concerned, amidst these transformations, the relation between civil society and state was redefined, as noted by Roniger (1994, p. 1):

> These processes have focused attention on the emasculation of the state and the parallel empowerment of civil society, which in turn has often involved a dual trend of disengagement from the state and the subsequent reshaping of participation in the public sphere.

Globalisation and technology advancement in the Anthropocene have become critical questions that touch deeply on the future of our ecosystem. Global warming, the rising fear of nuclear and biological warfare, intra-state brutality and murderous conflicts have become ongoing concerns in our time. Most conflicts in Africa are militarised because of the easy inflow of weaponry that is legally and illegally traded to fuel interethnic armed conflict and regional wars. Speaking of development suggests engaging in an ambivalent discourse that evokes both the devastating effects of technology on the environment and the quest for human welfare. Pieterse (2010, p. 1) observes:

> The classic aim of development, modernization or catching up with advanced countries, is in question because modernization is no longer an obvious ambition. Modernity no longer seems so attractive in view of ecological problems, the consequences of technological change and many other problems.

It appears that the field of development has been in crisis, not only as the result of confronting ideologies such as "neo-Marxism and dependency theory" on the one hand, and "Keynesianism and welfare politics" on the other, but mostly, as Pieterse (2010, p. 5) remarks:

> There have been plenty of critical positions but no coherent ideological response to the neoliberal turn. The crisis is further due to changing circumstances including development failures, the growing role of international financial institutions, and conflicts in developing countries.

In response to this crisis NGOs require a new mode of engagement that has to be differentiated into various developmental fields, namely: (1) the "multi-level negotiation and struggle among different stakeholders"; (2) "the relationship between power and knowledge in development"; and finally, in the areas of "globalization, sustainability, gender, diversity, poverty alleviation [...] [e]mergencies, such as humanitarian action, conflict prevention and post-conflict reconstruction" (Pieterse 2010, p. 11).

The issue of geographical context where NGOs set their interventions is impactful on development and peacebuilding. For this reason, a study on how NGOs develop their actions and determine the sites where they ought to intervene may have serious effects, for instance, on rural poverty (Bebbington 2004). Their target sites, such as rural areas which, in most of cases, remain underdeveloped and vulnerable to recurring militarised conflicts should get civil society approval for more openness and flexibility that could move NGOs and funding institutions beyond their policies and pre-established geographical areas of intervention.

In the context of South Kivu, due to the slow pace of the interventions by the Central DRC Government and the local/provincial government, efforts to build peace and accelerate developmental projects rely significantly upon civil society organisations via I-NGOs and locally-based non-governmental organisations (L-NGOs). As a matter of fact, there is an increasing number of I-NGOs and L-NGOs operating in the eastern part of the DRC; they are spread across all vital sectors of life (health, education, human rights, food security, etc.).

The center-periphery model is characterised as a continuum that deals with "the role played in the international market" (Grosfoguel 2000, p. 363); this is associated with financial relations between the donors and recipients to advance development projects. The weight on the balance of this relationship swings in favor of the funders who dictate their ideologies, philosophies and geographies to the recipient local organisations and their populations. According to Namkoong (1999) the scarcity of capital accumulation results in the connection of periphery and center; and here, dependence entails "the relations between centers and the periphery whereby a country is subjected to decisions taken in the centers" (Prebish 1980, cited in Namkoong 1999, p. 130).

20.3.2 Center-Periphery Theory

The *center-periphery* model is used in geography to explain a relationship of domination and dependency that exists between two types of places: the centers, which dominate and take advantage of this unequal relationship, and the peripheries, which are dominated and suffer (Encyclopædia Universalis 2021). For instance, Galtung depicts the center-periphery model in Fig. 20.1 as a complex web of relations between mainly the First World countries and the Third/Second World countries; and that is revealed in unequal living standards between cities, the rich, the elites and the multinationals on the one hand, and rural areas, workers, farmers, and the poor, on the other hand (Galtung n.d.).

A peripheral space is the inverse of the center. It is characterised by a lower standard of living and a more limited production and decision-making capacity (Encyclopædia Universalis 2021). Figure 20.1. represents the uneven type of relation and structure of the global society, the center portrayed as the *First world countries and the periphery as the Third world and since 1990, that has become the second world* (Galtung n.d.). This model (center-periphery) renders intelligible the uneven type of structure of the global society. Most of the beneficiaries of peacebuilding and development programs are in the periphery, in the Grand South whether in towns, cities or rural areas where we see an increasing need of external funds for peacebuilding ans sustainable development. The South is comprised of areas that are largely affected by underdelopment, poverty, precarious or inexisting infrastructure, and above all, by

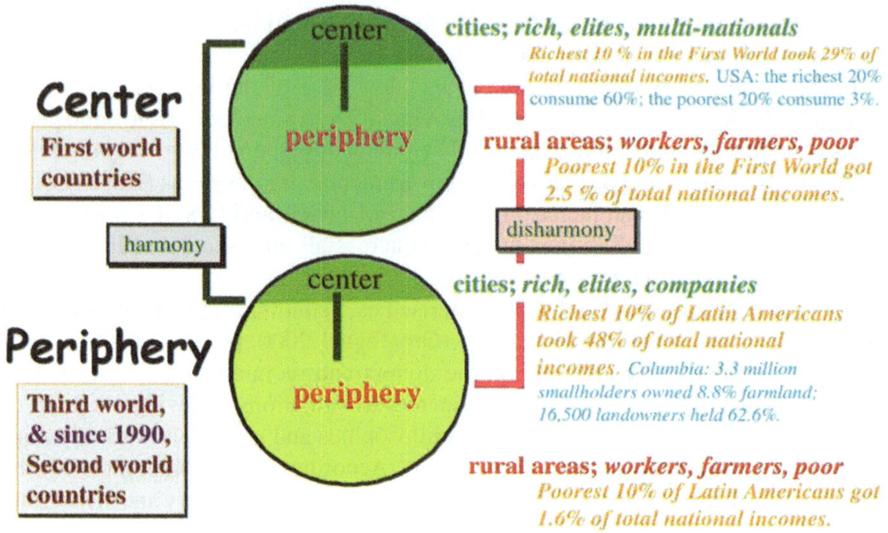

Fig. 20.1 Johan Galtung, Center-periphery model. *Source* https://image.slideserve.com/265453/galtung-s-center-periphery-model-l.jpg

militarised conflicts, and negative peace—described as absence of personal violence amidst structural violence also discernible in social injustice (Galtung 1969).

The relation between the center and the periphery is marked by dependency and imbalanced decision-making power in which the center always prevails. External partner organisations that finance and guide peace-building programs keep upper-hand on funding processes while local organisations that carry them out and the beneficiary communities are not represented in the deciding structures; they constitute the periphery, while external funders remain the center. Aid to the peripheries is guided and defined by ideological orientations of the centre for the implementation of funded projects.

Power-balance can be leveraged by civil society's intervention, possibly because:

> social movements, voluntary associations, and intermediate institutions of civil society could affect an overall reconstruction of the political centers and reformulation of community through a strong emphasis on participation and the endorsement of an egalitarian vision of rights and entitlements (Roniger 1994, p. 2).

This argument is equally applicable to interrelations between funders or external partners, L-NGOs, and beneficiary communities. Failure to redress the "unevenness and injustice" experienced in the geographies of funding development where civil society is sidelined in decision-making processes (Bebbington 2004), will further disadvantage communities that most need peace and social and economic growth. The involvement of civil society is crucial in rethinking a more equitable and balanced relation between the center and the periphery to better address the needs of the munities weakened by conflicts and violence; and this constitute opportunities for the center to intervene.

The section that follows discusses the design and methodological approach adopted to conduct this study.

20.4 Design and Methods

This inquiry is designed as a single case study limited to the South Kivu Province in the Democratic Republic of Congo (DRC). A case study offers an interesting advantage to research, namely its versatility to take in the philosophical position of the inquirer and "presents a unique platform for a range of studies that can generate greater insights into areas of inquiry" (Mills et al. 2017, p. n.p.). This model of investigation provides an ample body of explanatory insights into the complexities of civil society's participation in peacebuilding and development projects, particularly in a volatile environment that is still vulnerable to militarised conflicts, thus frustrating peacebuilders and discouraging donors and I-NGOs.

Empirical data were compiled through disengaged observation and semi-directive interview with thirteen non-governmental organisations (local civil society) between July and September 2015. According to Dockès/Kling-Eveillard (2006), the semi-directive method provides the person being investigated the latitude to convey his/her opinions by giving answers to open-ended questions. For Huntington (2000, p. 1271), "[t]he semi-directive interview is more a conversation than a question-and-answer session". The semi-directive interview method offers several research advantages to both the interviewer and the interviewee. On the one hand, it opens a level of freedom for the person from whom information is sought to freely express his/her views (Dockès/Kling-Eveillard 2006). On the other hand, it is a useful approach to investigate a problem with which respondents or participants may not feel comfortable, especially in dealing with direct questions; and it is similarly advantageous to the investigator when s/he cannot ascertain that the questions asked are comprehended by the respondent as intended (Huntington 2000).

The main sources of data utilised to conduct this inquiry being texts or qualitative data, content analysis was applied as method of analysis and interpretation of qualitative data gathered via semi-directive data and documentary sources. A Discussion section follows the display of the findings where the views expressed by the interviewees are analysed and interpreted via the *Three Approaches of Content Analysis* (Hsieh/Shannon 2005). These encompass Conventional Content Analysis (observation), and Directed Content Analysis (developing theory), and Summative Content Analysis (identifying keyword), followed by an "[a]n analysis of the patterns leads to an interpretation of the contextual meaning of specific terms or content" (Hsieh/Shannon 2005, p. 1286).

To this end, eighteen resource persons, comprising 10 men and 8 women, were interviewed. An average of 60 to 90 minutes was devoted to each interview conducted using a previously established protocol. The organisations involved in this study were the Women's Caucus, Heir to Justice, the Olame Centre, the Diocesan Development Office, the Catholic Diocese of Kivu Justice and Peace Commission, the Network of

Human Rights Associations in South Kivu (RADHOSKI), the Jeremiah Group, the Action for Children in Difficult Situations (AFESD), the Guinea Pig Village (VICO), Let Africa Live (LAV), the Women's Network, the Office for the Coordination of Civil Society, and the Inter-Community Baraza.

The sampling was purposive; participation in this study was consistent with ethical norms that require voluntary/consented participation, confidentiality, upholding the principles of beneficence, justice, and freedom to withdraw one's participation, etc. (Bless et al. 2013).

The interviews conducted provided the empirical basis for this study. The qualitative and quantitative data were processed through content analysis. For the intelligibility of analysis, reasoning constructs a dialectical relationship between the following variables: actors (structures), actions and strategies; logics and representations. Thus, the analysis attempts to detect the multifaceted contradictions of the logics and peace processes undertaken by non-governmental organisations in South Kivu. These contradictions do not mean inopportuneness or lack of necessity, but rather a contradictory way of producing the new material and cultural bases for collective self-determination.

20.5 Findings

The first research question consisted in establishing the nature and extent of L-NGOs involvement in peacebuilding in South Kivu. The findings are presented in themes that encompass specific objectives of investigated NGOs, in relation to the construction of peace, the beneficiaries, fields of action, and themes of intervention; organisational collaboration, namely monitoring and evaluation of projects; conflict apprehension; monitoring and evaluating peace building projects; and building peace for development, which focusses on ideological orientations, targets and program goals in intervention zones. Interpretation of these findings is simultaneously backed by the literature reviewed.

The actions of local peace-building organisations are guided by organisational strategies reflected in the definition and determination of the apprehension of peace; goals, beneficiaries, and themes; organisational collaboration and monitoring and evaluation methods.

20.5.1 Goals, Beneficiaries, and Themes

Emerging objectives of the interviews we conducted include raising awareness on peace issues, leading advocacy, and direct interventions in favor of beneficiaries, such as legal assistance and psycho-social assistance, as evidenced by the following words: "We raise awareness, advocacy and assist victims" (Interview of 27/8/2015 with a VICO facilitator in Bukavu). Goals such as program intervention areas are dictated from "outside" and no longer meet donors' visions.

According to our investigation, none of these organisations contacted received financial support from the Congolese state. This suggests that the Congolese state has no structured policy of partnership and promotion of national civil society working in the field of peacebuilding. The state plays neither a regulatory role, nor a restorative or ideological role. On the contrary, instead of playing its hegemonic role in the face of civil society, the Congolese state aligns itself with the beneficiaries of the actions of organisations through programs of institutional capacity building, material support, etc. given to certain state institutions such as the prosecutors, the provincial divisions of social affairs and justice, the territories, etc. As a result, through uncontrolled external financing from within, the Congolese state and national peacebuilding organisations are embarked on an alienating system defined by "a rigid external determinism", citing Touraine (1976, p. 59).

20.5.2 Recipients, Areas, and Themes of Intervention

The themes and areas chosen are diversified according to the need and interests of donors as shown in the Table 20.1.

Areas of intervention selected by funding institutions are mainly rural settings of South Kivu Province and the city of Bukavu, but these are primarily determined by external partners. Rural communities are most affected by the effects of conflict and violence. However, the actions of the organisations appear to be quite limited in relation to the need for peacebuilding. In addition, according to our observation, 55% of the organisations contacted concentrate certain activities that target rural populations in Bukavu. This forces the "beneficiaries" to move from their villages to the city of Bukavu to come and solicit the interventions of the organisations (legal support, health intervention, etc.). Other beneficiaries go through relay structures at

Table 20.1 NGO Areas and Topics for Intervention shows different themes and areas of interventions by NGOs. *Source* The authors

Areas	Themes
Justice	Human rights Sexual violence Violence against women
Social	Peaceful cohabitation Integration of children out of armed forces and groups Transformation of land and inter-communal conflicts
Policy	Civic education and peace education Election education Women's political participation
Economic	Microfinance

the base (Justice and Peace Commission, Mediation Framework, etc.) which have an intermediary role to refer certain problems to the central organisational structures far from the applicants' services.

20.5.3 Organisational Collaboration

All respondents (100%) reject opposition between organisations even when working in the same fields of intervention. Rather, they believe that there is collaboration, partnership, complementarity and even synergy among organisations working in the field of peacebuilding. According to the respondents, organisational networks (such as the Network of Human Rights Associations in South Kivu, the Peace and Reconciliation Council and the Office for the Coordination of Civil Society in South Kivu) would promote close collaboration between the organisations. This collaboration is manifested in the commitment to joint actions such as the defense of an actor or an organisation threatened by the political system, participation in meetings convened by networks, etc.

Nevertheless, from a nuanced point of view of six interviewees (33%), it appears that conflicts of interest sometimes occur between certain organisations and their respective networks. Such conflicts arise because of the misuse of the network without consulting all affiliates, non-transparent management of finances and a struggle for leadership. In this regard, the survey recorded the following statements: "Some network coordinators use us to spread their own organisations and not all affiliates" (Interview of 08/7/2015 with a member of VICO); "To run the network, logic disappears to the benefit of tribes, odds or money" (Interview of 15/7/2015 with a member of AFSD); "On several occasions, we were almost caught up in the mismanagement of civil society finances" (Interview of 03/8/2015 with a member of the Women's Caucus). Various networks are maintained and constitute a maintenance strategy for organisations involved in the construction of peace.

20.5.4 Monitoring and Evaluation of Peacebuilding Projects

Organisations recognise the importance of evaluation in the execution of any peacebuilding project. Evaluation is also a requirement of the funder. According to interviewees, follow-up is done throughout the execution of projects in order to comply with the guidelines. Evaluations are organised mid-term and at the end of projects. During the evaluation, respondents state that their organisations verify the progress of the project and its impact through the manipulation of qualitative and quantitative indicators. To do this, they use certain evaluation methods such as OPS, 3P and Canadian RPP. Only five organisations, or 28% of respondents, apply the objective-based assessment approach, i.e., the qualitative and quantitative measurement of achievements against pre-set targets. In fact, an evaluation of a project is not an impact analysis as identified by 72% of our respondents. In addition, it was noted that it was difficult for all respondents to identify these specific indicators for their various peace-building projects.

20.5.5 Apprehension of Conflict and Peace

According to our observation, conflict is understood to be about the various categories of disagreement, misunderstanding, opposition, divergence, uncertainty, violence, exclusion and injustice. It manifests itself in a relationship of tension, rejection, or violence between two or more people, between communities or between states. Fifty-six percent of respondents believe that conflict is not only negative, because it can lead to a desired change. On the other hand, almost half of them consider any conflict to be a lack of peace, and therefore negative. According to the latter opinion, organisations in collaboration with the state must put an end to conflicts within the communities of South Kivu. The most cited examples of these conflicts are land conflicts, rape, war, massacres and killings, and displacement of populations. These manifestations and consequences of conflict are derived from the context of conflict in the east of the D.R.C., and particularly in South Kivu.

The second question of this inquiry sought to determine the limitations CSOs in their interventions.

20.5.6 Ideological Orientations and Targets

Several local civil society organisations in South Kivu are working to build peace as demonstrated above. Their approach to peacebuilding, referred to here as an "organisational approach," is defined by a set of philosophy, strategies, means and goals to bring peace to communities whose socio-economic fabric has been weakened by cycles of conflict and violence.

CSO's interventions in the construction of peace is based on a goal of bringing peace to communities in conflict or post-conflict situations/states. Peace is both the leitmotif but also the philosophical orientation that determines the pacification programs put in place. According to the organisations contacted, peace is the prerequisite for the development of post-conflict zones. In this regard, all the subjects contacted said: "Our motivation is peace"; "Peace is the overriding need in our areas of intervention." This philosophical orientation is theoretically justified by the hegemonic mission of civil society. However, the actions generated by this vision do not address the complexity of peace because they are limited to certain areas for which each organisation has obtained funding from an external donor. To this end, the ideological basis of the pacification actions of local organisations suffers from a lack of autonomy, because their ideological orientation is dependent on ideologies directed/indicated by external partners. Indeed, international organisations are involved in the field of peacebuilding. Each carries its vision and mission, which it imposes on any local organisation seeking support, as evidenced by the following statement: "We guide our peace programs in the areas required by donors. Our organisation works in the field of ex-soldier children because that is what my partner is interested in this year until 2018" (Interview of 14/9/2015 with a member of the LAV staff).

Insufficient financial resources of the stakeholders were reported as cited here: "We cannot pretend to make peace at the level of the whole of Congo because we do not have the means" (interview of 28/8/2015 with a member of the Women's Caucus. Of course, civil society organisations do not have institutional assets to address peacebuilding at macro-social and/or mega-social levels. According to the respondents, the community level provides an in-depth understanding of the problems of individuals and communities and their needs sufficient to guide action. Generally, cyclical solutions include care for the sick or victims of sexual violence, the reintegration into society of unemployed children and ex-soldier children, advocacy, or legal assistance for victims of miscarriages of justice and social injustice, reconciliation between communities, etc.

The duration of program delivery is short and does not consider real time to provide sustainable solutions. According to our survey, the duration of the programs varies on average between four and six months. This time is that of the donor and not of the execution organisation. The organisations contacted acknowledge this lack of temporal realism in their peace-building programs. Interviewees stated that the duration of any program is imposed by funders: "The duration of a project is not up to us. It is the funder who sets the time according to the objectives he wants to achieve" (Interview of 23/8/2015 with a member of the staff of Heirs of Justice). For interviewees, the short duration imposed for the implementation of peace-building programs weakens the impact of interventions and does not allow for community ownership of philosophy and practice.

Peace-building programs carried out by civil society organisations appear to focus on cyclical solutions rather than structural changes in recipient communities. Structural change would mean communities benefiting from peace-building programs, changes in representations, practices, and structures to solve problems resulting from past conflicts and violence, but also the creation of resilient practices that can maintain social cohesion and collective economic progress. On the contrary, what is observed is the fragility of the beneficiary communities, even after the implementation of peace-building programs, and hence the permanence of non-peace. The intervening organisations contacted seem to be aware of this limitation, which is rightly or wrongly attributed to the inadequacy of financial resources as can be understood by the following statement: "We, as an NGO, are trying to solve certain problems such as the health of women raped or the microfinance in their favor, advocacy, etc., but unfortunately, we cannot do everything. Our villages are very fragile because of the conflicts and their consequences. True peace must be achieved by everyone with the state at the head" (Interview of 28/7/2015 with a member of the RFDP)[1].

20.5.7 Program Goals and Intervention Zones

Recipient populations are not involved in defining the objectives of the programs for which they are a beneficiary. Thus, the perceived needs that justify the objectives of the programs are transformed into expressed needs whose effect is the internal inappropriateness of philosophy and peace-building actions. The deployment of organisations to the implementing areas chosen by donors consists of explaining downstream to communities the merits of the action to obtain their membership and commitment. Organisational objectives hide the individual objectives of project designers and the ideological goals of external organisational and institutional partners.

The organisations' areas of intervention are in rural areas of South Kivu Province and the city of Bukavu. But they are previously determined by international partners. Rural communities are most affected by the effects of conflict and violence.

However, the actions of the organisations appear to be quite limited in relation to the need for peacebuilding. In addition, according to our observation, 55% of the organisations contacted concentrate certain activities, that target rural populations, in Bukavu. This forces the "beneficiaries" to move from their villages to the city of Bukavu to come and solicit the interventions of the organisations (legal support, health intervention, etc.). Other beneficiaries go through relay structures at the base (Justice and Peace Commission, Mediation Framework, etc.) which have an intermediary role to refer certain problems to the central organisational structures far from the applicants for services. Goals such as intervention areas are dictated from the outsiders and no funds are granted if the program for which the subsidisation is required no longer meet the donors' vision.

20.6 Discussion

The organisations contacted state that they provide judicial assistance to victims of various injustices such as rape, inheritance problems, land problems (mainly by the Heir to Justice), material and psychological assistance to victims of sexual violence and atrocities by armed groups. This was reported by the LAV, Women's Caucus, VICO, RFDP, etc. All the organisations contacted claim to have made a significant contribution to the building of peace through the actions they carried out.

---

[1] RFDP stands for a local civil society organisation, *the Réseau des Femmes pour la Défense des Droits et la Paix* [Women's Network for the Defense of Rights and Peace]. For more on this, see Cyril Musila, Bukavu, June 2006, Réseau des Femmes pour la Défense des Droits et la Paix (RFDP), http://www.irenees.net/bdf_fiche-acteurs-280_fr.html. Accessed 15 Oct 2021.

The organisations interviewed are interested in people of both sexes and of all ages, except for the Olame Centre and Women's Caucus, on the one hand, and Action for Children in Difficult Situations (AFESD) and LAV, on the other, which are aimed at women and children respectively. For specific groups of beneficiaries, a remarkable choice is reserved for female victims of sexual violence. Other beneficiaries identified are children in difficult situations, children exiting armed forces and groups, the poor, and vulnerable women. To these categories are added the political authorities, the judiciary, the police, and the Army through judicial police officers, magistrates, judicial police inspectors, police and military officers, customary authorities, mayors, and territorial administrators. They benefit mainly from various trainings to improve their governance capacities. On average, each organisation is interested in six categories of beneficiaries.

All the organisations surveyed claim to address the needs felt and expressed by the beneficiaries in relation to the construction of peace. They say they have knowledge about the areas of intervention and the needs of the beneficiaries. However, in most cases, knowledge about contexts is not the result of in-depth studies but of empirical observations by the leaders of organisations and relay structures at the grassroots level (local beneficiary community). Therefore, the knowledge that underpins the logics of intervention is imprecise, and can, in turn, reduce the effectiveness of actions. This observation shows that scientific research is not yet integrated into local organisational practices. However, the role of scientific research ahead of any intervention is unavoidable in view of the dynamics of conflicts in the East of the DRC. and in the Great Lakes region. Research can serve as the basis for peacebuilding actions by producing real knowledge of the socio-economic, political, and environmental contexts of the intervention zones.

Organisational actors in the construction of peace have different and even reductionist apprehensions and conceptions of conflict because of the identification of consequences, instead of the processes or dynamics that generate them. This shift in approach may be an obstacle to determining evidence of conflicting processes to be considered to induce structural changes rather than redressing the consequences. To this end, organisations would work more on the structural conditions that underlie the basis of rape, killings, exclusion between communities, etc. rather than only helping rape victims, ex-soldier children, and so on as is the case. The sociological apprehension of conflict emphasises processes. According to Bercovitch and Fretter (2004), the conflict is broad, as a process of interaction between two or more parties that seek to thwart, insult, or destroy their opponent because they perceive incompatible interests or goals. As for peace, it is considered by the respondents as a state of stability, of tranquility. For the organisations contacted, peace is general and not just the absence of war. It must be observed through the well-being of the people.

Moreover, the apprehension of peace is reductionist, especially since it essentially refers to the field or theme of intervention of each organisation, not allowing exploration of peace in all its complexity and in all its dimensions. Paradoxically, this reductionism is a strategic force, in that it directs organisations towards specific areas. One of the criteria of good health is that you do not feel your organs; the proper way of their proper functioning is that they go unnoticed, just as peace is mechanical, if not unconscious (Bouthoul 1970). In fact, social conflict is easier to grasp than peace.

20.7 Conclusion

This chapter intended to investigate the link between peacebuilding and development and used literary and empirical evidence to expound the inconsistencies found in I-NGOs' intervention policies, and the credibility and determination of L-NGOs to efficiently embark on peacebuilding and development projects funded by external institutions. It also investigated the relation between I-NGOs and L-NGOs in developing and implementing development and peacebuilding programs in a spirit of partnership which does not hinder each organisation's autonomy and respects the principle of subsidiarity.

Even when discussing the principle of subsidiarity in relation to financial aid needed by L-CSOs, one should not misinterpret the balance between what one institution can afford in terms of its financial self-reliance and the openness to receive needed backing funds to operate; subsidiarity entails that *Local First* is not antithetical to receiving external assistance; it is rather a call to make local actors self-sustainable. To illustrate this point, Hayman (2012), remarks: "[Local First] is based on the idea that aid should consciously and assertively assist countries to move to self-reliance through a self-help process that is locally led and determined" (2012, p. 13).

It was found that the model of cooperation to be forged between I-NGOs and L-NGOs should be overseen by civil society advocacy groups to contain any negative effects resulting from external NGOs' policies, ideologies, or geographies, and from L-NGOs' obstacles to act as peacebuilders. The mobilisation of L-NGOs is dictated by the logics of international actors (funders) who define, according to their own visions, the operationalisation of peace-building programs. Funding for peacebuilding programs, as well as their ideologies and actions, is beyond the reach of local actors and targeted populations. It is an uneven system that serves the ideologies of donors and the individual and collective ambitions of so-called "local civil society organizations".

Nevertheless, the permanence and re-emergence of stereotypes and violent conflicts between groups or communities in the areas of intervention of non-governmental conflict transformation organisations remain a real challenge that deserves institutional attention, an ideological revolution for a change of attitudes and a lasting peace. That is a *sine qua* non condition for development as well. Effective socio-economic development, the reconstruction of social capital and societal harmony in the South Kivu Province rely significantly upon balanced relations between all stakeholders, namely, donors and CSOs (I-NGOs, L-NGOs, community-based initiatives); and rethinking the conditions for financially backing peacebuilding and development projects in various regions of the globe.

The complexities of peacebuilding in South Kivu Province in particular, resulting from two decades of violent intercommunity conflict and civil wars, demand the restoration of the destroyed social capital—structures and infrastructure that maintained social cohesion—and accelerate the province's developmental agenda. Despite the contradictions, challenges, and limitations of the organisational approach to conflict transformation, it is, in the face of the failure of the Congolese state, a cyclical response to the fragility of communities affected by conflict and violence in South Kivu.

The intervention of organisations in the construction of peace is based on a goal of bringing peace to communities in conflict or post-conflict situations. Peace is both the *leitmotif* but also the philosophical orientation that determines the pacification programs put in place. According to the organisations contacted, peace is the prerequisite for the development of post-conflict zones. In this regard, all the subjects contacted said: "Our motivation is peace"; "Peace is the overriding need in our areas of intervention." This philosophical orientation is theoretically justified by the hegemonic mission of civil society. However, the actions generated by this vision do not address the complexity of peace, because they are limited to certain areas for which each organisation has obtained funding from an external donor.

To this end, the ideological basis of the pacification actions of local organisations suffers from a lack of autonomy, because their ideological orientation is dependent on ideologies crafted by external partners/outsiders. This situation needs to be remedied by civil society advocacy groups to ensure that this hurdle on the path of development and building durable peace is removed. I-CSOs' involvement in helping reimagine hardline ideologies, geographies, philosophy, and political economy of donors whose actions should not infringe the autonomy and performance of L-NGOs constitute the way forward to address the unevenness and unfair geography and targets of international civil society's donors.

References

Aembe, B., & Jordhus-Lier, D. (2017). Within, above, and beyond: Churches and religious civil society activism in South Kivu. *Journal of Civil Society, 13*(2), 149–165.

Bebbington, A. (2004). NGOs and uneven development: geographies of development intervention. *Progress in human geography*, 28(6), 725–745.

Bercovitch, J., & Fretter, J. (2004). *Regional guide to international conflict and management from 1945 to 2003*: CQ Press.

Bless, C., Higson-Smith, C., Sithole, S.L. (2013). *Fundamentals of social research methods: an African perspective* (5th ed.). Cape Town: Juta & Company.

Bouthoul, G. (1970). *Traité de polémologie: sociologie des guerres*. Lausanne: Payot. Broadbent, C.E. (2009). Civil Society Funding Mechanisms. https://gsdrc.org/publications/civil-society-funding-mechanisms/ 23 May 2021.

Broadbent, E. (2020). *Helpdesk Research Report: Civil Society Funding Mechanisms*. http://gsdrc.org/docs/open/hd633.pdf. Accessed 23 June 2021.

Dockès, A.C., & Kling-Eveillard, F. (2006). Farmers' and advisers' representations of animals and animal welfare. *Livestock science, 103*(3), 243–249.

DRC National Institute of Statistics (2015). *Statistical Yearbook 2014*.

Encyclopædia Universalis (2021). In (Vol. *Centre et périphérie*). Online: Universalis Junior.

Evans, M. (2013). The Principle of Subsidiarity as a Social and Political Principle in Catholic Social Teaching. Solidarity: *The Journal of Catholic Social Thought and Secular Ethics*, Article 4, 3(1), 42–60. https://researchonline.nd.edu.au/solidarity/vol3/iss1/4 Accessed 12 Dec 2021.

Galtung, J. (n.d.). Galtung's Center-periphery Model, https://image.slideserve.com/265453/galtung-s-center-periphery-model-l.jpg Accessed 23 June 2021.

Galtung, J. (1969). Violence, peace, and peace research. *Journal of peace research*, *6*(3), 167–191.

Grosfoguel, R. (2000). Developmentalism, modernity, and dependency theory in Latin America. *Nepantla: Views from South*, 1(2), 347–374.

Hayman, C. (2012). Introduction: Local first—A proposal for development in the twenty-first century. In K. McGuinness (Ed.), *Local first: Development for the twenty-first century* (pp. 13–43). London: Peace Direct.

Hayman, C. (2013). Local first in peacebuilding. *Peace Review*, *25*(1), 17–23.

Haynes, R., Ireland, V., Duke, J. (2019). Funding Civil Society Organisations & Networks Promising Approaches to Financing Development in The 21st Centuryresearch: Report May 2019. https://forus-international.org/en/resources/71. Accessed 23 June 2021.

Hsieh, H.F., & Shannon, S.E. (2005). Three approaches to qualitative content analysis. *Qualitative Health Research*, *15*(9), 1277–1288.

Huntington, H.P. (2000). Using traditional ecological knowledge in science: methods and applications. *Ecological Applications*, *10*(5), 1270–1274.

Mahoney, C., & Beckstrand, M. (2009). *Following the money: EU funding of civil society organizations*. Paper presented at the APSA 2009 Toronto Meeting Paper.

Mills, J., Harrison, H., Franklin, R., Birks, M. (2017). *Case study research: Foundations and methodological orientations*. Paper presented at the Forum Qualitative Sozialforschung/Forum: Qualitative Social Research.

Namkoong, Y. (1999). Dependency theory: concepts, classifications, and criticisms. *International Area Review*, *2*(1), 121–150. https://doi.org/10.1177/223386599900200106.

Paffenholz, T. (2015). Civil society and peacebuilding. *Development Dialogue*, *63*, 108–118.

Pieterse, J. N. (2010). *Development Theory Deconstructions/Reconstructions* (2nd ed.). Los Angeles: Sage. https://doi.org/10.4135/9781446279083.

Roniger, L. (1994). The comparative study of clientelism and the changing nature of civil society in the contemporary world. In L. Roniger & A. G. Ayata (Eds.), *Democracy, clientelism, and civil society* (pp. 1–18). Boulder: Lynne Rienner Publishers.

Stearns, J., & Vogel, C. (2015). *The landscape of armed groups in the Eastern Congo*. http://congoresearchgroup.org/essay-the-landscape-of-armed-groups-in-eastern-congo-2/ Acessed 25 Aug. 2021.

The World Bank (2007). *Civil Society and Peacebuilding: Potential, Limitations and Critical Factors*. Washington DC: The International Bank for Reconstruction and Development / The World Bank.

Touraine, A. (1976). *Les sociétés dépendantes*. Louvain-la-Neuve (Belgique) : Duculot.

United Nations (2020). Peacebuilding Fund Report of the Secretary-Genera. https://www.un.org/peacebuilding/sites/www.un.org.peacebuilding/files/documents/sg_report_2019.pdf. Accessed 23 June 2021.

Van Houten, K. (2018). The in-between: Local civil society organizations and their community-based structures in South Kivu, the Democratic Republic of the Congo. *Journal of Civil Society*, *14*(2), 116–134.

Verkoren, W., & Van Leeuwen, M. (2013). Civil society in peacebuilding: Global discourse, local reality. *International Peacekeeping*, 20(2), 159–172. https://doi.org/10.1080/13533312.2013.791560.

Vlassenroot, K. (2013). South Kivu: identity, territory, and power in the eastern Congo. London: Rift Valley Institute.

A Manifesto with Proposals for Action Towards a Peace Ecology and an Ecological Peace Policy: A Global Research Partnership Between Organisations in Sub-Saharan Africa and Countries in the Global North

Introduction

Based on Chap. 2 on "Peace Ecology in the Anthropocene for Africa", this manifesto draws general conclusions with the goal of improving both the possibilities and prospects for an integrated analysis and assessment of both global environmental challenges and their impact on national and international peace and security in the social sciences and for strategies and action in the framework of an ecological peace policy.

Background

The countries in Sub-Saharan Africa face a difficult dilemma while

- they have been affected by multiple types of *postcolonial conflicts*,
- they already face severe *impacts of global environmental changeand climate change*,
- they have experienced a high number of *fatalities and economic damage from hydro-meteorological hazards and disastersand anthropogenicclimate change* as well as physical and societal impacts from epidemics including the Covid 19 pandemic caused by the virus SARS-CoV-2.

These countries lack:

- the *resources needed to build up the infrastructure* and the *capacities for adaptation, mitigation and resilience-building* with regard to climate change and conflict risks, and
- the *scientific capacities* to contribute, be accepted and have their research results published in peer-reviewed peace research, environmental studies and the

J. C. K. Kiyala and G. T. Harris (eds.), *Civil Society and Peacebuilding in Sub-Saharan Africa in the Anthropocene*, The Anthropocene: Politik—Economics—Society—Science 34,
https://doi.org/10.1007/978-3-030-95179-5

proposed peace ecology publications, because of several policies and measures such as:

- the present *major citation indexes* (e.g. the Web of Science and Scopus) only accept publications in English and have selected primarily research results from English-speaking countries;
- the insistence of *scientific funders in Organisation for Economic Co-operation and Development (OECD) countries on 'open access' publications* by their funding recipients. The national deals that have been agreed upon in the meantime further privilege scholars from those countries that can offer their researchers an easier access to a subsidised or free open access publishing.

Disadvantages for Researchers in the Global South: A Diagnosis

The preceding discussion shows elements that discriminate against researchers and scholars from the global South and especially from Sub-Saharan Africa. Therefore, this memorandum makes the following recommendations for developing a 'peace ecology approach' for Africa by African scholars and those who have specialised in both peace and security issue areas pertaining to Sub-Saharan Africa from a peace research perspective as well as those working on environmental studies, ecological concerns on sustainabilityissues, sustainable development goals and on problems of sustainability transition.

A simultaneous analysis from a combined peace research and ecological perspective matters for Sub-Saharan Africa, a continent that has suffered both from violent conflicts and from the physical and societal consequences of anthropogenic global environmentaland climate change impacts.

Many African universities and research centres have lacked both financial resources for a research infrastructure (including library resources) and access to relevant academic information resources in peer-reviewed academic journals and books. Most African scholars and researchers on peace and ecological issues cannot afford to attend those conferences in North America or in Europe where many peace researchers from the global North are meeting to present and discuss their research results.

Difficulty in Attending Scientific Conferences Outside Sub-Saharan Africa

Economic Constraints

Scholars from African universities and research centres and institutes often lack travel funding to attend international scientific conferences. In addition, some countries – not only in the global North – have been reluctant to grant visa for scholars from

Sub-Saharan Africa to participate in international conferences, even if they received travel grants with a return ticket to their countries.

Visa Constraints

Occasionally conference participants and speakers from sub-Saharan countries have difficulties in attending conferences in another country of the global South. Many scholars and young researchers have lacked this opportunity to share research results and get inspired by the research results of others. Science – and also the social sciences and peace and ecological studies – requires a free flow of ideas, information and research results.

Lack of Resources and Human Capacities

Lack of Resources

Many African institutions of higher education lack the financial resources to subscribe to scientific journals and to purchase books and other research resources. They also lack the equipment – especially specialised instruments in the natural sciences – both at the level of high or grammar schools, and in undergraduate and graduate studies. Only the least developed countries are granted free access to some scientific journals by scientific publishers in the global North.

Programmes exist by which major global publishers make scientific journals and books available to academic institutions in the global South. Among them are: *Research4Life*[1] and its specific programmes: *Hinari* (Research for Health), *AGORA*[2] (Research in Agriculture)[3]; *OARE* (Research in the Environment),[4] *ARDI*, *GOALI*[5] (Research for Global Justice).[6] No similar support programme exists for the Social Sciences and for peace, security and development programmes.

---

[1] See *Research4Life*, at: https://www.research4life.org/about/programs/ (19 April 2020).

[2] See *Hinari*, at: https://www.who.int/hinari/en/ (19 April 2020).

[3] See *AGORA*, at: http://www.fao.org/agora/en/ (19 April 2020).

[4] See *OARE*, at: https://www.unep.org/explore-topics/environment-under-review/what-we-do/information-management/online-access-research (19 April 2020).

[5] See *ARDI*, at: https://www.wipo.int/ardi/en/ (19 April 2020).

[6] See GOALI, at: http://goali.ilo.org/content/en/partners.php.

Lack of Human Capacities

Whilesome universities in Asia require their PhD candidates to get their highest academic degree in institutions in the global North, this is not the case for many universities and research centres in Sub-Saharan Africa as these countries cannot afford to send their gifted students abroad. Some may also fear that they will stay overseas and thus contribute to a brain-drain, and that these experts will be lacking for the development of their own country.

Opportunities

Frameworks and support mechanisms have been developed

on a *global level* in the framework of the United Nations system supported by

- the World Bankand International Monetary Fund (IMF);
- specific UN programmes (UNDP, UNEP);
- on an *interregional level*between the African Union (AU) and the European Union (EU): see Sect. 4.4.1 below:
- on a *bilateral level* between individual OECD countries (USA, Canada, Japan, UK, France, Germany, Sweden, Netherlands, Switzerland etc.) and individual African countries or a selected group of African countries: see 4.5 below for the initiatives launched by Germany;
- on a bilateral levelbetween China and …in the framework of *China'sOne Belt One Road Strategy.*

Existing Frameworks of Cooperation Between the European Union and African Union Countries in Sub-Sahara Africa

The*Association of African Universities* (AAU) collaborates with the *African Academy of Sciences* (AAS) to strengthen the continent's science granting councils under the *Science Granting Councils Initiative* (SGCI):

> The SGCI … will strengthen the capacities of Africa's science granting councils to play their essential roles effectively and efficiently in research, knowledge generation and innovation through training, technical support and peer-to-peer learning. It is a multi-funder project supported by the Swedish International Development Cooperation Agency, the United Kingdom's Department for International Development, Canada's International Development Research Centre, South Africa's National Research Foundation and the German Research Foundation (DFG).
>
> The AAU announced [its decision] … to change the situation in its 2020–2025 Strategic Plan, prioritising closer engagement with national policy systems in African countries, while the

> AAS Strategy emphasises strengthening of African institutions to become excellent and adaptable to proven needs. … Building on the success of the first phase of the SGCI, the Research Management Project is being implemented in 15 Sub-Saharan African countries, specifically, Ghana, Burkina Faso, Côte d'Ivoire, Senegal, Botswana, Malawi, Uganda, Kenya, Tanzania, Mozambique, Zimbabwe, Zambia, Ethiopia, Namibiaand Rwanda. The project will support the development of research and evidence-based policies that will contribute to economic and social development on the continent. …This may change over the coming years as national agendas evolve to reflect local priorities, increased Japanese and Chinese supported activity, and with climate change, ICT and energy gaining momentum.[7]

The European Union (EU) and the African Union (AU)

The *European Union* (EU) and the *African Union* (AU) have entered into an inter-regional cooperation in the framework of the *EU-AU High-Level Policy Dialogue on Science, Technology and Innovation* "with its investments reaching more than €1.8 billion. It will support African scientists and their capacity to contribute with cutting-edge research, reinforce their potential to benefit from EU research and innovation instruments, as well as enhance collaboration on equal footing with the European continent."[8] On 8 December 2020 the EU announced the launch of the *African Research Initiative for Scientific Excellence* (ARISE) Pilot Programme with a contribution of €25 million that will support:

> 40 young and emerging African research talents, early career scientists and their ideas, will have the opportunity to be recognised and offered further development opportunities with grants. Beneficiaries will be selected on the basis of scientific excellence in a continent-wide open competition. [After] a rigorous grant and award process for researchers with two to seven years of experience after obtaining their PhD, [t]he ARISE grantees will be awarded up to €500,000 in grants over a period of up to five years to start their own independent research teams and projects. The ARISE grantees and their research teams will be hosted in African higher education or research institutions in 40 African countries. …

The focus of these

> Collaborative R&I actions will be on the four priority areas of public health, *green transition*, innovation & technology, and capacities for science. The common thread of the collaborative activities is to speed-up the translation of scientific advances into tangible impact.

In line with the

> *2030 Agenda for Sustainable Development*, the role of science, technology and innovation is a vital driver of sustainability. The EU and the AU equally recognise that a *sustainable transition to knowledge-based economies needs to rely on science, research and innovation.* Investments in research and innovation ensure a sustainable and inclusive future. Specifically

[7] See Francis Kokutse: "Continental bodies collaborate to boost science granting councils", in: *University World News, Africa Edition* (25 August 2020); at: https://www.universityworldnews.com/post.php?story=20200825074812179 (26 July 2021); see: *Association of African Universities* (AAU); at: https://aau.org/.

[8] See at: https://ec.europa.eu/international-partnerships/news/arise-programme-open-applications_en.

> in times of the coronavirus pandemic, research and innovation are best placed to accelerate the green and digital transitions, strengthen resilience and crisis preparedness, and support global competitiveness.[9]

This modest initial funding of the European Union for young African scholars at leading African scientific institutions on green and sustainable transition seems to be limited to the natural sciences and engineering and excludes the social sciences and humanities. However, it could offer a framework for subsequent funding that includes combined research programmes of peace ecologyin peace studies and ecology.

Initiatives of the Federal German Government with African Partner Countries

In2015, the German Federal Ministry for Education and Research announced a programme for partnerships for sustainable solutions with Sub-Saharan Africa for measures for research and integrated postgraduate education.[10] This funding has focused on bio-economics, sustainable urbanisation and resource management. The cooperation involved: Angola, Benin, Burkina Faso, Côte d'Ivoire, Gambia, Ghana, Cameroon, Kenya, Namibia, Nigeria, Rwanda, Senegal, Somalia, South Africa, Tanzania, Togo, the Central African Republic and the regions of South, West and East Africa.

The *Maria Sibylla Merian Institute for Advanced Studies in Africa* (MIASA) that is located at the University of Ghana in Accra started the main phase in September 2020 with nearly 12 million Euros for the following six years (2020–2025). The goal of this international research institute is to support the internationalisation of the humanities and cultural and social sciences and to regularly invite fellows from African institutions. Main partners in Germany have been the Universities of Freiburg, Frankfurt (with its Centre on Interdisciplinary Research on Africa), the Max-Weber Foundation and the Institute of African Affairs of the German Institute of Global and Area Studies in Hamburg. Topical research themes are migration, democratisation and ecological transformation and also sustainable governance.

So far, the German Foundation on Peace Research (DSF) in Osnabrück has not been actively involved in these collaborative research activities with African institutions due to its mandate to fund only German research institutions.[11]

---

[9] See: "European Unionand African Union join forces to empower a new generation of African research talents", at: https://ec.europa.eu/commission/presscorner/detail/en/ip_20_2324 (19 April 2020).

[10] See: "Neuer Ansatz für Ausbau der Wissenschaftskooperation mit Afrika"; at: https://www.kooperation-international.de/aktuelles/nachrichten/detail/info/neuer-ansatz-fuer-ausbau-der-wissenschaftskooperation-mit-afrika/ (19 April 2020).

[11] See: "The Funding Concept of the DSF"; at: https://bundesstiftung-friedensforschung.de/wp-content/uploads/2021/01/Funding_Concept_EN.pdf (19 April 2021).

The existing networks of scientific publishers, in making scientific information accessible for medium- and low-income countries, and the new funding programmes of the European Commission and the German *Federal Ministry on Education and Science* (BMBF) are presently supporting research on sustainability transition that is an inherent part of the proposed peace ecology framework and research programme. The German BMBF has since 2018 launched several research and network projects with African partners, among them:

- On 4 March 2021 a preparatory phase for an African-German Research network on innovation in the health sector in Sub-Saharan Africa was announced
- On 21 April 2021 the BMBF announced guidelines on sustainable land management in Sub-Saharan Africa with the goal of enhancing food production with research areas on digitalisation (of farming) and strengthening of governance for sustainable development
- On 22 July 2021: *Construction of a Climate Competence Centre on Climate Changeand Adapted Land Use* in Quagadougou (Burkina Faso) with training of 400 young African students at 12 graduate schools and support for a Master's programme on green hydrogen.

Other Bilateral Funding Opportunities by OECD Member Governments in Support of African Research Institutions

Other national funding agencies, such as U.S. AID, the United Kingdom's Department for International Development (DFID), the French Development Agency (AFD), the Japanese, Canadian, Dutch (Institute for Social Sciences [ISS] in The Hague), Swedish, Norwegian and Finnish Development Ministries have in the past funded specific collaborative research programmes with African countries.

Research Partnerships by Peace Research Centres in the Global North

The*Stockholm International Peace ResearchInstitute* (SIPRI)[12] maintains a research programme on peace and development that addresses: (a) *Climate changeand risk* with a special emphasis on: (i) *Climate change and security*, (ii) *Climate change and food security* and *Climate-related peace and securityrisks*; (b) *Environment of peace*; (c) *Governance and society on sustainable peace.*

> SIPRI looks at the long-term causes of insecurity to understand how societies identify and navigate paths to sustainable peace.

[12] See SIPRI; at: https://www.sipri.org/ (20 April 2021).

> Developing and sustaining peace requires an understanding of the root causes of conflict and insecurity. SIPRI looks at what fuels conflict and what drives long-term, positive change by analysing economic, social, political and environmental factors. We contribute to both research and dialogue in order to inform policy and practice to forge a path to peace.
>
> Our work in this field is broad, with topics ranging from corruption in the security sector, inclusive peacebuilding to climate change. Conflicts are rarely caused by one single factor; often several issues play a role to reinforce and exacerbate each other. Our analysis is also multidisciplinary and applies a variety of social sciences and methods. SIPRI's research reflects the complexity of conflict drivers and the relationships between them, as well as how they differ across contexts.
>
> A crucial part of SIPRI's work on peace and development is *the Stockholm Forum on Peace and Development*. Held every spring, the Stockholm Forum brings together senior researchers, policymakers and practitioners to discuss and address the most pressing peacebuilding issues.[13]

However, thus far no specific cooperative training and research programme exists to support the infrastructure and regional networking of African scholars with colleagues in the global North in the areas of peace research(conflict resolution, peacebuilding, post-conflict reconstruction), and on peace ecology (linking e.g. climate change adaptation and conflict avoidance, sustainable development and transition strategies towards sustainable development) that could train a generation of African students on coping capacities and resilience activities to mitigate the impact of climate change, such as extreme weather events, recurring epidemics and pandemics.

Initiatives by the Global Peace Research Community Towards a Peace Ecology

The global peaceresearch community is fragmented and underfinanced. The peace scholars from the global North usually meet within the *Peace StudiesSection* (PEACE) during the annual conferences of the *International Studies Association* (ISA) in North America which junior scholars studying and working at institutions in the global South find it impossible to attend due to a lack of financial resources and scholarships and difficulties in obtaining entry visas. The oldest global peace research organisation IPRA (founded in 1964) has met since 2014 only in the global South: *2014: Istanbul(Turkey*; Brauch/Oswald Spring/Bennett/Serrano Oswald 2016; Oswald Spring/Brauch/ Serrano Oswald/Bennett 2016); *2016: Freetown* (Sierra Leone; Brauch/Oswald Spring/Collins/ Serrano Oswald 2018); *2018: Ahmedabad*(India; Oswald Spring/Brauch 2021) and in *2021 in Nairobi* (Kenya) with only few scholars from the global North attending.

Working on research issues of an emerging peace ecology requires knowledge-sharing, joint networking and conferences of scholars working at institutions in

[13] See the SIPRI Research on 'peace and development'; at: https://www.sipri.org/research/peace-and-development (20 April 2021).

the global North and global South and assisting scholars from the global South in publishing in global peer-reviewed journals and in peer-reviewed books with major scientific publishers. Moving from knowledge to action necessitates the joint development of ecological peace policies and setting the agenda of policy programmes and activities to cope with the impact of climate change, and enhance adaptation, mitigation, resilience and ecological peacebuilding.[14]

For Africa, Wikipedia lists the following ecological peacebuilding initiatives:

> In Southern Africa, several trans-boundary peace parks have been established with the support of the Peace Parks Foundation since the 1990s, such as the *Great Limpopo Transfrontier Park* (Mozambique, South Africa, Zimbabwe), the *Kgalagadi Transfrontier Park* (Botswana, South Africa), the *Maloti-Drakensberg Transfrontier Conservation and Development Area* (Lesotho, South Africa) and the *Lubombo Conservancy* (Eswatini, Mozambique, South Africa).[15] Assessments of these projects often highlight several negative aspects, such as the dominance of South Africa, the exclusion of local populations vis-a-vis state and business interests, and a low impact on peaceful international relations.[16] Conservation cooperation between the DR Congo, Rwandaand Uganda in the Virunga region, by contrast, is reported to have yielded some conservation and peace benefits even though it took place in an environment characterized by political instability and low economic development.[17] There are also bottom-up dynamics of environmental cooperation amidst conflict between pastoralists and farmers in several African countries, including Ghana[18]and Kenya.[19]

[14] Tobias Ide, Carl Bruch, Alexander Carius, Ken Conca, Geoffrey D. Dabelko, Richard Matthew, Erika Weinthal, 2021: "The past and future(s) of environmental peacebuilding", in: *International Affairs*, 97,1 (January): 1–16, https://doi.org/10.1093/ia/iiaa177 (4 August 2021); open access: https://academic.oup.com/iaarticle/97/1/1/6041492; "Environmental peacebuilding", at: https://en.wikipedia.org/wiki/Environmental_peacebuilding.

[15] *Peace Parks Foundation, 2019:* "Peace Parks Foundation" (19 February).

[16] *Ramutsindela, Maano, (January 2017): "Greening Africa's borderlands: The symbiotic politics of land and borders in peace parks", in: Political Geography, 56 (January): 106–113;* 10.1016/j.polgeo.2016.11.012.

[17] *Martin, Adrian; Rutagarama, Eugene; Cascão, Ana; Gray, Maryke; Chhotray, Vasudha, 2011:* "Understanding the co-existence of conflict and cooperation: Transboundary ecosystem management in the Virunga Massif", *in: Journal of Peace Research, 48,5 (september): 621–635,* 10.1177/0022343311412410. ISSN 0022-3433.

[18] *Bukari, Kaderi Noagah; Sow, Papa; Scheffran, Jürgen, 2018:* "Cooperation and Co-Existence Between Farmers and Herders in the Midst of Violent Farmer-Herder Conflicts in Ghana", *in: African Studies Review, 61,2 (July): 78–102;* 10.1017/asr.2017.124. ISSN0002-0206.

[19] *Adano, Wario R; Dietz, Ton; Witsenburg, Karen; Zaal, Fred, 2012: "Climate change, violent conflictand local institutions in Kenya's drylands", in: Journal of Peace Research, 49,1 (January): 65–80;* 10.1177/0022343311427344. hdl:1887/18493. ISSN 0022-3433.

Proposals for International Research Cooperation on Peace Ecology Issues

Knowledge Sharing and Scientific Networking

The development of a peace ecology in the Anthropocene for Africa requires overcoming the constraints and impediments to a global and intercontinental knowledge sharing, and discussion and debate of research trends, especially among peace researchers, environmental specialists and ecologists from both the global North and the global South. For many decades, the *International Peace Research Association* (IPRA) used to be the forum for this controversial debate of different theories and empirical results (Brauch 2021).

Increased Scientific Cooperation of Universities and Research Centres

An increased scientific cooperation and debate between scholars and researchers in the global South and in OECD countries, especially on case studiesin Sub-Saharan Africa, is highly advisable for global peace and security, and in meeting the economic and environmental challenges facing Earth in the Anthropocene. This requires a longer-term vision as well as funding by both international institutions (IGOs, INGOs) and globally-oriented foundations).

Student Exchange from the Global North with Universities in Sub-Saharan Africa

In several countries, national ministries and private foundations have supported the exchange of students: for example, among EU countries through the Erasmus programme that focuses primarily on its 27 member countries but has also permitted specific programmes with selected countries in the global South. The most recent ERASMUS programme (2021 to 2027) offers a framework for funding opportunities for the support of exchanges among students and trainees which universities in the 27 EU countries can apply for.[20] Besides the EU, the US and many OECD countries offers scholarships for students and young professionals in the global South:

---

[20] See: https://eucalls.net/blog/new-erasmus (31 July 2021): "First of all, the new Erasmus+ programme 2021 will have more resources to support education, training, youth and sport. The Erasmus 2021–27 budget is estimated at €26.2 billion, almost double (€14.7 billion) the budget compared with the programme's previous finances for 2014-2020. Out of this budget, it is expected

- Erasmus Mundus Scholarships: European Scholarships for Developing Countries 2021/22;
- List of Scholarships for International Students 2021–2022[21];
- 2021 Grants for NGOs and Organisations: Funding Grants for NGOs in Developing Countries.[22]

Internships with International Organisations in Sub-Saharan Africa

Many universities in the global North require or encourage their Bachelor and Master's students to engage in internships with international organisations, development agencies or NGOs in the global South:

- List of Paid Internships in Africa for International Students[23];
- Stockholm Environment Institute (SEI) Africa: Research internship opportunities[24];
- US Aid: Bureau for Africa Internship.[25]

Stipends for Junior Scholars

In selected countries, national academic exchange agencies and universities in the global North offer stipends, grants and fellowships for PhD candidates and recent Post Docs in high-quality institutions which are working on peace researchand environmental studies, and ecological concerns – the two components of any peace ecology for guest scholars from countries in the global South, particularly for the countries in Sub-Saharan Africa.

Fellowships for Post Docs from and in Africa

Some academic institutions offer fellowships for Post Doctoral Fellows from all parts of the world, including countries in Sub-Sahara Africa, among others:

---

that 70% will be dedicated to supporting mobility, and the remaining 30% will go towards cooperation projects and policy development activities. Finally, it is estimated that approximately 10 million Europeans of all ages and backgrounds will be benefited from Erasmus+ 2021–2027.

[21] See at: https://www.advance-africa.com/List-of-Scholarships-for-International-Students.html.

[22] See at: https://www.advance-africa.com/Grants-for-NGOs-and-Organisations.html (31 July 2021).

[23] See at: https://ugfacts.net/list-paid-internships-africa-international-students/

[24] See at: https://www.sei.org/people/jobs/internships-africa/ (31 July 2021).

[25] See at: https://www.usaid.gov/work-usaid/careers/student-internships/bureau-africa-internship.

- German Academic Exchange Service[26];
- AESA-RISE Postdoctoral Fellowship Programme to support the training of postdoctoral researchers in Africa[27];
- Opportunities for Africans[28];
- African Research Universities Alliance (ARUA) Postdoctoral fellowships on Mobility and Sociality in Africa's emerging Urban.[29]

Support for Cooperative Research Projects

A few Research and Training Institutions in Europe and North America have launched joint research projects with junior and senior researchers in the global South and in Sub-Saharan Africa; among them are:

- Volkswagen Stiftung; Knowledge for tomorrow[30];
- German Research Foundation (DFG): Working Together with Africa and the Middle East[31];
- Research Africa Network (RAN).[32]

An Open Access Partnership between the EU and the AU

Theexisting open-access programmes have so far excluded publications by scholars from the global South in the social sciences and humanities. These programmes should be extended. Publications in open-access journals in the global North should be facilitated for scholars from the global South from a special fund that covers their open access fees. The institutions with whom they collaborated during the research project should cover their open access fee to avoid discrimination against scholars from the global South whose institutions are unable to pay the high open access fees.

---

[26] See: https://www2.daad.de/deutschland/stipendium/datenbank/en/21148-scholarship-database/?detail= 57191387 (31 July 20221).

[27] See: https://www.aasciences.africa/aesa/programmes/aesa-rise-postdoctoral-fellowship-programme.

[28] See at: https://www.opportunitiesforafricans.com/ (31 July 20221).

[29] See at: African Research Universities Alliance (ARUA) Postdoctoral fellowships on Mobility and sociality in Africa's emerging urban (31 July 2021).

[30] See at: https://www.volkswagenstiftung.de/en/funding/our-funding-portfolio-at-a-glance/knowledge-for-tomorrow-%E2%80%93-cooperative-research-projects-in-sub-saharan-africa (31 July 2021).

[31] See at: https://www.dfg.de/en/dfg_profile/international_cooperation/international_activities/africa_middle_east/index.html (31 July 2021).

[32] See at: http://resanet.org/ (31 July 2021).

Challenges

Lack of a Systematic Global Research Survey on Peace Research

Duringthe Cold WarUNESCO regularly monitored the institutional arrangements in global peace studies with the publication of a "World Directory of Peace Research and Training Institutions" (1988, 1991, 1995) and the provision scholarships for junior scholars to attend IPRA conferences and workshops where both established scholarsand innovative thinkers from North and South met and discussed common global research themes. Since the early 21st century it was increasingly the case that promising young peace researchers met annually only at ISA in North America, while young scholars from the Global South met at conferences in the Global South in the IPRA framework supported by few travel grants – primarily supported by the IPRA Foundation in the USA and only very few partial grants by ISA.

Policy Challenges: Addressing the Challenges via a Peace Ecology Approach and Aiming at an Ecological Peace Policy in Sub-Saharan Africa

Given the multiple environmental challenges and threats to the peace, security and economic policies of many Sub-Saharan countries during the 21st century, Chap. 2 in this volume has suggested an integrated peace ecology approach and strategies aiming at an ecological peace policy for the continent. This requires an institutional strengthening of the research infrastructure and for African peace, security, development and environmental capabilities to be able to address, face and cope with these challenges and thus to avoid negative impacts on peace and security.

For the African peace research community, this requires improved access to digitalised scientific publications (to eBooks and to open access journals) and the of African scholars to publish their results in open access publications, for which high fees usually apply.

Conclusion

If OECD and AU member countries actively pursue the goal of preventing violent conflicts being triggered by issues of global environmental changeand climate change in the Anthropocene (see several German initiatives during its membership in the UN

Security Council)[33] there is a need to develop an integrated assessment and understanding of multiple environmental and climate-related threats for local, regional and continental crises, conflicts and wars (Brauch 2002, 2009, 2011, 2012, 2014). These countries should invest in the knowledge base which analyses these linkages in comparative case studiesfor Sub-Saharan Africa within a peace ecology approach that integrates both knowledge provided by peace research institutions on violent conflicts, and on ecological conflictprevention and peace building by environmental specialists and political and social ecologists.

The peace research community in the global North should be encouraged to initiate longer-term cooperative research projects and to actively network with social scientists in the global South, most particularly from Sub-Saharan Africa, and to co-publish their joint results in open-access publications supported by the national open-access agreements in OECD countries.

References

Brauch, Hans Günter, 2002: "Climate Change, Environmental Stress and Conflict: AFES-PRESS Report for the Federal Ministry for the Environment, Nature Conservation and Nuclear Safety", in: Federal Ministry for the Environment, Nature Conservation and Nuclear Safety (Ed.): *Climate Change and Conflict: Can Climate Change Impacts Increase Conflict Potentials? What is the Relevance of this Issue for the International Process on Climate Change?* (Berlin: Federal Ministry for the Environment, Nature Conservation and Nuclear Safety, 2002): 9–112; at: http://www.afes-press.de/pdf/Brauch_ClimateChange_BMU.pdf.

Brauch, Hans Günter, 2009: "Securitizing Global Environmental Change", in: Brauch, Hans Günter; Oswald Spring, Ursula; Grin, John; Mesjasz, Czeslaw; Kameri-Mbote, Patricia; Behera, Navnita Chadha; Chourou, Béchir; Krummenacher, Heinz (Eds.), 2009: *Facing Global Environmental Change: Environmental, Human, Energy, Food, Health and Water Security Concepts* (Berlin – Heidelberg – New York: Springer-Verlag): 65–102.

Brauch, Hans Günter, 2011: "Global Climate Change Impacts for the Mediterranean in the 21st Century: Challenges for Human and Environmental Security", in: Hans Günter Brauch, Úrsula Oswald Spring, Czeslaw Mesjasz, John Grin, Patricia Kameri-Mbote, Béchir Chourou, Pal Dunay, Jörn Birkmann (Eds.): *Coping with Global Environmental Change, Disasters and Security – Threats, Challenges, Vulnerabilities and Risks* (Berlin – Heidelberg – New York: Springer-Verlag, 2011): 485–524.

Brauch, Hans Günter, 2012: "Environmental Security", in: Juergensmeyer, Mark; Anheier Helmut K. (Eds.): *Encyclopedia of Global Studies* (London: Sage, 2012).

Brauch, Hans Günter, 2014: "From Climate Change and Security Impacts to Sustainability Transition: Two Policy Debates and Scientific Discourses", in: Oswald Spring, Úrsula; Brauch, Hans Günter; Tidball, Keith G. (Eds): *Expanding Peace Ecology: Peace, Security, Sustainability, Equity and Gender* (Cham: Springer International Publishing): 33–62.

Brauch, Hans Günter, 2021: "Peace Ecology in the Anthropocene", in: Oswald Spring, Úrsula; Brauch, Hans Günter (Eds.): *Decolonising Conflicts, Security, Peace, Gender, Environment and Development in the Anthropocene* (Cham: Springer Nature Switzerland).

Brauch, Hans Günter; Oswald Spring, Úrsula; Bennett, Juliet; Serrano Oswald, Serena Eréndira (Eds.),2016: *Addressing Global Environmental Challenges from a Peace Ecology Perspective* (Cham – Heidelberg – New York – Dordrecht – London: Springer International Publishers).

[33] See Brauch/Scheffran 2012; APESS vol. 33: Hardt/Harrington/Simpson/von Lucke/Estève, 2022.

Brauch, Hans Günter; Oswald Spring, Ursula; Collins, Andrew E.; Serrano Oswald, Serena Eréndira (Eds.), 2019: *Climate Change, Disasters, Sustainability Transition and Peace in the Anthropocene* (Cham: Springer International Publishing).

Brauch, Hans Günter; Oswald Spring, Úrsula; Grin, John; Scheffran; Jürgen (Eds.), 2016: *Handbook on Sustainability Transition and Sustainable Peace* (Cham – Heidelberg – New York – Dordrecht – London: Springer): 187–236.

Brauch, Hans Günter; Scheffran, Jürgen, 2012: "Introduction", in: Scheffran, Jürgen; Brzoska, Michael; Brauch, Hans Günter; Link, Peter Michael; Schilling, Janpeter (Eds.): *Climate Change, Human Security and Violent Conflict: Challenges for Societal Stability* (Berlin – Heidelberg – New York: Springer-Verlag).

Hardt, Judith Nora; Harrington, Cameron; Simpson, Nick; von Lucke, Franziskus; Estève, Adrien, 2022: *Climate Security in the Anthropocene: Exploring the Approaches of United Nations Security Council Member-States* (Cham: Springer Nature Switzerland).

Oswald Spring, Úrsula; Brauch, Hans Günter (Eds.), 2021: *Decolonising Conflicts, Security, Peace, Gender, Environment and Development in the Anthropocene* (Cham: Springer Nature Switzerland).

Oswald Spring, Úrsula; Brauch, Hans Günter; Serrano Oswald, S.E.; Bennett, Juliet (Eds.), 2016: *Regional Ecological Challenges for Peace in Africa, the Middle East, Latin America and Asia Pacific Perspective* (Cham: Springer International Publishers).

UNESCO, 1988: *World Directory of Peace Research and Training Institutions* (Paris – Oxford: UNESCO – Berg).

UNESCO, 1991: *World Directory of Peace Research and Training Institutions* (Paris: UNESCO).

UNESCO, 1995: *World Directory of Peace Research and Training Institutions* (Paris: UNESCO).

PD Dr. Hans Günter Brauch chairman of AFES-PRESS; chairman of the board of the Hans Günter Brauch Foundation on Peace and Ecology in the Anthropocene (HGBS), and editor of this Anthropocene (APESS) book series; Email: brauch@onlinehome.de. The author is grateful for comments and suggestions by Dr. Jean Chrysostome K. Kiyala, a coeditor of this book, and to Ms. Elizabeth Norton.Amor for language-editing.

On Durban University of Technology

DUT

DURBAN UNIVERSITY OF TECHNOLOGY
INYUVESI YASETHEKWINI YEZOBUCHWEPHESHE

With approximately 23 000 students, the Durban University of Technology (DUT) is the first choice for higher educationin KwaZulu-Natal (KZN). It is located in the beautiful cities of Durban and Pietermaritzburg (PMB). As a University of Technology, it prioritises the quality of teaching and learning by ensuring its academic staff possess the highest possible qualification that they can get.

The Durban University of Technology is a result of the merger in April 2002 of two prestigious technikons, ML Sultan and Technikon Natal. It was named the Durban Institute of Technology and later became the Durban University of Technology in line with the rest of the universities of technology.

DUT, a member of the International Association of Universities, is a multi-campus university of technology at the cutting edge of higher education, technological training and research. The university aspires to be a "preferred university for developing leadership in technology and productive citizenship", and to "making knowledge useful".

As a butterfly develops from a pupa, so have the students at our institution. From the moment they register as green freshers, to their capping at the hallowed graduation ceremony, our students undergo an intellectual evolution. Website: http://www.dut.ac.za/.

J. C. K. Kiyala and G. T. Harris (eds.), *Civil Society and Peacebuilding in Sub-Saharan Africa in the Anthropocene*, The Anthropocene: Politik—Economics—Society—Science 34,
https://doi.org/10.1007/978-3-030-95179-5

On International Centre of Nonviolence (ICON)

ICON is one of DUT's ten focus areas. Its mission is to make strategic interventions in education that challenge structural violence and help develop a culture of nonviolence.

ICON seeks to make nonviolence real through analysis of the local and global context, through research and through reflective practice. These inform its actions and strengthen its capacity in the field of nonviolence.

Accordingly, ICON is active in the teaching nonviolence to undergraduates as part of in DUT's general education programme, using highly participatory learning methods. In addition, it has a large postgraduate programme in peacebuilding, with some 60 PhD and 15 master‘s students as at early 2017, drawn from many African countries. All postgraduate students follow an action research approach in their thesis work.

Website: http://www.icon.org.za/current.html.

J. C. K. Kiyala and G. T. Harris (eds.), *Civil Society and Peacebuilding in Sub-Saharan Africa in the Anthropocene*, The Anthropocene: Politik—Economics—Society—Science 34,
https://doi.org/10.1007/978-3-030-95179-5

On University of Bandundu

UNIBAND

DUC IN ALTUM

The **University of Bandundu (UNIBAND)** is a public university in the Democratic Republic of the Congo (DRC), located in Bandundu, the capital city of the province of Kwilu. At its inception, UNIBAND—then called *Centre Universitaire the Bandundu* (C.UB.) [Bandundu University Centre]—created on 1 October 2004, was an appendage of the University of Kinshasa (UNIKIN). By 2012, UNIBAND had 800 students spread through its six faculties. It became autonomous in 2010 following Ministerial Decree No. 157/MINESU/CABMIN/ EBK/PK/2010 27 September 2010. At present, UNIBAND hosts six faculties: Law, Medicine, Psychology and Sciences of Education, Economics and Agronomy. Based on Law No. 86/025 of 3 October 1981 and the Framework Law No. 89/005 of 22 September 1986 on the general organisation of higher learning institution and universities, which sets the objectives of universities and higher learning institutions in the DRC, UNIBAND, like

J. C. K. Kiyala and G. T. Harris (eds.), *Civil Society and Peacebuilding in Sub-Saharan Africa in the Anthropocene*, The Anthropocene: Politik—Economics—Society—Science 34,
https://doi.org/10.1007/978-3-030-95179-5

any other university, has the following main mission: to ensure the training of intellectuals, researchers, academics, and managers in the diverse domains of life. As such, it provides training and lectures in various fields that encourage the emergence of new ideas and the development of professional skills: to organise basic and applied scientific research aimed at meeting the many challenges of social, scientific, economic, and political life. Because of its integrated academic scheme, UNIBAND's students enjoy ongoing formation provided by a qualified staff who pursues the goals of building a harmonious and integral development of the human person. UNIBAND's academic programmes bridge traditional training and innovative learning which promote research and community engagement, and train leaders capable of meeting the demanding criteria required to make society a better place to live in.

On the Editors

Jean Chrysostome K. Kiyala is a Senior Lecturer at the International Centre of Violence, Durban University of Technology, South Africa; Associate Professor and visiting lecturer at the University of Bandundu, Catholic University of Bukavu, Evangelical University in Africa (DR Congo). He holds Diplomas and BAs in Philosophy, Theology and Mission Studies from Tangaza College and the Pontifical Urbaniana University, a BA Honours in Religious Studies from the Catholic University of Eastern Africa, and an M Phil (2010) from the St Augustine College of South Africa. He was conferred M Tech in Public Management by DUT in 2012 and was awarded a Doctoral Degree in Philosophy, specializing in Public Administration, in 2016 from the same institution.

Chrys is the founder of the A Hipfunaneni [Let us help each other] Community Project that provided psychosocial assistance to Zimbabwean and Mozambican refugees in Limpopo Province (South Africa); the Youth Advocacy and Peacebuilding Organisation, an NGO committed to training young people in nonviolent approaches to dispute resolution in KwaZulu-Natal Province (South Africa); and the Youth Fraternity for Peacebuilding and Reconciliation, based in North Kivu in the Democratic Republic of Congo, which aims to accompany former child soldiers and train them to become peacebuilders. He has specialized in the child-soldiering phenomenon; restorative justice, post-conflictreconciliation and nation-building; civil society and peacebuilding; and is developing an interest in peace ecology and ecojustice. His publications include:

Responding to Epistemic Injustice against Child Soldiers. *The International Journal of Children's Rights, 2021*, 29(3), 662–700. 10.1163/15718182-29030006 Kiyala, J. C. K. (2020). Dynamics of Child soldiers' Psychosocial Rehabilitation and Well-being: Perspectives from Bioecological Systems Theory in

J. C. K. Kiyala and G. T. Harris (eds.), *Civil Society and Peacebuilding in Sub-Saharan Africa in the Anthropocene*, The Anthropocene: Politik—Economics—Society—Science 34,
https://doi.org/10.1007/978-3-030-95179-5

the Democratic Republic of Congo. *Child Psychiatry & Human Development*, 52(3), 376–388. https://doi.org/10.1007/s10578-020-01022-4. The impact of globalisation on child soldiering: challenges and opportunities in the context of the Democratic Republic of Congo. *International Social Science Journal*, 66(221–222), 271–284. 10.1111/issj.12132. *Child Soldiers and Restorative Justice : Participatory Action Researchin the Eastern Democratic Republic of Congo*. Cham ZG: Springer International Publishing. https://doi.org/10.1007/978-3-319-900 71-1. Combining social justice with transitional and restorative justice: an agenda for an integrative justice model in transitional societies. In J. Evans (Ed.), *Restorative justice and Transitional Justice: perspectives, progress and considerations for the future* (pp. 82–177). New York: Nova Publishers. Reintegrating former child soldiers in the eastern Democratic Republic of Congo. In S. Kaye & G. Harris (Eds.), *BuildingPeace via Action Research: African Case Studies* (pp. 223–236). Addis Ababa: United Nations University for Peace (UPEACE). Kiyala, J. C. (2016). Utilising a traditional approach to restorative justice in the reintegration of former child soldiers in the North Kivu Province, Democratic Republic of Congo. *Africa Insight*, 46(3), 33–50. Challenges of reintegrating self-demobilised child soldiers in North Kivu Province: prospects for accountability and reconciliationvia restorative justicepeacemakingcircles. *Human RightsReview*, 16(2), 99–122. 10.1007/s12142-015-0361-7 Kiyala, J. C. K., & Afagbegee, G. L. (2008). Amos Programme: Bible sharing and social issues. In M. A. Adekambi (Ed.), *Bible and Social Issues in Africa: Acts of the 6th Continental Workshop of Biblical Apostolate Coordinators in Africa, Dar Es Salaam, 30th July–3rd August 2007* (pp. 87-102). Accra: Bicam Publications/Les Publications du Cebam.

Address: Dr. Jean Chrysostome K. Kiyala, International Centre of Nonviolence (ICON), 292 Varsity Drive, Reservoir Hills, Durban, P.O. Box 65075, Reservoir Hills 4091, Durban KwaZulu-Natal, South Africa;

Email: kljeanchrysostome@gmail.com and JeanK@dut.ac.za.

Website: https://www.linkedin.com/in/kiyala-chrysostome-82268a9b/ and http://afes-press-books.de/html/APESS_34.htm and http://afes-press-books.de/html/SAB_01.htm; URL: www.theophilsapiens.co.za; Researchgate: https://www.researchgate.net/profile/Jean_Kiyala ORCID: https://orcid.org/0000-0002-9587-0230.

Geoff Thomas Harris (South Africa, Australia) is a Professor in the International Centre for Nonviolence, Durban University of Technology, where he set up its postgraduate Peacebuilding Programme, which had in 2021 some 70 doctoral and 20 master's students. His current research interests include restorative justice, the reintegration of prisoners and demilitarisation. He was appointed to a chair in economics at the then University of Natal, South Africa, in 1999. Prior to this he held posts at La Trobe University, Australia (1969–71), the University of Papua New Guinea (1972–78), Lincoln University, New Zealand (1978–80), the University of New England, Australia (1980–99) and the University of KwaZulu-Natal, South Africa (1999–2011). He edited *Recovery from armed conflictin developing countries: an economic and political analysis* (London: Routledge, 1999), *Achieving security in sub-Saharan Africa: cost effective alternatives to the military* (Pretoria: Institute for Security Studies, 2004) and co-edited, with Sylvia Kaye, *Building peace via action research. African case studies* (Addis Ababa: University for Peace, 2017) and, with Mediel Hove: *Infrastructuresfor Peace in Sub-Saharan Africa* (Cham: Springer International Publishing, 2019).

Address: Prof. Geoff Harris, International Centre of Nonviolence (ICON), Durban University of Technology, South Africa.

Email: geoffreyh@dut.ac.za

On the Contributors

Each contribution underwent a thorough peer-review process, initially by the editors and subsequently by two anonymous reviewers.

Ashton Murwira holds a PhD in Peace Studies from the Durban University of Technology. He is currently a lecturer in the Department of Political and Administrative Studies at the University of Zimbabwe. His research interests are in peace and security, governance and foreign policy issues. Email: murwirashie@gmail.com.

Denis Mukengere Mukwege is a world-renowned gynaecologist, human rights activist, and Nobel Peace Prize laureate from Eastern Congo. He has become a world leader in the treatment of the victims of wartime sexual violence. He is also an advocate of human rights, and a campaigner against sexual violence and the use of rape as a "weapon of war". Email: denismukwege@hotmail.com.

Dorothy Moyo is a PhD student in Peace Studiesat Durban University of Technology. She holds a BSc Sociology degree and an MSc in Peace Leadership and Conflict Resolution. Dorothy is a Commissioner and Deputy Chairperson in the Zimbabwe Human Rights Commission. She is also a Facilitator and Trustee at the Alternatives to Violence Project Zimbabwe (AVPZ). Email: dorothyhopemoyo@gmail.com.

Dumisani Maqeda Ngwenya is a peacebuilding practitioner and academic who has been working in the field of conflict transformation and trauma healing since 2003. He holds a Masters in Peace and Conflict studies (University of KwaZulu-Natal) and a PhD in Peace Studies from the Durban University of Technology. He is a co-founder and executive director of Grace to Heal, a faith-based peacebuilding and community healing organisation, working in Matebeleland, Zimbabwe. He authored *Healing the Wounds of Gukurahundi in Zimbabwe: a Participatory Action ResearchProject*, published by Springer in 2018. Email: dumiemngwenya@gmail.com.

Geoff Thomas Harris (South Africa, Australia) is a Professor in the International Centre for Nonviolence, Durban University of Technology, where he set up its post-graduate Peacebuilding Programme, which had in 2021 some 70 doctoral and 20

J. C. K. Kiyala and G. T. Harris (eds.), *Civil Society and Peacebuilding in Sub-Saharan Africa in the Anthropocene*, The Anthropocene: Politik—Economics—Society—Science 34,
https://doi.org/10.1007/978-3-030-95179-5

master's students. His current research interests include restorative justice, the reintegration of prisoners and demilitarisation. He was appointed to a chair in Economics at the then University of Natal, South Africa, in 1999. Prior to this he held posts at La Trobe University, Australia (1969–71), the University of Papua New Guinea (1972–78), Lincoln University, New Zealand (1978–80), the University of New England, Australia (1980–99) and the University of KwaZulu-Natal, South Africa (1999–2011). Email: geoffreyh@dut.ac.za.

Georges Bomino Bosakaibo is a Catholic priest born in the Democratic Republic of Congo in 1972. He joined the Catholic Divine Word Missionary congregation (SVD) in 1995. After a BA in Philosophy jointly from the University of Saint Augustine in Kinshasa and the Pontifical Urbaniana university in Rome in 2000, he obtained Master's and PhD degrees in International Development. He is presently an assistant professor in the Department of Policy Studies, Faculty of Policy Studies, Nanzan University, Nagoya, Japan. Email: gbbilbos@gmail.com.

Gérard Mpang'de Bisambu is a Congolese trained in Philosophy and Anthropology (at the Catholic University of Congo and National University of Kinshasa). He holds a Masters in Electoral Administration from the Electoral Training School of Central Africa (EFEAC). He has some 28 years of experience in civil society, with emphasis on the themes of democracy and good governance. He was a member of the mediation team at the Political Dialogue facilitated by the Catholic Bishops of the Democratic Republic of Congo in 2016. Currently, he serves as the Secretary General of the Non-Governmental Organization (NGO) *Agir pour les Elections Transparentes et Appaisées* [Act for Transparent and Peaceful Elections] (AETA). Email: philobis1991@gmail.com.

Hans Günter Brauch (Germany), Dr, was until 2012 Adjunct Professor at the Free University of Berlin. Since 1987 he has been chairman of Peace Research and European Security Studies (AFES-PRESS). In 2020 he set up the Hans Günter Brauch Foundation on Peace and Ecology in the Anthropocene (HGBS), of which he is the board chairman. He is editor of the Hexagon book series on Human and Environmental Security and Peace; Springer Briefs on Environment, Security, Development and Peace (ESDP); Springer Briefs on Pioneers in Science and Practice and Pioneers in Arts, Humanities, Science, Engineeringand Practice; and of The Anthropocene: Politik – Economics – Society – Science. He was guest Professor of International Relations at the universities of Frankfurt am Main, Leipzig, Greifswald, and Erfurt, a research associate at Heidelberg and Stuttgart, and a research fellow at Harvard and Stanford Universities. In 2013/2014 he was guest professor at Chulalongkorn University in Bangkok. Address: PD Dr Hans Günter Brauch, Alte Bergsteige 47, 74821 Mosbach, Germany. Email: brauch@afes-press.de.

Jean Chrysostome K. Kiyala is a Senior Lecturer, International Centre of Violence, Durban University of Technology, South Africa; Associate Professor and visiting lecturer the University of Bandundu, Catholic University of Bukavu and Evangelical University in Africa based in Bukavu (DR Congo), visiting professor at the Catholic University of Bukavu (CUB), the Evangelical University in Africa (UEA) and the Catholic University of Goma-La Sapientia in the DRC. He has a PhD in Management Sciences: Peace Studies from the Durban University of Technology and is developing a research programme in peace ecologyand environmental justice. Email: kljeanchrysostome@gmail.com.

Josephine Kimanu Mauwa is a PhD student in Peace Studies at the Durban University of Technology. She holds a Master's degree in Peace and Development and works as a junior lecturer in the Department of Peace and Conflict Transformation in the Faculty of Social Sciences at the Evangelical University in Africa. Her field of studies is gender- and peacebuilding-oriented. Email: mauwakimanu@yahoo.fr.

Lawrence Mhandara, asenior lecturer in the Department of Political and Administrative Studies at the University of Zimbabwe, holds a PhD in Peace Studies from the Durban University of Technology. His latest publication 'Community-based Reconciliation in Practice: Lessons for the National Peace and Reconciliation of Zimbabwe' appears in the *African Journal on Conflict Resolution*, Volume 20, Issue June 2020. Email: lmhandara@gmail.com.

Michael Aeby is a Visiting Academic at the University of Edinburgh and a Research Associate of the University of Cape Town's Institute for Democracy, Citizenship and Public Policy in Africa. His publications focus on the African Peace and Security Architecture, civil societyinclusion in peace processes, mediation, agreement implementation, and transitional governance and power-sharing, with a focus on Southern Africa. He holds a PhD from the University of Basel, worked at the Graduate Institute Geneva and Swisspeace, and was a visiting fellow at the Universities of the Western Cape, Witwatersrand and Zimbabwe. Email: michael.aeby@ed.ac.uk.

Moses Monday John is an associate lecturer on conflict management and nonviolence at the School of Public Service, University of Juba and Catholic University of South Sudan respectively. He holds a PhD in Public Administration -Peace Studies from Durban University of Technology, South Africa. He co-founded Organization for Nonviolence and Development (ONAD) based in Juba, South Sudan. He is an author of several book chapters including Nonviolence: The Tactics and Strategies of Grassroots Social Change In: Connecting Contemporary African-Asian Peacemaking and Nonviolence. From Satyagraha to Ujamaa edited by Vidya Jain and Matt Meyer, Published by Cambridge Scholars Publishing 2018. Email: mosesjoa@gmail.com.

Norman Chivasa is a Postdoctoral Research Fellow at the Durban University of Technology, South Africa. He holds a PhD and Masters in Conflict Resolutionand Peace Studies from the University of KwaZulu-Natal, Durban, South Africa. He is currently a Post-Doctoral Research Fellow in the International Centre of Nonviolence at Durban University of Technology. He was formerly a senior lecturer at the War & Strategic Studies Unit, Department of History, University of Zimbabwe, Harare. His research focuses on community peacebuilding and informal infrastructures for peace. He has long been involved in community peacebuilding efforts in Seke district, Zimbabwe, where he has been working closely with villagers in creating local peace committees. His consulting specialty involves designing, implementing and evaluating local infrastructures for peace at village level. Email: normanchivasa@gmail.com.

Ntombizakhe Moyo-Nyoni is a lecture in Peace Studies at the Midlands State University in Zimbabwe. Besides teaching, she is a peacebuilding practitioner working with communities as a facilitator of Alternative to Violence Project and a consultant in community trauma processing session. Her research interest is including peace education, violent prevention and restorative justice.

Paul Kariuki serves as the Executive Director of the Democracy Development Programme (DDP), based in Durban, South Africa. Academically, he holds a PhD in Public Administration from the University of KwaZulu-Natal, where he is also a research fellow at the School of Management, IT and Governance. His main research interests include public governance, public participation, citizenship, electronic governance as well as monitoring and evaluation in the public sector. Email: paulk@ddp.org.za.

Philippe Mulume-Oderhwa Kaganda is a Doctor of Sociology and Professor at the Faculties of Social Sciences of the Official University of Bukavu (UOB) and the Evangelical University in Africa (UEA). His scientific and consulting research focuses on conflict dynamics and peacebuilding. He is scientific director of CRECOPAX-GL (Centre de Recherche et d'Etude sur les Conflits et la Paix dans la Région des Grands Lacs). Email: kamuldoudou@yahoo.fr.

Sylvia Blanche Kaye is a senior lecturer in the Peacebuilding Programme at the Durban University of Technology. Her PhD study was on women entrepreneurs in the informal sector in Botswana, and she has been actively involved in community-based and educational peacebuilding efforts in Botswanaand South Africa. A major focus area of her research interest is on the need for women to be included at all levels of peacebuilding efforts. Email: SylviaK@dut.ac.za.

Index

A
Abbasi, M. A., 398
Abdela, L., 392
Abdrabo, M. A., 37
Abij Ahmed, 27
Abrahamic, 128, 131, 135, 136
Abrams, B. E., 320
Abu-Nimer, M., 101
Accountability of former child soldiers, 423, 431, 432, 458
Acidifying oceans, xv
Act for Transparent and Peaceful Elections, 271, 272, 554
Action research, 54, 145, 147, 159–162, 164, 170, 173, 183, 225, 291, 297, 308, 355, 358–362, 364, 365, 371, 372, 374–379, 427, 545, 550, 551, 553
Action Support Centre, 388
Adan, M., 384, 385, 388, 390
Adano Wario, Roban, 49, 535
Addis Ababa Peace Agreement, 356
Adekeye, A., 217, 218
Adeosun, H. O., 195
Adetula, Victor, 478
Adger, W. N., 46, 68
Advocacy organisations, 7, 331
Aeby, Michael, 465–467, 470–478, 480–494, 555
Aembe, B., 510
Aerosol spray cans, 23
Aertsen, I., 448
Afako, B., 455–457
Afghan, xx
Afghanistan, 384, 390
Africa Confidential, 489
African Alternative Dispute Resolution, 440
African Case Studies, 550, 551
African Centre for the Constructive Resolution of Disputes (ACCORD), 50, 177, 343–346, 348, 471, 476, 483, 488
African civil society organisations, 465
African morality, 435, 439, 459
African National Congress (ANC), 27, 320, 321, 340
African Non-Governmental Organisations (NGOs), 466, 467, 471, 472, 476, 477, 480, 487, 490, 492
African palaver, 424, 434, 436–439
African palaver and peacebuilding, 438
African Partner Countries, 532
African partner NGOs, 471
African Peace and Security Architecture, 42, 465, 555
African philosophy, 423, 431, 434
African Research Initiative for Scientific Excellence, 531
African Solidarity Network, 346
African traditional religions, 105, 128, 135, 136, 163
African Union (AU), 41, 42, 54, 149, 152, 153, 298, 344, 465–468, 470–478, 480, 485, 488–492, 530–532, 538, 540
Agence Française de Développement, 341
Agency, 84–88, 102, 113–115, 145, 150, 151, 154, 155, 183, 208, 210, 216, 219, 223, 224, 275, 326, 340,

J. C. K. Kiyala and G. T. Harris (eds.), *Civil Society and Peacebuilding in Sub-Saharan Africa in the Anthropocene*, The Anthropocene: Politik—Economics—Society—Science 34,
https://doi.org/10.1007/978-3-030-95179-5

342–347, 349, 393, 434, 469, 501, 508–510, 530, 533, 537
AGORA, 529
Agreement on the Resolution of Conflict, 356
Agricultural land, 11, 17
Agriculture, 30, 48, 49, 87, 109, 166, 214, 225, 289, 504, 529
Ahmed, F., 96
Ahmed, J. U., 281
Aid to the Church in Need, xxxi
Aid, Trade and Security in an Unequal World, 31
Aker, D., 384
Albert Lutuli, 27
Aldehoff, Larissa M., 54
Alexander, J., 320, 322
Alexander, K., 237
Alier, A., 356
Aliu, Y., 258
Allen, J. A., 281
Alliance for Peacebuilding, 237
Allison, S., 359
Alternatives to Violence Project, 308, 553
Amazon, xvi
America, 26, 27, 33, 35, 55, 132, 370, 528, 534, 538, 539
Amnesty International, 7, 209, 285
Amr, Mohamed Sameh, 474
Amster, Randall, 21, 51, 56
Analytical methodologies, ix
Ancient Greek, 6, 97
Anderson, D. M., 391
Andreevska, E., 276
Anglo-Egyptian Condominium colonial rule, 355
Angola, 19, 320, 532
Angom, S., 225
Animation Civique, 294
Antarctica, 74, 99
Anthropos, xi, xvi
Anthropocene, 5, 11–13, 17–27, 30–32, 34, 35, 40, 42, 47, 50–54, 56, 57, 69–72, 74–78, 81, 84, 86, 88, 95–97, 99, 105, 107, 108, 110, 112, 120–122, 512, 527, 536, 540, 554
Anthropocene Epoch, 26, 50–52, 69, 70, 72, 96, 108, 120
Anthropocene Working Group, 18, 23, 70
Anthropogenic, 5, 21–23, 25, 26, 30, 31, 51, 53, 56, 68, 70–74, 78, 84, 86, 95, 99, 101, 110, 115, 119, 128, 527
Anthropogenic atmospheric disruptions, 85
Anthropogenic global environmental, 11, 19, 20, 24, 528
Anthropogenic threats, 17, 19, 20, 23
Antimilitarism, 75
Antje, D., 403
Anyanya One Movement, 356
Apartheid ideology, xiii
Apostle Paul, 101, 103, 112
Apuuli, Kasaija Philip, 457, 476
Arato, A., 6
Arbour, L., 276
Armed conflict, 3, 6, 7, 9, 11, 29, 42, 43, 46–48, 148, 150, 152, 154, 156, 193, 194, 205–208, 210, 216, 229, 230, 238, 240, 255, 261, 264, 266, 273, 274, 276, 293, 356, 358, 424–426, 429, 431–433, 436, 439, 445, 448, 449, 455, 456, 458, 460, 503, 512, 551
Armed Conflict Survey, 42, 44
Armed groups, 7, 150, 205, 209, 214, 216, 220, 224, 227, 286, 288, 324, 423, 424, 429, 430, 432, 433, 440, 444–449, 453, 459, 469, 504, 521
Arthur, Peter, 475
Arts, K., 424
Artur, L., 68
Ashworth, J., 357
Association of African Universities, 530
Associations, 6, 83, 85, 98, 105, 136, 150, 154, 156, 193, 194, 197, 209–211, 219, 221, 226, 227, 229, 237, 241, 248, 253, 254, 265, 282–284, 286–289, 347, 349, 357, 376, 397, 398, 400, 401, 407–409, 411–414, 416, 434, 509, 514, 516, 518, 534, 543
Asuelime, L. E., 403
Atomic bomb, xi
Atmosphere, 13, 21, 24, 52, 56, 70, 71, 75, 80, 87, 96, 105, 134, 200, 272, 297
Atmospheric carbon dioxide concentration, xv
Atmospheric chemistry, 105
Atmospheric conditions, 68–70, 72, 75, 84, 86, 97, 110, 111, 121
Atmospheric transformations, 71, 84, 88
Atom bomb, 17, 18, 20
Atomic weapons, 19
AU Continental Early Warning System, 475
Audu, B., 339
AU Liaison Offices and Special Representatives, 477

Aulin, Jenny, 469
AU Mediation Support Handbook, 473
AU Panel of the Wise, 471, 475
AU Policy Framework, 472, 473
Augenstein, Karoline, 52
Austin, Z., 402
Autesserre, S., 210, 211, 221
Azadi, Hossein, 50
Azadi, J., 50

B

Ba'al taschit, 132
Babatunde, A. O., 444
Babcicky, P., 77
Babunga, B., 278
Baines, E. K., 276
Balbo, Andrea L., 50
Bali, Ranjula, 50
Bangura, A. K., 455, 457
Banpasirichote, C., 337
Baraza, 5, 214, 263, 423–425, 434, 445–449, 455, 458, 459, 516
Barbie, E., 159
Barndt, W., 253
Barnes, Catherine, 150, 193, 235, 239, 468–470
Barnett, M., 9, 425
Barnhoorn, A., 68
Barry, T. L., 99
Bartkowski, J., 356
Bashingantahe, 443, 444
Baškarada, S., 400
Bauckham, R., 112
BBC bitesize, 130
BBC News, 72
Beare, D. J., 87
Bebbington, A., 512, 514
Beckstrand, M., 507
Behera, Navnita Chadha, 31
Beidelman, T. O., 414
Beijing Platform, 146
Belgium, 251, 271
Bell, Christine, 469, 470
Bello, M. M., 255
Belloni, Roberto, 98, 150, 255, 272, 275, 276, 468
Benchley, P., 99
Benedick, Richard Elliot, 23
Benin, 532
Benner, Susanne, 18
Bennett, E. M., 3, 87
Bennett, G. D., 133
Bennett, Juliet, 26, 534
Benoit, D., 299
Bentahar, M., 279, 280
Bercovitch, J., 100, 101, 106, 522
Berg, B., 359
Berghof Foundation, 470
Besheer, M., 451
Beyond Scarcity, 31
Bhargava, V., 337
Bickler, G., 73, 74
Biggs, R., 68, 73, 79, 84, 89
Bignami, F., 403
Bilak, A., 220
Bilal, M. G., 135
Bilateral level, 530
Biodiversity, 20, 28, 33, 41, 67, 69, 110
Biological, 25, 29, 104, 130, 133, 402, 434, 512
Biosphere, 18, 33, 34, 75, 80, 82, 99
Birhane, E., 50
Birks, M., 515
Birmingham, David, 19
Bless, C., 260, 516
Bloch, N., 365, 368, 370, 371
Bloor, M., 360
Bob, Urmilla, 50, 67, 68, 75, 77
Boesch, D. M., 77
Böhmelt, T., 68, 83
Böhner, J., 50
Borsenberger, M., 341
Bosakaibo Bomino Georges, 205
Bosmans, M., 218
Botswana, 322, 531, 535, 556
Bottom-Up, 195, 197, 200, 201, 217, 221, 225, 229, 231, 275, 280, 337, 338, 342, 343, 346, 349, 351, 352, 453, 479, 506, 535
Boulding, Elise, 21, 51, 79, 228
Boulding, J. Russell, 22
Boulding, Kenneth E., 9, 21, 51
Bourantonis, Dimitris, 19
Bouthoul, G., 523
Bown, P. R., 99
Bradbury, H., 159
Branton, C. J., 128–130
Bratton, M., 236, 258, 259, 276
Brauch, Hans Günter, 5, 11, 12, 17, 18, 21, 24–26, 31, 51, 52, 55–57, 68, 70, 71, 84, 96, 97, 99, 108, 534, 536, 538, 540, 554
Braun, V., 310
Brennan, A., 81
Brennan, R. J., 205

Brett, R., 430
Britton, 79, 82, 93
Brisman, A., 5, 68
Broache, M., 398
Broadbent, E., 506, 507
Bronkhorst, Salome, 67, 68, 77, 78, 80, 81, 86, 99, 176, 180
Brounéus, K, 436
Brown, G. K., 195
Brulle, R. J., 68
Brundtland Commission, 29, 30
Bryman, A., 360
Brzoska, Michael, 18, 25
Buhaug, Halvard, 47
Building Resilience, 31, 39, 80
Building sustainable peace, 28
Building the capacity of civil society, 555
Building True Peace, 241, 324
Burkina Faso, 55, 287, 531–533
Burns, J., 342
Burton, B., 201, 308
Burundi, 152, 153, 260, 261, 272, 384, 386, 390, 423–425, 430, 442–444, 504
Burundians, 272, 430, 443, 444
Bushingantahe, 5, 423–425, 434, 442–444, 458, 459
Business associations, 7
Bussy, J. M., 287
Buszka, A. J., 444
Butedi, F., 446

C
Cabot, Charlene, 49, 50
Caelesta, B., 342
Cahen, M., 430
Cairns Jr, J., 81
Cameron, H., 192, 532
Campbell, B. M., 87
Canada, 297, 298, 530
Caniglia, B. S., 68
Cantrill, D., 99
Carbon-cutting ambitions, 69
Carbon dioxide, 21, 35, 42, 54–56, 70, 71, 75, 87, 99
Carbon-intensive processes, 24
Carbon-neutral, xvi
Cardinal Laurent Monsengwo Pasinya, 271
Carius, A., 54, 77, 535
Carney, D., 170
Carothers, T., 253, 468
Carpenter, S. R., 67, 73, 79, 84, 89
Case studies, 13, 48, 95, 97, 106, 210, 217, 251, 252, 259, 281, 285, 355, 366, 374, 400, 402, 425, 467, 507, 515, 536, 540
Cashdan, B., 350
Catholic Bishops, 108, 242, 245, 324, 554
Catholic Church, 101, 104, 106, 133, 212–214, 226, 227, 277, 278, 295, 301, 324
Catholic Commission for Justice and Peace, 324
Catholicism, 96, 97, 104, 106, 108
Catholic News Service, 96, 113
Causal explanations, 21
Causes of climate change, 68, 119
Cele, S., 387
Central Africa, 48, 49, 291, 299, 554
Central African Republic, 532
Centre d'Etudes, 294
Centre for Conflict Management, 242
Centre for the Study of Violence and Reconciliation, 471
Centre on Interdisciplinary Research on Africa, 532
Centre-periphery, 501, 503, 510, 511, 513, 514
Chandhoke, N., 150, 253, 258
Chan, E., 341
Chan, J, 341
Chapin III, F. S., 67
Chataway, C. J., 10
Chazan, N., 237
Cheema, G. S., 150
Chemical weapons, 29
Chen, Baiping, 19
Chen, K., 424
Chenoweth, E., 368, 373
Cheruiyot, K., 341
Chesterman, Simon, 469
Chibhamu, A., 244
Chick, K., 299
Chigas, D., 379
Children born from rape, 397–400, 403, 404, 406–416
Children's maltreatment, 397
Child Soldiers International, 424, 445
Chimurenga violence, 146
China, 23, 33, 36, 530
Chinoputsa, L., 386
Chitando, A., 146
Chitiga Rudy, 471
Chivasa, N., 158, 247, 384, 555
Chiweshe, K. M., 385, 386
Chlorofluorocarbons, 23, 24
Chourou, Béchir, 31

Christensen, D., 412
Christensen, J. E., 130, 131
Christian Council, 347, 485, 490
Christianity, 100, 110, 128, 131, 132, 134, 441
Christian, R. R., 77
Christos N. Kyrou, 21
Cinner, J., 77
Citizens and Diaspora Directorate, 474
Citizens' movements, 252, 261, 266, 271, 273, 274, 282, 286–288, 298
City of God, 106
City of the Organisation of African Unity, 298, 300
Civil society, 5–10, 13, 33, 49, 67–70, 72, 79, 83–89, 95, 97–100, 102, 104, 106, 107, 110, 116, 120, 121, 147–151, 153, 156, 191–193, 198, 201, 205–228, 230, 231, 235–249, 251–261, 263–267, 271–275, 277–292, 294, 295, 300–302, 323, 337–339, 342, 343, 345–348, 350–352, 355–357, 359, 366, 374, 376, 379, 426, 441, 443, 447, 453, 456, 457, 466–470, 472, 473, 475, 478–494, 501–503, 506, 508–512, 514, 515, 517, 518, 523, 524, 549, 554, 555
Civil society actors, 150, 207–210, 216, 218, 223–225, 228, 229, 251, 257, 262, 265, 272, 299, 340, 369, 457, 466–469, 475, 477, 480, 483, 485, 491–494, 506, 507, 509
Civil Society and Violence Protection, 403
Civil Society Areas of Action, 282
Civil society groups, 209, 210, 218, 221, 223, 241, 246, 327, 339, 340, 444, 467, 489
Civil Society Inclusion in Peace-making, 469, 470
Civil Society Inclusion in SADC Mediations, 489
Civil society organisations, 5, 76, 84, 102, 120, 145–147, 150, 152, 156, 191, 205, 206, 220, 235, 237, 239, 242–244, 248, 251, 252, 254, 262, 271, 272, 277, 282–286, 289–293, 296, 297, 307, 325, 326, 331, 337–340, 347, 348, 350, 352, 355, 357, 375, 391, 397, 398, 407, 427, 433, 465, 470, 505, 509, 510, 513, 519–521
Civil society participation, 208, 261, 339, 466–473, 478, 481, 482, 486, 489, 490
Civil wars, 7, 27, 76, 205, 207, 208, 245, 251, 256, 264, 284, 286, 287, 290, 294, 302, 356, 385, 425, 428, 430, 431, 437, 439, 440, 442, 443, 450, 454, 456, 457, 478, 511, 523
Clapham, Christopher, 19
Clarke, R., 69
Clark, V., 310
Clément, P. M., 446
Climate change, 3, 5, 11–13, 17, 19, 20, 22–25, 28, 29, 31, 36, 37, 39–42, 45–51, 53, 54, 56, 57, 67–69, 72, 73, 75–78, 80–86, 88, 89, 95, 96, 99, 101, 102, 108, 114, 115, 121, 127, 128, 132, 134–136, 138, 139, 166, 527, 528, 531, 533–535, 540
Climate change vulnerability, 48
Climate conflicts, 76, 77
Climate-management risk policy, 76
Climate models, 31, 35, 36, 56
Climate-neutral world between, 19
Climate-related environmental scarcities, 11, 17
Climate-related security risks, 76
Climate-sensitive adaptation, 67, 78
Clowes, W., 255, 288
CO_2 emissions, 19, 55, 87
Coalition groups, 334
Coalition to Stop the Use of Child soldiers, 7
Coastal wetlands, 77
Coe, A. L., 99
Coe, Brooke, 466
Coghlan, B., 205
Cold War, 19, 24, 51, 100, 539
Coleman, J., 226, 227
Collective agency, 84–86, 121
Collective security system, 18
Colombo, A., 272
Colonial dependencies, 19
Colonial rule, 19, 192, 240, 241
Coltart, D., 148
Commissioner for Peace and Security, 475
Commitment and engagement, vi
Common geostory, 85
Common Market for Eastern and Southern Africa, 487
Communalist ethic, 435
Communitarianism, 5, 434
Community-based associations, 357

Community building, 235, 243, 246, 247, 249, 349
Community mobilisation, 10, 428
Community of Life, 114
Community ownership, 10, 428, 502, 520
Community war-crime victimisation, 397
Comprehensive Peace Agreement National Congress Party, 356
Compulsive consumerism, 114, 119
Conca, Ken, 4, 19, 21, 23, 54, 535
Conflict Barometer, 42, 44
Conflict mediation, 75, 343
Conflict prevention, 27, 45, 77, 154, 177, 244, 425, 465–471, 476, 478–480, 482, 484–486, 490, 494, 503, 512, 540
Conflict-related outcomes, 47
Conflict resolution, 7, 28, 100, 101, 149, 165, 177, 179, 180, 199, 201, 207, 211, 221, 225, 231, 246, 247, 256, 261–263, 267, 275, 343, 387, 389, 425, 434, 437, 439–442, 445–447, 450, 453, 534, 553, 555
Conflict risks, 45–47, 49, 475, 481, 482, 527
Conflict-sensitive, 50, 75, 77, 78, 163, 256
Conflict-sensitive mechanisms, 67
Conflict transformation, 146, 147, 156–158, 161, 263, 403, 469, 479, 523, 524, 553, 555
Confucianism, 128, 130
Congo Brazza-Ville, 434
Congo Catholic Bishops' Conference, 288
Congo Forum, 254, 283, 284
Conscientious stewardship, 110–112
Conservation clubs, 137
Conservationists, 138
Constitutive Act, 471–474, 491
Consumerism, 102, 103, 114–116, 120
Consumption preferences, 20
Copeland, T. F., 252
Coral bleaching, 77
Cornell, S. E., 3, 84, 89
Correa, D. D. S., 281, 425
Corruption Watch, 340
Côte d'Ivoire, 483, 531, 532
Council of the Wise, 478
Court, J., 236
COVID-19 pandemic, 338, 351, 352
Coward, H., 130
Creator, 98, 100, 105, 111, 120, 133, 135, 136, 138, 449
Creator-Redeemer, 134
Creswell, J. W., 402
Crutzen, Paul J., 11, 17, 22–24, 26, 51, 56, 70–72, 77, 84, 96, 97, 99
Cultural violence, 168
Cupchik, G., 259
Curle, A., 274
Cvetek, N., 278, 282

D

Da'esh, 424
Daiber, F., 282
Dalby, Simon, 12, 13, 25, 52, 80, 99, 105
Daley, P., 235, 236
Daneel, M. L., 105, 128, 136–138
Daniel Wegner, 485
D'Aoust O., 272
Daudi, Faruma, 49
Davies, M. B., 259, 260
Davis, Ian, 42
Day, J. W., 77
Deane-Drummond, C., 12, 107
Decarbonisation of the economy, 19
De Carvalho, Gustavo, 476, 492
Decision-making, 4, 41, 53, 82, 151, 152, 225, 256, 265, 293, 334, 343, 349, 377, 383, 390, 411, 451, 466, 474, 475, 481, 486, 491, 492, 507, 513, 514
Decision-policy-makers, viii
Deen, M. Y. I, 12, 105
De Klerk, F.W., 27
Del Bene, D., 78, 79
Delivery institutions, 7
Demaria, F., 78, 79
Democracy Development Program, 347, 348, 556
Democratic activism, 235
Democratic legitimacy, 469
Democratic peace, 75
Democratic Republic of Congo, 18, 145, 205, 251, 253, 254, 271, 284, 285, 288, 289, 291, 292, 300, 386, 397, 398, 423, 424, 445, 501, 504, 515, 549, 550, 554
Demographic projection, 35
Denis Mukengere Mukwege, 27, 397, 553
Desmond Tutu, 27
Desportes, I., 179
Detges, Adrien, 54
Developmental projects, 181, 513
Development service, 7
Development theory, 503

De Vries, W., 79, 84, 89
DeWiel, B., 6
De Wit, C. A., 3
Diamond, J. L., 238
Dickes, P., 341
Dieter, N., 403
Dill, A., 8
Diplomacy, 5, 10, 23, 24, 30, 31, 54, 56, 75, 426, 466, 475, 479, 482, 487, 488, 492, 494
Direct democracy, 257, 258
Dirigisme, 253
Disciplinary boundaries, 47, 52
Dockès, A. C., 515
Dolamo, R., 436
Dominant feature, 170
Donais, T., 194, 201
Dörner, W., 404
Do Rosário, D., 430
Dorothy Moyo, 307, 553
Douma, N., 398
Drivers of climate change, 32
Drivers of conflict, 10, 40, 46, 47, 81, 426
Drumbl, M. A., 424
Dube, T., 128
Duffy, Rosaleen, 25
Duke, J., 507
Dumisani Maqeda Ngwenya, 553
Dunning, T., 424
Durable peace, 9, 13, 80, 153, 155–157, 183, 191, 193, 195, 197, 198, 201, 255, 259, 261, 264, 275, 292, 424, 427, 436, 441, 455, 458, 459, 510, 524
Durban University of Technology, 95, 145, 251, 261, 271, 307, 355, 362, 383, 397, 423, 501, 543, 549, 551, 553–556
Duvenage, P., 436–438

E

Eagle, G., 312
Early warning analysis, 472
Earth, 3, 9, 11, 12, 18, 19, 24, 26, 32, 51, 52, 56, 67, 68, 70–75, 77–80, 82, 84–87, 89, 95–97, 99–101, 103–106, 108, 111–115, 117, 119–122, 127–129, 131–133, 135, 138, 217, 536
Earth Charter, 110, 114, 115, 119, 122
Earthkeeping, 105, 127, 128, 136–139
Earth system, 3, 5, 12, 18, 21, 54, 56, 67, 69, 71–73, 77, 79, 80, 84–86, 88, 95–97, 113
Earth Systems Science, 12, 24, 26, 52, 56, 57, 80
Eastern religions, 129, 130
Eaton, H., 12, 103–105, 110–112
Echo-deontology, 82
Ecoeducation, 67, 81, 82
Eco-engineering, xvi
Ecoethics, 67, 81, 82, 89
Ecogeomorphology, 77
Ecojustice philosophy, 79, 81, 82
Ecological conflicts, 28, 29, 540
Ecological conversion, 101, 109, 112, 113
Ecological distribution conflicts, 78
Ecological education, 69, 89, 113, 116, 119, 120
Ecological Integrity, 115
Ecological Peace Policy, 17, 20, 21, 26, 27, 29–31, 36, 40, 45, 50, 54–57, 527, 535, 539
Ecological refugees, 78
Ecological sustainability, 67, 78, 89
Ecological threats to peace, 28
Ecological violence, 75, 80
Ecologists, 105, 136, 536, 540
Economic Community of West African States, 7
Economic Community of West African States (ECOWAS), 7, 76, 466–468, 470, 471, 478–483, 490–493
Economic Social and Culture Council (ECOSOCC), 471–474, 485, 490, 491
Eco-pedagogy, 82
Ecosystems, 4–6, 12, 22–24, 34, 38, 39, 41, 49, 51, 67, 69, 70, 72, 73, 75, 80–82, 87, 88, 96, 99, 102–104, 109, 112, 113, 119, 121, 139, 512, 535
Eco-theology, 82, 132
ECOWAS' Civil Society Interface, 480
ECOWAS Council of the Wise, 482
ECOWAS Policy Framework, 478
ECOWAS Treaty civil society and states, 479
Ecumenical Church Leader's Forum, 387
Edinger-Schons, L. M., 342
Edwards, B., 208, 339
Egotism, xi
Ehrhart, H. -G., 9, 425, 427
Ehteshammajd, S., 50
Einarsdóttir, J., 401
Ekiyor, T., 207
Eklöw, K., 68, 76
Election-related violence, 193, 197, 272

Ellen Johnson Sirleaf, 27
Ellis, C., 217, 219, 222, 225
Elo, S., 402
Elsobki, M. S., 50
Empirical independence, 235, 249
Empirical research, 40, 47
Encyclical Letter, 95–97, 104, 106, 107, 109, 110
Engelke, Peter, 18
Engendering Development, 146
Enlightenment, 253
Environmental change, 28, 49
Environmental conflict, 3, 51, 67, 69, 76–78, 80, 81, 86, 99, 121
Environmental conservation, 27, 28, 104, 127, 139
Environmental cooperation, 54, 76, 535
Environmental education, 115, 120
Environmental instability, 67, 89
Environmentalists, 27, 89, 129
Environmental justice, 5, 11, 12, 25, 67, 68, 78, 79, 82, 87, 89, 95–97, 99, 121, 122, 555
Environmental Law Institute, 76
Environmental peace, 5, 49, 51, 67, 68, 87, 89, 95, 97, 102, 103, 105, 106, 118, 121, 122
Environmental peacemaking, 4, 5, 12, 54, 69, 70, 78–80, 102
Environmental policy, 12, 28, 30, 78, 82, 83, 376
Environmental risks, 12, 78, 120
Environmental sciences, 55
Environmental stability, vi
Environmental stress, 29
Environmental studies, 17, 20, 21, 26, 27, 52, 56, 79–81, 108, 527, 528, 537
Environment Institute, 340, 537
Environment preservationists, v
Environment programme, 23, 76
Epistemic violence, xix
Erasmus, V., 10, 428, 536
Erlingsson, C., 402, 403
Estève, Adrien, 540
Ethical principles, 8, 114, 294
Ethics, 37, 69, 81, 82, 89, 96, 97, 101, 104, 108, 110, 120, 121, 155, 260, 291, 294, 333, 423, 431, 434–436, 459
Ethiopia, 27, 36, 55, 256, 356, 377, 531
Ethnic cleansing in Matabeleland, 325
Etkin, D., 82
Ettang, D., 155
Eugene Stoermer, 24, 51, 70, 535
European Centre for Conflict Prevention, 404
European Commission, 19, 30, 87, 533
European Union, 19, 30, 297, 298, 507, 508, 530–532
Evadzi, P. I. K., 50
Evans, M., 509
Evans Ogden, 28
Everman, M., 451

F

Fæste, Lars, 19
Faith-Based Organisations, 7, 12, 79, 87, 96, 97, 102, 104–106, 120
Fambul Tok, 5, 423–425, 434, 454, 455, 458, 459
Fanning, A. L., 74
Fargas-Malet, M., 402
Feldman, D.L., 104, 112
FemWise mediators, 476–478, 483, 492, 493
Ferrara, P., 107, 108, 115
Fetzer, I., 3, 68, 73, 74, 79, 84, 89
FFKM, 489
Fifita, S., 69
Filimbi, 287, 288
Finkelhora, D., 401
Fischer, M., 403
Fisher, S., 163, 181
Floribert Chebeya, 285
Foley, M. W., 339
Food insecurity, 12, 39, 48, 81, 87
Food supplies, 11, 17
Forster, P., 69
Fossil fuels, 21, 56, 87, 88, 367
Franchi, L., 107
Francis of Assisi, 107, 109, 112, 113, 121
Franklin, R., 515
Freedman, J., 225
Freeman, J., 384
Fretter, J., 522
Freund, Bill, 19
Frogh, W., 392
Funding Partners, 5, 503, 504, 508
Funk, N. C., 197–199
Furini, G., 128

G

Gacaca, 5, 423–425, 432, 434, 442, 451–455, 458, 459
Gade, C. B., 436, 437
Gaffney, Owen, 24

Gale, A., 99
Gallez, E., 445, 446
Galtung, Johan, 22, 148, 156, 194, 237, 513, 514
Gambia, 532
Gazibo, M., 302
Gender-based violence, 149, 151, 152, 165, 218, 230, 289, 316, 375, 378
Gender violence, xix
Genocidal violence, 4
Genocide, 325, 326, 430, 451, 452, 455
Geoengineer, xvi
Geo-graphy, 85
Geological epoch, 17–20, 23, 32, 70, 71, 84, 86
Geo-morphology, 85
Geophysical effects, 21
Geo-physics, 85
Geo-physiology, 85
Geopolitical, 352
Geopolitical conflicts, 11, 17
Geo-politics, 85
Geosphere-Biosphere, xviii
Gerard Mpang'de Bisambu, 271
German Advisory Council on Climate Change, 44
German Corporation for Development Cooperation, 467
German Foundation on Peace Research, 532
German government announced, 19
Ghana, 27, 55, 272, 507, 531, 532, 535
Gibbard, P., 99
Gibson, J. L., 455
Gibson, W. E., 134
Ginanjar, W. R., 83
Ginzburg, V., 69
Global and Inclusive Agreement, 300
Global cooperation, 20
Global environmental change, 20, 22, 24, 29, 35, 36, 51–53, 55, 73, 527, 539
Globalisation, 128, 263, 430, 431, 440, 450, 511, 512, 550
Global North, 5, 13, 528–530, 534, 536–538, 540
Global Peace, 4, 5, 77, 88, 89, 99, 438, 460, 501, 534, 536, 539
Global population, xii
Global Research Partnership, 527
Global response, 78, 83
Global South, 13, 27, 33, 80, 383, 528, 529, 534–540
Global temperature, 68, 71, 88
Global Water Crisis, 31
Gnanguênon, Amandine, 481
GOALI, 529
Gomes Porto, João, 475, 476, 492
Göpel, Maja, 53
Gordon, C., 50
Gottlieb, R. S., 12, 96, 102–104, 112
Gouzou, J., 214
Government of Sudan, 356
Grace to Heal Trust, 145, 147, 159
Graham, T. L., 87, 88
Grassroots leadership, 10, 72, 157, 275, 283, 401
Graybill, L. S., 432, 442, 455
Great Britain, 297, 298
Green Belt Movement, 27
Greenhouse, 21, 24, 37, 51, 52, 54, 56, 71, 87
Greenhouse gas emissions, 21, 46, 71, 87
Greenpeace, v
Greenpeace movements, v
Green Teen Team, v
Gregory, F. J., 99
Greta Thunberg, 68, 84
Grinevald, J., 75, 84, 99
Grin, John, 24, 31
Grosfoguel, R., 511, 513
Guinn, A. S., 414
Gukurahundi Era, 321
Gukurahundi violence, 146
Gulliksen, C., 308
Gumbu, M., 400
Gupta, M. D., 414
Gyekye, K., 434, 435
Gyimah-Boadi, E., 193

H

Hahne, H. A., 112
Haile, D., 452, 453
Haile, M., 50
Hall, A., 276
Hallen, B., 435
Hallman, D. G., 134
Hall-Spencer, J. M., 87
Hallward, M., 358
Halpert, J., 131
Hambyb, S., 401
Hamuli Kabarhuza, B., 284, 285
Handa, C., 69
Hapanyengwi-Chemhuru, O., 385
Harber, C., 308
Harbeson, J., 237

Hardlife, Z., 385
Hardt, Judith Nora, 68, 99, 540
Harpviken, K. B., 195
Harrington, Cameron, 540
Harris, Geoff, 18, 146, 161, 307, 360, 551, 553
Harrison, H., 515
Hartman, Henrik, 487
Hayman, C., 509, 510, 523
Haynes, R., 507
Hazards and disasters, 20, 27, 527
Healey, J., 435
Healing, 5, 10, 86, 100, 127, 128, 139, 146, 151, 167, 236, 243, 275, 311, 319, 323, 325, 326, 333, 334, 385, 423, 424, 426, 432, 434, 436, 437, 439, 440, 442, 450, 453–455, 458, 459, 489, 503, 553
Healing nature, 128, 136
Heal Zimbabwe Trust, 247, 387
Heathershaw, John, 469
Heaton, L., 284
Hegel, 6, 236
Heidelberg Institute for International Conflict Research, 42
Helena, Kavsek, 485
Hellmüller, Sara, 468–470
Hendricks, Cheryl, 486
Hertig, E., 50
Higher education, 30, 529, 531, 543
Higson-Smith, C., 260
Hilhorst, D., 68, 398
Hinduism, 101, 110, 128, 130
Hittinger, R., 106, 109
Hoffman, A. J., 101, 107
Hoffman, E., 454, 455
Hoger, T., 358
Holistic approaches, 3, 26, 28–30, 52, 272
Hölkeskamp, K. -J., 278
Holocene, 18, 70, 71, 73, 97, 99
Homo Sapiens, 70
Horn of Africa, 68, 78
Hoseini, B. L., 398
Houghton, D. D., 75
Hounslow, M. W., 99
Houten, K. V., 398, 509, 510
Hove, M., 551
Hovhannes, H., 98
Howell, J., 226, 227
Hsieh, H. -F., 69, 161, 400, 403, 515
Huberman, A. M., 281
Huggett, Richard John, 25
Hulse, Merran, 485
Human actions, 12, 24, 72, 80, 99, 360
Human agency, 67, 69, 71, 79, 81, 84, 89, 95, 100, 110
Human beings, 21, 23, 33, 56, 70, 75, 81, 88, 103, 104, 111, 113, 114, 119, 128–136, 138, 179, 279, 280, 312, 356, 376, 435
Human capacities, 529, 530
Human civilization, 99, 130
Human development, 20, 31–35, 48, 109, 365, 550
Human health, 69
Human-induced disasters, 17
Humanism, 5, 434, 435
Humanitarian, 7, 10, 42, 48, 193, 194, 215, 216, 221, 223, 231, 252, 256, 275, 276, 283, 285, 323, 356, 424, 426, 432, 433, 440, 454, 457, 458, 477, 479, 482, 506, 510, 512
Humanitarian affairs, 7, 471
Humanitarian relief, 100, 472
Human life, 82, 136
Human progress, 31
Human resilience, v
Human rights, 7, 8, 24, 27, 106, 146, 149, 154, 168, 192, 194, 209, 214–216, 218, 223, 224, 227, 230, 235–237, 239, 241–247, 261, 264–267, 276, 277, 283–289, 292, 295, 297, 323, 328, 331, 349, 366, 378, 411, 415, 416, 424, 426, 432, 437, 440, 456, 457, 460, 479, 513, 516–518, 550, 553
Human rights advocacy groups, 7
Human rights violations, 7, 156, 165, 209, 213, 241, 243, 244, 252, 255, 272, 287, 298, 301, 322, 434, 437, 439, 443, 449, 459
Human Rights Watch, 7, 215, 224, 277, 398, 486
Human security, 3, 31, 40, 45–48, 75, 80, 81, 167, 209, 479–481, 484, 486, 487, 490, 492
Human solidarity, 31
Human-triggered, 20
Hünicke, B., 50
Huntington, H. P., 515
Hybrid knowledge-management system, 477
Hyden, G., 236

I

Ibuot, E. J., 413, 414

Idealistic values, 20
Ideological orientations and targets, 516, 519
Ide, Tobias, 54, 535
Ijumba, N., 78
Ikenberry, G. John, 18
Ikhonaindaba blogspot, 325
Ilker Etikan, I., 400
Imbovane Yamahlabezulu, 325
Imwalle, Lawrence E., 19
In-between, 503, 509, 510
Inclusion in mediations, 465, 493
Inclusion in peace processes, 473, 491, 555
Inclusion of local civil society, 472, 483
Inclusion of non-dominant groups, 469
Inclusion of powerful political and military elites, 469
Inclusive peace-making, 469
Inclusive peace process designs, 467, 470, 477, 492, 494
Income-generating projects, 169, 170
Independence of Sudan, 355
Independence of the Democratic Republic of Congo, 264
India, 23, 36, 534
Indigenous, 5, 10, 81, 83, 96, 137, 157, 195, 198, 221, 256, 263, 274, 275, 283, 357, 423, 424, 426, 427, 434, 439, 440, 442, 444, 447, 456, 458, 459, 501
Indigenous Disputes Resolution Institution, 443
Indigenous knowledge, 37
Indigenous knowledge system, 191, 192, 195, 196, 198
Indigenous peacebuilding approaches, 193
Individual agency, 68, 84, 349
Individual changes, 7, 97, 450
Individualism, 115, 116, 119, 120
Industrialisation, 71, 110, 128, 504, 505
Industrial revolution, 72, 73, 95, 99
Informal peace committees, 158, 247, 383, 384, 390
Infrastructure, 7, 10, 20, 41, 48, 53, 68, 80, 180, 194, 251, 263, 289, 302, 308, 315, 326, 383–385, 426, 471, 484, 487–490, 501, 502, 513, 524, 527, 528, 534, 539, 551, 555
Ingram, J. S., 87
Initiatives of the Federal German Government, 532
Institute for Justice and Reconciliation, 148, 167, 169, 170, 467, 471
Institute for Peace and Security Studies (IPSS), 466, 471, 476
Institute for Security Studies, 49, 471, 551
Institute of African Affairs, 532
Institute of Statistics, 504
Intellectual solidarity, xxii
Interactive conflict resolution programs, 10
Inter-Congolese Dialogue, 206, 284
Interest groups, 7, 85, 86, 239, 331, 338
Intergovernmental organisations, 467, 468, 470, 480
Intergovernmental Panel on Climate Change (IPCC), 25, 26, 31, 37, 39, 40, 45, 57, 68, 83, 84
Intergroup social cohesion, 8, 208, 210, 227, 266
Internal Displacement Monitoring Centre, 33
International actors as peacebuilders, 469
International Alert, 216, 217, 219, 220, 228, 508
International and local peace actors, 469
International Centre of Nonviolence, 95, 145, 251, 261, 271, 383, 423, 501, 545, 549–551, 554, 555
International-Civil Society Organisations, 155, 501
International community, 7, 146, 147, 150, 152, 207, 212, 220, 255, 276, 278, 298, 326, 454, 455, 460
International Cooperation at a Crossroads, 31
International Criminal Court, 333, 429
International Crisis Group, 7, 205, 213
International Institute for Strategic Studies, 42
International Monetary, 530
International Non-Governmental Organisations (NGOs), 7, 86, 210, 220, 222, 228, 229, 323, 473, 482, 501
International Panel on Climate Change, 68
International relations, 20, 25, 26, 56, 107, 108, 486, 535, 554
International sovereignty, 271
International Union of Geologic Sciences, 70
Interventions for decision-makers, 295, 297
Iovan, M., 81
Iraq, 180
Ireland, V., 507
Irene, O., 308
Isike, C., 154, 183

IsiZulu, 436
Islam, 100, 101, 105, 110, 128, 131, 135, 441
Islam environmental ethics, 105
Islamic eco-theology, 135
Islamic rule of Mahdiyya, 355
Islamic sultanates, 355
Ivankova, N. V., 359
Izzi, V., 208

J
Jackson, Simon, 19
Jacobsen, Rune, 19
Jacob Zuma, 340
Janette Yarwood, 475
Jan Weinreich, 485
Jan Zalasiewicz, 70, 475
Japan International Cooperation Agency, 357
Jaramillo, F., 87
Jayakumar, K., 424
Jenneker, M., 385
Jeong, H-W., 275
Joanna, S., 359
Johansen, J., 359
Johansen, R. C., 456
John, M., 355, 357, 368, 555
John, V., 308
Joint Mediation Team, 489
Jordhus-Lier, D., 510
Jörgel, M., 430
Joseph, G., 368
Josephine Kimanu Mauwa, 397, 399, 555
Joseph Kabila, 276–278, 284, 287, 288, 291, 294, 295, 299, 302
Joshua Nkomo, 320, 324
Jos Lelieveld, 70
Judaism, 101, 110, 128, 131, 133, 441
Judicial structure, 7, 443, 449, 452, 453
Juma, M., 308
Juvenile conscription, 424

K
Kääriäinen, M., 402
Kache, C., 147
Kacuol, A., 357
Kadayifci-Orellana, A. S., 100, 101, 106
Kagame, H E Paul, 408, 473
Kambale, V. A., 287
Kameri-Mbote, Patricia, 31
Kaminer, D., 312
Kamta, Frederic, 50
Kanste, O., 402
Kantian policy goals, 20
Ka pahla, 423–425, 434, 450, 458, 459
Kapinga Yvette Ngandu, 475, 476
Kapundu, A. K., 350
Karbo, Tony, 453, 466
Katema, W., 195
Kathryn Nash, 466
Katumba, S., 341
Kawulich, B. B., 402, 403
Kaya, H., 78
Kaye, S., 161, 360, 550, 551, 556
Keal, 18
Keane, J., 253
Keasley, A., 196, 199
Kennedy, M., 101
Kenya, 27, 49, 55, 272, 386, 391, 531, 532, 534, 535
Kenyon, E. W., 413
Kerber, G., 134
Kerr, A. C., 99
Keverian, S., 87, 88
Key players, 263, 278, 293, 295–297, 300
Khademi, G., 398
Khadiagala, Gilbert M., 484
Khartoum government, 356
Khartoum regime, 356, 357
Kheshgi, H., 69
Khodary, Y. M., 154, 180
Kigali Amendment, 23
Killander, M., 146
Kim, H., 9
King, M., 368
Kioko, E. M., 391
Kissinger, Henry, 18
Kivu, 205, 206, 214, 220–231, 286, 511, 515
Kiyala, Jean Chrysostome K., 252, 263, 430–433, 440, 446, 447, 459, 549, 554
Kjellman, K. E., 195
Klein, S., 403
Kling-Eveillard, F., 515
Klot, J. F., 146, 147, 151–155, 181, 183
Knowledge-management system, 477
Knox, R., 99
Kobayashi, S., 69
Kode, D., 243
Kofi Annan, 27, 483
Kogoui Kamta, 50
Kohlhagen, D., 443
Koko, S., 212–214
Kolk, M., 160

Kollmeier, K., 424
Kopstein, P., 131, 132
Koselleck, R., 403
Kouassi, C., 283
Kraan, Carolien M., 46
Krampe, Florian, 28, 68, 76
Krause, S., 259
Krippendorff, K., 69, 402, 403
Kühne R. W., 68, 86
Kurtenbach, S., 150
Kwame Nkrumah, 434
KwaZulu-Natal, 308, 337, 344–348, 388, 543, 549–551, 553–556
Kyngäs, H., 402
Kyrou, Christos, 5

L
Laakso, L., 192
Lack of a Systematic Global Research Survey on Peace Research, 539
Lack of Human Capacities, 530
Lack of Resources, 529
Laitin, D. D., 412
Lalbahadur, A., 445
La LUCHA, 287
Lambin, E.., 68
Lamb, W. F., 74
Lamontagne, C., 87, 88
Landon E. Hancock, 469
Larkin, E., 402
LaRocco, A., 276
Last, D. M., 98, 248, 255, 274
Latour, Bruno, 29, 85, 86
Laudani, R., 98
Laudato Si', 95–97, 100, 104, 106–108, 110, 113, 115–117, 120–122, 132
Laurent Kabila, 254
Lauterbach, F., 272
Lawrence Mhandara, 235, 555
Lay Coordination Committee, 288
Laymah Gbowee, 27
Leadership in peace processes, 469
Learner Support Agents, 316
Le Billon, Philippe, 25
Le Cordeur, M., 340
Lederach, J. P., 10, 156, 157, 167, 194, 238, 255–257, 263, 275, 401, 413, 426–428
Lee, C. -S., 403
Leeuwen, M. V., 208, 502
Lefko-Everett, K., 342
Legitimacy to negotiations and agreements, 469
Lehman, H. P., 338
Lemanski, C., 387, 388
Lennard, C., 37
Lenton, Timothy, 24
Leoschut, L., 308
Leroux, S., 277
Lesley, 28
Lesotho, 487, 489, 490, 535
Lewis, P. M, 350
Lewis, S., 398
Liberal peacebuilding model, 191
Liberia, 27, 215, 223, 444
Libya, xxi
Life and Peace Institute partners, 471
Lisa Gürth, 485
Lisk, F., 68
Living beings, 96, 111, 119, 129, 235
Livingstone Formula, 472, 474, 475
Local and international advocacy, 7
Local and international monitoring, 7
Local and international protection, 7
Local First approach, 509
Local Non-Governmental Organisations, 501
Local ownership of peace processes, 469
Locham, R., 50
Locke, F., 359
Locust plagues, 11
London, J. K., 68
LoWilla, E., 357
Low-onset sea-level rise, 26, 56
Lo, Y.-S., 81
Lubunga, E., 412
Lucky E. Asuelime, 403
Ludin, J., 163, 181
Lu, Dylan, 19
Lupane District, 145, 147, 148
Lwizi, Gladyse, 474

M
Maas, A., 77
Mac Ginty, R., 386
Machakanja, P., 385, 386
Mach, Katharine J., 46
Macro-infrastructures for Peace, 385
Madalina, B., 342
Madhuku, J., 146
Maguwu, F., 241
Mahlobo, D., 340
Mahoney, C., 507
Mahony, L., 214, 215
Maiese, M., 200, 201, 401

Maina, W. M., 439
Mainstream peacebuilding, 127, 157, 384
Making Peace with Nature, 28
Malan, M., 205
Malan, N., 343
Malawi, 55, 531
Manby, Bronwen, 471, 474
Mandikwaza, E., 356
Maneno, B. K., 212
Manifesto, 527
Mapara, J., 195
Maphosa S. B., 196, 199
March of Hope, 277
Maria Sibylla Merian Institute for
Advanced Studies in Africa, 532
Marie-Joëlle Zahar, 50
Marie, M., 50
Marimbe, D., 247
Marina Ottaway, 468
Marine resources, 69
Marongwe, N., 168
Marshall, J., 99
Martinez-Alier, J., 78, 79
Martin Luther King, 287
Martin, M. L., 238
Martin, S., 398, 414
Maruta, S., 385
Marxist ideology, 207
Mashingaidze, T. M., 192, 193
Mashonaland, 146, 320, 386, 387
Masolo, D. A., 435
Masunungure, E., 148, 192
Matabeleland, 145, 148, 161, 192,
319–332, 386
Mathematics, xiii
Mato oput, 5, 456, 457
Matovic, V., 163, 181
Matthews, H. D., 87, 88
Mawarire, T., 243
Max-Weber Foundation, 532
Maxwell, A. M., 158
Mayeda, D., 284
Mazarr, Michael J., 18
Mazrui, A. A., 431, 439
Mbayo, A. S., 207
Mbombo, J-M. K., 287
McClanahan, T. R., 77
McGregor, J., 320, 322
McKay, S., 151, 154
McKnight, J., 391
McLaverty, P., 264
McNeely, J. A., 68
McNeill, John R., 18
Mease, K., 236
Mediation agendas and agreements, 465,
467, 494
Mediation and Security Council, 481
Mediation Facilitation Division, 482
Mediation Handbook, 473, 479
Mediation Reference Group, 487
Mediation structures, 485
Mediation support, 466, 472, 473, 476, 482,
490, 492, 493
Mediation Support Unit, 476, 487, 488
Melting ice, xv
Memorandum of Understanding, 349, 472
Mende, G., 342
Mennonite Central Committee, 308
Merrick, M. T., 414
Methodological approaches, 281, 515
Mexico, 18
Mhandara, L., 386, 555
Michael Aeby, 465, 555
Michelle Ndiaye, 466, 474–476, 489, 492
Middle East, 28, 424, 538
Middle-range leadership, 10
Midlands, 192, 319–322, 324, 325, 328,
386
Miles, M. B., 281
Military alliances UN Charter, 18
Military conflict, 29
Military infrastructure, 20
Millennia, 71
Miller, C., 372
Mills, J., 515
Missbach, A., 332
Mitchell, C. R., 469
Mnangagwa, 243, 319, 327–329, 333, 334
Mncube, V., 308
Mobilising the Population, 298
Mobjörk, Malin, 50, 68, 76
Mobjörk, M., 50, 68, 76, 128
Mobutu Sese Seko, 276, 445
Modelling, 105
Molutsi, P., 237
Molvaer, R. K., 78
Monsen, E. R., 359
Montreal Protocol, 23
Moral Authority, 290
More, M. P., 438
Morier-Genoud, E., 430
Morris, J., 424
Morris, Odhiambo, 471
Morrison, Mary Lee, 22
Morton, S., 73, 74
Moseley, L., 104, 112

Moses Monday John, 355
Mother Earth, 107, 109, 121, 122
Motsamai, Dimpho, 485, 490
Moulaert, F., 341
Moules, N., 310
Moustakas, C., 281
Mouton, J., 159, 360
Moyer, J., 112
Moyo, A., 390, 392
Moyo, J., 236, 237
Moyo-Nyoni, N., 145
Mozambique, 55, 68, 320, 423–425, 430, 440, 442, 449, 510, 531, 535
Mrics, P. B., 372
Mthwakazi Action Group, 325
Mubarrok, A. Z., 83
Muchemwa, C., 386
Muchiri, S., 444
Mudumbula, M. G., 290
Mueller, M. P., 82
Mugumo, F. M., 284, 285
Mulegwa, P., 277
Mulla, R., 356
Multi-ethnic societies, 29
Multitrack mediations, 478, 479, 492
Munhande, C., 192
Murambadoro, R., 146, 386
Murekasenge, J., 444
Murithi, Tim, 466, 471
Mushagalusa, A. K., 398
Mushi, M. F., 284
Mushongera, D., 341
Musingafi, M., 146
Mutisi, M., 155, 183, 453
Muzavazi, F., 385, 386
Mwale, N., 136
Myths, 80, 102, 120, 154
Mzola Central Ward, 145, 147, 149, 156, 158, 161, 162, 165, 166, 168, 171–173, 176, 180, 182, 183

N

Nagler, M., 368
Nahimana, T., 444
Nakayi, R., 457
Namibia, 486, 531, 532
Namkoong, Y., 513
Nathan, Laurie, 466, 473–476, 489, 492
National Academy of Sciences, 398
National Agreement Monitoring Council, 298
National Constitutional Assembly, 241
National consultations, 278, 298
National Order of Electoral Experts, 291
National reconstruction, 7
National Science Board, 52, 53
National Stabilization and Reconstruction Program, 216
National Technical Secretariat of Civil Society, 289
Natural disasters, 11, 48, 54, 55, 82, 505
Nature Lovers, v
Navigating Channels, 494
Nciizah, E., 192
Ncube, C., 237, 238, 240–242
Ndlovu-Gatsheni, S.J., 148, 320, 325
Ndlovu, P., 244
Nduwimana, F., 149, 151–153, 155, 183
Neenan, J., 413
Negara, K., 432
Nelson Mandela, 27, 287, 434
Newell, P., 68
New Tribe, 357, 371, 379
New Zimbabwe National Army, 321
Nganje, F., 388, 389
NGO networks, 467, 470, 479, 484, 487, 493
Ngoy, P., 205
Ngwenya, D., 146–148, 320–323, 325, 330–332
Ngwerume, T. E., 386
Niang, I., 37
Nigeria, 36, 55, 272, 532
Nilsson, D., 256, 257, 272, 469, 470
Nindorera, L. -M., 444
Ninian, A., 258, 263
Njozela, L., 342
Nkala, S., 329
Nkomo, J. M., 321, 322
Noagah Bukari, Kaderi, 50, 535
Nobel Peace, 27, 553
Noll, D., 98, 433
Noma, E., 384
Non-elitist Diversity, 391
Non-formality of Peace Committees, 391
Non-governmental societal organisations, 339
Non-political, 282
Non-public action, 7
Non-state, 6, 83, 84, 98, 102, 106, 150, 156, 193, 235, 238, 253, 360, 455, 483, 488
Nontshokweni, S., 342

Nonviolence, 115, 130, 314, 315, 357–360, 362, 363, 367–370, 373, 374, 378, 379, 545, 551
Noone, K., 68
Noori, S. M., 135
Norman Chivasa, 383, 555
Norris, J., 310
North Africa, 44, 48, 49
North Atlantic, xvi
North Korean instructors, 322
Not-profit platforms, 282
Novy, A., 341
Nowell, L., 310
Noyes, Alexander, 475
Nsapu, P., 288
Nuclear arms control, 23
Nuclear war, 23, 29, 431
Nuclear winter, 23
Nyaba, P., 356
Nyang'oro, J., 238
Nyathi, M., 146
Nyerere, J. K., 435
Nzisabira, S. C., 444
Nzongola-Ntalaja, G., 212

O
Oates, M., 99
Obenga, T., 438
Oberschall, A., 255
Oborji, F. A., 438
O'Brien, Karen, 22, 51
O'Brien, R., 159–161, 164
Observations, 148, 156, 161–165, 170, 171, 173, 174, 176–178, 180, 236, 248, 281, 293, 295, 297, 310, 355, 359, 400, 515, 517, 519, 521, 522
Oceans, 4, 70, 84
Odendaal, A., 388, 389
Odigie, Brown, 482
O'Donnell,M., 5
O'Driscoll, D., 155, 180, 181
Ogunsanya, K., 155
Okumu-Alya, F., 392
Olugbemi Jaiyebo, 478
O'Malley, Alanna, 19
O'Neill, D.W., 74
Open Access Partnership, 538
Open Society Initiative for Southern Africa, 293
Organisational mandate, 484
Organisations in Sub-Saharan Africa, 537
Organised crime, xix
Organization for Security and Co-operation in Europe, 403, 404
Orjuela, C., 150, 235, 238
Orthodox churches, 132
Ortiz, R., 87
Osamba, J., 200, 201
Osbahr, H., 68
Osmer, R. R., 385
Oswald Spring, Ú. O., 21, 22, 24–26, 31, 51, 52, 56, 71, 534
Otto, I. M., 70, 84–86
Outcome evaluation, 307
Outdoor pollution, 41
Overseas non-governmental organisations, 323
Owens, W. R., 281, 425
Owned the process, 469
Oxfam South Sudan, 378
Oxford Research Encyclopaedia of Climate Science, 48
Ozone layer, 23, 74, 120
Ozone layer depletion, 23
Ozone-depleting substances, 89

P
Pacifism, 75
Padgham, J., 37
Paffenholz, T., 6–8, 10, 98, 106, 149–151, 156–158, 193, 194, 206–208, 222–225, 227, 235, 236, 238, 247, 248, 252, 253, 255, 256, 272, 274, 275, 403, 426, 428, 470, 508, 509, 511
Paladini Adell, Borja, 470
Palaver, 423, 424, 434–436, 438, 439, 445, 453, 459
Panafricanism, 434
Panel of the Wise, 475
Panzi General Reference Hospital, 400
Paradigms, 4, 5, 12, 52, 53, 68, 78, 79, 82, 89, 97, 106, 110, 111, 115, 138, 148, 156, 157, 161, 249, 274, 281, 285, 286, 290, 400, 425, 442, 466–470, 479, 484, 494
Paris, 68, 69, 429, 431, 433, 445, 447, 449
Paris, R., 194, 208
Park, A. S., 455
Parker, R., 442
Parmar, Inderjeet, 18
Participation in Organisational Structures, 491
Partnership, 5, 72, 83, 102, 152, 181, 208, 222, 235, 236, 242, 248, 249, 263,

284, 289, 299, 345, 346, 348, 351, 352, 458, 471, 472, 477, 480, 482, 492, 503, 505, 506, 508, 509, 517, 518, 523, 532, 533
Paterson, M., 217, 218
Patriarch Bartholomew, 96, 101, 103, 104, 112, 113, 116, 118–120, 132–134
Pattberg, P., 84, 85, 102
Paul J. Crutzen, 11, 72, 99
Paul Kariuki, 337, 556
Peace agents, 307, 309, 314, 316
Peace and security, 12, 19–23, 27–29, 42, 50, 56, 75–77, 80, 83, 151–154, 207, 264, 302, 466, 467, 471, 472, 474, 475, 478–481, 484, 485, 491, 527, 528, 533, 539, 553
Peace and Security Council, 466, 472
Peace and Security Department, 472
Peace architecture, 307
Peace Brigades International, 7, 208
Peacebuilding actors, 11, 256, 275
Peacebuilding Commission, 76, 153, 155
Peacebuilding funding, 506
Peacebuilding in North-East Africa, 471
Peacebuilding non-governmental organisations, 7
Peacebuilding, 3, 5, 7, 9, 10, 12, 13, 28, 69, 75–78, 97–101, 106, 118, 127–129, 136, 138, 139, 145–159, 161, 163, 166, 168–176, 179–185, 191–202, 205–208, 210–212, 216–218, 220–225, 227, 230, 231, 235–243, 245–249, 251, 252, 255–257, 259, 262, 263, 267, 272–275, 308, 331–334, 339, 357, 358, 360, 378, 379, 384, 386, 389–392, 424–429, 431, 434–437, 440–442, 444, 447, 453, 455, 456, 458–460, 466, 468, 469, 472, 476, 477, 480, 489, 501–504, 506, 508–513, 515–521, 523, 534, 535, 545, 549, 551, 553–556
Peace Club Learners are Pace Agents and Role Models, 310
Peace ecology, 4, 5, 11–13, 17, 18, 20–23, 25, 27, 45, 50, 51, 54, 56, 67–69, 78–80, 86, 89, 96, 106, 120, 121, 527, 528, 532–534, 536, 537, 549, 555
Peace ecology approach, 17, 21, 23, 24, 26, 31, 52, 54, 56, 57, 528, 539, 540
Peace ecology research, 24, 26, 51, 57
Peacekeeping, 5, 156, 274, 466, 471, 472, 478
Peacemaking, 5, 8, 9, 42, 100, 156, 217, 550
Peace-Making and Mediation Support, 467
Peace-making efforts, 466, 467, 471, 479
Peace-Making in the African Peace, 465
Peace negotiations, 10, 154, 206, 208, 225, 228, 466, 468, 477, 510
Peace networks, 7, 227
Peace prize, 3, 397
Peace process, 8, 10, 42, 101, 145, 149–152, 154, 156, 176, 178, 181, 183, 210, 218, 220, 221, 228, 231, 235, 248, 255–257, 261, 272, 290, 386, 390, 416, 426, 451, 453, 465–470, 472, 473, 477–480, 484, 487, 493, 494, 516
Peace research, 17, 18, 20, 21, 26, 27, 31, 40, 42, 45, 47, 49, 52, 56, 57, 358, 468, 527, 528, 533–535, 537, 539, 540, 554
Peace studies, 4, 11, 20, 25, 67, 78, 79, 89, 145, 307, 355, 397, 532, 534, 553, 555
Pearce, J., 226, 227, 430
Pearl Harbor, 30
Pearson, P., 99
Peinado, M. M., 194, 195
Pelczynski, Z., 6
Pencheon, D., 73, 74
Perera, S., 255, 287
Persson, Å., 67
Pettersson, Therese, 50
Pew Research Center, 102, 117
Phenomenological reduction, 281
Philippe Mulume-Oderhwa Kaganda, 501
Phiri, K., 128
Physical effects, 19, 26, 29, 31, 55–57
Piccolino, G., 333
Pieterse, J. N., 512
Pihkala, P., 68
Pike, J., 430
Pius Yanda, P., 50, 68, 77, 78
Pkalya, R., 384, 385, 388, 390
Planetary boundaries, 3, 72–74, 87
Planetary boundaries framework, 3, 69, 72, 73, 89
Planetary consequences, 20
Planetary home, xii, xvi
Planet Earth, 24, 35, 96, 116
Pogodda, S., 208
Polarisation, 192, 197, 206, 242, 249, 391, 393

Policy challenges, 539
Policy frameworks, 88, 89, 467, 472, 473, 478, 484, 490
Political dialogue, 205, 212, 213, 298, 366, 490, 554
Political environment, 252, 280, 297, 300
Political geo-ecology, 20, 25, 52
Political independence, 19
Political issues, 19, 20, 49, 173, 331, 412
Political transition, 5, 151, 252, 258, 259, 263, 271–273, 276, 277, 282, 292, 294–296, 302, 388
Pölkki, T., 402
Poor service delivery, 338, 339, 344
Pope Francis, 68, 75, 84, 85, 87, 95–98, 100, 101, 103, 104, 106–121, 132
Popovski, V., 424
Population growth, 11, 17, 29, 35, 117
Porto Gomes, João, 487
Porto, J., 205
Post-Cold War, 24, 51, 235
Postcolonial order, 19
Postcolonial violence, 4
Post-conflict, 9, 10, 54, 76, 148, 150, 274, 384, 398, 403, 415, 416, 425, 439, 444, 447, 454, 456, 501, 519, 524, 549
Post-conflict peacebuilding, 9, 51, 151, 383, 384, 426, 502
Post-conflict reconstruction, 8, 152, 217, 440, 451, 472, 512, 534
Post-Genocide's Justice Experiments, 451
Post-Mugabe Period, 242
Post-War, 28, 275, 427, 432, 440, 442
Pouligny, B., 150, 222, 235, 239
Powell, J., 99
Power relations, 7, 25, 48, 49, 358, 366
Power-sharing pacts, 466, 469
Practical perspective, 469
Praxis, 76, 89, 299
Pre-industrial, 40, 68, 69, 71, 88
President Joseph Kabila, 276–278, 284, 287, 291, 298
Press releases, 7, 209, 247
Prevention and peacebuilding, 479
Priebe, Miranda, 18
Primary data, 251, 259, 339, 355
Primary schools, 308–310
Principe, M., 357
Problem solving, 12, 349, 360
Production processes, 20
Program Goals and Funding, 504
Program Goals and Intervention Zones, 521
Promote development, 470, 502
Proposals for Action Towards a Peace Ecology, 527
Protecting and nurturing children, 398, 416
Protection and monitoring, 235, 243–245, 249
Protocol on the Establishment, 466, 472
Protracted rebellious attitudes of children, 397
Proxy conflicts and wars, 19
Prunier, G., 211
Psychological violence, 410
Public campaigns, 8, 245
Public participation in peace-making, 469
Purdue, D., 207, 208
Pyramid of peacebuilding, 11, 426

Q

Quadruple-level democratisation monitoring process, 295

R

Rabbi Green, A., 132
Rabbi Waskow, A., 132
Radin, Andrew, 18
Radio Tarmazuj, 375
Raeymaekers, T., 211
Rafferty, John P., 70
Rahmstorf, Stefan, 24
Rainforest, xvi
Rakkhit, P., 129
Rakover, N., 131
Raleigh, C., 128
Ramankutty, N., 87
Ranger, T., 320, 322
Rape in South Kivu, 398, 416
Rape in wartime, 397
Rape survivors, 218, 398–401, 404–406, 409, 410, 416
Rawson, P., 99
Realist geopolitics, 20
Realist logic, 20
Reason, P., 159
Recommendations, 7, 50, 77, 148, 155, 182, 183, 200, 209, 212, 217, 252, 261–263, 290, 296, 300, 324, 339, 346, 351, 399, 414, 415, 443, 472, 473, 476, 491, 528
Reconciliation, 5, 10, 113, 146, 147, 156, 157, 192–195, 207, 210, 220, 228, 235, 236, 238, 240, 242, 256, 263, 275, 276, 323, 326, 334, 385, 386,

423, 424, 426, 427, 432, 434–444, 447, 450, 452–459, 468, 471, 489, 520, 549, 550, 555
Redie Bereketeab, 478
Reducing vulnerabilities, 31, 39
Reflective Level of Action, 295
Regina, A. L., 403, 404
Regional Early Warning Centre, 486
Regional Economic Communities, 465
Regional intergovernmental organisations, 76
Regions of south, 532
Relational models of IGOs and CSOs, 470
Relationships, 12, 21, 22, 24, 42, 46, 47, 50, 56, 78, 79, 81, 82, 98, 108, 119, 129–131, 134, 136, 138, 155, 156, 162, 165–167, 170, 173, 178, 179, 181, 196, 200, 208, 209, 212, 218, 220, 222, 227, 228, 230, 231, 241, 243, 246, 275, 314, 320, 345, 347, 349, 352, 357, 358, 371, 378, 393, 398, 410, 411, 413, 414, 416, 430, 434–436, 439–441, 450, 457, 469, 485, 503, 508, 512, 513, 516, 519, 534
Relief Web, 398, 400
Religions, 96, 97, 100–107, 110, 112, 113, 115–122, 127–129, 131, 136, 138, 341, 439–441
Renewable forms of energy, 41
Renugadevi, R., 104
Repair harm, 132
Repair the earth, 132
Representative and liberal democracy, 258
Repressive violence, 148
Research community, 47, 534
Researchers and research institutions, 7
Researchers in the Global South, 528, 536, 538
Research for Global Justice, 529
Research priorities, 47
Resilience, 39, 48, 75, 87, 89, 95, 264, 350, 432, 532, 534, 535
Resource-induced instability, 76
Resource Scarcity, 31, 49, 52, 56
Respondent, 162, 164, 192, 214, 215, 239, 259, 260, 262, 360, 362, 363, 365, 366, 373, 378, 400, 401, 508, 515, 518–520, 522
Restorative justice, 98, 263, 388, 433, 436, 458, 549–551, 554
Rettig, M., 453
Reuter, T. A., 135
Reverend Ndabaningi Sithole, 320
Rey, Cdl., 151, 154
Reychler, L., 252, 428
Reyntjens, F., 443, 451, 452
Richardson, Katherine, 25, 68, 73, 79, 84, 89
Richmond, O. P., 208
Right2Know Campaign, 340
Rio Earth Summit, 23, 25
Rising sea levels, 77
Risks management, 67
Risks of natural resource management, 76
Ritimo network, 287
Rival superpowers, 19
River basins, 29, 48, 49
Rivers, M., 244
Robert Gerenge, 490
Robert Mugabe, 137, 192, 320
Robinson, C., 402
Rockström, Johan, 24, 25, 68, 73, 89
Rohwerder, B., 398
Rolston, H., 106, 108
Roman civilisations, 6
Ronan, M., 132
Roniger, L., 511, 514
Rossy Tshimanga, 255, 288, 298
Rothchild, D., 237
Rothe, Delf, 50
Rowe, C. J., 258
Rowlatt, J., 68, 69
Rubbers, B., 445, 446
Ruddiman, W. F., 71, 72
Rudo, Chitiga, 471, 474
Rudolph, L., 402
Rukuni, T., 146
Ruppel, O. C., 37
Russia, xx
Ruun, H., 357
Rwanda, 408, 411, 412, 423–425, 430, 432, 442, 451–453, 504, 531, 532, 535
Rwandan Genocide, 451, 452
Ryan, M., 357
Ryan, S., 237

S

Sachikonye, Lloyd, 192–194, 237, 240, 241, 486
Sacred Union of the Radical Opposition, 212
SADC architecture, 485, 486
SADC Mediation Support Unit, 488
SADC National Committees, 484, 485

SADC Panel of Elders, 487
SADC Regional Early Warning Centre, 486
SADC Summit and Organ Troika, 486, 491
Saeidi, M., 398
Sahel Region, 68
Saint Sylvester's Peaceful Transition Deal, 301
Saki, O., 195
Sala, S., 67, 77
Salinger, J., 131, 132
Sangqu, S., 389, 390
Saniotis, A., 105
Saunders, D. G., 484
Savage, T., 445
Save the Children, 7
Save Zimbabwe Campaign, 242
Sayer, J. A., 87
Scarnecchia, T., 323
Schaffner, L., 77
Scheffran, Jürgen, 12, 24, 25, 47–49, 68, 96, 99
Scheidel, A., 78
Scheidler, S., 342
Schellnhuber, Hans Joachim, 105
Schilling, Janpeter, 47, 49
Schirch, L., 368, 370
Schlosberg, D, 5, 11, 12, 68, 78, 79
Schnabel, A., 425, 427
Schneidewind, Uwe, 53
School Peace Clubs, 307, 308
School violence, 307
Sciences and Technology Institutions, v
Scientific capacities, 4, 527
Scientific Conferences Outside Sub-Saharan Africa, 528
Sean Molloy, 470
Secondary, 152, 156, 176, 192, 206, 251, 259, 262, 271, 308–310, 313, 315, 339, 402
Second Liberian Civil War, 27
Security Architecture, 42, 465, 478
Security realm, 18
Security risk, 40, 44, 48, 76
Security studies, 20, 49
Self-amnesty, 294
Senegal, 531, 532
Serrano Oswald, S. E., 534
Service delivery, 7, 8, 151, 206, 208, 210, 222, 228–231, 239, 256, 261, 266, 338, 351, 355, 357, 403
Seto, D., 87, 88
Sexual violence, 171, 217, 218, 225, 398, 446, 448, 449, 517, 520–522, 553
Shaaban, M., 50
Shale, Victor, 490
Shannon, S. E., 69, 161, 400, 403, 515
Shanyisa, W., 146
Shattucka, A., 401
Shaw, R., 455
Shearing, C., 384, 385, 390
Shepler, S., 432
Sherry, D., 402
Shindell, D., 87
Shippee, R. S., 441
Shomrei Adamah, 132
Shona idiom, 195, 198
Shuku, N. Y., 284, 285
Sibanda, E. M., 320
Sibanda, P., 195
Sikhism, 101
Silverman, S., 359
Simpson, Nick, 540
Singer-Brodowski, Mandy, 52
Singer, P.W., 423, 424, 432
Sister, 107, 108, 113, 121, 181, 405, 407, 434
Sitea, L., 5
Sithole, S. L., 260, 516
Skolimowski, H., 82
Smith, A., 99
Smith, D., 68
Smith, E., 68
Smith, T. J., 87, 88
Social and Economic Justice, 115, 276
Social and political movements, 7
Social capital, 7, 8, 146, 181, 195, 210, 218, 227, 248, 263, 265, 454, 503, 523, 524
Social change, 7, 20, 157
Social cohesion, 7, 10, 151, 158, 206, 208, 210, 227, 228, 230, 231, 246, 256, 265, 272, 337–339, 341–352, 426, 427, 444, 502, 503, 520, 524
Social consciousness, 325
Social ills, 316
Social justice, 22, 25, 73, 79, 108, 167, 182, 218, 280, 324, 348
Social media, 7, 209, 246, 328, 332–334, 372, 374
Social movements, 12, 79, 85, 86, 89, 183, 222, 357, 379, 514
Social organisation, 29
Social Science Research for Global Change, 53
Social system, 21, 56, 427
Socialisation action, 8

Socialisation In-group Identity, 226
Socialism, 5, 434, 435
Societal harmony, 149, 435, 449, 450, 458, 459, 523
Societal peace, 87
Socioeconomic development initiative, 9
Socio-economic impact, viii
Socio-economic outcomes, 19
Socio-environmental conflicts, 78
Socio-political development, 337
Söderbaum, Fredrik, 471
Solar system, xv
Solomon, N., 50
Somalia, 68, 428, 532
Sörlin, S., 3, 67
South Africa, 27, 34, 38, 44, 49, 54, 55, 134, 153, 206, 240, 260, 284, 293, 294, 307, 308, 321, 322, 337–343, 345, 348–350, 352, 384, 386–390, 436, 437, 439, 442, 532, 535, 551, 553–556
South Sudan, 355–358, 362, 365–367, 369, 370, 373, 375, 378, 379, 386, 387, 389, 428, 555
Southern Africa Trust, 485
Southern African Development Community (SADC), 149, 153, 298, 466, 484
Southern African Faith Communities, xxxv
Southern Sahel of Sudan, 78
Sow, Papa, 50, 535
Specht, I., 430
Spirduso, W., 359
Spoilers, 469, 502
Spurk, C., 6, 98, 106, 150, 156, 193, 235–239, 253, 255, 256, 272, 274, 275, 403
Sriram, Chandra Lekha, 466, 469
St Anthony, 107, 111
State-centric liberal peacebuilding, 468
Status quo, 26, 57, 182, 224, 301, 392
Staudt, R. J., 108, 111
Stauffer, C. S., 148
Stearns, J., 504
Stearns, J. K., 398
Stefanovic, I. L., 82
Steffen, Will, 3, 19, 25, 67, 72–75, 79, 84, 89, 97, 99
Steinberger, J. K., 74
Stephen, M., 70, 217, 219, 220, 222, 228–230, 368
Stephan, M. J., 373
Sterck, O., 272
Stewardship, 32–34, 79, 82, 84, 86, 88, 95, 100, 103, 108–113, 118–121, 132, 137
St Francis of Assisi, 109, 112
Stick, S. L., 359
Stigmas, 397, 400
Stockholm International Peace Research Institute, 42
Stoermer, Eugene F., 24, 51, 70, 535
Stöger-Eising, V., 435
Stone, Madeleine, 19
Stone, P., 99
St Pope John XXIII, 106
Strategic policy plans, 472, 490
Stripple, J., 84, 85, 102
Structure of the Book, 13
Stuth Cevallos, Astrid, 18
Sub-Sahara, 530, 538
Sub-Saharan, 529
Sub-Saharan Africa, 5, 11, 17, 35, 36, 40, 44, 48, 121, 207, 301, 355, 356, 358, 379, 424, 429, 439, 440, 442, 465, 466, 511, 527–530, 532, 533, 536–538, 540, 551
Sudan People's Liberation Movement, 355, 356
Sulaiman Abubakar, M., 400
Sulfate aerosol emissions, 87, 88
Sümer, Vakur, 54
Sun City, 277, 278, 284, 294
Sunusi Alkassim, R., 400
Support mechanisms, 530
Survival of humanity, 68, 73, 88, 103, 112
Survival of humankind, 19, 23, 80
Susanti, E., 146
Sustainability, 11, 19, 22, 23, 51, 53, 67, 73, 74, 78, 79, 82, 110, 194, 198, 201, 222, 229, 238, 247, 275, 358, 368, 371, 383, 479, 491, 509, 512, 528, 531
Sustainability transition research network, 53
Sustainability transition, 24, 53, 54, 528, 533
Sustainable biodiverse, 89
Sustainable development goals, 30, 41, 54, 73, 528
Sustainable governance, 532
Sustainable peace, 9, 10, 24, 49, 76, 191, 194, 195, 199, 202, 205, 206, 231, 238, 247, 249, 252, 253, 263, 286, 344, 371, 379, 383, 385, 426, 444, 455, 458, 466, 469, 533

Sustaining human progress, 31
Sutton, J., 402
Swain Anders, Themnér, 50
Swain, Ashok, 50
Swiatek, D. C., 341
Swilling, Mark, 52
Swiss National Science Foundation, 467
Sybertz, D., 435
Sygna, Linda, 22, 51
Sylvia Blanche Kaye, 397
Syria, xxi
Szasz, A., 68
Sze, J., 68
Szpak, A., 444

T
Taheri, F., 50
Takawira Musavengana, 486
Takeuchi, S., 432, 453
Talja, S., 161
Talmudic laws, 132
Tanzania, 18, 36, 320, 507, 531, 532
Tänzler, D., 77
Tarif, M., 276
Tarusarira, J., 135, 136
Teachers, 307–312, 314–316
Techno-economic paradigm, 114
Teck Kong, Ming, 19
Teddlie, C., 400
Telecommunication, 375
Temper, L., 78, 79
Temporal violence, xxi, xxii
Thabo Mbeki, 340, 487
Theoretical framework, 12, 79, 148, 156, 252, 273, 510
Thermonuclear war, 29
Think tanks, 25, 52, 72, 89, 262, 295, 474, 475, 484, 487, 492
Thomas, D. S., 68
Thompson, C., 402
Thomson, K., 414
Through Gender Equality in Rights, 146
Thunberg, G., 86
Tidball, Keith, 21, 26, 56
Tikkun Olam, 132
Timothy Murithi, 466
Tippins, D., 82
Todd, J. G., 322
Togo, 434, 532
To, H.-P., 341
Tokyo, 76
Top leadership, 10, 256, 284
Torabi, M., 135
Toroitich, I. K., 134
Touraine, A., 508, 517
Towheed, S., 69, 425
Track I diplomacy, 9
Track II diplomacy, 10
Traditional Christian, 137
Training for Peace programme, 480
Tramblay, Y., 50
Transdisciplinarity, 52
Transformative cornerstones, 53
Transformative research programme, 26, 52, 53
Transformative role of peace committees, 391
Transformative science, 23, 52, 53
Transformative strategy, 17, 20
Transitional Justice Policy, 471
Transnational Climate Governance, 85, 86, 102, 106
Tree-planting Eucharist, 137
Trefon, T., 211
Troika, 486, 491
Trust and confidence, 7
Truth and reconciliation commission, 252, 294, 445, 453, 454
Tshuma, N., 384, 385, 390
Tsitsi Dangarembga, 3, 4
Turko-Egyptian occupation of Sudan, 355
Turnera, H., 401
Twilley, R. R., 77
Twyman, C., 68

U
Ubuntu, 5, 199, 423, 424, 434–439, 444, 453, 458, 459
Udo, E. J., 437, 438
Uganda, 308, 377, 384, 386, 390, 423–425, 428, 430, 432, 434, 442, 455–457, 531, 535
Ujamaa, 5, 434, 435
Ukase, P., 339
Umkonto WeSizwe, 321
UN Secretary-General, 27, 45, 77, 424, 483
UN Security Council, 45, 150, 153, 218, 432, 481, 540
UNESCO, 81, 168, 539
United Nations, 7, 45, 84, 87, 149, 151–153, 155, 183, 209, 215, 218, 247, 276, 280, 294, 297, 346, 351, 429, 431, 432, 445, 447, 451, 452, 454, 458, 459, 470, 506, 530, 550

United Nations Development Programme (UNDP), 9, 20, 31–35, 155, 297, 425, 489, 504, 508, 530
United Nations Framework Convention on Climate Change, 83
United Nations Office on Drugs and Crime, 307
United States Conference of Catholic Bishops, 108
United States Institute for Peace, 127, 452, 456
Universities of Freiburg, 532
University of Bandundu, 547, 549, 554
Unsustainable development, 29, 30
Unwanted pregnancies, 397, 407
Uppsala, 42
Uppsala University, 18, 28, 42
Urbanisation, 41, 69, 71, 532
Urmilla Bob, viii
Urquhart, P., 37
Urusaro Karekezi, A., 432
USA, 18, 272, 280, 294, 297, 298, 441, 530, 539
US National Science Board, 52, 53
Utas, M., 430
Utriainen, K., 402
Uzodike, U. O., 100, 154, 183

V
Valentova, M., 341
Van Baalen, Sebastian, 50, 128
Vandeginste, Stef, 443, 451, 452, 466
Van De Goor, L., 68
Vandermindena, J., 401
Van Houten, K., 509, 510
Van Leeuwen, M., 208, 502
Van Nieuwkerk, Anthoni, 484
Van Reybrouck, D., 302
Vanspauwen, K., 445
Van Staden, M., 436
Van Tongeren, P., 384, 389, 390, 392
Variations in precipitation, 26, 56
Vautravers, A. J., 423
Verena Stauber, 485
Verkoren, W., 502
Vienna Convention for the Protection of the Ozone Layer, 23
Vilariño, M. V., 69
Villellas, M., 149, 157
Violence-makers, 469
Violent climate conflicts, 11, 17
Violent conflicts, 5, 17, 21, 22, 27, 40, 46–51, 78, 79, 86, 96, 128, 138, 139, 205, 222, 251, 252, 256, 295, 356, 369, 385, 387, 389, 428, 438, 441, 444, 458, 470, 501, 523, 528, 535, 540
Vircoulon, T., 276
Vlassenroot, K., 211, 503
Vogel, C., 398, 504
Volgy, Thomas J., 19
Von Lucke, Franziskus, 540
Von Soest, Christian, 50
Von Uexkull, Nina, 47, 50

W
Walters, C., 99
Wamba, L., 440, 451, 459
Wangarĩ Maathai, 27
War and peace, 29
War-affected communities, 8, 454, 456
Warming water, 77
War Resisters' International, 372
Water scarcity, 48, 69, 80, 81
WBGU, 40, 44, 45, 53
Weapons and wars on the environment, 51
Weapons of mass destruction, 19, 29
Werner, K., 67, 96, 99
Wessells, M. G., 440
Westad, Odd Arne, 19
West African Civil Society Forum, 480
West African Network for Peacebuilding (WANEP), 471, 475, 479–483, 491–493
West and East Africa, 532
Western psychology, 440, 451
Westing, Arthur H., 51
Westra, L., 5, 11, 78
Whetho, A., 100
White, D., 106
Wiedermann, M., 84, 86
Wielenga, C., 386
Wiener, Jarrod, 19
Wildfires, 78, 101
Wildlife, 41, 137, 376
Williams, M., 71, 75, 99
Willms, D. G., 67, 96, 99
Wilmot, P., 357
Wilson, Edward O, 21
Wilson, G., 68
Wilson, J., 358
Wiredu, K., 434, 435, 437, 439, 459
Wisdom, J., 359
Witt, Antonia, 489
Wolfgang, D., 403

Wolf, Johanna, 22, 51
Women empowerment, 147–149, 152, 161
Women of Liberia Mass Action for Peace, 27
Women's resiliency, 146
Women's rights, 27, 217, 225
Wood, E. M., 6, 360
Wood, F., 360
Woollett, N., 414
World Bank, 6, 9, 55, 146, 150, 151, 181, 183, 208, 426, 469, 502–504, 506, 508–511, 530
World Bank's Post-Conflict, 504
World Council of Churches, 127, 134
World Health Organization, 413
World Meteorological, 75
World Meteorological Organization, 75
Worldometer, 118
World War II, 18, 24, 26, 30, 51, 52, 54, 205, 253

Y

Yahaya, J. U., 255
Yahia Zoubir, 474–476, 492
Yanda, Pius, 50, 68, 75, 77, 78
Yáñez-Arancibia, A., 77
Yin, R. K., 400, 402
Yirga, F., 50
Yonekawa, M., 276
Young, N. J., 391
Yŏ, D. S., 129
Yu, F., 400

Z

Zabondo, G. D., 279, 280
Zacharie, A., 280
Zagzebski, L. T., 291
Zaire, 284, 434
Zalasiewicz. J., 70, 71, 75, 99
Zambara, W., 236
Zambia, 308, 320, 531
Zanker, F., 228, 469
Zaremba, N., 357
Zehr, H., 98, 99, 433
Zembe, S., 386
Zengenene, M., 146, 168
Zhou, G., 385
Zigomo, K., 193
Zimbabwe African People's Union, 192, 320
Zimbabwe Christian Alliance, 242
Zimbabwe Civic Education Trust (ZIMCET), 386, 387, 391
Zimbabwe Peace and Security Programme, 242
Zimbabwe People's Revolutionary Army, 320
Zimbabwe's National Peace and Reconciliation Commission, 235
Zimbabwe, 105, 127, 136, 145, 146, 148, 149, 157, 158, 161, 168, 179, 183, 191–194, 196, 201, 235, 236, 239–241, 243–249, 272, 319–321, 324, 328, 331, 383–387, 389–392, 434, 485–487, 489, 531, 535, 553, 555
Zongwe, D. P., 446
Zorita, E., 50
Zuber-Skerritt, O., 159

The manufacturer's authorised representative in the EU is Springer
Nature Customer Service Centre GmbH, Europaplatz 3, 69115 Heidelberg,
Germany. If you have any concerns regarding our products, please
contact ProductSafety@springernature.com

Printed and bound by CPI Group (UK) Ltd, Croydon, CR0 4YY
30/09/2026
02225251-0017